# Nano Drug Delivery Systems for Colon Cancer

## Innovations, Challenges, and Applications

# Nano Drug Delivery Systems for Colon Cancer

## Innovations, Challenges, and Applications

Edited by

**Arun Kumar Singh**

*University School of Pharmaceutical Sciences, Rayat Bahra University, Mohali, Punjab, India*

**Vipan Kumar**

*School of Pharmaceutical Sciences, CGC University, Mohali, Punjab, India*

**Pranav Kumar Prabhakar**

*Department of Biotechnology, School of Engineering and Technology, Nagaland University, Kohima, Nagaland, India*

**Meenakshi Dhanawat**

*Amity Institute of Pharmacy, Amity University Haryana, Gurugram, Haryana, India*

**Neeraj Mishra**

*Department of Pharmaceutics, Amity Institute of Pharmacy, Amity University, Gwalior, Madhya Pradesh, India*

ELSEVIER

ACADEMIC PRESS

An imprint of Elsevier

Academic Press is an imprint of Elsevier
125 London Wall, London EC2Y 5AS, United Kingdom
50 Hampshire Street, 5th Floor, Cambridge, MA 02139, United States



For accessibility purposes, images in electronic versions of this book are accompanied by alt text descriptions independently provided by Elsevier, generated with the aid of AI. For more information, see https://www.elsevier.com/about/accessibility or Radarweg 29, 1043 NX Amsterdam, Netherlands.

Books and Journals published by Elsevier comply with applicable product safety requirements. For any product safety concerns or queries, please contact our authorised representative, Elsevier B.V., at productsafety@elsevier.com.

Publisher's note: Elsevier takes a neutral position with respect to territorial disputes or jurisdictional claims in its published content, including in maps and institutional affiliations.



**Notices**
Knowledge and best practice in this field are constantly changing. As new research and experience broaden our understanding, changes in research methods, professional practices, or medical treatment may become necessary.

Practitioners and researchers must always rely on their own experience and knowledge in evaluating and using any information, methods, compounds, or experiments described herein. In using such information or methods they should be mindful of their own safety and the safety of others, including parties for whom they have a professional responsibility.

To the fullest extent of the law, neither the Publisher nor the authors, contributors, or editors, assume any liability for any injury and/or damage to persons or property as a matter of products liability, negligence or otherwise, or from any use or operation of any methods, products, instructions, or ideas contained in the material herein.

ISBN: 978-0-443-40443-6

For Information on all Academic Press publications
visit our website at https://www.elsevier.com/books-and-journals

*Publisher:* Megan Ball
*Acquisitions Editor:* Linda Buschman
*Editorial Project Manager:* Harshita Bathla
*Production Project Manager:* Fahmida Sultana
*Cover Designer:* Mark Rogers

Printed in England by CPI Group (UK) Ltd
Croydon CR0 4YY
United Kingdom

Typeset by MPS Limited, Chennai, India

# Contents

# List of contributors


**Lalit Mohan Aggarwal**
Department of Radiotherapy and Radiation Medicine, Institute of Medical Sciences, Banaras Hindu University, Varanasi, Uttar Pradesh, India

**Ajinkya Aher**
Department of Pharmacology, Arnold and Marie Schwartz College of Pharmacy and Health Sciences, Long Island University, Brooklyn, NY, United States

**Adil Ali**
Department of Biotechnology, Parul Institute of Applied Sciences and Research and Development Cell, Parul University, Vadodara, Gujarat, India

**Molakpogu Ravindra Babu**
Department of Pharmacy, School of Medical and Allied Sciences, Galgotias University, Greater Noida, Uttar Pradesh, India

**Abdul Basit**
Department of Biological Sciences, National University of Medical Sciences (NUMS), Rawalpindi, Punjab, Pakistan

**Swetha Belavadi**
Department of Pharmacology, College of Pharmaceutical Sciences, Dayananda Sagar University, Bengaluru, Karnataka, India

**Sanheeta Chakrabarty**
Department of Pharmaceutical Engineering & Technology, Indian Institute of Technology (BHU), Varanasi, Uttar Pradesh, India

**Subhendu Chakroborty**
Department of Chemistry, School of Basic Sciences, Chandigarh University, Unnao, Uttar Pradesh, India

**Pramila Chaubey**
Department of Pharmaceutics, College of Pharmacy, Shaqra University, Shaqra, Kingdom of Saudi Arabia

**Shikha Baghel Chauhan**
Amity Institute of Pharmacy, Amity University, Noida, Uttar Pradesh, India; Centre of Industrial Pharmacy, Uttar Pradesh, India

**Ananda Chettupalli**
Department of Pharmaceutical Technology, School of Health and Medical Sciences, Adamas Knowledge City, Adamas University, North 24 Parganas, Kolkata, West Bengal, India

**Sunil Choudhary**
Department of Radiotherapy and Radiation Medicine, Institute of Medical Sciences, Banaras Hindu University, Varanasi, Uttar Pradesh, India

**Melroy D'Sa**
Department of Pharmaceutics, OES's Oriental College of Pharmacy, Navi-Mumbai, Maharashtra, India

**Meenakshi Dhanawat**
Amity Institute of Pharmacy, Amity University Haryana, Gurugram, Haryana, India

**Charmi Dudhat**
Department of Pharmaceutics, Institute of Pharmacy, Nirma University, Ahmedabad, Gujarat, India

**Manoj Gadewar**
Department of Pharmacology, JSS University, Greater Noida, Uttar Pradesh, India

**Rajesh Gavit**
Herbanext Laboratories Inc, Bacolod, Negros Occidental, Philippines

**Madiha Ghazanfar**
Department of Biological Sciences, National University of Medical Sciences (NUMS), Rawalpindi, Punjab, Pakistan

**Prabhanjan Giram**
Department of Pharmaceutical Sciences, University at Buffalo, The State University of New York, Buffalo, NY, United States

**Dhanashree Guram**
Department of Pharmaceutics, OES's Oriental College of Pharmacy, Navi-Mumbai, Maharashtra, India

**Yash Jadhav**
Department of Pharmaceutics, OES's Oriental College of Pharmacy, Navi-Mumbai, Maharashtra, India

**Shreyans Jain**
Department of Pharmaceutical Engineering & Technology, Indian Institute of Technology (BHU), Varanasi, Uttar Pradesh, India

**Sunil Kumar Kadiri**
Department of Pharmacology, College of Pharmaceutical Sciences, Dayananda Sagar University, Bengaluru, Karnataka, India

**Darpan Kalra**
M.M College of Pharmacy, Maharishi Markandeshwar (Deemed to be University), Ambala, Haryana, India

**Princy Kashyap**
Department of Pharmacy, Guru Ghasidas Vishwavidyalaya, Bilaspur, Chhattisgarh, India

**Mrunalini Kulkarni**
Deparment of Pharmaceutical Chemistry, Vishwakarma University, Pune, Maharashtra, India

**Manish Kumar**
Department of Pharmaceutics, ISF College of Pharmacy, Moga, Punjab, India; KIET School of Pharmacy, Krishna Institute of Engineering & Technology (KIET), Ghaziabad, Uttar Pradesh, India

**Manoj Kumar**
Department of Pharmacy, Guru Ghasidas Vishwavidyalaya, Bilaspur, Chhattisgarh, India

**Rahul Kumar**
Faculty of Pharmacy, Vidya University, Meerut, Uttar Pradesh, India

**Vipan Kumar**
School of Pharmaceutical Sciences, CGC University, Mohali, Punjab, India

**Arun Kumar Singh**
University School of Pharmaceutical Sciences, Rayat Bahra University, Mohali, Punjab, India

**Mehul Luhar**
Department of Pharmaceutics, Vidhyadeep Institute of Pharmacy, Vidhyadeep University, Surat, Gujarat, India

**Garima Malik**
Chitkara College of Pharmacy, Chitkara University, Rajpura, Punjab, India

**Uma Shanker Maurya**
Goel Institute of Pharmacy and Sciences, Lucknow, Uttar Pradesh, India

**Neeraj Mishra**
Department of Pharmaceutics, Amity Institute of Pharmacy, Amity University, Gwalior, Madhya Pradesh, India

**Dikshitkumar Modi**
Pharmacy, S. S. Agrawal Institute of Pharmacy, Navsari, Gujarat, India

**Satyajit Mohanty**
Department of Pharmacy, School of Medical and Allied Science, Galgotias University, Greater Noida, Uttar Pradesh, India

**Muhammad Naeem**
Department of Biological Sciences, National University of Medical Sciences (NUMS), Rawalpindi, Punjab, Pakistan

**Sakshi Patel**
Faculty of Pharmacy, Vidya University, Meerut, Uttar Pradesh, India

**Mansi Patil**
Department of Pharmaceutical Chemistry, TSSM's Jayawant Institute of Pharmaceutical Sciences and Research Bavdhan, Pune, Maharashtra, India

**Ankit Paul**
Department of Biotechnology, Parul Institute of Applied Sciences and Research and Development Cell, Parul University, Vadodara, Gujarat, India

**Sampriti Paul**
Department of Pharmacology, College of Pharmaceutical Sciences, Dayananda Sagar University, Bengaluru, Karnataka, India

**Dipti G. Phadtare**
Department of Pharmaceutics & Pharmaceutical Chemistry, KCT, R. G. Sapkal Institute of Pharmacy, Nashik, Maharashtra, India

**Pranav Kumar Prabhakar**
Department of Biotechnology, School of Engineering and Technology, Nagaland University, Kohima, Nagaland, India

**Shaik Abdul Rahaman**
Department of Pharmacy, School of Medical and Allied Sciences, Galgotias University, Greater Noida, Uttar Pradesh, India

**Devyani Rajput**
Department of Pharmacognosy, Nims Institute of Pharmacy, Nims University, Jaipur, Rajasthan, India

**Manali Rupareliya**
Department of Pharmaceutics, Shobhaben Pratapbhai Patel School of Pharmacy and Technology Management, SVKM's NMIMS, Mumbai, Maharashtra, India

**Ashok Kumar Sah**
Department of Medical Laboratory Sciences, College of Applied & Health Sciences, A' Sharqiyah University, Ibra, North Sharqiyah, Oman

**Mohit Shah**
Department of Pharmaceutics, Institute of Pharmacy, Nirma University, Ahmedabad, Gujarat, India

**Pranav Shah**
Dr. Chunibhai Vallabhbhai Patel College of Pharmacy, UKA Tarsadia University (NAAC Accredited), Bardoli, Surat, Gujarat, India

**Hitesh V. Shahare**
Deparment of Pharmaceutical Chemistry, SNJB's Shriman Sureshdada Jain College of Pharmacy, Nashik, Maharashtra, India

**Shweta H. Shahare**
Department of Pharmaceutics & Pharmaceutical Chemistry, KCT, R. G. Sapkal Institute of Pharmacy, Nashik, Maharashtra, India

**Shivam Sharma**
Department of Pharmacy Practice, Guru Gobind Singh College of Pharmacy, Yamunanagar, Haryana, India

**Shivangi Sharma**
Department of Applied Chemistry, Amity School of Applied Science, Amity University, Gwalior, Madhya Pradesh, India

**Debayan Sil**
Department of Quality Assurance, ISF College of Pharmacy, Moga, Punjab, India

**Deepak B. Somavanshi**
Dept. of Pharmacognosy, K. B. H. S. S. Trust Institute of Pharmacy, Malegaon, Maharashtra, India

**Shruti Srivatava**
Directorate of Research, Innovation & Consultancy, Chandigarh University, Unnao, Uttar Pradesh, India

**Prashant Tiwari**
Department of Pharmacology, College of Pharmaceutical Sciences, Dayananda Sagar University, Bengaluru, Karnataka, India

**Paramvir Singh Tomar**
Department of Pharmacy, School of Medical and Allied Sciences, Galgotias University, Greater Noida, Uttar Pradesh, India

**Arya Tripathi**
Department of Pharmacy, School of Medical and Allied Sciences, Galgotias University, Greater Noida, Uttar Pradesh, India

**Muneeb Ullah**
Department of Manufacturing Pharmacy, College of Pharmacy, Pusan National University, Busan, Republic of Korea

**Tarun Kumar Upadhyay**
Department of Biotechnology, Parul Institute of Applied Sciences and Research and Development Cell, Parul University, Vadodara, Gujarat, India

**Pitchaikannu Venkatraman**
Department of Neurosurgery and Gamma Knife Unit, Neurosciences Centre, All India Institute of Medical Sciences, New Delhi, India

**Jyotsna Vitore**
Department of Pharmaceutics, National Institute of Pharmaceutical Education and Research, Ahmedabad, Gujarat, India

**Shikha Yadav**
Department of Pharmacy, School of Medical and Allied Sciences, Galgotias University, Greater Noida, Uttar Pradesh, India

**Jin-Wook Yoo**
Department of Manufacturing Pharmacy, College of Pharmacy, Pusan National University, Busan, Republic of Korea

# About the editors

**Arun Kumar Singh** has completed his Master of Pharmacy (M.Pharm.) in pharmaceutics from Galgotias University, Greater Noida, Uttar Pradesh, India. Currently, he is serving as an assistant professor in the Department of Pharmacy at Vivekanand Global University, Jaipur, Rajasthan, India, where he is actively engaged in both teaching and research. His research interests encompass a wide range of emerging and interdisciplinary fields, including nano-formulation, blockchain technology, the Internet of Things (IoT), machine learning, cancer biology, artificial intelligence, big data analytics, and neuroscience. Demonstrating a strong commitment to academic excellence, Mr. Singh has made significant contributions to the scientific community. He has authored one book with IOP Publishing and has contributed five book chapters in the field of big data with the prestigious River Publishers, Denmark. Furthermore, he has published 28 review articles in reputed journals, including two high-impact papers in *Biochimica et Biophysica Acta (BBA)—Reviews on Cancer*. His research work reflects a cumulative impact factor of 95.6, underlining the quality and influence of his scholarly contributions. In addition to journal publications, Mr. Singh has authored and edited 58 books with internationally recognized publishers such as IOP Publishing, Elsevier, CRC Press, and Wiley, demonstrating his broad expertise and dedication to advancing knowledge in pharmaceutical sciences and related areas. Mr. Singh is known for his exceptional research skills, innovative thinking, leadership qualities, effective decision-making, and positive outlook. His diligent work ethic and unwavering commitment to academic and research excellence set him apart as an extraordinary professional in his field. With a passion for driving innovation and fostering interdisciplinary collaboration, Arun Kumar Singh continues to contribute to the growth of pharmaceutical sciences and aims to make a lasting impact in healthcare and technology-driven research.

**Dr. Vipan Kumar** is an associate professor at the School of Pharmaceutical Sciences, CGC University, Mohali, Punjab, India. He received his PhD in Pharmaceutical Sciences from Maharshi Dayanand University, Rohtak, Haryana, India, where he was awarded fellowship support from the University Grants Commission (UGC), New Delhi, India. Dr. Kumar's research focuses on medicinal chemistry, nanotechnology-based drug delivery, and the development of novel therapeutic agents for cancer and metabolic disorders. Dr. Kumar has published over 35 research and review articles in peer-reviewed national and international journals and has authored and edited 4 national and international books, along with several book chapters. He also holds four Indian patents in the areas of drug discovery and nanomedicine. His academic achievements have been recognized with several honors, including the Backbone Award (2025–26) from CGC University, Mohali, Punjab, India, and the Devi Lal Memorial Award (2020–21) from JCD Vidyapeeth, Sirsa, Haryana, India, for excellence in teaching and research. Dr. Kumar serves on the editorial boards of multiple international journals and acts as a reviewer for several reputed publications. He has delivered more than 30 oral and poster presentations at national and international scientific meetings, and he actively contributes to academic governance and policy development through his roles in ethical committees, inspection teams, examination committees, and university Boards of Studies.

**Dr. Pranav Kumar Prabhakar** is currently working as a professor and head at the Department of Biotechnology, School of Engineering and Technology, Nagaland University, Kohima, Nagaland, India. He is among the world's top 2% scientists (list published by Stanford University, United States, 2021, 2022, 2023, 2024, and 2025). He completed his PhD in biotechnology from IIT Madras.

His research focuses on elucidating molecular mechanisms and strategies for oral insulin delivery and mimicking signaling pathways in metabolic disorders (diabetes) using natural products. Dr. Pranav is a member of the Royal Society of Chemistry, London, United Kingdom, and the Asia-Pacific Chemical, Biological & Environmental Engineering Society. He serves as an editorial board member and reviewer for many reputed national and international journals. His honors include a travel grant from IIT Madras and the Council for Scientific and Industrial Research (CSIR) to attend ATTD 2009 in Greece, approved by the Department of Science and Technology (DST). He has published over 160+ research articles, authored/edited 24 books and 48 book chapters, and delivered 9 oral and poster presentations at scientific meetings.

**Dr. Meenakshi Dhanawat** is currently working as an associate professor at the Amity Institute of Pharmacy, Amity University Haryana, Gurugram, Haryana, India. She has a background in pharmaceutical sciences with a specialization in medicinal chemistry. She has done a master's at NIPER, Mohali, Punjab, India, and the project was based on glycosylamino acids. She completed her PhD from IIT-BHU, Varanasi, Uttar Pradesh, India, and she worked on CNS drug delivery targeting the 5HT1A/2A receptor. Apart from that, she has vast experience in the extraction of herbal drugs. Her areas of interest are synthetic chemistry, heterocyclic chemistry, nanotechnology, drug delivery, prodrug and co-drug drug design complexation, polymer chemistry, and their application in drug delivery. She has 17 years of teaching and research experience. She has attended several national and international conferences in India and abroad and is the recipient of the travelship award in the DDD section of AAPS-2010, National Biotechnology Conference, California, United States. She has received the international travel grants from DBT and DST to attend the International Conference in California, United States, and Munich, Germany, respectively. She has also received many other fellowship awards, including the Rajiv Gandhi National Fellowship. She has published more than 85 research/reviews along with 6 patents, 6 books, 22 book chapters, edited 2 books, and 1 copyright.

**Dr. Neeraj Mishra** is working as a professor and head of the Department of Pharmaceutics at the Amity Institute of Pharmacy, Gwalior, Madhya Pradesh, India, since July 2019. He has around 20 years of teaching and research experience. He has more than 140 publications of international and national repute in recent concepts of novel drug delivery systems, localized drug delivery, and targeted and controlled drug delivery of nanocarriers for the treatment of breast and colon cancer and neurodegenerative disorders. Dr. Mishra has supervised 30 M.Pharm. and 3 PhD students. Dr. Mishra has more than 4500 citations with an H-index of 37. The total cumulative impact factor of his published papers is more than 500 (as per SCOPUS). He has also written 25 books and 45 book chapters in Elsevier, Bentham, Springer, Wiley Publication, etc. He has also been granted three international patents and two Indian patents. Dr. Mishra is the recipient of the "Distinguished Professor Award" 2019 from DST-NSTMIS, SPAICS, Indore, Madhya Pradesh, India, in September 2019. Dr. Mishra has also received the "Outstanding Scientist Award," 2020, Sixth International Scientist Awards on Engineering, Science, and Medicine, June 20–21, 2020, Chennai, India. Dr. Mishra has received the "Distinguished Professor Award" 2021 from the Indian Pharmaceutical Association, Madhya Pradesh State Branch, Indore, Madhya Pradesh, India. Dr. Mishra has received the Best Academician Award in the Indian Pharmacy Graduate Association, M.P. State International Conference held on May 06, 2023, at DAVV Auditorium, Indore, India. Recently, he was honored to be in the world's top 2% of scientists by prestigious Stanford University, CA, United States, and Elsevier in 2025.

# Preface

Colon cancer remains a significant global health challenge, ranking among the leading causes of cancer-related morbidity and mortality. Despite notable progress in early detection and conventional treatments such as surgery, chemotherapy, and radiotherapy, its effective management is often limited by late diagnosis, systemic toxicity, drug resistance, and insufficient targeting of cancer cells. These challenges emphasize the urgent need for more efficient, precise, and patient-friendly therapeutic strategies. In recent years, nanotechnology has emerged as a powerful tool in oncology, offering innovative solutions to overcome the limitations of traditional therapies. Nano drug delivery systems enhance drug solubility, improve pharmacokinetics, and enable targeted delivery to tumor tissues, thereby increasing therapeutic efficacy while minimizing adverse effects. Diverse nanocarriers—including liposomes, polymeric nanoparticles, dendrimers, and solid lipid nanoparticles—have shown considerable promise in improving treatment outcomes for colon cancer. *Nano Drug Delivery Systems for Colon Cancer: Innovations, Challenges, and Applications* presents a comprehensive and up-to-date overview of advancements in nanomedicine for colon cancer therapy. It begins with an exploration of colon cancer pathophysiology and the shortcomings of existing treatments, followed by detailed discussions on the design, formulation, and characterization of various nanocarriers. Particular emphasis is placed on targeting strategies, including passive and active targeting, as well as stimuli-responsive systems that adapt to the tumor microenvironment for enhanced drug delivery. This book also addresses key challenges in clinical translation, such as biocompatibility, toxicity, large-scale manufacturing, regulatory considerations, and cost-effectiveness. By highlighting both opportunities and barriers, it provides a balanced perspective essential for advancing research into clinical applications. Furthermore, emerging approaches such as theranostics, combination therapies, gene delivery, and personalized nanomedicine are explored, reflecting the growing integration of nanotechnology with precision medicine. Designed for researchers, clinicians, academicians, and students, this book aims to inspire innovation and collaboration, ultimately contributing to the development of safer and more effective treatments for colon cancer.

**Arun Kumar Singh, Vipan Kumar, Pranav Kumar Prabhakar, Meenakshi Dhanawat, and Neeraj Mishra**

# Acknowledgments

The author gratefully acknowledges the invaluable support and encouragement provided by the Honorable CEO of Vivekananda Global University, Jaipur, Rajasthan, India. Sincere thanks are also extended to the entire team of the Faculty of Pharmaceutical Sciences and Nursing for their continued cooperation and contributions that have significantly facilitated the research and development activities associated with this work. Lastly, my deepest and most heartfelt gratitude goes to my beloved family, especially my father, Mr. Kshetra Pal Singh, whose unwavering love, patience, and encouragement have been my greatest source of strength.

**~ Mr. Arun Kumar Singh**

We would like to express our sincere gratitude to the Honorable S. Rashpal Singh Dhaliwal, Founder Chancellor, and Mr. Arsh Dhaliwal, Managing Director, CGC University, Mohali, Punjab, India, for their visionary leadership and continued commitment to foster academic excellence and research-driven initiatives, which provided an encouraging environment for the successful completion of this edited book. We would like to sincerely thank the respected Vice Chancellor of CGC University, Mohali, Prof. (Dr.) Vinay Goyal, for his visionary leadership, constant encouragement, and unwavering support in promoting academic and research excellence. His guidance has played a significant role in the successful completion of this book. I am also grateful to Prof. (Dr.) Anish Gupta, Pro Vice-Chancellor—Academics and International Affairs at CGC University, Mohali, Punjab, India, for his valuable academic support and continued encouragement. My heartfelt appreciation is extended to Dr. Sandeep Singh, Dean of Research, for his consistent support throughout the drafting of this book. I also acknowledge Prof. (Dr.) Deepak Kapoor, Dean, School of Pharmaceutical Sciences, and Prof. (Dr.) Vikas Bansal, Head, School of Pharmaceutical Sciences, CGC University, Mohali, Punjab, India, for their institutional guidance and constructive support. Finally, I extend my deepest gratitude to my family members for their patience, understanding, and unwavering encouragement throughout this journey.

**~ Dr. Vipan Kumar**

I would like to express my sincere gratitude to the Honorable Vice Chancellor of Nagaland University, Prof. Jagadish Kumar Patnaik, for his visionary leadership, constant encouragement, and unwavering support in strengthening academic and research excellence. His guidance and motivation have played a significant role in the successful completion of this edited volume. I am equally thankful to Prof. N. Vehun, Pro Vice-Chancellor, Nagaland University, Kohima Campus, Kohima, Nagaland, India, for his continuous encouragement and academic support, which have greatly inspired our scholarly endeavors. My heartfelt appreciation is extended to Prof. Sujata Dash, Dean, School of Engineering and Technology, Nagaland University, Kohima Campus, Kohima, Nagaland, India, for her valuable guidance, institutional support, and encouragement throughout the preparation of this book. Her commitment to academic growth and research excellence has been deeply motivating. I also acknowledge with gratitude the support and cooperation of our colleagues, faculty members, and staff of Nagaland University, whose assistance and encouragement contributed meaningfully to this work. Finally, we express our deepest gratitude to our family members for their unwavering love, patience, and constant support. Their understanding and encouragement have been the cornerstone of our academic journey, and this book stands as a testament to their enduring faith in us.

**~ Prof. Pranav Kumar Prabhakar**

We sincerely thank Honorable Founder-President Dr. Ashok K. Chauhan, Ritnand Balved Education Foundation (RBEF), and Dr. Aseem Chauhan, Chancellor, Amity University Haryana, Gurugram, Haryana, India, for their visionary leadership and inspiration. We gratefully acknowledge Honorable Pro-Chancellor Lt. Gen. V. K. Sharma, AVSM (Retd.), and Vice Chancellor Prof. (Dr.) P. B. Sharma for their constant guidance and support. We extend our appreciation to Pro Vice-Chancellor Prof. Vikas Madhukar for fostering a strong research environment and to Prof. Satish Sardana, Head of Institution (HOI), Amity Institute of Pharmacy, Amity University Haryana, Haryana, India, for their valuable guidance during this work. We also thank the faculty and staff of Amity Institute of Pharmacy, Amity University Haryana, Gurugram, Haryana, India, for their support.

**~ Dr. Meenakshi Dhanawat**

We would also like to express our sincere gratitude to Honorable Founder-President Dr. Ashok K. Chauhan, Ritnand Balved Education Foundation (RBEF), and Dr. Aseem Chauhan, Chancellor, Amity University Madhya Pradesh, Gwalior, India. We are thankful to our honorable pro-chancellor Lt. Gen. V. K. Sharma, AVSM (Retd.), Amity University Madhya Pradesh, Gwalior, India; Vice Chancellor Prof. (Dr.) R.S. Tomar, Amity University Madhya Pradesh, Gwalior, India; and Pro Vice-Chancellor (Research) Prof. (Dr.) M. P. Kaushik, Amity University Madhya Pradesh, Gwalior, India, for their continuous encouragement. We are also grateful to Prof. (Dr.) Manoj Kumar Mishra, Director, Amity Institute of Pharmacy, Amity University Madhya Pradesh, Gwalior, India, for her support during the writing of this book. Finally, we would like to express our wholehearted appreciation to the faculty and staff members of Amity Institute of Pharmacy, Amity University, Madhya Pradesh, Gwalior, India.

**~ Prof. Dr. Neeraj Mishra**

CHAPTER

# Etiology and pathophysiology of colon cancer

1

**Shivam Sharma[1], Darpan Kalra[2], Meenakshi Dhanawat[3], Garima Malik[4] and Pramila Chaubey[5]**

[1]*Department of Pharmacy Practice, Guru Gobind Singh College of Pharmacy, Yamunanagar, Haryana, India,*
[2]*M.M College of Pharmacy, Maharishi Markandeshwar (Deemed to be University), Ambala, Haryana, India,*
[3]*Amity Institute of Pharmacy, Amity University Haryana, Gurugram, Haryana, India,*
[4]*Chitkara College of Pharmacy, Chitkara University, Rajpura, Punjab, India,*
[5]*Department of Pharmaceutics, College of Pharmacy, Shaqra University, Shaqra, Kingdom of Saudi Arabia*

## 1.1 Introduction

Colon cancer (Fig. 1.1) is an ailment that occurs explicitly in the colon and is brought on by the aberrant development of glandular epithelial cells. Colon cancer originates from specific polyps or neoplasms in the colon's mucosal layer (Gajendran et al., 2019; Xiao et al., 2019). Healthcare practitioners possess screening techniques that may identify precancerous polyps before they transform into cancerous tumors (Smith et al., 2019). The three main categories of colorectal cancer (CRC) are sporadic, inherited, and colitis-related. The incidence of colon cancer cases is steadily rising worldwide. Both hereditary and environmental factors act as determinants in the risk of developing CRC. Furthermore, the likelihood of acquiring CRC in individuals with chronic ulcerative colitis and Crohn's disease rises as they become older. Various studies have shown that risk factors for CRC include nutrition and chronic illness, genetic background, and lifestyle (Edwards et al., 2010; Menon & Cagir, 2025). The most efficient approach to prevent colon cancer and decrease mortality rates associated with it is to screen individuals who are at an average risk. Consequently, several European nations, Canada, and some areas in America, both north and south, Oceania, and Asia, have implemented population-based screening programs (Navarro et al., 2017). Participation in CRC screening is based on age and place of residence. Furthermore, population-based screening seeks to identify hidden diseases within the general population at an average risk, allowing for early treatments and decreasing the risk to individuals and communities (Harford, 2006). Screening is especially suitable for CRC because of its high prevalence and the belief that it follows a steady progression from adenoma to carcinoma (Morson, 1984). Moreover, the development of CRC may be averted by eliminating colorectal adenomas (Winawer et al., 1993). Additionally, the likelihood of patient mortality decreases if CRC is identified earlier. Interventions targeting the adenoma-carcinoma pathway have a beneficial influence on treatment outcomes. Additional effective tactics involve the identification and surveillance of high-risk populations, such as people with hereditary CRC syndrome, those with inflammatory bowel disease, and those with a family history indicating a genetic susceptibility to CRC but lacking identifiable genetic markers, as well as individuals displaying phenotypic characteristics indicative of heightened risk. The metabolome facilitated the identification of crucial biological processes influenced by genetic

Nano Drug Delivery Systems for Colon Cancer. DOI: https://doi.org/10.1016/B978-0-443-40443-6.00009-X

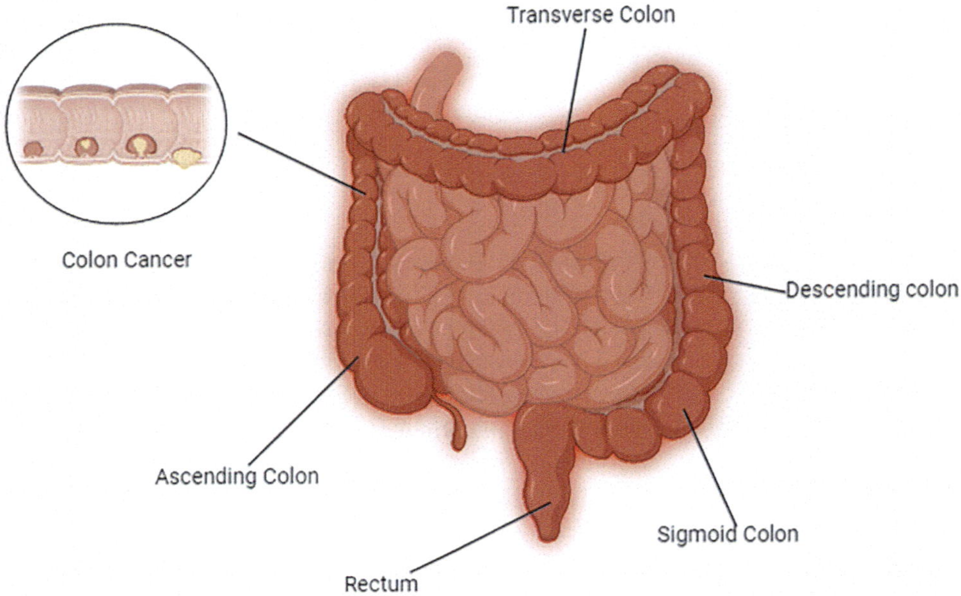


**FIGURE 1.1**

Diagrammatic representation of colon cancer anatomically by myoclonic.

*Created with BioRender.com.*

variation (Garcia-Etxebarria et al., 2021). The two most often used procedures for CRC screening are fecal occult blood tests (FOBTs) and lower endoscopy (Keum & Giovannucci, 2019). The formation of CRC requires a significant amount of time, usually spanning many years. Typically, it takes around 10–15 years to develop a polyp into a cancerous tumor. Frequent screening, early polyp identification, and removal are essential to avoid colon cancer. The current diagnostic methods have a limited ability to identify just 40% of CRC cases in the early stages. Additionally, there is a possibility of colon cancer recurring after undergoing surgery and postsurgical therapy ("Cancer Facts & Figures, 2020" American Cancer Society. Accessed February 3, 2025). Chemotherapeutic drugs selectively target cancer cells while simultaneously causing collateral damage to healthy cells in the surrounding environment. Most colon cancer patients develop resistance to modern chemotherapies, which reduces the effectiveness of anticancer drugs and eventually results in chemotherapy failure. Hence, it is essential to thoroughly examine potential risk factors, epidemiology, and prevention strategies for colon cancer using the latest information based on available evidence to address future CRC-related issues (Hossain et al., 2022a).

## 1.2 Epidemiology

According to the American cancer society (ACS), there will be an estimated 152,810 new instances of CRC in 2024. Estimates indicate that there will be 81,540 cases among males and 71,270 cases among

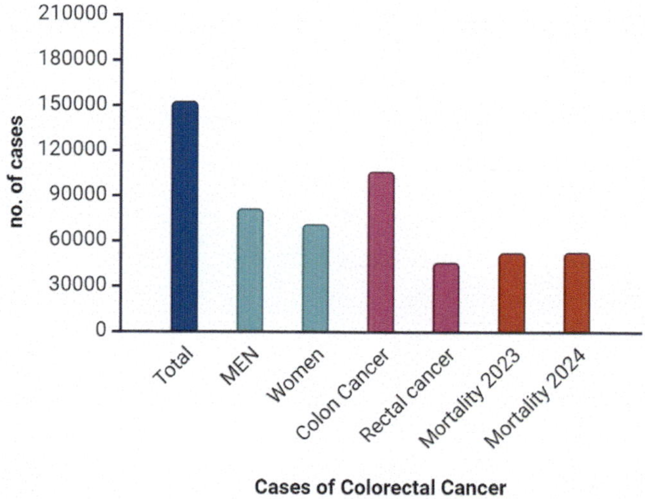


**FIGURE 1.2**

Cases of colorectal cancer (CRC) according to ACS releases CRC estimates for 2024.

*Created with BioRender.com.*

women. Out of the total, 106,590 instances will be screened for cancer of the colon, while 46,220 cases will be screened for rectal cancer. CRC is the second most significant contributor to mortality, which is cancer-associated mortality in the United States. It is projected that there will be around 53,010 fatalities due to CRC in 2024, which represents a slight rise compared to the predicted 52,550 mortalities in the previous year (Fig. 1.2).

It is estimated that there are over 608,000 fatalities globally associated with CRC, which represents 8% of all mortalities related to cancer. This makes CRC the fourth leading cause of deaths related to cancer. In India, the yearly occurrence rate for colon and rectal cancer in males is 4.4 and 4.1 per 100,000 individuals, respectively. The age-adjusted rate (AAR) for women with colon cancer is 3.9 per 100,000 individuals. Among males, cancer of the colon is ranked eighth, while cancer of the rectum is ranked ninth. Rectal cancer is not among the top 10 malignancies for women, but colon cancer is ranked ninth. The annual incidence of colon cancer in India is estimated to be around 70,038 new cases per year. Colon cancer is ranked as the second most common cancer in women and the third most common in men in India. Urban-rural disparities: the prevalence is more remarkable in urban and rural areas ("Cancer Facts & Figures, 2020." American Cancer Society. Accessed February 3, 2025).

## 1.3 Etiology

A complicated illness, CRC is impacted by several variables, including environmental exposures (such as nutrition), inflammatory diseases of the digestive system, hereditary factors, etc. (Fig. 1.3).

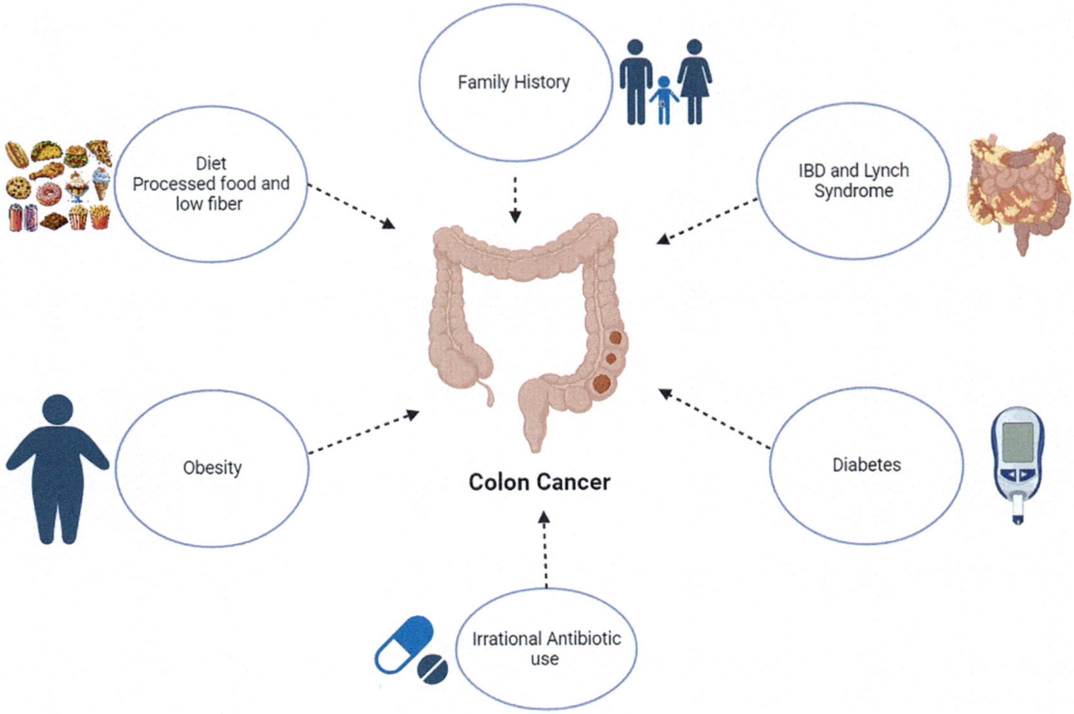


**FIGURE 1.3**

Etiology of colon cancer.

*Created with BioRender.com.*

While there is still much to learn about the genetics of CRC, current research suggests that genetic factors have the strongest association with the disease. In particular, a hereditary mutation of familial adenomatous polyposis (FAP), a disorder caused by the adenomatous polyposis coli (APC) gene, significantly raises the chance of developing colon cancer before the age of 40.

## 1.3.1 Hereditary nonpolyposis colon cancer syndrome (Lynch syndrome)

Hereditary nonpolyposis colorectal cancer (HNPCC), also known as Lynch syndrome, carries a lifetime risk of around 40% for developing CRC. Additionally, people with this condition are more likely to develop endometrial cancer, urothelial carcinoma, and other less common types of cancer. Lynch syndrome is defined by a lack of proper mismatch repair (dMMR) caused by a hereditary mutation in one of the mismatch repair genes, including hMLH1, hMSH2, hMSH6, hPMS1, hPMS2, and possibly more genes that have not yet been identified. HNPCC accounts for around 6% of all cases of colon cancer. Research conducted by Burn et al. discovered that the use of resistant starch and aspirin, either separately or together, did not have any impact on the occurrence of CRC in individuals with Lynch syndrome, even though aspirin has been shown to lower the chance of colorectal neoplasia in some groups (Burn et al., 2008).

### 1.3.2 Dietary factors

Researchers are now conducting extensive examinations of dietary issues (Meyerhardt et al., 2007). Epidemiological studies have shown a correlation between a heightened susceptibility to CRC and a diet characterized by an elevated consumption of fat, which is of animal origin, such as goat/red meat, and a diet low in fiber with insufficient intake of vegetables and fruits. Research conducted by Aune et al. revealed that consuming a significant amount of fiber was linked to a decreased likelihood of developing CRC. Specifically, the study showed that cereal fiber and whole grains were efficacious (Aune et al., 2011). Pala et al. discovered that taking a significant quantity of yogurt is associated with a reduced likelihood of developing CRC. Longitudinal research conducted by Tabung et al., which tracked 121,050 persons over 26 years, discovered that people of both genders who consumed proinflammatory diets (rich in specifically processed meat, red meat, and animal organ origin, among other things) had a considerably increased likelihood of acquiring CRC (Pala et al., 2011). The risk was particularly elevated in overweight and obese males and, interestingly, in lean females. Additionally, the risk was heightened in both males and females who abstain from alcohol use (Tabung et al., 2018).

Consuming an excessive amount of drinks that contain high-fructose corn syrup (HFCS) is associated with a higher likelihood of developing CRC. A study was conducted on mice with APC mutations, which make them more likely to develop intestinal tumors. The mice were given a daily dose of 20 g of weight-adjusted HFCS, equivalent to one soda per day. This led to a substantial increase in polyps, quickly progressing into advanced, high-grade dysplastic lesions. Carbon tagging demonstrated the absorption of fructose by the intestinal tumors themselves. Fructose was metabolized into fructose-1-phosphate inside the tumors, which then triggered the activation of glycolysis and promoted the production of fatty acids. These fatty acids provide the necessary resources for the development of the tumor (Goncalves et al., 2019).

### 1.3.3 Obesity and lifestyle

Being overweight and having unhealthy habits, such as drinking alcohol, tobacco use, and sedentary behavior, are known to raise the risk of CRC. A metaanalysis of prospective cohort studies found that current smokers had a slight but significantly increased risk of CRC, with the risk being higher in men and for rectal cancers compared with colon cancers and still present in former smokers.

### 1.3.4 Family history and alcohol intake

Cho and colleagues conducted a comprehensive investigation and found that those with a familial history of CRC who drink high amounts of alcohol had a higher chance of developing the illness. The correlation was only significant for those who consumed 30 g or more of alcohol per day. No significant linear trend was observed. This was compared to those who do not drink and are not prone to CRC in their family. A relative risk of 2.80 for colon cancer exists for those who consume 30 g/d or more and have a family history of CRC (Cho et al., 2012).

### 1.3.5 Your age, racial background, and gender

As you age, your chance of CRC increases. Although it can strike younger persons, it is far more prevalent beyond the age of 50. Uncertainty surrounds the cause of the increase in CRC in those under 50. The most significant rates of CRC in the United States are seen in the Indian population living in America and Native individuals of Alaska, followed by African American men and women. Of all the ethnic groups in the world, Ashkenazi Jews, or Jews of Eastern European ancestry, have some of the highest rates of CRC. Compared to women, men are more likely to die from CRC. The causes are not entirely evident. Right-sided colon cancer is more common in women with CRC, especially if they are no longer menstruating (postmenopausal) (Zhou et al., 2022).

### 1.3.6 Diabetes

Current screening guidelines advise clinicians to consider the heightened risk of this cancer in affected people who smoke or have above-normal body weight. However, these guidelines do not specifically emphasize the elevated risk in diabetic individuals. Diabetes is a distinct risk factor for colon and rectal cancer based on a thorough review of case-control and cohort studies. Further analysis of specific subgroups within these studies has confirmed that these results were consistent across several demographics. This new information may have implications for updating screening guidelines and developing more accurate risk models for CRC (Yuhara et al., 2011).

### 1.3.7 Body mass index and colorectal adenomas

Previous research has shown an association between a person's body mass index (BMI) and the risk of colon cancer and adenomas. However, only some studies have had a sufficient sample size for detailed analysis based on different subgroups. Jacobs et al. conducted a metaanalysis using statistics and facts from 8213 people in seven prospective studies. They discovered a substantial correlation between BMI and significant histologic features of metachronous adenomas in males, not females. The researchers determined that body size may have an impact on the development of CRC, especially in males, during the early stages of the disease (Jacobs et al., 2009).

The loss of APC often leads to the initiation of the Wingless-Type MMTV Integration Site Family (WNT) signaling pathway, which is critical for the progression of CRC. The WNT pathway is primarily regulated by CTNNB1 (β-catenin), a significant mediator. The WNT-CTNNB1 signaling pathway is also implicated in metabolic diseases such as type II diabetes, obesity, and glucose metabolism. Therefore, Morikawa and colleagues proposed that the relationship between obesity, physical activity, and the risk of CRC may vary depending on the specific subtypes of tumors based on CTNNB1 status (Bailly et al., 2020).

By utilizing a molecular pathological epidemiology database, the researchers established a clear correlation between an elevated risk and a higher BMI of CTNNB1-negative cancer. Conversely, they also observed that engaging in A decreased risk was linked to increased physical activity levels for

CTNNB1-negative cancer. However, no significant relationship was found between either physical activity level or BMI and the risk of CTNNB1-positive cancer (Morikawa et al., 2013).

### 1.3.8 Gastric comorbidities

Individuals with ulcerative colitis and Crohn's disease, classified as inflammatory bowel diseases, face a heightened risk of colorectal adenocarcinoma. The probability of developing CRC escalates proportionately with the duration of inflammatory bowel disease and the extent of colon involvement.

### 1.3.9 Irrational use of antibiotics

A case-control investigation conducted in the United Kingdom from 1989 to 2012 demonstrated a notable association between the administration of oral antibiotics and an elevated susceptibility to colon cancer, especially in the proximal segment of the colon. This association was observed when individuals were exposed to antibiotics more than 10 years prior to being diagnosed with colon cancer. The risk of developing colon cancer was found to depend on the dosage of antibiotics, and even a single course of antibiotics was associated with an elevated risk. Furthermore, the use of antianaerobic antibiotics was found to pose the most significant risk. The authors of the study suggest that these antibiotics disturb the gastrointestinal (GIT) microbiota, which is primarily composed of anaerobic bacteria, and this disruption may contribute to the disturbances in colon microbiota that promote the growth of cancer cells (Zhang et al., 2019).

### 1.3.10 A documented personal medical history of exposure to radiation in the abdominal or pelvic regions

Your risk of CRC is elevated if you have previously survived cancer and were exposed to radiation therapy in the abdomen and pelvis, where your colon is located, as part of your treatment. You may need to begin screening for CRC earlier in life and be examined more frequently if you have had radiation to the abdomen or pelvis, particularly during childhood.

Studies reveal that rectal tissue is capable of absorbing a fraction of radiation during treatment, resulting in an elevated risk of rectal cancers among men treated with radiation for prostate cancer. The cohort primarily examined in this research consists of males who received treatment during the 1980s and 1990s, characterized by less precise radiation techniques than contemporary standards. The implications of modern radiation methodologies on the risk of rectal cancer remain ambiguous.

### 1.3.11 Familial adenomatous polyposis

APC gene mutations, which are inherited from one's parents, are the cause of FAP. FAP accounts for around 1% of all colorectal malignancies.

The most prevalent kind of FAP commonly begins between the ages of 10 and 12 and involves hundreds or thousands of polyps growing in the colon and rectum. As early as age 20, one or more of these polyps often become cancerous. Almost all FAP patients will have colon cancer by the time they are 40 if their colon removal has not been done for prevention. Additionally, there is a higher chance of stomach, small intestine, pancreatic, liver, and several other organ cancers in people with FAP.

There are three subtypes of FAP

1 Less than 100 polyps are seen in people with attenuated FAP (AFAP), and CRC typically strikes later in life (in the 40s and 50s).
2 A form of FAP known as Gardner syndrome also results in noncancerous tumors of the bones, soft tissues, and skin.
3 APC gene mutations in Turcot syndrome increase the likelihood of developing numerous adenomatous polyps, colon cancer, and medulloblastoma, a specific type of brain cancer.

#### 1.3.11.1 Rare inherited conditions associated with colorectal cancer

1. **Peutz-Jeghers syndrome (PJS)**: freckles around the mouth (and occasionally on the hands and feet) and a particular kind of polyp in the digestive system called hamartomas are common in people with this hereditary disorder. These individuals have a significantly increased risk of CRC, in addition to other malignancies such as pancreatic, ovarian, and breast cancers. They are typically identified at a younger age than is customary. The STK11 (LKB1) gene is mutated in this condition.
2. **MUTYH-associated polyposis (MAP)**: people with this disease form numerous colon polyps. If not thoroughly monitored with regular colonoscopies, they tend to develop into cancer. These individuals are also more susceptible to thyroid, breast, ovarian, bladder, and other GI (gastrointestinal) tract malignancies. This disease, which frequently results in cancer at an earlier age, is caused by mutations in the MutY DNA Glycosylase (MUTYH) gene, which is involved in "proofreading" the DNA and correcting any errors.
3. **Cystic fibrosis (CF)**: the cells in some bodily organs produce mucus that is thicker and stickier than usual in people with CF, a hereditary disorder. Health issues may result from this, particularly in the pancreas and lungs. It has been evident that individuals with CF are also more susceptible to CRC, which often manifests at a significantly earlier age than in those without the illness, since improved medical treatment has allowed them to live longer. Those who have received an organ transplant, such as a lung transplant, are considerably more at risk for CRC. Mutations in the cystic fibrosis transmembrane conductance regulator (CFTR) gene cause CF (Colorectal Cancer Risk Factors: Hereditary Colorectal Risk Factors, 2025).

## Other risk factors

A systematic review and metaanalysis conducted by Zhou et al. found that there is a correlation between higher adult-attained height and a greater risk of developing CRC and adenoma. The study comprised 47 observational studies, which included a total of 280,644 cases of CRC and 14,139 instances of colorectal adenoma. The research revealed that individuals in the highest percentile of height had a 24% greater chance of developing CRC compared to those in the lowest percentile of height. Each additional 3.9-inch (10-cm) increment in height corresponded to a 14% elevated likelihood of having colon cancer and a 6% increased probability of having adenoma. Zhou et al. suggest that height should be considered a risk factor for CRC screening (Zhou et al., 2022).

## 1.4 Pathophysiology

The evolution from normal colonic epithelium to dysplastic modifications entails a steady rise in genetic changes, ultimately giving rise to carcinoma. Colon malignancy may emerge via three predominant genetic mechanisms: chromosomal instability (CIN), mismatch repair (dMMR), and the CpG island methylator phenotype (CIMP). These genetic mechanisms are not mutually exclusive but demonstrate considerable interrelation. The pathophysiology of colon cancer entails an intricate interaction between genetic and epigenetic factors, which collectively subsidize the onset and advancement of the disease (Grady & Carethers, 2008).

It includes:

- Heterogeneity of CRC with three main molecular pathways: CIMP, microsatellite instability (MSI)/MMR (DNA mismatch repair), and CIN.
- Genetic mutations in critical genes such as APC, Kirsten-ras, p53, and MutY DNA Glycosylase (MYH).
- MSI associated with Lynch syndrome and sporadic CRC (Kocher et al., 2007).
- Epigenetic changes include CIMP and DNA hypermethylation.
- Serrated pathway characterized by B-Raf Proto-Oncogene, Serine/Threonine Kinase (BRAF) mutations and promoter hypermethylation (Kambara et al., 2004).
- Involvement of the WNT signaling pathway and tumor suppressor genes, like p53 (Kuhnert et al., 2004).
- Progression from benign neoplasms to malignant tumors through recognized models like the adenoma-carcinoma sequence and the serrated pathway (Cadigan & Liu, 2006).

### 1.4.1 Chromosomal instability pathway

Previously designated as the classical adenoma-carcinoma progression, the CIN pathway is characterized by a mutation gain that precipitates an imbalance between oncogenes and tumor suppressor genes. Common genetic alterations identified include APC, Kirsten Rat Sarcoma Viral Oncogene Homolog (KRAS), and TP53. Variations in the APC gene typically initiate neoplasia and can be observed in nearly 60% of CRCs. The conventional APC protein interacts with β-catenin, modulating the WNT signaling pathway. Alterations in the APC gene disrupt this regulatory mechanism, resulting in cellular proliferation, apoptosis, and differentiation anomalies. Mutations in KRAS and BRAF are observed in CIN tumors, yet they are not confined exclusively to this pathway (Wielandt et al., 2020).

### 1.4.2 dMMR pathway

dMMR refers to the genetic DNA mismatch repair system, which is often altered in MLH1, MSH2, MSH6, or PMS2 genes. These mutations lead to a gradual buildup of errors during DNA replication. Tissue samples can be analyzed to detect MSI related to dMMR. Germline mutations in dMMR are associated with HNPCC. However, most tumors exhibiting high microsatellite instability (MSI-high) are typically sporadic. Lesions with MSI-high characteristics are often found on the right side, have poorly differentiated or mucinous histological features, and exhibit high resistance to 5-fluorouracil (5-FU). These tumors respond better to immunotherapy than those with stable microsatellites (Mármol et al., 2017).

### 1.4.3 CpG island methylator phenotype pathway

The CIMP pathway is involved in about 15% of colon malignancies and is characterized by excessive methylation of CpG islands. CIMP-related tumors often develop in serrated polyps rather than traditional adenomas and are frequently linked to KRAS and BRAF mutations. When seen with MSI, the prognosis is much more favorable than that of hypermethylated non-MSI-high tumors (Huang et al., 2021). Recently, a comprehensive analysis was conducted on several gene-expression-based categorization approaches, identifying and categorizing four key molecular subtypes (Guinney et al., 2015).

- CMS1 is characterized by being MSI-immune (14%), having a high load of hypermutation, dMMR, MSI, and substantial immune activation.
- CMS2 (canonical, 37%) is characterized by high CIN, mostly in epithelial cells, and shows significant activation of WNT and myelocytomatosis proto oncogene (MYC) signaling pathways.
- CMS3 (metabolic, 13%) is characterized by epithelial cells and exhibits apparent metabolic dysregulation, along with a KRAS mutation.
- The characteristics of CMS4 include mesenchymal subtype (23%), CpG hypermethylation, significant activation of transforming growth factor-β, stromal invasion, and angiogenesis. This categorization enables a sophisticated understanding of the interrelated expression of genes; however, it is not yet prevalent in clinical use (Fig. 1.4).

## 1.5 Clinical manifestation

Colon cancer can exist without showing any signs. If you do experience symptoms, you might not be sure whether changes in your body signal colon cancer. This uncertainty arises because certain symptoms of colon cancer can mimic those of less serious conditions. Typical indications of colon cancer include:

- Hematochezia: consult a healthcare professional if you observe blood in the bowl or on toilet paper after defecation, or if your stool appears dark or bright red. It is crucial to bear in mind that the presence of blood in feces does not always indicate colon cancer. Various factors, such as hemorrhoids, anal tears, and consumption of beetroots, might alter the appearance of your stool. However, it is advisable to see a healthcare professional if you observe blood in or on your stool.
- If you have persistent alterations in your bowel habits, such as chronic constipation or diarrhea, or if you have a lingering sensation of needing to defecate even after using the toilet, it is advisable to consult a healthcare professional.
- Abdominal discomfort, often known as belly pain, should be discussed with a healthcare physician if it is persistent, severe, and has no identifiable cause. Several factors might contribute to abdominal pain. However, it is advisable to see a healthcare professional if you have uncommon or recurring abdominal discomfort.
- Bloating in stomach: various factors might contribute to the sensation of a distended abdomen, similar to abdominal discomfort. Suppose you have persistent bloating for over a week, worsening symptoms, or other signs such as vomiting or blood in your stool. In that case, it is advisable to see a healthcare professional.

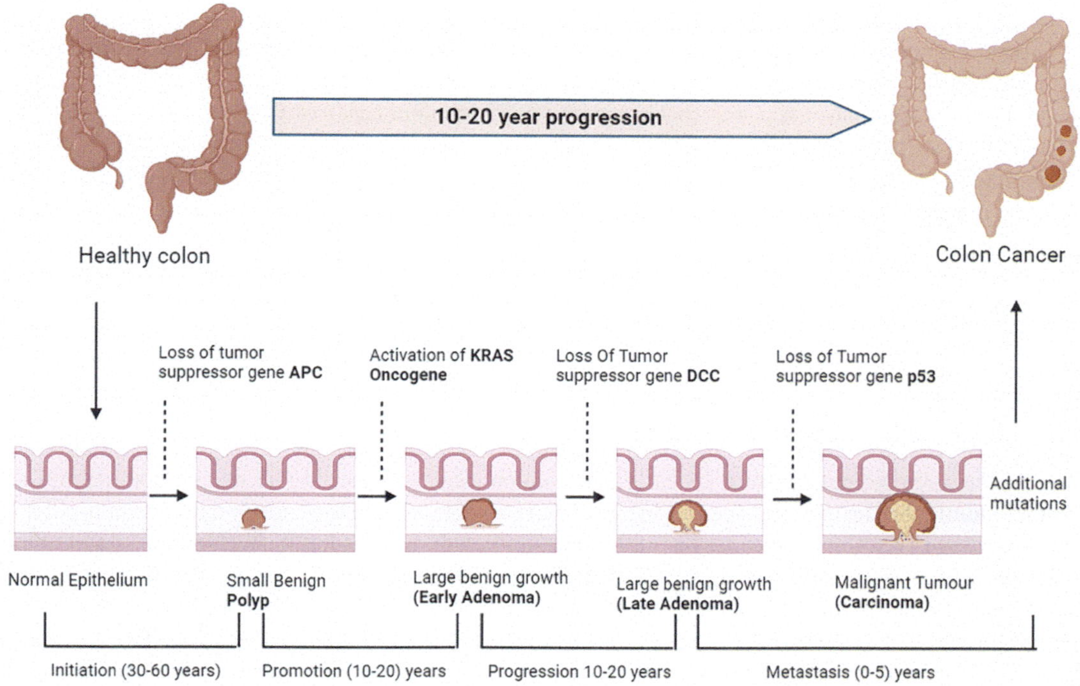


**FIGURE 1.4**

Colorectal cancer (*CRC*) stages and development. There are four stages in the development of CRC carcinogenesis: initiation, promotion, progression, and metastasis.

*Hossain, Sanower, M., Karuniawati, H., Jairoun, A.A., Urbi, Z., Ooi, D.J., John, A., Lim, Y.C. et al. (2022). Colorectal cancer: A review of carcinogenesis, global epidemiology, current challenges, risk factors, preventive and treatment strategies.* Cancers, 14, *(7). https://doi.org/10.3390/CANCERS14071732; created with BioRender.com.*

- Significant weight loss that occurs without deliberate effort is referred to as unexplained weight loss.
- Vomiting: consult a healthcare professional if you have recurrent vomiting without a discernible cause, or if you vomit excessively during 24 hours.
- Fatigue and dyspnea are indicative of anemia. Anemia may serve as an indication of colon cancer ("Colorectal Cancer." Centers for Disease Control and Prevention. Accessed February 3, 2025).
- Lesions on the right side of the body are often seen in younger individuals. The typical symptoms that are frequently observed in these cases include bleeding and diarrhea. Tumors located on the left side of the body are more typically seen in elderly individuals and often result in bowel obstruction symptoms. Research using data from England's Clinical Practice Research Datalink revealed that abdominal pain was the predominant symptom among patients under 50. This demographic is experiencing an increase in CRC incidence. In comparison to other age groups, these younger patients had the lowest proportion of common indicators and symptoms (such as abdominal bulk, diarrhea, altered bowel habits, anemia, and rectal hemorrhage). On the other

hand, these individuals were more likely to have sought medical care from their primary healthcare practitioner during the year before diagnosis due to vague symptoms (Arhi et al., 2019).

- Similarly, case-control research that compared 5075 incident cases of CRC in patients under the age of 50 revealed specific clinical symptoms as red-flag indicators in this age range. These symptoms included iron deficiency anemia, diarrhea, stomach pain, and rectal bleeding that occurred three months to two years prior to the diagnosis. Each clinical manifestation was linked to a 1.94 times greater risk, having any two was connected with a 3.59 times greater risk, and having three or more was related to a 6.52 times greater risk. These correlations were particularly pronounced in younger age groups (Fritz et al., 2023).

## 1.6 Diagnostic considerations

Due to the asymptomatic nature of early-stage colon cancer, screening is crucial for diagnosing treatable malignant lesions and identifying precancerous lesions, such as adenomatous colon polyps. The reduction in CRC incidence and mortality rates in recent decades has been chiefly ascribed to the extensive use of screening practices ("Cancer Facts & Figures, 2024" American Cancer Society. Accessed February 3, 2025). Screening recommendations advocate for various tests and methods that identify adenomatous polyps and cancer or predominantly focus on cancer detection. The American College of Gastroenterology recommends that a colonoscopy be performed every decade, starting at age 50, as the optimal screening approach (Shaukat et al., 2021). A rectal examination, colonoscopy, and biopsy of any concerning lesions are required if CRC is suspected. All patients diagnosed with CRC under the age of 70 should be tested for HNPCC (Lynch syndrome), according to the National Comprehensive Cancer Network. Patients 70 years of age and older should only be tested if they meet the updated Bethesda guidelines for HNPCC. Following tissue diagnosis confirmation, laboratory tests aim to measure tumor burden and assess organ function (liver, kidneys) in advance of diagnostic and treatment measures. For staging, thorough imaging of the abdomen and chest should be obtained, ideally prior to surgery.

### 1.6.1 Blood studies

Laboratory investigations evaluate patients' organ function (liver, kidneys) to prepare for diagnostic and therapeutic interventions and measure tumor burden. Research may include the following:

- Total number of blood cells
- Chemicals in serum
- Tests of liver function
- Tests of renal function
- The level of serum carcinoembryonic antigen (CEA)

A preoperative baseline CEA level should be obtained because of its prognostic significance, as markedly increased levels may suggest more advanced, disseminated illness. Elevated blood CEA levels have been linked to a worse prognosis in individuals with resectable CRC; nevertheless, this biochemical measure has not yet been incorporated into CRC staging criteria (Thirunavukarasu et al., 2011).

### 1.6.2 Imaging studies

Comprehensive imaging of the thoracic and abdominal regions should be conducted for staging, ideally before surgical intervention. To accurately identify problems involving the abdomen and liver, various imaging methods, such as abdominal and pelvic CT, enhanced ultrasound imaging, and MRI, may be used. Imaging evaluations may encompass a chest radiograph, a chest CT scan, and an abdominal barium swallow study to delineate the primary lesion with greater precision before surgical procedures.

Positron emission tomography (PET) scanning is invaluable for staging and evaluating colorectal malignancies. The latest enhancement, a fusion PET-CT scan, facilitates the identification of metastatic deposits and incorporates the tissue-based resolution of CT. Notably, some histologies, particularly the mucinous signet-ring cell type of CRC, may not be well visible on a PET scan.

### 1.6.3 Screening tests

Detecting and treating precancerous conditions, particularly adenomatous colon polyps and early cancerous formations that can be treated, are the main goals of CRC screening. Although the importance of detecting and removing colorectal polyps early to prevent the development of invasive cancer is primarily supported by indirect findings, it has been confirmed by the findings of many research studies. In the United States, a guideline was established by the American Cancer Society, the American College of Radiology, and the US Multi-Society Task Force on CRC. According to the recommendation, those without symptoms, irrespective of gender, should begin screening for adenomatous polyps and CRC at age 50. Based on the evaluated individual risk factors for the development of CRC, the advice describes suitable screening approaches, their reasons, and their frequency. The American Cancer Society has modified its recommendation to start screening for CRC at age 45 for individuals deemed to be at average risk due to the rising number of incidences in younger populations (Kumar & Lewis, 2024).

#### *1.6.3.1 Colonoscopy*

Screening methodologies encompass examinations that ascertain the presence of adenomatous polyps and malignancies, in addition to diagnostic procedures that concentrate solely on identifying cancer. Each of these methodologies may be employed for screening purposes. Examinations that detect adenomatous polyps and malignancies, along with their recommended intervals for administration, comprise the following:

Flexible sigmoidoscopy is performed every five years. A colonoscopy is carried out every 10 years. Double-contrast barium enema is conducted every five years. Computed tomographic colonography is executed every five years. Examinations primarily intended for cancer detection, accompanied by their recommended frequency, include the annual guaiac-based fecal occult blood test (gFOBT), demonstrating commendable sensitivity for cancer identification; annual fecal immunochemical test (FIT), showcasing substantial sensitivity for cancer detection; and stool DNA test, revealing significant sensitivity for cancer, with an undefined interval (Littlejohn et al., 2012).

The initiation of CRC screening protocols should occur at a younger age, accompanied by a significant increase in the frequency and thoroughness of these screenings for individuals categorized as high risk, including those with a documented history of polyps, prior CRC, family instances of this malignancy, and established inflammatory bowel disorders.

In addition, individuals who have received genetic screenings and are either recognized or believed to possess hereditary family syndromes, including HNPCC or FAP, ought to be regarded as at an increased risk for cancers affecting the colon and rectum, thereby requiring more stringent monitoring and preventive interventions (Holme et al., 2018).

#### *1.6.3.2 Other tests for screening*

- FIT: this diagnostic procedure facilitates the detection of hidden blood in fecal matter.
- Medical pathologists carefully examine fecal specimens for occult blood that is invisible to the unaided eye.
- The gFOBT identifies blood in stool that may remain undetected. Fecal DNA test: this analytical method identifies the genetic mutations and hematological components of fecal samples.
- Healthcare experts utilize a flexible tool called a sigmoidoscope to examine the inner workings of the lower colon and rectum.
- By employing X-ray technology, virtual colonoscopy enables the visualization and identification of polyps, neoplasms, and ulcerations in the colon and rectum (Levin et al., 2008).

### 1.6.4 Staging

1. **Primary tumor**
2. **Definition of regional lymph nodes (N)**
3. **Definition of distant metastasis (M)**

The terms pM0 and Mx do not constitute recognized classifications within the tumor node metastasis (TNM) classification system. The M category used for clinical categorization may be cM0, cM1, or pM1. The pathological stage classification may employ any of these classifications (cM0, cM1, or pM1).

Sources for staging criteria (Amin et al., 2017).

## 1.7 Prevention

### 1.7.1 Screening for colon cancer

Around age 45, doctors advise individuals with an average risk of colon cancer to start screening for the disease. However, those at higher risk should consider beginning screening earlier. Individuals with a family history of colon cancer are at a higher risk.

### 1.7.2 Lifestyle changes to reduce the risk of colon cancer

- Changing daily habits can lower the risk of colon cancer. To reduce the chance of colon cancer:
- Consume a variety of whole grains, fruits, and vegetables. Antioxidants, fiber, vitamins, and minerals found in fruits, vegetables, and whole grains may help prevent cancer. A selection of fruits and vegetables may provide a range of vitamins and nutrients.
- If you drink alcohol, do so sparingly. Limit your alcohol intake to no more than one drink per day for women and two for men if you decide to drink.
- Quit smoking. Discuss quitting with your medical team.
- Most days of the week, work out. Try to get in at least 30 minutes of activity on most days. If you have not been active, gradually increase to 30 minutes. Additionally, before starting an exercise regimen, consult a healthcare provider.
- Keep your weight in check. If you are at a healthy weight, try to maintain it by eating well and exercising every day. If you need to lose weight, ask your healthcare provider about healthy approaches to your weight-loss goals. Reduce your calorie intake and increase your physical activity to lose weight gradually.

### 1.7.3 Colon cancer prevention for individuals at high risk

Certain medications can lower the risk of colon cancer or colon polyps. For example, there is evidence that frequent use of aspirin or aspirin-like medications reduces the incidence of colon cancer and polyps. However, what dosage and duration would be required to lower the risk of colon cancer is unclear. There are hazards associated with using aspirin regularly, such as stomach bleeding and ulcers.

Those with a high risk of colon cancer are typically the only ones eligible for these alternatives. For those with an average risk of colon cancer, there is insufficient evidence to suggest these medications. Talk to your healthcare provider about your risk factors if you have an elevated risk of colon cancer to find out if taking preventive medication is safe for you ("Colon Cancer." Mayo Clinic, December 19, 2024).

## 1.8 Conclusion

To sum up, colon cancer is a complicated, multifaceted illness that requires a thorough comprehension of its genesis, pathophysiology, signs and symptoms, and diagnostic factors. An inclusive exploration of the epidemiology, risk factors, and preventive strategies of colon cancer is necessary to adequately address future problems related to CRC. Early identification and screening are also critical for preventing colon cancer and lowering death rates. The TNM and American Joint Committee on Cancer (AJCC) staging systems are often used for colon cancer staging, which is essential for assessing prognosis and available treatments. The colon may develop cancer following the atypical increase in glandular epithelial cell numbers. On a worldwide scale, there is an ongoing upward trend in CRC cases linked to inherited genetics and external conditions that determine the risk of acquiring the illness. Screening individuals identified as being at average risk is seen as the most potent tactic for combating colon cancer and curtailing mortality figures, paving the way for the adoption of population-oriented screening initiatives throughout various European countries, Canada, and multiple regions in North and South America, along with some areas in Asia and Oceania.

Numerous variables, such as genetics, environmental exposures, and inflammatory states of the digestive system, contribute to the etiology of colon cancer. A lifetime risk of around 40% for CRC is associated with HNPCC (Lynch syndrome). Dietary variables related to an elevated risk of CRC include food items such as heavy red meat, animal fat, low fiber, fruits, and vegetables. The risk of CRC is also increased by obesity and lifestyle choices, such as drinking alcohol, smoking cigarettes, and leading a sedentary lifestyle.

Three main genetic pathways can give rise to the pathophysiology of colon cancer: the CIN, mismatch repair (dMMR), and the CIMP. These mechanisms include the slow accumulation of genetic changes that eventually lead to carcinoma. Though hematochezia, modifications in digestive tract activity, inner unease, bloating in the stomach, unexplained weight loss, vomiting, weariness, and trouble breathing are common symptoms, colon cancer can present clinically as asymptomatic (Tables 1.1 and 1.2).

The methodologies for screening encompass colonoscopy, flexible sigmoidoscopy, double-contrast barium enema, computed tomographic colonography, gFOBT, fecal immunochemical examination, and stool DNA analysis, all of which represent critical diagnostic variables that warrant consideration. For staging purposes, imaging tests such as abdominal/pelvic computed tomography, abdominal/pelvic magnetic resonance imaging, and contrast ultrasound of the abdomen/liver are also utilized. The American College of Gastroenterology recommends a colonoscopy every ten years, starting at age 50, as the best detection procedure for CRC and adenomatous polyps. The American Cancer Society advises that screening for these conditions in asymptomatic males and females should begin at age 50.

Additionally, the prognosis and available treatments for colon cancer depend heavily on the stage of the disease. The TNM approach, which considers the size and spread of the primary tumor (T), the existence of distant metastases (M), and the presence of lymph node metastases (N), is frequently used for staging colon cancer (Table 1.3).

**Table 1.1 Classification of Tumor (T) category according to the AJCC TNM Staging System for colorectal cancer. Specifically, these descriptions detail the anatomic depth of invasion through the layers of the bowel wall, ranging from Tx/T0 up to T4b.**

| TX | The primary tumor is unassessable |
|---|---|
| T0 | No indication of a primary neoplasm. |
| Tis | cancer in situ: intraepithelial or intramucosal cancer (involvement of the lamina propria without penetration into the muscularis mucosa). |
| T1 | The tumor infiltrates the submucosa, penetrating the muscularis mucosa but not extending into the muscularis propria. |
| T2 | The tumor infiltrates muscularis propria. |
| T3 | The tumor infiltrates the muscularis propria and extends into the pericolorectal tissues. |
| T4 | The tumor infiltrates the visceral peritoneum or invades or clings to neighboring organs or structures. |
| T4a | The tumor infiltrates the visceral peritoneum, resulting in significant intestinal perforation and ongoing invasion via inflamed regions to the visceral peritoneum's surface. |
| T4b | The tumor immediately infiltrates or adheres to adjacent organs or tissues. |

**Table 1.2 AJCC TNM staging system: Pathological classification and criteria for regional lymph node (N) involvement in colorectal cancer.**

| NX | Regional lymph nodes are unassessable |
|---|---|
| N0 | Absence of regional lymph node metastasis. |
| N1 | Metastasis occurs in one to three regional lymph nodes (tumor in lymph nodes measuring ≥0.2 mm) or any number of tumor deposits, but all detected nodes remain negative. |
| N1a | In one local lymphatic node, metastases. |
| N1b | Two to three regional lymph node metastases. |
| N1c | Malignant neoplastic formations located within the subserosal layer, mesenteric tissue, or nonpersonalized pericolic and perirectal/mesorectal structures are characterized by the absence of regional lymph node metastasis. |
| N2 | Four or more lymph node metastases. |
| N2a | Four to six local lymph nodes have been affected. |
| N2b | At least seven or more regional lymph nodes have exhibited signs of involvement. |

**Table 1.3 AJCC TNM staging system: Clinical (M) and Pathological (pM) classification criteria for distant metastasis in colorectal cancer.**

| Category | Criteria |
|---|---|
| M0 | No evidence of distant metastasis was identified through imaging techniques or additional assessments, and there was no evidence of tumor manifestations in distant sites or organs. Pathologists do not recognize this classification. |
| M1 | There have been observations of metastasis in solitary or multiple distant locations or organs, and inside the peritoneal cavity. |
| M1a | Metastasis confined to a solitary organ or location is identified without peritoneal metastasis. |
| M1b | The presence of metastatic dissemination to two or more anatomical sites or organs is observed without peritoneal metastasis. |
| M1c | Metastatic dissemination occurs solely to the peritoneal surface or concurrently with metastatic spread to alternative sites or organs. |
| Pertonial (p) M1 | The phenomenon of metastasis, which may occur in one or multiple remote sites, including various organs or the peritoneal cavity, is identified and verified through microscopic examination. |
| pM1a | The presence of metastasis to a solitary site or organ is identified without peritoneal involvement and is substantiated through microscopic examination. |
| pM1b | Metastatic dissemination to two or more anatomical sites or organ systems is identified without peritoneal involvement, with validation achieved through microscopic examination. |
| pM1c | Metastatic dissemination to the peritoneal surface may occur either as an isolated event or concomitantly with metastases to additional anatomical sites or organs, and this phenomenon is confirmed by microscopic examination. |

## References

Amin, M. B., Greene, F. L., Edge, S. B., Compton, C. C., Gershenwald, J. E., Brookland, R. K., Meyer, L., Gress, D. M., Byrd, D. R., & Winchester, D. P. (2017). The eighth edition AJCC cancer staging manual: Continuing to build a bridge from a population-based to a more 'personalized' approach to cancer staging. *CA: A Cancer Journal for Clinicians, 67*(2), 93–99. https://doi.org/10.3322/CAAC.21388.

Arhi, C. S., Ziprin, P., Bottle, A., Burns, E. M., Aylin, P., & Darzi, A. (2019). Colorectal cancer patients under the age of 50 experience delays in primary care leading to emergency diagnoses: A population-based study. *Colorectal Disease : The Official Journal of the Association of Coloproctology of Great Britain and Ireland, 21*(11), 1270–1278. https://doi.org/10.1111/CODI.14734.

Aune, D., Chan, D. S. M., Lau, R., Vieira, R., Greenwood, D. C., Kampman, E., & Norat, T. (2011). Dietary fibre, whole grains, and risk of colorectal cancer: Systematic review and dose-response meta-analysis of prospective studies. *British Medical Journal, 343*(7833), 1082. https://doi.org/10.1136/BMJ.D6617.

Bailly, L., Roxane, F., Christian, P., & Antonio, I. (2020). Colorectal cancer risk following bariatric surgery in a nationwide study of french individuals with obesity. *JAMA Surgery, 155*(5), 395–402. https://doi.org/10.1001/JAMASURG.2020.0089.

Burn, J., Bishop, D. T., Mecklin, J.-P., Macrae, F., Möslein, G., Olschwang, S., Bisgaard, M.-L., et al. (2008). Effect of aspirin or resistant starch on colorectal neoplasia in the lynch syndrome. *New England Journal of Medicine, 359*(24), 2567–2578. https://doi.org/10.1056/nejmoa0801297.

Cadigan, K. M., & Liu, Y. I. (2006). Wnt signaling: Complexity at the surface. *Journal of Cell Science, 119*(Pt 3), 395–402. https://doi.org/10.1242/JCS.02826.

Cancer Facts & Figures. (2020). American Cancer Society. *Cancer facts & figures 2020.* https://www.cancer.org/research/cancer-facts-statistics/all-cancer-facts-figures/cancer-facts-figures-2020.html.

Cancer Facts & Figures. (2024). American Cancer Society. https://www.cancer.org/research/cancer-facts-statistics/all-cancer-facts-figures/2024-cancer-facts-figures.html.

Cho, E., Lee, J. E., Rimm, E. B., Fuchs, C. S., & Giovannucci, E. L. (2012). Alcohol consumption and the risk of colon cancer by family history of colorectal cancer. *The American Journal of Clinical Nutrition, 95*(2), 413–419. https://doi.org/10.3945/AJCN.111.022145.

Colon Cancer. (2024, December 19). Mayo Clinic. https://www.mayoclinic.org/diseases-conditions/colon-cancer/symptoms-causes/syc-20353669.

Colorectal Cancer Risk Factors: Hereditary Colorectal Risk Factors. (2025). *Hereditary colorectal risk factors* |. American Cancer Society. Accessed February 3 https://www.cancer.org/cancer/types/colon-rectal-cancer/causes-risks-prevention/risk-factors.html.

Edwards, B. K., Elizabeth, W., Kohler, B. A., Eheman, C., Zauber, A. G., Anderson, R. N., Jemal, A., et al. (2010). Annual report to the nation on the status of cancer, 1975-2006, featuring colorectal cancer trends and impact of interventions (Risk Factors, Screening, and Treatment) to reduce future rates. *Cancer, 116*(3), 544–573. https://doi.org/10.1002/cncr.24760.

Fritz, C. D. L., Otegbeye, E. E., Zong, X., Demb, J., Nickel, K. B., Olsen, M. A., Mutch, M., Davidson, N. O., Gupta, S., & Cao, Y. (2023). Red-flag signs and symptoms for earlier diagnosis of early-onset colorectal cancer. *Journal of the National Cancer Institute, 115*(8), 909–916. https://doi.org/10.1093/JNCI/DJAD068.

Gajendran, M., Loganathan, P., Jimenez, G., Catinella, A. P., Nathaniel, N., Umapathy, C., Ziade, N., & Hashash, J. G. (2019). A comprehensive review and update on ulcerative colitis. *Disease-a-Month : DM, 65*(12), 100851. https://doi.org/10.1016/j.disamonth.2019.02.004.

Garcia-Etxebarria, K., Marc, Clos-Garcia, Oiana, T., Beatriz, N., Cristina, A., Marta, I.-L., Andre, F., et al. (2021). Interplay between genome, metabolome and microbiome in colorectal cancer. *Cancers, 13*(24). https://doi.org/10.3390/CANCERS13246216.

Goncalves, M. D., Changyuan, L., Tutnauer, J., Hartman, T. E., Hwang, S. K., Murphy, C. J., Pauli, C., et al. (2019). High-fructose corn syrup enhances intestinal tumor growth in mice. *Science (New York, N.Y.), 363*(6433), 1345–1349. https://doi.org/10.1126/SCIENCE.AAT8515.

Grady, W. M., & Carethers, J. M. (2008). Genomic and epigenetic instability in colorectal cancer pathogenesis. *Gastroenterology, 135*(4), 1079. https://doi.org/10.1053/J.GASTRO.2008.07.076.

Guinney, Rodrigo Dienstmann, J., Wang, X., De Reyniès, A., Schlicker, A., Soneson, C., Marisa, L., et al. (2015). The consensus molecular subtypes of colorectal cancer. *Nature Medicine, 21*(11), 1350–1356. https://doi.org/10.1038/NM.3967.

Harford, W. V. (2006). Colorectal cancer screening and surveillance. *Surgical Oncology Clinics of North America, 15*(1), 1–20. https://doi.org/10.1016/j.soc.2005.11.003.

Holme, Magnus Løberg, Ø., Kalager, M., Bretthauer, M., Hernán, M. A., Aas, E., Eide, T. J., et al. (2018). Long-Term effectiveness of sigmoidoscopy screening on colorectal cancer incidence and mortality in women and men: A randomized trial. *Annals of Internal Medicine, 168*(11), 775–782. https://doi.org/10.7326/M17-1441.

Hossain, Sanower, M., Karuniawati, H., Jairoun, A. A., Urbi, Z., Ooi, D. J., John, A., Lim, Y. C., et al. (2022a). Colorectal cancer: A review of carcinogenesis, global epidemiology, current challenges, risk factors, preventive and treatment strategies. *Cancers, 14*(7). https://doi.org/10.3390/CANCERS14071732.

Huang, J., Soupir, A. C., Schlick, B. D., Teng, M., Sahin, I. H., Permuth, J. B., Siegel, E. M., Manley, B. J., Pellini, B., & Wang, L. (2021). Cancer detection and classification by CpG Island hypermethylation signatures in plasma cell-free DNA. *Cancers, 13*(22). https://doi.org/10.3390/CANCERS13225611.

Jacobs, E. T., Ahnen, D. J., Ashbeck, E. L., Baron, J. A., Greenberg, E. R., Lance, P., Lieberman, D. A., et al. (2009). Association between body mass index and colorectal neoplasia at follow-up colonoscopy: A pooling study. *American Journal of Epidemiology, 169*(6), 657–666. https://doi.org/10.1093/AJE/KWN401.

Kambara, T., Simms, L. A., Whitehall, V. L. J., Spring, K. J., Wynter, C. V. A., Walsh, M. D., Barker, M. A., et al. (2004). BRAF mutation is associated with DNA methylation in serrated polyps and cancers of the colorectum. *Gut, 53*(8), 1137–1144. https://doi.org/10.1136/GUT.2003.037671.

Keum, N. N., & Giovannucci, E. (2019). Globalncer: Emerging trends, risk factors and prevention strategies. *Nature Reviews. Gastroenterology & Hepatology, 16*(12), 713–732. https://doi.org/10.1038/S41575-019-0189-8.

Kocher, H. M., Mears, L., Lea, N. C., Raj, K., & Mufti, G. J. (2007). Sleisenger and fordtran's gastrointestinal and liver disease. *Gut, 56*(8), 1175. https://doi.org/10.1136/GUT.2007.121533.

Kuhnert, F., Davis, C. R., Wang, H. T., Chu, P., Lee, M., Yuan, J., Nusse, R., & Kuo, C. J. (2004). Essential requirement for wnt signaling in proliferation of adult small intestine and colon revealed by adenoviral expression of dickkopf-1. *Proceedings of the National Academy of Sciences of the United States of America, 101*(1), 266–271. https://doi.org/10.1073/PNAS.2536800100.

Kumar, R., & Lewis, C. R. (2024). Colon cancer screening. *StatPearls, September*. https://www.ncbi.nlm.nih.gov/books/NBK559064/.

Levin, B., Lieberman, D. A., McFarland, B., Smith, R. A., Brooks, D., Andrews, K. S., Dash, C., et al. (2008). Screening and surveillance for the early detection of colorectal cancer and adenomatous polyps, 2008: A joint guideline from the american cancer society, the us multi-society task force on colorectal cancer, and the american college of radiology. *CA: A Cancer Journal for Clinicians, 58*(3), 130–160. https://doi.org/10.3322/CA.2007.0018.

Littlejohn, C., Hilton, S., MacFarlane, G. J., & Phull, P. (2012). Systematic review and meta-analysis of the evidence for flexible sigmoidoscopy as a screening method for the prevention of colorectal cancer. *The British Journal of Surgery, 99*(11), 1488–1500. https://doi.org/10.1002/BJS.8882.

Mármol, I., Cristina, S., Dieste, A. P., Cerrada, Elena, & María, J. R. Y. (2017). Colorectal carcinoma: A general overview and future perspectives in colorectal cancer. *International Journal of Molecular Sciences, 18*(1). https://doi.org/10.3390/IJMS18010197.

Menon, G., & Cagir, B. (2025). *Colon cancer*.

Meyerhardt, Donna Niedzwiecki, J. A., Hollis, D., Saltz, L. B., Hu, F. B., Mayer, R. J., Nelson, H., et al. (2007). Association of dietary patterns with cancer recurrence and survival in patients with stage iii colon cancer. *The Journal of the American Medical Association, 298*(7), 754–764. 10.1001/JAMA.298.7.754.

Morikawa, Aya Kuchiba, T., Lochhead, P., Nishihara, R., Yamauchi, M., Imamura, Y., Liao, X., et al. (2013). Prospective analysis of body mass index, physical activity, and colorectal cancer risk associated with β-Catenin (CTNNB1) status. *Cancer Research, 73*(5), 1600–1610. https://doi.org/10.1158/0008-5472.CAN-12-2276.

Morson, B. C. (1984). The evolution of colorectal carcinoma. *Clinical Radiology, 35*(6), 425–431. https://doi.org/10.1016/S0009-9260(84)80033-1.

Navarro, M., Nicolas, A., Ferrandez, A., & Lanas, A. (2017). Colorectal cancer population screening programs worldwide in 2016: An update. *World Journal of Gastroenterology, 23*(20), 3632–3642. https://doi.org/10.3748/wjg.v23.i20.3632.

Pala, Sabina Sieri, V., Berrino, F., Vineis, P., Sacerdote, C., Palli, D., Masala, G., et al. (2011). Yogurt consumption and risk of colorectal cancer in the italian european prospective investigation into cancer and nutrition cohort. *International Journal of Cancer, 129*(11), 2712–2719. https://doi.org/10.1002/IJC.26193.

Shaukat, A., Kahi, C. J., Burke, C. A., Rabeneck, L., Sauer, B. G., & Rex, D. K. (2021). ACG clinical guidelines: Colorectal cancer screening 2021. *The American Journal of Gastroenterology, 116*(3), 458–479. https://doi.org/10.14309/AJG.0000000000001122.

Smith, R. A., Andrews, K. S., Brooks, D., Fedewa, S. A., Manassaram-Baptiste, D., Saslow, D., & Wender, R. C. (2019). Cancer screening in the United States, 2019: A review of current american cancer society guidelines and current issues in cancer screening. *CA: A Cancer Journal for Clinicians, 69*(3), 184–210. https://doi.org/10.3322/caac.21557.

Tabung, F. K., Li, L., Wang, W., Fung, T. T., Wu, K., Smith-Warner, S. A., Cao, Y., et al. (2018). Association of dietary inflammatory potential with colorectal cancer risk in men and women. *JAMA Oncology, 4*(3), 366–373. https://doi.org/10.1001/JAMAONCOL.2017.4844.

Thirunavukarasu, P., Sukumar, S., Sathaiah, M., Mahan, M., Pragatheeshwar, K. D., Pingpank, J. F., Zeh, H., Bartels, C. J., Lee, K. K. W., & Bartlett, D. L. (2011). C-Stage in colon cancer: Implications of carcinoembryonic antigen biomarker in staging, prognosis, and management. *Journal of the National Cancer Institute, 103*(8), 689–697. https://doi.org/10.1093/JNCI/DJR078.

Wielandt, Claudia Hurtado, A. M., Moreno C, M., Villarroel, C., Castro, M., Estay, M., Simian, D., et al. (2020). Characterization of chilean patients with sporadic colorectal cancer according to the three main carcinogenic pathways: Microsatellite instability, cpg island methylator phenotype and chromosomal instability. *Tumor Biology, 42*(7). https://doi.org/10.1177/1010428320938492.

Winawer, S. J., Zauber, A. G., Ho, M. N., O'Brien, M. J., Gottlieb, L. S., Sternberg, S. S., Waye, J. D., et al. (1993). Prevention of colorectal cancer by colonoscopic polypectomy. The national polyp study workgroup. *The New England Journal of Medicine, 329*(27), 1977–1981. https://doi.org/10.1056/NEJM199312303292701.

Xiao, J.-B., Ai-Min, L., Yan-Qiu, Z., Zhen, W., Jian, H., & Guan-Nan, Y. (2019). CUEDC2: Multifunctional roles in carcinogenesis. *Frontiers in Bioscience (Landmark Edition), 24*(5), 935–946. https://doi.org/10.2741/4759.

Yuhara, H., Craig Steinmaus, S. E., Cohen, D. A., Corley, Y., Tei, P. A., & Buffler (2011). Is diabetes mellitus an independent risk factor for colon cancer and rectal cancer? *The American Journal of Gastroenterology, 106*(11), 1911–1921. https://doi.org/10.1038/AJG.2011.301.

Zhang, J., Charles, H., Watson, A. J. M., Hart, A. R., Platt, M. J., Pardoll, D. M., Cosgrove, S. E., Gebo, K. A., & Sears, C. L. (2019). Oral antibiotic use and risk of colorectal cancer in the United Kingdom, 1989-2012: A matched case-control study. *Gut, 68*(11), 1971–1978. https://doi.org/10.1136/GUTJNL-2019-318593.

Zhou, Lin Wang, E., Santiago, C. N., Nanavati, J., Rifkin, S., Spence, E., Hylind, L. M., et al. (2022). Adult-attained height and colorectal cancer risk: A cohort study, systematic review, and meta-analysis. *Cancer Epidemiology, Biomarkers & Prevention : A Publication of the American Association for Cancer Research, Cosponsored by the American Society of Preventive Oncology, 31*(4), 783–792. https://doi.org/10.1158/1055-9965.EPI-21-0398.

CHAPTER

# Current trends of targeted nanoparticulate carriers for the treatment of colon cancer

# 2

**Princy Kashyap[1], Manoj Kumar[1], Arun Kumar Singh[2] and Vipan Kumar[3]**

[1]*Department of Pharmacy, Guru Ghasidas Vishwavidyalaya, Bilaspur, Chhattisgarh, India,*
[2]*University School of Pharmaceutical Sciences, Rayat Bahra University, Mohali, Punjab, India,*
[3]*School of Pharmaceutical Sciences, CGC University, Mohali, Punjab, India*

## 2.1 Introduction

Cancer is a condition in which abnormal cells can divide without control and can invade other tissues. If genes turn cells inside out, changing a controlled process of cell division into misshapen, uncontrollable growth, the cells start to proliferate abnormally. Cancer cells may form a tumor. There are essentially two types of tumors: malignant and benign. There are more than 100 clear cancer types, which can include breast, myeloma, leukemia, lymphoma, lung, prostate, and colon. The third most diagnosed cancer globally is colon cancer among these (Araghi and Soerjomataram, 2019). It has about 1.36 million cases around the world, making it the fourth most common cause of cancer deaths. Colon cancer has a high incidence in Western countries, such as Australia, Europe, and the Americas. It is relatively rare in Africa, Asia, and Central and South America (May & Anandasabapathy, 2019). Colon cancer signs and symptoms can include diarrhea or constipation, a change in bowel habits, loss of appetite, weight loss, nausea, vomiting, and bloating, as well as abdominal cramping after meals. It occurs when cancerous tumors develop in the colon, the last section of the digestive tract, which is almost five feet long. The colon forms and moves feces into the rectum for elimination, and it also absorbs water, electrolytes, and vitamins. It is also a repository of fecal matter before evacuation (Renzi and Lyratzopoulos, 2019). It should be noted that colon cancer develops over the course of time. The process typically begins as a "polyp" in the inner lining of the colon. Polyps are typically benign growths, but some may develop into cancer. Two types of polyps can occur in the colon. These include (1) hyperplastic and inflammatory polyps, which are not usually considered to be at risk of becoming cancerous, and (2) adenomas or adenomatous polyps that may develop into colorectal cancer if untreated. The cancer cells first start to grow in the wall of the colon, then begin to spread into the nearby blood and lymph. The cells then migrate to the adjacent lymph nodes and proceed to disseminate to distant organ sites (Rajekar, 2024).

Colonic carcinomas have been linked to many lifestyle factors such as physical inactivity, low fiber or high fat intake, alcohol abuse, smoking, obesity, etc. Individuals with type two diabetes or insulin resistance and those susceptible to chronic inflammatory diseases, including ulcerative colitis and Crohn's disease, have been identified as being at high risk of developing colon cancer. It is estimated that 2% of cases of colon cancer per year occur because of inflammatory bowel disease.

Nano Drug Delivery Systems for Colon Cancer. DOI: https://doi.org/10.1016/B978-0-443-40443-6.00003-9

Colon cancer can result from mutations in DNA. These mutations can cause oncogenes to become activated or tumor suppressor genes to be turned off, which causes cells to grow uncontrollably. Some of the mutations are inherited by offspring and passed down through the family. These mutations are called germline gene mutations. This mutation leads to hereditary nonpolyposis colon cancer (HNPCC, or Lynch syndrome), which accounts for approximately 3% of all cases of colorectal cancer, with 1% that is highly linked to Gardner's syndrome and familial adenomatous polyposis (FAP) (Antelo and Golubicki, 2019). Metastasis accounts for close to 90% of colon cancer-caused mortalities. Lifestyle interventions to prevent colon cancers include physical activity, maintaining a normal body mass index (BMI), and consuming high quantities of fruits and grains.

Several tests may be used to screen for colon cancer. These tests can be generally classified into two categories. (1) Stool-based tests: These tests examine the sample of stool for signs of colon cancer or polyps. Some of the screening tests in this category include guaiac-based fecal occult blood test (gFOBT), fecal immunochemical test (FIT), and stool DNA tests. (2) Visual (structural) tests: These tests examine the structure of the inside of the colon and rectum for abnormal areas that could indicate cancer or polyps. These include colonoscopy, CT colonography, and flexible sigmoidoscopy. Colon cancer can also be diagnosed with a fecal occult blood test (FOBT) and urinary volatile organic compounds. The diagnosis of colon cancer can be made with the help of common tumor markers such as CEA or CA 19–9. The CT scan can measure how far the cancer has spread outside the colon. Depending on the location of the lesion, it may also be "confirmed" by colonoscopy for colon cancer. It is a safe biopsy in which biopsied samples are removed and examined for histopathological changes, including cell types and grades (Burt, 2000). Generally, the lesion of a tumor is characterized by a relatively abnormal tubular structure with multiple lumina and decreased stroma. Mucus is occasionally produced by tumors. Immunostaining can also be employed to verify antigens in colon cancer cells.

For early-stage colon cancers, surgery is frequently the primary therapy. The type of operation depends on the stage and size of the colon cancer, as well as the kind of tumor it is. Most polyps can be removed (e.g., during a colonoscopy) when they are in stage 0 and some early-stage I tumors (early colon cancer). In a polypectomy, small growths in the colon lining are removed. Local excision can also remove a portion of the colon. In a total colectomy, the entire colon is removed. Even with surgery, cancer might not be fully cured. In these cases, treatments such as surgery, radiotherapy, and chemotherapy can be used to alleviate the advancement of colon cancer. Radiation consists of X-rays generated for radio-therapeutic killing of tumor cells (Deleon & Stocchi, 2022). Radiation therapy can impart side effects such as skin irritation, nausea, hair loss, fatigue, and sexual malfunction. With chemotherapy, radiation treatment is generally administered as well. Chemotherapeutic bioactives may also directly kill tumor cells and enhance the efficacy of radiation therapy. However, another treatment modality is chemotherapy, in which anticancer bioactives could be applied to inhibit or decelerate the proliferation of bona fide cancer cells. Many bioactives, such as capecitabine, oxaliplatin, 5-fluorouracil (5-FU), and irinotecan, are used to treat colon cancer. However, chemotherapy is often accompanied by adverse side effects such as nausea and vomiting, bone marrow depression, alopecia (i.e., loss of hair), anorexia, and mucositis. To circumvent this limitation, different nanoparticle carrier systems have been tailored to the right size and surface properties to optimize the biodistribution of bioactives and their circulation time in the bloodstream (You and Kang, 2016). This chapter will focus on trends in research on nanoparticle-based treatment for colon cancer.

## 2.2 Molecular biology of colon cancer

Colon cancer is a progressive disease, originating from the accumulation of epigenetic as well as genetic changes that enable normal colonic epithelial cells to progress into colon cancer cells. This sequence of colon carcinogenesis, referred to as the polyp-carcinoma sequence, generally takes 10–15 years and involves concomitant histological and molecular changes. The net result of these epigenetic and genetic changes is the molecular biology of tumor cells, which are replaced not only by critical biological markers but also by cancerous factors that drive malignancy. From the molecular genetics of colon cancer, it is evident that colon cancer evolves, as cancer is a multistage process. The molecular biology of colon cancer is based on the fact that it arises from genomic instability secondary to a loss of DNA repair capabilities within cells. Thus, genetic or epigenetic mutations are present. The genetic and epigenetic mutations involved in this form of genomic instability work together to promote the development and progression of colon cancer (Fig. 2.1) (Nguyen & Duong, 2018).

- **Adenoma carcinoma sequence:** The development of adenocarcinoma from normal epithelial cells typically occurs through a sequence of histological changes and accompanying genomic and epigenomic changes. These epigenetic alterations and genetic mutations confer a growth advantage to mutant cells. The process leads to the proliferation of these mutated cells. This cycle underpins the progression of adenomas to adenocarcinomas through the acquisition of epigenetic and genetic changes that lead to clonal heterogeneity. The only type of polyp shown so far to have a potential for malignancy is adenomatous. However, only a minority of adenomas develop into cancer, and the progression most likely requires years or decades (Bragg and Breaux, 2023). Colonic cancer seems to follow its pathway and sequence, the hyperplastic polyp-serrated adenoma-adenocarcinoma pathway, which occurs with mitosis (Fig. 2.1).

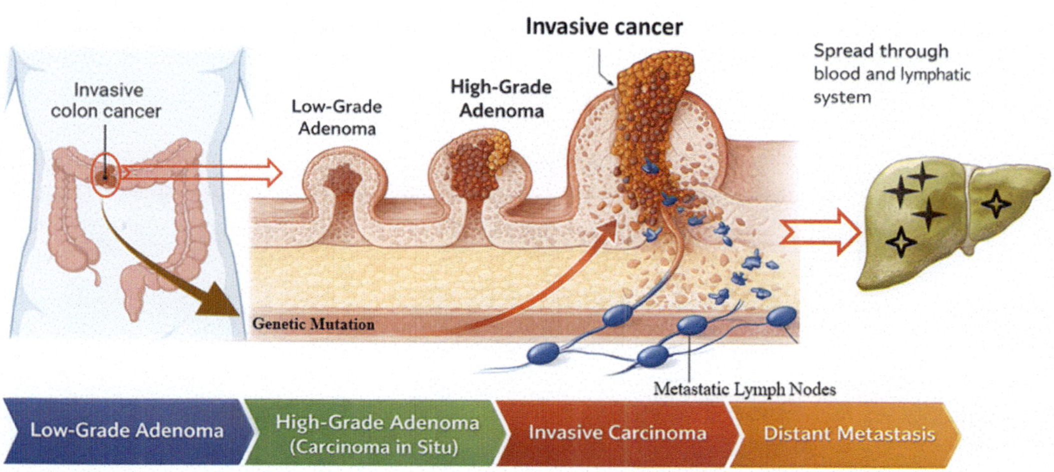


**FIGURE 2.1**

Schematic representation of the stepwise progression of colon cancer from low-grade adenoma to high-grade adenoma, invasive cancer, and distant metastasis, driven by cumulative genetic mutations.

**Genetic mutations:** The molecular world of colon cancer has advanced considerably since the 1990s, when the first genetic model for the disease was developed. Heterogeneous mutations in all four genes—APC, K-ras, P53, and DCC—evidence the progression from normal mucosa to adenoma and carcinoma cells. All of these are thought to be critical but not uniformly accounted for in all colon cancers. The early lesion in colonic carcinogenesis is the aberrant crypt foci (ACF). Its actual neoplastic potential is currently unknown. However, with these lesions, some may seem to progress and become adenocarcinoma. APC gene mutations are frequently detected in dysplastic ACF and are presumed to be the most likely precursors of progression to colon cancer. Thus, alterations in APC that lead to overactivity of the Wingless/Wnt signaling cascade are likely to stimulate tumorigenesis in the colon. Later mutations in the other genes are involved in tumor development and other malignant properties, such as tissue invasiveness and metastatic capacity (Roper & Hung, 2013).

## 2.3 Biomarkers used in cancer

A cancer biomarker or molecular marker refers to a process or entity that indicates the presence of a tumor/cancer in the body. It could be a molecule that causes tumors or a specific response in the body to having cancer. Ideally, these biomarkers are measured in biofluids such as blood or serum. They are relevant for early diagnosis of colon cancer, its therapy, and the patients' prognosis. These biomarkers can be loosely classified as diagnostic, predictive, and prognostic (Fig. 2.2) (Peluso and Incollingo, 2017).

1. **Diagnostic biomarkers**: These biomarkers can be used in the early detection of colon cancer.
   (a) **Genomic instability:** a changing signature of colon cancer. Genomic instability is a characteristic of most colon cancers. This results from genomic changes during cell division. Cancer arises when several genes that regulate cell division and tumor suppression become damaged (Tables 2.1 and 2.2). It has been well documented that various mediators tightly regulate genome integrity, including the mitotic checkpoint, the DNA damage checkpoint,

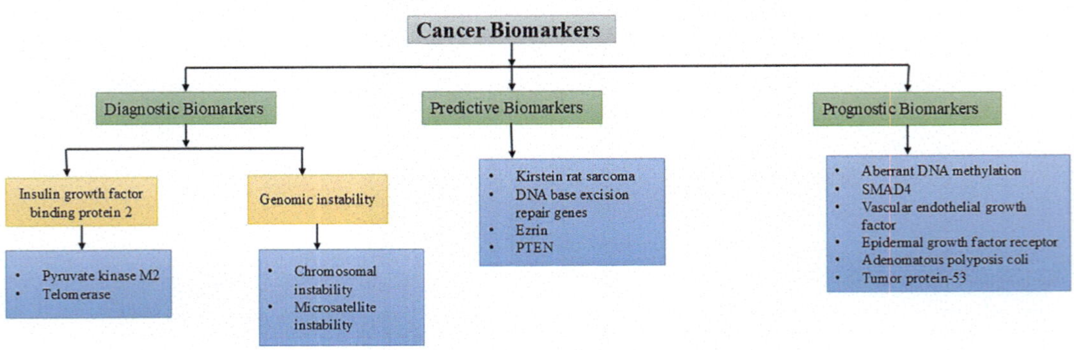


**FIGURE 2.2**

Biomarkers used in cancer detection.

**Table 2.1 Several critical genes linked to the tumorigenesis of colon cancer.**

| Mutation type | Gene | Location on chromosomes | Function of gene product |
|---|---|---|---|
| Point mutations (most frequent V600E) | BRAF | 7q34 | Cell proliferation and survival |
| Point mutation (codons 12 and 13 of exon 2) | K-RAS | 12p21 | Cell proliferation and survival |
| Gene amplification | MYC | 8q24 | Cell proliferation and survival |
| Gene amplification | EGFR | 7p21 | Cell proliferation and survival |
| Germline mutation | MLH1 (associated with microsatellite instability) | 3p22.2 | Cell proliferation and survival |

**Table 2.2 List of essential genes involved in the suppression of tumors in colon cancer.**

| Gene | Location on chromosomes | Mutation type | Function of gene product |
|---|---|---|---|
| TP53 | 17q13 | Point mutation (missense), allele loss | Apoptosis, autophagy induction, and cell-cycle arrest |
| APC | 5q21 | Point mutation, frameshift mutation, allele loss leading to truncated protein deletion | Wingless signaling pathway inhibition |
| SMAD4 | 18q21 | Allele loss, nonsense, missense | Intracellular mediator of TGF-β pathway |
| PTEN | 10q23 | Nonsense, deletion | Inhibition of the phosphatidylinositol-3 kinase pathway |
| BAX | 19q13 | Frameshift mutation | Apoptotic activator |
| DCC | 18q21 | Point mutation | Triggers tumor cell apoptosis, cell surface receptor for nectrin-1 |

and the DNA repair machinery. Dysregulation of any one of these mechanisms often results in genomic instability, which is prone to malignant transformation. The integrity of the genome is required for proper cellular function to avoid errors caused by endogenous genotoxic stress, such as reactive oxidative species (ROS) from cellular metabolism, DNA replication, and exogenous carcinogens, including ultraviolet radiation, ionizing radiation, and chemicals that damage DNA. The loss of genomic stability allows for the acquisition of numerous mutations required for colon cancer progression (Markowitz & Bertagnolli, 2009). Various types of events contribute to genomic instability, including chromosomal instability, microsatellite instability (MSI), abnormal DNA methylation, and defective DNA repair. A comprehensive genome-wide survey of gene mutations in colon cancers has identified acquired somatic mutations in hundreds of genes.

(b) **Insulin-like growth factor binding protein 2:** Introduction Insulin-like growth factor binding protein 2 (IGFBP2) is an important regulator of IGF and IGF-1. The upregulation of their levels in colon tumors is driven by an overabundance of the corresponding mRNAs. In colorectal cancer and advanced disease, researchers observe elevated plasma or serum IGFBP2 levels compared to early-stage disease (Liou and Shun, 2010).

2. **Predictive biomarkers:** This biomarker is beneficial to forecast the response to treatment. The patient can be selected for a particular treatment based on a positive response. They could even be applied for finding the proper dose of bioactives and to prevent their toxicity (Taylor, 2018; Won and Lee, 2017).
3. **Prognostic biomarkers:** These help determine the course of the disease. They are molecules with key roles in various biological events, including angiogenesis, cellular proliferation, invasion, differentiation, and metastasis (Feng and Chang, 2019; Galuppini and Pennelli, 2017).

## 2.4 Conventional treatment options for colon cancer

There are many traditional treatment options for colon cancer, which include surgical resection, radiation therapy, chemotherapy, immunotherapy, and targeted therapy. These can be used as monotherapy, depending on the stage at which the disease is detected and diagnosed, or they can be combined (Desantis and Lin, 2014).

(a) **Surgical resection:** Surgical resection of precursor or cancer lesions can be curative for the patient. This could be a preferable and more effective alternative for small, focal (localized) cancer growths. Polyps can be surgically removed through a procedure called polypectomy. Recurrence is a possibility; however, for radiation-resistant solid tumors and chemotherapy-resistant tumors (e.g., pancreatic carcinoma), surgical treatment is the only option. In addition, comprehensive geriatric assessment (CGA) may identify vulnerable patients at very high risk of severe postoperative complications (Kristjansson and Nesbakken, 2010).

(b) **Radiation therapy**: Radiation therapy is treatment with high-energy X-rays that can destroy cancer cells. Whereas chemoradiation forms part of adjuvant or neoadjuvant treatment for rectal cancers, chemotherapy alone is more frequently used in colon cancer (Marshall, 2008). But radiation also has its share of side effects. Potential side effects of radiation treatment for colon cancer may include nausea, tiredness/fatigue, changes to the skin, ranging from redness to blistering and peeling, irritation to the bladder, and difficulty with wound healing. Before radiation therapy, surgery may cause scarring fibrosis (stiffening), adhesions, or bands of tissue that form between organs and cause them to adhere (stick) to one another.

(c) **Chemotherapy:** After proper treatment, colon cancer patients undergo chemotherapy. It is a therapy that uses cancer-killing bioactives to destroy colon cancer cells. This is also the principal mode of action in cancer therapy. It contains bioactives, for example, alkylating agents, plant alkaloids, antitumor antibiotics, enzymes, hormones, etc., which can kill or slow the growth of cancerous cells or make them unable to divide any longer. However, there are no bioactives that can selectively kill malignant cells without harming normal tissues, and in many cases, the side effects are dangerous. Cancer and metastasis are treated mainly with chemotherapy, which is a form of bioactivity for the body (effect) on tumor cells, followed by direct cytotoxicity and regression, either indirectly or directly. These chemotherapeutic bioactives inhibit cell division processes, such as DNA replication and chromosome segregation, but are not specific to tumor cells. While reaching all the proliferating cells in the body, only a fraction of bioactives reach the target tissue. This can damage healthy cells, especially those with a high turnover rate (such as the gastrointestinal lining and immune cells). However, these cells usually heal themselves when chemotherapy is stopped (Cunningham and Atkin, 2010).

### 2.4.1 Targeted therapy

(a) **Antibody-based therapies:** Conventional chemotherapy is nonselective. Since the chemotherapy does indeed conjugate with cells, it can and does kill off normal tissues. Tumor cells release growth factors that trigger neovascularization during tumorigenesis, suggesting the possibility of hypothesis-driven, more cancer-specific treatment options. As an example of protein pro-angiogenesis overexpression in cancer cells, vascular endothelial growth factor (VEGF) is one of the most commonly studied targets (Chu, 2012). Bevacizumab (marketed under the brand name Avastin) became the first FDA-approved monoclonal anti-VEGF bioactive. It is combined with standard chemotherapy for treatment of metastatic colon cancer. It has an antiangiogenic effect, disrupting vascularization and new blood vessel growth (Van Cutsem and Lambrechts, 2011). Another potential target is EGFR, which is frequently overexpressed in colon cancer cells. Panitumumab is a monoclonal antibody that blocks EGFR on the outside of cells and stops it from becoming active. Cetuximab is also a monoclonal antibody that works by binding to EGFR. It is a brake on cell proliferation in cancers that harbor EGFR mutations. Preclinical studies have shown that cetuximab in combination with small-molecule chemotherapeutic agents, particularly irinotecan, is synergistic. Panitumumab is usually given along with cetuximab to treat metastatic colon cancer. Thus, the incorporation of chemotherapy with these targeted agents has indeed led to a median survival exceeding two years (Intriago-Baldeón and Pérez-Coral, 2026).

Targeted drug delivery aims to increase the concentration of anticancer bioactives in cancer tissue while reducing the amount of this type of bioactive used in neighboring tissue (Gulbake and Jain, 2016). Nanoparticles have attracted a great deal of interest because of their physicochemical properties, their ability to be conjugated or coupled to attractive functional surface groups, and their biocompatibility, biodegradability, nontoxicity, and high surface-to-volume ratios. Nanoscale materials, including biodegradable biopolymeric, metal oxide, metal, and polymeric substrates, have contributed to the successful development of nanodrug formulations for targeted therapy. Nano-bioconjugation is an important factor in the development of targeted colon cancer therapy (Singh and Waghambare, 2022). It is defined by the linkage of cancer-binding receptors, antibodies, targeting moieties, cytotoxic anticancer bioactives, and immunotoxins through a covalent bond. Nanoparticulate delivery systems may deliver anticancer bioactives to targeted tumor sites via active or passive targeting, thereby increasing therapeutic efficacy and reducing side effects. These carriers are of general interest worldwide for their ability to seek cancer cells (at worst, for treatment, the rare, undesired side effects). They can be easily coated with bioactives via conjugation/coupling to fluorescently labeled cancer-specific markers (You and Kang, 2016).

## 2.5 Nanoparticles: the modern trends in the treatment of colon cancer

Nanomaterials, or nanoparticles, have emerged as promising therapeutic agents for cancer treatment, helping to develop superior treatments. One of the prominent features of nanoparticles is their biocompatibility and bioavailability, which make them more promising for drug delivery, diagnostics, and cancer therapy. Advances and discoveries in the field of intestinal drug delivery have stimulated worldwide interest in nanotechnology, particularly for its potential in diagnosing and treating colon cancer. Chemotherapy treatment has been reported to effectively deposit drugs within cancerous tissues, including regions where large molecules cannot directly penetrate (Schütz and Staedler, 2014). Drug molecules should be small enough to successfully penetrate cancerous cells. The advantages of nanoparticles are that they are nanosized and thereby penetrate easily into cancer cells (Ahmed and Dong, 2013). The nanoparticle size and geometry are important for the successful transport of anticancer agents. Various-shaped (spherical, nanotube, nanorod, nanoflower, nanostar, and nanoprism) nanoparticles can be prepared for efficient drug delivery/systemic therapy. The shapes of nanoparticles (NPs) are vital to their efficient targeting and penetration into cancer cells (Schütz and Staedler, 2014). The spherical morphology of nanoparticles is active in penetrating deeply into cancer cells. Apart from shape, the size of nanoparticles is also very important for efficient therapeutic response, as nanoparticle size has a direct correlation with the cellular uptake mechanism, which is primarily based on endocytosis.

There is another interesting fact that nanoparticles from 10 to 400 nm can cross the cancer cell membrane. Nanoparticles above 500 nm, by contrast, can be taken up into cancer cells via the mechanism of endocytosis (Pelka and Gehrke, 2009). Meanwhile, nanoparticle size also affects the kinetics of drug release and cellular uptake (Miethling-Graff and Rumpker, 2014). The cancer cells contain quantum dots, iron oxide, gold nanoparticles, platinum nanoparticles, silver nanoparticles, and cerium oxide nanoparticles of 10–100 nm in size. At the same time, even polymeric nanoparticles, polypeptide nanoparticles, and dendrimers with a size range of 150–350 nm could also be easily infiltrated into cancer cells (Ma and Chen, 2017). Nanoformulations—a cocktail of different nanoparticles—have been employed in the diagnosis and therapy for colon cancer (Table 2.3) (Huang and Wei, 2015).

New avenues for the early diagnosis and treatment of colon cancer are now available using nanoparticle- and nanotechnology-based targeted drug-delivery vehicles. Various types of masks and other nanoparticles have been designed for use as delivery vehicles in diagnostic and therapeutic applications. These nanoparticles are coupled to targeting moieties to enable selective therapy and reduce toxicity. For passive targeting, the enhanced permeability and retention (EPR) effect of tumors and extravasation is dependent, whereas for active tumor targeting, recognition by colon cancer tissues, such as folate, mAbs, VEGF, aptamers, and membrane-penetrating peptides, is assumed to be the binding moiety. Furthermore, advanced carriers have gradually been transferred to nanotechnology-based drug-delivery systems. These carrier systems are designed to passively accumulate at the tumor site and release encapsulated bioactive contents selectively in response to specific

**Table 2.3 List of some nanomaterial-based products for the treatment of colon cancer.**

| S. No. | Drug | Product name | Types of nanoparticles | Indication |
|---|---|---|---|---|
| 1. | Doxorubicin | Doxil | Liposome | Ovarian cancer, breast cancer, multiple myeloma |
| 2. | Irinotecan | Onivyde | Liposome | Pancreatic cancer |
| 3. | Paclitaxel | Abraxane | Nanoparticle-bound albumin | Breast cancer, pancreatic cancer, nonsmall-cell lung cancer |
| 4. | Cytarabine | DepoCyt | Liposome | Neoplastic meningitis |
| 5. | Vincristine | Marqibo | Liposome | Acute lymphoid leukemia |
| 6. | L-asparaginase | Oncaspar | Polymer-protein conjugate | Leukemia |

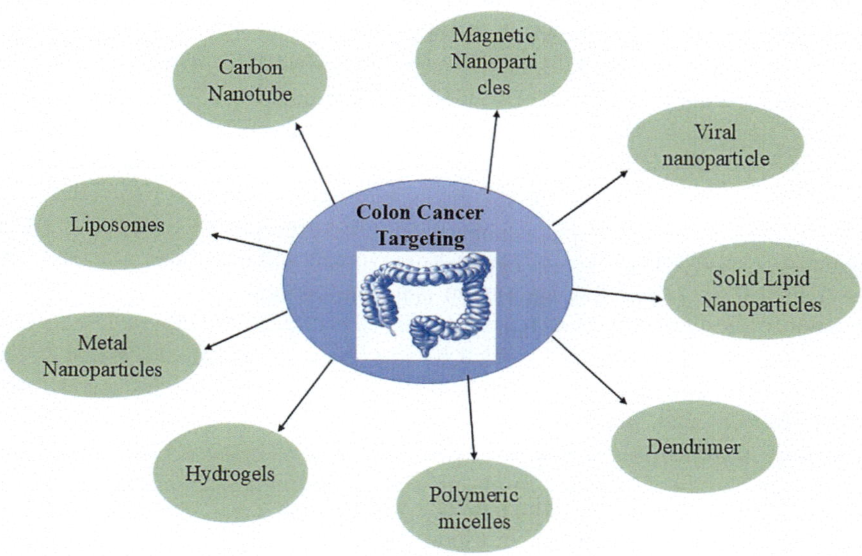


**FIGURE 2.3**

Schematic diagrams of various nanocarriers for targeted colon cancer treatment.

alterations in the bioenvironment, resulting in high bioactive concentrations at the tumor site (Wathoni and Nguyen, 2020). Various nanodrug delivery carriers in pharmaceutical approaches for targeted drug delivery to cancerous colon tissues are discussed here (Fig. 2.3).

**(a)** **pH-responsive nanoparticles:** These colloidal particles present an extracellular acidic environment and a changed pH gradient across their cell compartments. pH-responsive nanoparticles are promising carriers for delivering bioactives to the colon. The anticancer nanoparticles described above were composed of a corona and a core. One or both of these modulate the extracellular pH to alter their solubility/insolubility or state. Nanoparticles with a corona that is either positively charged or becomes soluble, thereby exposing targeting groups and enabling binding at extracellular pH on tumor cells, are developed for enhanced cellular targeting and internalization. Nanoparticles with a core that are soluble or undergo structural changes to release the carried bioactive(s) at the extracellular pH of tumors (or lysosomal pH) have also been designed to rapidly release bioactives into the extracellular fluid or cytosol. These pH-sensitive nanoparticles have several advantages over their conventional pH-insensitive counterparts (Shen and Liu, 2019). Prajakta and Ratnesh, 2009 developed pH-responsive curcumin-laden nanoparticles to deliver the drug to colorectal carcinoma cells. Eudragit S100, which dissolves at pH in the colon, was incorporated into the formulation to target the colon. This results in the selective delivery of entrapped bioactives to the colon. The HT-29 human colon adenocarcinoma cell line (Caco-2) was used for a cytotoxicity test methyl thiazolyl tetrazolium assay (MTT). Equal molar amounts of the nanoparticle formulation and curcumin were compared for efficacy. Results showed that nanoparticles inhibited cell viability by 91.22% after 24 hours of treatment, which is significantly higher ($P$ <0.05) than that observed with curcumin alone (63.85%; 100 μM). The enhanced therapeutic efficacy can be associated with greater cellular uptake, which is size-dependent, and might result in a reduction in the dose requirement (Prajakta and Ratnesh, 2009). Ahmadi and Zarghami, 2019 synthesized pH-sensitive biodegradable nanoparticles for the codelivery of HT/DOX, which degrade by intrinsic mechanisms in hyperthermia (HT)-29 colon cancer cells. Preparation of poly(lactide-co-glycolic acid-co-acrylic acid) (PLGA-co-PAA)[96]: PLGA-co-PAA was synthesized by radical AA telomerization using mercaptoethanol (ME). This is followed by ring-opening polymerization of lactide and glycolide in the presence of PAA-OH as a chain-transfer agent. The measured $IC_{50}$ of the hydroxytyrosol- and doxorubicin-loaded nanoparticles was lower than that of their free analogs at this time concentration. The free HT- and DOX-loaded nanoparticles had weaker cytotoxicity than the respective OM- or DD-loaded nanoparticles. Uptake of the HT- and DOX-loaded NPs by the cells was about 94%. 4',6-diamidino-2-phenylindole (DAPI) staining and cell-cycle analysis demonstrated increased apoptosis of HT-29 cells after treatment with the two-drug-loaded nanoparticles compared to formulas containing only one drug or free drugs. The HT- and DOX-loaded nanoparticles could inhibit gene expression even better than isolated forms, as well. These findings indicated the potential use of the combination drug protocol described in this report for HT-29 colon cancer cell lysis (Ahmadi and Zarghami, 2019). These nanocarrier systems should be further explored in animal models and in clinical trials.

**(b)** **Liposomes: A** liposome is a spherical vesicle that consists of one or more phospholipid bilayers and cholesterol, with sizes ranging from 30 nm to several micrometers. These vesicular systems are of interest because they allow the encapsulation of hydrophilic and lipophilic bioactives. The interior of the aqueous cavity of liposomes is effectively camouflaged by lipid bilayers, which form an envelope to trap hydrophilic species, while the hydrophobic sections/moieties in some lipid bicelles facilitate the intercalation of hydrophobic entities. These vesicular systems mimic natural cell membranes. They have been intensively employed as drug carriers for a long time, and their high loading capacity, safety profile, and biocompatibility make them quite attractive. Additionally, the surface of these nanocarriers is easy to modify with vesicle-specific properties by conjugating polymers and ligands. Moreover, targeting liposomes could provide site-specific delivery of bioactives to solid tumors, leading to a slower release of bioactives at the site of action within the tumor compared with surrounding tissues (Malam and Loizidou, 2009). Xiong and Lei, 2017 reported the fabrication of mannosylated liposomes to enhance the therapeutic performance of paclitaxel in colon cancer models. The mannosylated liposomes demonstrated higher cellular uptake in CT26 (colon cancer cells) and a higher tumor inhibitory ratio due to their targeting of the mannose receptor (Xiong and Lei, 2017). Ektate and Munteanu, 2018 exploited ultrasound and Salmonella-loaded temperature-sensitive liposomes (thermobots) for colon cancer therapy. They developed thermobots that would carry membrane-bound temperature-sensitive liposomes (LTSL) actively within colon cancer cells. This results in the release of doxorubicin. Meanwhile, the fabricated carrier system further polarized macrophages to the M1 phenotype by high-intensity focused ultrasound (HIFU) treatment (42°C). Biocompatibility test results indicated that the resultant thermobots could effectively load LTSL without compromising their viability. The in vitro study indicated that the thermobots exhibited effective intracellular targeting, high doxorubicin accumulation, and increased expression of proinflammatory cytokines in C26 murine colon cancer cells. In vivo, the thermobot combined with HIFU heating (B30 minutes) increased M1 polarization of macrophages and therapeutic efficacy in C26 murine colon cancer cells. The data suggest that thermobots and focused ultrasound therapy could enhance the treatment of colon cancer (Ektate and Munteanu, 2018). Korani and Ghaffari, 2019 designed bortezomib-loaded nanoliposomal formulations and investigated their anticancer effects in mice bearing C26 colon carcinoma cells. The $IC_{50}$ values of the developed bortezomib formulation against the C26 colon cancer cell line were compared with those of free bortezomib. The $IC_{50}$ values obtained for the free drug and the liposomal preparation were $0.0072 \pm 0.00075\ \mu g\ ml^{-1}$ (free drug) and $0.05 \pm 0.001\ \mu g\ ml^{-1}$ (liposomal preparation) against the C26 colon carcinoma cell line, respectively. An in vivo study of the liposomal formulation showed significant inhibition of tumor growth and an increase in median survival time (MST) in mice. These results also suggested that the liposomal formulation of bortezomib has great potential to improve the therapeutic index of bortezomib. This nanocarrier system could be considered a promising drug-delivery system in clinical practice for the future (Korani and Ghaffari, 2019).

(c) **Polymeric nanoparticles**: Nanosized colloidal particles in a size range from 10 nm to 1 μm, which are of a solid nature. The small sizes and unique properties of these nanoparticulate systems have brought tremendous attention in recent years. They use a matrix of synthetic or natural, biodegradable, and biocompatible polymers, which allows them to entrap/encapsulate bioactives, physically adsorb them on the matrix surface, or chemically bond them to its surface. Biodegradable polymers have certain advantages, as they are fully metabolized by natural metabolic pathways. Natural polymers are generally biodegradable and suitable for the human body. However, they are limited by the irregular properties of carbon black 8 in each batch. On the other hand, synthetic polymers can be synthesized with precisely controlled chemical composition (Parveen & Sahoo, 2008). The most frequently employed synthetic polymers are polyacrylates, polycaprolactones, polylactide, and polylactide-polyglycolide copolymers. The lactide-glycolide copolymer is well studied. Among the different natural polymers, alginate, albumin, and chitosan are mainly studied. Polymeric nanoparticles can, in fact, be obtained, depending on the method of preparation used: nanospheres or nanocapsules. Nanospheres are a matrix system in which an active is homogeneously dispersed, whereas in nanocapsules, an active is located inside a cavity, which is surrounded by a polymeric wall. The polymeric nanoparticles are prepared in three ways: (1) preparation of nanoparticles from preformed polymer dispersion using nanoprecipitation, emulsification, solvent evaporation, dialysis, salting out, and supercritical fluid; (2) the polymerization of monomers using emulsion, miniemulsion, microemulsion, and controlled radical polymerization or interfacial polymerization methods; and (3) coacervation of hydrophilic polymers or ionic gelation. The advantages of polymeric nanoparticles are that (1) bioactives can be controllably released, (2) therapy can be coupled with imaging (i.e., theranostics), (3) protection is provided, and (4) precise targeting of bioactives is facilitated. Polymeric nanoparticles are characterized by their many desirable features, including small size, biodegradability, long stability during storage, and water solubility. These properties make them an ideal candidate for the delivery of bioactives, proteins, DNA, or genes to precisely targeted tissues/organs. So, these are used in cancer therapy, vaccines, and gene therapy for crossing the blood–brain barrier as well as in the diagnosis of diseases (Liechty & Peppas, 2012).

**(d) Nanocapsules:** Nanoencapsulation, as a modern approach, has recently offered new opportunities in drug delivery. Nanocapsules are one of the submicro-drug-delivery systems. They consist of an oily or aqueous core encapsulated in a polymer shell. Their size ranges from 5 to 1000 nm, with most falling within 100–500 nm. Advantages of nanocapsules as bioactive carriers. The core cavity contains bioactives protected from rapid degradation. Frequently, the wall of nanocapsules is formed from a biodegradable polymer. Such shells are typically formed by tightly wrapping around the spherical interfaces of particles or droplets or by self-assembly of amphiphilic polymer molecules. They are prepared using various techniques, such as nanoprecipitation, emulsion diffusion, double emulsification, and polymer coating (Creixell & Peppas, 2012). Capsules are formed from aqueous solutions via a sequential deposit of polymer layers onto a sacrificial template. Various interactions, such as hydrogen bonding, electrostatic interactions, and covalent interactions, can drive the assembly of a multilayer shell. Decomposition of the sacrificial template leads to the generation of hollow capsules. The nanocapsules are similar to vesicular drug-delivery systems, which entrap bioactives in their cavities, whether as a liquid core or a polymer matrix, encapsulated by a polymeric membrane. The bioactives are present in the cavity as a molecular dispersion in either liquid or solid form. Hollow nanocapsules such as these have been demonstrated to encapsulate catalysts, dyes, bioactives, and biopolymers, including nucleic acids and proteins. In fact, the same can even offer a shielded environment (protective covering) to peptides, hormones, proteins, enzymes, and bioactives/metabolites from both chemical and biological degradation. They also manifest site specificity, which could enhance the stability and bioavailability of bioactives. These vehicles are likewise able to regulate cutaneous bioactive permeation/penetration and serve as physical sunscreens due to their light-reflecting properties. The permeability of such capsules can be adjusted by changing the pH and by incorporating a temperature-sensitive, redox-sensitive, or pH-sensitive polymer. Moreover, targeted delivery of proteins into cells allows for the localization of proteins to treat tumor cells. Consequently, these intelligent polymeric nanocapsules have broad potential for the targeted delivery of intracellular degradable bioactives because of their flexible biofunctionality (Santos and Ponte, 2013).

Tsakiris and Papavasileiou, 2019 fabricated nanocapsules for colon cancer therapy. The authors entrapped six anticancer bioactives, namely SN38, doxorubicin hydrochloride, oxaliplatin, 5-fluorouracil, irinotecan, and regorafinib in lipid nanocapsules. They were about 50 nm in size. In vitro assays demonstrated that the toxicity of lipid nanocapsules loaded with the drug was slightly lower compared to the free drug. However, there was no noticeable difference in the $IC_{50}$ values between the free drug and the encapsulated drug within liposomes. Among the tested formulations, SN38-lipid nanocapsules were among the most potent formulations on the CT26 colon cancer cell line in clonogenic assays. However, they were found to be cytotoxic at lower doses. Therefore, this formulation was used for the in vivo investigations. In the in vivo study, a synchronous therapy of SN38 lipid nanocapsules with regorafinib lipid nanocapsules retarded the growth and enhanced the MST of Balb/C mice implanted with CT26 murine colon cancer (Tsakiris and Papavasileiou, 2019). Oyarzun-Ampuero and Rivera-Rodríguez, 2013 developed hyaluronan nanocapsules containing docetaxel to treat colon cancer. The nanocapsules were prepared by a solvent displacement method. The average diameter of the nanocapsules was about 200 nm, and they were spherical. To investigate the therapeutic effects of docetaxel-loaded hyaluronan nanocapsules, a cell viability assay was conducted using the H460 lung cancer cell line. These cells were selected because they overexpress CD44 receptors, which are known to interact strongly with hyaluronic acid. The cytotoxicity assay showed that the docetaxel-loaded hyaluronan nanocapsules were significantly more cytotoxic against H460 lung cancer cells than free docetaxel. The nanocarrier formulations were 3.6-fold ($P$=.005) higher than the free drug when comparing $IC_{50}$ values. These nanocomposites are potentially applicable to intracellular drug-delivery systems (Oyarzun-Ampuero and Rivera-Rodríguez, 2013).

**(e)** **Nanospheres:** Nanospheres 10–200 nm in diameter are formed of a polymeric matrix composite of spherical shape. An abiotically active substance is trapped, dissolved, encapsulated, or bonded to the polymeric matrix composite. The morphology of nanospheres can be crystalline or amorphous. Such nanocarriers are designed to preserve bioactives from chemical and enzymatic degradation. The bioactives are distributed and dispersed evenly in the polymeric matrix. There are principally two kinds of nanospheres: biodegradable and nonbiodegradable. Biodegradable nanospheres include albumin, gelatin, and polypropylene dextran nanospheres; modified-starch nanospheres; polylactic acid (PLA); poly(D-, L-glycolide) (pLG); poly(lactic-co-glycolic acid) (PLGA); and poly(cyanoacrylate) (PCA). Among these, only PLA has been granted permission as a controlled-release agent for nonbiodegradable nanospheres. In addition, magnetic and immune nanospheres are also presented in the following paragraphs. Nanospheres that encapsulate and transport immune components, such as antibodies or antigens, provide in-transit protection for the bioactive(s) during circulation. Further, they may also provide a slow release of the bioactive(s) over an extended period. Furthermore, they can also be directed specifically (Meng et al., 2025). The magnetic nanospheres are coated with magnetite ($Fe_3O_4$)/ferrite ($Fe_2O_3$) and the drug. Thus, they exhibit magnetic responsiveness. Accordingly, when these structures are placed under the applied magnetic field, they can be localized to a specific site or organ within the patient's body, and the drug can be released in a controlled manner depending on the intensity of the external magnetic field (Kumar et al., 2025; Sharifi-Azad et al., 2022).

Xu and Sun, 2019 prepared bufalin-loaded targeted nanospheres for colon cancer therapy. In the present study, a facile room temperature route was developed to prepare bufalin-loaded CaP/1,2-bis(diphenylphosphino)ethane polyethylene glycol (CaP/DPPE-PEG) nanospheres with covalently coupled epidermal growth factor (CaP/DPPE-PEG-EGF); it was employed as a carrier of bufalin for specific antitumor therapy. For the cell line study, HCT116 colon cancer cells were used. The in vitro cellular uptake studies indicated that the developed nanocarrier formulation could be efficiently taken up by HCT116 colon cancer cells via receptor-mediated endocytosis mediated by epidermal growth factor. Cytotoxicity of the bufalin-loaded nanospheres was greater than that of bufalin in HCT116 cells. The cytotoxicity test showed that HCT116 cells were more affected by bufalin-loaded nanospheres than by free bufalin. In animal studies, the nanosphere preparation was observed to be more readily captured by the liver and spleen owing to the extensive phagocytic activity of the reticuloendothelial system (RES). It also showed much stronger luminescence than free bufalin in the tumor (Xu and Sun, 2019). Ayub and Chiu, 2019 prepared paclitaxel-loaded self-aggregated disulfide-cross-linked sodium alginate-derived nanoparticles for the treatment of colon cancer. The encapsulation efficiency of 77.1% and the cumulative percentage drug release were observed to be 45.1% for the developed formulation, as reported. The pH response of the CC was followed by a physicochemical pathway involving a volumetric transition (105.15% increase in diameter) between pH 1 and 7 in the CC nanocarrier formulation. HT-29 cells were subjected to a cell viability assay. The viability was 86.7% for the developed nanosphere formulation at a concentration of 0.8 $\mu g\ mL^{-1}$. Over 70% of the paclitaxel-loaded nanospheres were found in HT-29 cells. These results demonstrated the efficient uptake of the synthesized nanospheres by the cancer cells. Thus, this may be a hopeful and feasible nanocarrier system for drug delivery in colon cancer (Ayub and Chiu, 2019).

**(f) Solid lipid nanoparticles (SLNs):** SLNs are one of the promising new generation submicron-sized colloidal carrier systems, which are solid at both room and body temperature. These are surfactant-stabilized liposomes. They combine the advantages of polymeric nanoparticles, emulsions, and liposomes, while avoiding some of their disadvantages. Generally, they comprise a hydrophobic solid matrix core coated with phospholipid. The hydrophobic tail portion of the phospholipid is inserted into the core matrix. For this reason, the core is purely hydrophobic, and it is presumed that SLNs should have better encapsulation efficiency of hydrophobic bioactives into the core than conventional liposomes (Yassin et al., 2010). Orally, parenterally, ophthalmically, dermally, by rectum, and nasally, SLNs have been investigated for a number of routes of administration, that is, intravenous. The most appropriate form for the administration of SLNs is the intravenous route. Such particles can be homogeneously dispersed in aqueous and aqueous surfactant solutions. SLNs are composed of solid lipids (e.g., waxes, steroids, purified triglycerides, and fatty acids), emulsifiers, and water for preparation. Various techniques are employed for the preparation of SLNs, such as hot homogenization, cold homogenization, solvent emulsification/evaporation, and solvent emulsification/diffusion. SLNs possess the following advantages: simple scale-up, nontoxicity, resistance to coalescence or aggregation, protection of the entrapped bioactive agent, sustained release of bioactives from the matrix, biodegradability, and low-cost production. Numerous studies have demonstrated that, when these nanoparticles are applied as bioactive transport systems for chemotherapeutics, particularly for colon cancer and malignant melanoma, promising results can be achieved (Ngwuluka and Kotak, 2017).

Shen and Liu, 2019 formulated orally administered SLNs with a polysaccharide coating to target colon cancer. The emulsion method was used to prepare SLNs coloaded with doxorubicin and superparamagnetic iron oxide nanoparticles (SPIONS). The resultant nanoformulation was surface-decorated with folic acid (FA)/D-α-tocopheryl polyethylene glycol (PEG) 1000 succinate and octadecanol-modified dextran using the layer-by-layer technique. The hydrodynamic diameter of SLNs was 100 nm. The prepared formulation was tested for cell uptake using mouse colon CT26 cancer cells. The IC50 value of the prepared formulation was lower than that of free doxorubicin (58 μM) on CT26 cells. An in vivo antitumor efficacy study showed that primary colon tumors and peritoneal metastasis were significantly suppressed, associated with decreased ascites, tumor volume, size, and the number of tumor nodules in the mouse model after oral administration of SLN therapy (Shen and Liu, 2019). Campos and Fernandes, 2019 formulated and optimized nimuslide-encapsulated SLNs. The in vitro drug release profile of the nimuslide-loaded SLNs exhibited a sustained release pattern, with 30% of the drug released within 24 hours. Both plain and nimuslide-loaded SLNs were noncytotoxic at the highest concentration (100 μg $mL^{-1}$) against Caco-2 cells up to 48 hours. The cell viability was 80%. To date, such nanocarrier systems hold the potential for the efficacious treatment of colon cancer (Campos and Fernandes, 2019).

**(g) Metallic nanoparticles:** Metallic nanoparticles have gained considerable attention nowadays, in comparison with their bulk material, because of their unusual properties. Researchers have become interested in synthesizing metallic nanoparticles, including gold, iron oxide, and silver, due to their potential applications in biotechnology, diagnostic imaging, magnetic separation, and targeted drug delivery. They can be synthesized and surface-modified using several chemical functionalizations, enabling this class of nanocarriers to be combined with targeting ligands. These carriers are particularly favorable in tumor therapy since the nanocarrier's charge, size, shape, and surface modification can be manipulated. The increased density of these nanocarriers can be better internalized by the cells, which is advantageous for cancer vaccination strategies compared to other nonmetal-based drug carriers. They also have unique optical properties (Desai and Momin, 2021). Han and Jiang, 2019 studied the anticarcinogenic properties of gold nanoparticles with respect to colon cancer. The nanocarrier was formulated with an extract of *Trichosanthes kirilowii*. The synthesized gold nanoparticles exhibited an average particle diameter of 50 nm. Their cytotoxic potential was assessed using the MTT assay in HCT116 colorectal cancer cells, where a 50% reduction in cell viability was observed at a concentration of 15.5 μg/mL. (Han and Jiang, 2019). Majeed and Aripin, 2019 prepared bovine serum albumin-coated silver nanoparticles for the treatment of colon cancer. The nanoparticle size ranged from 11.26 to 23.85 nm. The $IC_{50}$ value against HCT116 colon cancer cells from a cytotoxicity assay was 60 μg $mL^{-1}$. An apoptotic assay showed chromatin condensation caused by ROS production and membrane blebbing using ethidium bromide and acridine orange double staining. These data suggest that the constructed delivery system might offer a promising method for colon cancer targeting (Majeed and Aripin, 2019; Shrivastava and Sharma, 2020).

**(h) Viral nanoparticles:** These NP formulations are virus-derived and have promising prospects for nanomedicine. They are naturally occurring nanoparticles. They might serve as general-purpose building blocks (Licciardi and Scialabba, 2019). In medical research, virus-derived materials are increasingly interesting. Specific examples of novel viruses include, but are not limited to, infections caused by bacteria, as well as those in mammals and plants (Cho and Yoon, 2010). These are self-assembled and genetically programmed, discrete, monodisperse structures with specific sizes and shapes. These structures have been known to possess atomic resolution and have been realized at the atomic scale for some of the virus-based systems. So far, with artificial nanoparticles, you cannot achieve this level of quality control; it is just structural engineering. For targeted applications, they may be engineered using several strategies: (1) Bioconjugate chemistries/approaches can be implemented to conjugate bioactives, targeting moieties, and contrast agents on the surface of the outer or inner capsid shell; (2) disassembly/reassembly protocols can be employed to encapsulate artificial payloads (or cargoes), that is, bioactives and/

or contrast agents; and (3) genetic engineering approaches allow precise and reproducible alterations/modifications to be made so that identical particles can be produced in mass with such functionalities as presenting targeting moieties for binding ligands or unique ligation handles for subsequent modification via selected functionalization reactions. Deo and Kato, 2016 produced and constructed viral nanoparticles surface-modified with recombinant single-chain variable fragment (rscfv) for targeting colon cancer tumors. Specifically, rscfv binds to tumor-associated glycoprotein, which is highly expressed on the surface of colorectal cancer cells. Calcein, as a representative bioactive compound, was used to focus on the colon cancer cells. The nanocarrier formulation was found to have a size of 50 nm. Chemotaxis chamber assays showed that the formulated nanoparticles attracted macrophages and adhered to colonic cancer cells (HU1197 human monocytic cells). The research found that macrophages secreted tumor necrosis factor-α, a cytokine necessary to kill cancer cells. This result suggests the potential of viral nanoparticles as smart nanobiomaterials for colon cancer targeting (Deo and Kato, 2016).

(i) **Polymeric micelles:** There have been marked developments and improvements in delivering bioactives, while an appropriate delivery system for the delivery of bioactives remains a formidable challenge to scientists despite progress. Polymeric micelles are λ,τ-structured nanomolecules. Normally, they are self-assembled amphiphilic block copolymers in aqueous conditions. The latter nanoparticles are characterized by a hydrophilic shell and a lipophilic core architecture, favoring the embedding of lipophilic bioactives. They are also considered drug-delivery vehicles because of their specificity toward targets and their controlled delivery of lipophilic anticancer bioactives. They have also emerged as carriers in drug delivery owing to their interesting characteristics, such as tailorability, nontoxicity, controlled drug release, high cargo capacity, and high stability. Polymeric micelles can improve the delivery of several poorly aqueous-soluble anticancer bioactives through combined or sequential regimens, presenting an interesting concept that may lead to new approaches for incorporating drugs in cancer treatment (Shrivastava et al., 2020). Le and Kim (2019) synthesized polymeric micelles to deliver curcumin into colon cancer cells. Hydrophilic graft-like O-(2-aminoethyl) PEG and hydrophobic octadecylamine were attached to a polysuccinimide backbone for micelle formation. Folic acid was chosen as the targeting ligand. Curcumin drug release is pH-sensitive and is linked via an acid-cleavable hydrazone bond. The cytotoxicity assay (MTT assay) studies showed that cancer cell viability exposed to the prepared nanocarrier formulation was significantly reduced compared with free curcumin at the same concentration of 0.25 μg $mL^{-1}$. Western blot analysis showed that the expression of cyclin D1, c-myc, and downstream genes in plain curcumin was lower than in curcumin-loaded polymeric micelles at 0.5 μg $mL^{-1}$. It was deduced from these results that curcumin-loaded folic acid-conjugated polymeric micelles should be a promising agent for colon cancer-targeting therapy by suppressing the Wnt/β-catenin signaling (Le & Kim, 2019).

(j) **Hydrogel:** A hydrogel (also known as aquagel) is a 3D network of hydrophilic polymers that absorbs water. The 3D polymer network of an aquagel is retained as an elastic solid. Hydrogels, by definition, are systems that contain at least 10% of the total weight of water. Hydrogels are nanocarriers and demand specific and potent characteristics for targeted drug-delivery systems in cancer therapy. Cancer is a life-threatening disease for human health worldwide, with various shortcomings of the conventional anticancer therapies. However, hydrogels offer new and improved opportunities for cancer treatment without systemic cytotoxicity to healthy cells or tissues. In addition, their morphology and high biocompatibility facilitate the design of hydrogels.

Because of these features, hydrogels have been the focus of attention and can also be a potent tool for nanomedicines. Zheng and Wang, 2019 constructed a chitosan-capable temperature-sensitive hydrogel for the treatment of colon cancer. The formulation was prepared by dissolving bioactive molecules, that is, doxorubicin and $MoS_2/Bi_2S_3$-PEG (MBP) (a photothermal material) nanosheets, into the hydrogel. Moreover, the hydrogel system could efficiently prevent MBP nanosheets and an anticancer drug (DOX) from entering the bloodstream and injuring healthy tissues. At a low concentration of MBP (0.5 mg/mL$^{-1}$), under the first near-infrared light, the developed hydrogel formulation exhibited a photothermal efficiency of 22.18%, and under the second, a thermal effect ratio of 31.42%. Doxorubicin release was sustained and controlled from the hydrogel. In addition, the formulated preparation has antibacterial effects. The fabricated nanocomposite hydrogel is anticipated to serve as a channel for effective cancer therapy, as it can selectively absorb light with different penetration depths (the first n**ear-infrared** (NIR) light and the second NIR light) (Zheng and Wang, 2019).

**(k) Polymerosomes:** Like liposomes, polymerosomes have a vesicular hollow-shell structure and are suitable for bioactive delivery. Polymerosomes are constructed from amphiphilic copolymers containing covalently linked homopolymer blocks. This is an interesting example of a versatile and robust drug-delivery technology that relies on block copolymers with amphiphilic properties. Further, their properties, such as encapsulation efficiency, release profile, and drug loading, can be modified. Biodegradable polymerosomes have been shown to harness the robust membrane of block copolymer vesicles and their aqueous core for drug sequestration, delivery, pH-induced release in endolysosomes, and transfer into the cytosol of bioactive blends. Biodegradable and tumor-specific polymers disintegrate under acidic conditions, and bioactives are activated in the endosomes of cancerous cells. A polymerosome is similar to a liposome, which has two layers of phospholipids, in that it consists of two synthetic polymers. For the transport of hydrophilic bioactives, polymerosomes, namely artificial vesicles, have emerged as propitious pharmaceutical carriers (Shrivastava and Sharma, 2020).

Paclitaxel-loaded folate-conjugated polymeric nanoparticles for the treatment of colon cancer by Pan and Gong, 2019. The formulation (polymerosomes) exhibited biphasic drug release in simulated gastric and intestinal fluids. A cytotoxicity (MTT) assay was performed with the colon cancer cell line Caco-2. The viability of colon cancer cells dropped to 20% for polymerosomes at a high concentration (0.4 μg/mL$^{-1}$). By fluorescence microscopy, the cell uptake was shown to be much higher for polymerosomes than for free paclitaxel. Pharmacokinetic studies have shown that the area under the curve of folate-conjugated polymerosomes was higher than that of the unconjugated ones. These results indicate that the ligand-conjugated polymerosomes may be prospective for anticancer bioactive drug delivery systems (Pan and Gong, 2019).

**(l) Carbon nanotubes:** These nanocarriers have specific tubular or barrel-like structures with a diameter of the nanoscale. The nanocarriers are formed as 10–100 concentric carbon shells with an adjacent taper of separation between two neighboring shell layers. This network of concentric shells correlates with the honeycomb arrangement of carbon atoms in graphite-like sheets. The exceptional electronic and mechanical behavior of these nanocarriers results from their graphite-like carbon atoms that form the shell and the quasione-dimensional structure. Carbon nanotubes may be semiconductor or metallic types, depending on their structural parameters. Furthermore, they may be single or multiwalled, with a high inner-to-outer volume ratio. The outer shell of these nanocarriers could be conveniently tailored (Ahmed and Dong, 2013).

They present a bioactive carrier capability that results from their high aspect ratio, tunable structure and morphology, adaptable surface chemistry, and CT-SR GER. For targeting colon cancer, multi-walled carbon nanotube (MWCNTs) were synthesized and surface-functionalized with hyaluronic acid (Liechty & Peppas, 2012). Gemcitabine was encapsulated in the prepared nanocarrier. The ligand (hyaluronic acid) was grafted onto the external surface of the pegylated MWCNTs. Gemcitabine is released faster at an acidic pH than under physiological conditions (PBS, pH 7.4), as demonstrated by the in vitro drug release study. The hemolytic toxicity of the resulting nanocarrier system was 7.73% and 60.4%, far lower than that of plain drug, which was 18.71% and 60.44%. The cytotoxicity assay indicated that, compared with RAW264.7, the HA-functionalized MWCNTs were more cytotoxic to HT-29 (a colon cancer cell line). In tumor-carrying rats, in vivo animal testing with these products showed higher AUMC, AUC, and MST than in rats loaded with gemcitabine-loaded hyaluronic acid-conjugated MWCNTs. The *P*-value was 0.0001. These investigations suggest that the synthesized surface-modified MWCNTs represent a potential, effective, and safe nano-vehicle for colon cancer targeting (Ma and Chen, 2017).

## 2.6 Conclusion

In conclusion, targeted cancer therapies and nanoparticle-based drug-delivery systems are effective in interacting with specific molecules implicated in tumor growth, dissemination, and progression. Nanoparticle-based formulations are a suitable option for treating colon cancer because they increase tumor cell apoptosis (cell death) and selectivity. However, a number of obstacles still need to be removed in order to develop the best possible treatment for various forms of colon cancer. Only a small number of nanoparticle formulations have been licensed and are available for the treatment of cancer, despite significant research efforts. Additionally, new methods for particle targeting and stabilization are needed.

## References


Ahmadi, E., Zarghami, N., Jafarabadi, M. A., Alizadeh, L., Khojastehfard, M., Yamchi, M. R., & Salehi, R. (2019). Enhanced anticancer potency by combination chemotherapy of HT-29 cells with biodegradable, pH-sensitive nanoparticles for co-delivery of hydroxytyrosol and doxorubicin. *Journal of Drug Delivery Science and Technology, 51*. http://www.editionsdesante.fr/category.php?id_category=48, https://doi.org/10.1016/j.jddst.2019.03.003.

Ahmed, S. R., Dong, J., Yui, M., Kato, T., Lee, J., & Park, E. Y. (2013). Quantum dots incorporated magnetic nanoparticles for imaging colon carcinoma cells. *Journal of Nanobiotechnology*. http://www.jnanobiotechnology.com/content/11/1/28, https://doi.org/10.1186/1477-3155-11-28, Japan.

Antelo, M., Golubicki, M., Roca, E., Mendez, G., Carballido, M., Iseas, S., Cuatrecasas, M., Moreira, L., Sanchez, A., Carballal, S., Castells, A., Boland, C. R., Goel, A., & Balaguer, F. (2019). Lynch-like syndrome is as frequent as Lynch syndrome in early-onset nonfamilial nonpolyposis colorectal cancer. *International Journal of Cancer, 145*(3). https://doi.org/10.1002/ijc.32160.

Araghi, M., Soerjomataram, I., Jenkins, M., Brierley, J., Morris, E., Bray, F., & Arnold, M. (2019). Global trends in colorectal cancer mortality: Projections to the year 2035. *International Journal of Cancer, 144*(12). https://doi.org/10.1002/ijc.32055.

Ayub, A. D., Chiu, H. I., Mat Yusuf, S. N. A., Abd Kadir, E., Ngalim, S. H., & Lim, V. (2019). Biocompatible disulfide cross-linked sodium alginate derivative nanoparticles for oral colon-targeted drug delivery. *Artificial Cells, Nanomedicine and Biotechnology, 47*(1). http://www.tandfonline.com/loi/ianb20#, https://doi.org/10.1080/21691401.2018.1557672.

Bragg, M. A., Breaux, W. A., & M'Koma, A. E. (2023). Inflammatory bowel disease-associated colorectal cancer: Translational and transformational risks posed by exogenous free hemoglobin alpha chain, a by-product of extravasated erythrocyte macrophage erythrophagocytosis. *Medicina (Lithuania)*. https://www.mdpi.com/journal/medicina, https://doi.org/10.3390/medicina59071254.

Burt, R. W. (2000). Colon cancer screening. *Gastroenterology, 119*(3). https://doi.org/10.1053/gast.2000.16508, http://www.journals.elsevier.com/gastroenterology/.

Campos, J. R., Fernandes, A. R., Sousa, R., Fangueiro, J. F., Boonme, P., Garcia, M. L., Silva, A. M., Naveros, B. C., & Souto, E. B. (2019). Optimization of nimesulide-loaded solid lipid nanoparticles (SLN) by factorial design, release profile, and cytotoxicity in a human colon adenocarcinoma cell line. *Pharmaceutical Development and Technology,* 616–622. https://doi.org/10.1080/10837450.2018.1549075, http://www.tandfonline.com/loi/iphd20.

Cho, Y. S., Yoon, T. J., Jang, E. S., Soo Hong, K., Young Lee, S., Ran Kim, O., Park, C., Kim, Y. J., Yi, G. C., & Chang, K. (2010). Cetuximab-conjugated magneto-fluorescent silica nanoparticles for in vivo colon cancer targeting and imaging. *Cancer Letters, 299*(1). https://doi.org/10.1016/j.canlet.2010.08.004.

Chu, E. (2012). An update on the current and emerging targeted agents in metastatic colorectal cancer. *Clinical Colorectal Cancer, 11*(1). https://doi.org/10.1016/j.clcc.2011.05.005.

Creixell, M., & Peppas, N. A. (2012). Co-delivery of siRNA and therapeutic agents using nanocarriers to overcome cancer resistance. *Nano Today, 7*(4). https://doi.org/10.1016/j.nantod.2012.06.013.

Cunningham, D., Atkin, W., Lenz, H. J., Lynch, H. T., Minsky, B., Nordlinger, B., & Starling, N. (2010). Colorectal cancer. *The Lancet, 375*(9719). https://doi.org/10.1016/S0140-6736(10)60353-4, http://www.journals.elsevier.com/the-lancet/.

Van Cutsem, E., Lambrechts, D., Prenen, H., Jain, R. K., & Carmeliet, P. (2011). Lessons from the adjuvant bevacizumab trial on colon cancer: What next? *Journal of Clinical Oncology, 29*(1). https://doi.org/10.1200/jco.2010.32.2701.

Deleon, M. F., & Stocchi, L. (2022). Elective and emergent surgery in the ulcerative colitis patient. *Clinics in Colon and Rectal Surgery, 35*(6). http://www.thieme-connect.com/ejournals/toc/ccrs, https://doi.org/10.1055/s-0042-1758134.

Deo, V. K., Kato, T., & Park, E. Y. (2016). Virus-like particles displaying recombinant short-chain fragment region and interleukin 2 for targeting colon cancer tumors and attracting macrophages. *Journal of Pharmaceutical Sciences, 105*(5). http://www.interscience.wiley.com/jpages/0022-3549, https://doi.org/10.1016/j.xphs.2016.02.011.

Desai, N., Momin, M., Khan, T., Gharat, S., Ningthoujam, R. S., & Omri, A. (2021). Metallic nanoparticles as a drug delivery system for the treatment of cancer. *Expert Opinion on Drug Delivery, 18*(9). http://www.tandfonline.com/loi/iedd20, https://doi.org/10.1080/17425247.2021.1912008.

Desantis, C. E., Lin, C. C., Mariotto, A. B., Siegel, R. L., Stein, K. D., Kramer, J. L., Alteri, R., Robbins, A. S., & Jemal, A. (2014). Cancer treatment and survivorship statistics, 2014. *CA Cancer Journal for Clinicians*. https://doi.org/10.3322/caac.21235, http://onlinelibrary.wiley.com/journal/10.3322/(ISSN)1542-4863.

Ektate, K., Munteanu, M. C., Ashar, H., Malayer, J., & Ranjan, A. (2018). Chemo-immunotherapy of colon cancer with focused ultrasound and Salmonella-laden, temperature-sensitive liposomes (thermobots). *Scientific Reports*. http://www.nature.com/srep/index.html, https://doi.org/10.1038/s41598-018-30106-4.

Feng, Q., Chang, W., Mao, Y., He, G., Zheng, P., Tang, W., Wei, Y., Ren, L., Zhu, D., Ji, M., Tu, Y., Qin, X., & Xu, J. (2019). Tumor-associated macrophages as prognostic and predictive biomarkers for postoperative adjuvant chemotherapy in patients with stage II colon cancer. *Clinical Cancer Research, 25*(13). https://doi.org/10.1158/1078-0432.ccr-18-2076.

Galuppini, F., Pennelli, G., Loupakis, F., Lanza, C., Vianello, L., Sacchi, D., Mescoli, C., Salmaso, R., Agostini, M., Lonardi, S., Farinati, F., Rugge, M., & Fassan, M. (2017). BRAF p.V600E-specific immunohistochemical assessment in colorectal cancer endoscopy biopsies is consistent with the mutational profiling. *Histopathology, 71*(6), 1008–1011. https://doi.org/10.1111/his.13315.

Gulbake, A., Jain, A., Jain, A., Jain, A., & Jain, S. K. (2016). Insight into drug delivery aspects for colorectal cancer. *World Journal of Gastroenterology, 22*(2), 582–599. http://www.wjgnet.com/esps/DownLoadFile.aspx?Type=Digital&SubType=1&DOI=10.3748%2fwjg.v22.i2.582&FilePath=Pub%5c10.3748%5cv22%5ci2%5cWJG-22-582.pdf, https://doi.org/10.3748/wjg.v22.i2.582.

Han, X., Jiang, X., Guo, L., Wang, Y., Veeraraghavan, V. P., Krishna Mohan, S., Wang, Z., & Cao, D. (2019). Anticarcinogenic potential of gold nanoparticles synthesized from *Trichosanthes kirilowii* in colon cancer cells through the induction of the apoptotic pathway. *Artificial Cells, Nanomedicine and Biotechnology, 47*(1). https://doi.org/10.1080/21691401.2019.1626412, http://www.tandfonline.com/loi/ianb20#.VmugQbfovcs.

Huang, R. F. S., Wei, Y. J., Inbaraj, B. S., & Chen, B. H. (2015). Inhibition of colon cancer cell growth by nanoemulsion carrying gold nanoparticles and lycopene. *International Journal of Nanomedicine, 10*. https://doi.org/10.2147/IJN.S79107, http://www.dovepress.com/getfile.php?fileID=24525.

Intriago-Baldeón, D. P., Pérez-Coral, E. S., Armas Samaniego, M. I., Romero, V. I., Pozo Palacios, J. C., & Bigoni-Ordóñez, G. D. (2026). Decoding breast cancer: Emerging molecular biomarkers and novel therapeutic targets for precision medicine. *Ecuador International Journal of Molecular Sciences, 27(1)*. https://doi.org/10.3390/ijms27010138, http://www.mdpi.com/journal/ijms.

Korani, M., Ghaffari, S., Attar, H., Mashreghi, M., & Jaafari, M. R. (2019). Preparation and characterization of nanoliposomal bortezomib formulations and evaluation of their anti-cancer efficacy in mice bearing C26 colon carcinoma and B16F0 melanoma. *Nanomedicine: Nanotechnology, Biology, and Medicine, 20*. https://www.journals.elsevier.com/nanomedicine-nanotechnology-biology-and-medicine, https://doi.org/10.1016/j.nano.2019.04.016.

Kristjansson, S. R., Nesbakken, A., Jordhøy, M. S., Skovlund, E., Audisio, R. A., Johannessen, H. O., Bakka, A., & Wyller, T. B. (2010). Comprehensive geriatric assessment can predict complications in elderly patients after elective surgery for colorectal cancer: A prospective observational cohort study. *Critical Reviews in Oncology/Hematology, 76*(3), 208–217. https://doi.org/10.1016/j.critrevonc.2009.11.002, http://www.elsevier.com/locate/critrevonc.

Kumar, A., Shahvej, S. K., Yadav, P., Modi, U., Yadav, A. K., Solanki, R., & Bhatia, D. (2025). Clinical Applications of Targeted Nanomaterials. *Pharmaceutics, 17*(3), 379. https://doi.org/10.3390/pharmaceutics17030379.

Le, T. T., & Kim, D. (2019). Folate-PEG/Hyd-curcumin/C18-g-PSI micelles for site-specific delivery of curcumin to colon cancer cells via the Wnt/β-catenin signaling pathway. *Materials Science and Engineering C, 101*. https://doi.org/10.1016/j.msec.2019.03.100.

Licciardi, M., Scialabba, C., Puleio, R., Cassata, G., Cicero, L., Cavallaro, G., & Giammona, G. (2019). Smart copolymer coated SPIONs for colon cancer chemotherapy. *International Journal of Pharmaceutics, 556*. https://doi.org/10.1016/j.ijpharm.2018.11.069.

Liechty, W. B., & Peppas, N. A. (2012). Expert opinion: Responsive polymer nanoparticles in cancer therapy. *European Journal of Pharmaceutics and Biopharmaceutics, 80*(2). https://doi.org/10.1016/j.ejpb.2011.08.004.

Liou, J. M., Shun, C. T., Liang, J. T., Chiu, H. M., Chen, M. J., Chen, C. C., Wang, H. P., Wu, M. S., & Lin, J. T. (2010). Plasma insulin-like growth factor-binding protein-2 levels as a diagnostic and prognostic biomarker of colorectal cancer. *Journal of Clinical Endocrinology and Metabolism, 95*(4). https://doi.org/10.1210/jc.2009-2668, http://jcem.endojournals.org/cgi/reprint/95/4/1717.

Ma, L., Chen, Q., Ma, P., Han, M. K., Xu, Z., Kang, Y., Xiao, B., & Merlin, D. (2017). IRGD-functionalized PEGylated nanoparticles for enhanced colon tumor accumulation and targeted drug delivery. *Nanomedicine:*

*Nanotechnology, Biology, and Medicine, 12*(16). https://doi.org/10.2217/nnm-2017-0107, http://www.futuremedicine.com/loi/nnm.

Majeed, S., Aripin, F. H. B., Shoeb, N. S. B., Danish, M., Ibrahim, M. N. M., & Hashim, R. (2019). Bioengineered silver nanoparticles capped with bovine serum albumin and their anticancer and apoptotic activity against breast, bone, and intestinal colon cancer cell lines. *Materials Science and Engineering C, 102*. https://doi.org/10.1016/j.msec.2019.04.041.

Malam, Y., Loizidou, M., & Seifalian, A. M. (2009). Liposomes and nanoparticles: Nanosized vehicles for drug delivery in cancer. *Trends in Pharmacological Sciences, 30*(11). https://doi.org/10.1016/j.tips.2009.08.00.

Markowitz, S. D., & Bertagnolli, M. M. (2009). Molecular basis of colorectal cancer. *New England Journal of Medicine, 361*(25). http://content.nejm.org/cgi/reprint/361/25/2449.pdf, https://doi.org/10.1056/NEJMra0804588.

Marshall (2008). Managing potentially resectable metastatic colon cancer. *Gastrointestinal Cancer Research: GCR, 2*.

May, F. P., & Anandasabapathy, S. (2019). Colon cancer in Africa: Primetime for screening? Gastrointestinal. *Endoscopy, 89*(6). http://www.elsevier.com/inca/publications/store/6/2/3/1/3/7/index.htt, https://doi.org/10.1016/j.gie.2019.04.206.

Meng, D., Cui, Z., Li, L., Shang, Q., & Chen, X. (2025). Matrine in cancer therapy: Antitumor mechanisms and nano-delivery strategies. *Frontiers in Pharmacology, 16, 16*. https://doi.org/10.3389/fphar.2025.1656862.

Miethling-Graff, R., Rumpker, R., Richter, M., Verano-Braga, T., Kjeldsen, F., Brewer, J., Hoyland, J., Rubahn, H. G., & Erdmann, H. (2014). Exposure to silver nanoparticles induces size- and dose-dependent oxidative stress and cytotoxicity in human colon carcinoma cells. *Toxicology in Vitro, 28*(7), 1280–1289. https://doi.org/10.1016/j.tiv.2014.06.005.

Nguyen, H. T., & Duong, H. Q. (2018). The molecular characteristics of colorectal cancer: Implications for diagnosis and therapy (review). *Oncology Letters, 16*(1), 9–18. https://doi.org/10.3892/ol.2018.8679, http://www.spandidos-publications.com/ol/16/1/9/download.

Ngwuluka, N. C., Kotak, D. J., & Devarajan, P. V. (2017). Design and characterization of metformin-loaded solid lipid nanoparticles for colon cancer. *AAPS PharmSciTech, 18*(2). https://doi.org/10.1208/s12249-016-0505-3, http://www.springerlink.com/content/1530-9932/.

Oyarzun-Ampuero, F. A., Rivera-Rodríguez, G. R., Alonso, M. J., & Torres, D. (2013). Hyaluronan nanocapsules as a new vehicle for intracellular drug delivery. *European Journal of Pharmaceutical Sciences, 49*(4). https://doi.org/10.1016/j.ejps.2013.05.008.

Pan, X. Q., Gong, Y. C., Li, Z. L., Li, Y. P., & Xiong, X. Y. (2019). Folate-conjugated pluronic/polylactic acid polymersomes for oral delivery of paclitaxel. *International Journal of Biological Macromolecules, 139*, 377–386. https://doi.org/10.1016/j.ijbiomac.2019.07.22, http://www.elsevier.com/locate/ijbiomac.

Parveen, S., & Sahoo, S. K. (2008). Polymeric nanoparticles for cancer therapy. *Journal of Drug Targeting, 16*(2), 108–123. https://doi.org/10.1080/10611860701794353.

Pelka, J., Gehrke, H., Esselen, M., Türk, M., Crone, M., Bräse, S., Muller, T., Blank, H., Send, W., Zibat, V., Brenner, P., Schneider, R., Gerthsen, D., & Marko, D. (2009). Cellular uptake of platinum nanoparticles in human colon carcinoma cells and their impact on cellular redox systems and DNA integrity. *Chemical Research in Toxicology, 22*(4). http://pubs.acs.org/doi/pdfplus/10.1021/tx800354g, https://doi.org/10.1021/tx800354g.

Peluso, G., Incollingo, P., Calogero, A., Tammaro, V., Rupealta, N., Chiacchio, G., Sandoval Sotelo, M.L., Minieri, G., Pisani, A., Riccio, E., Sabbatini, M., Bracale, U.M., Dodaro, C.A., & Carlomagno, N. (2017). Current tissue molecular markers in colorectal cancer: A literature review. https://doi.org/10.1155/2017/2605628, http://www.hindawi.com/journals/biomed/.

Prajakta, D., Ratnesh, J., Chandan, K., Suresh, S., Grace, S., Meera, V., & Vandana, P. (2009). Curcumin-loaded pH-sensitive nanoparticles for the treatment of colon cancer. *Journal of Biomedical Nanotechnology, 5*(5). https://doi.org/10.1166/jbn.2009.1038.

Rajekar, H. (2024). Recent advances in the treatment of liver metastases from colorectal cancer. A comprehensive review. *Journal of Clinical Oncology and Cancer Research, 2*(1). https://doi.org/10.61148/2836-2888/GHR/049.

Renzi, C., Lyratzopoulos, G., Hamilton, W., & Rachet, B. (2019). Opportunities for reducing emergency diagnoses of colon cancer in women and men: A data-linkage study on pre-diagnostic symptomatic presentations and benign diagnoses. *European Journal of Cancer Care, 28*(2), e13000. https://doi.org/10.1111/ecc.13000.

Roper, J., & Hung, K.E. (2013). Molecular mechanisms of colorectal carcinogenesis. pp. 25–65). https://doi.org/10.1007/978-1-4614-8412-7. New York, United States.

Santos, I. S., Ponte, B. M., Boonme, P., Silva, A. M., & Souto, E. B. (2013). Nanoencapsulation of polyphenols for a protective effect against colorectal cancer. *Biotechnology Advances, 31*(5). https://doi.org/10.1016/j.biotechadv.2012.08.005.

Schütz, C. A., Staedler, D., Crosbie-Staunton, K., Movia, D., Bernasconi, C. C., Kenzaoui, B. H., Prina-Mello, A., & Juillerat-Jeanneret, L. (2014). Differential stress reaction of human colon cells to oleic-acid-stabilized and unstabilized ultrasmall iron oxide nanoparticles. *International Journal of Nanomedicine, 9*(1). https://doi.org/10.2147/IJN.S65082, http://www.dovepress.com/getfile.php?fileID=20942.

Sharifi-Azad, M., Fathi, M., Cho, W. C., Barzegari, A., Dadashi, H., Dadashpour, M., & Jahanban-Esfahlan, R. (2022). Recent advances in targeted drug delivery systems for resistant colorectal cancer. *Cancer Cell International*. https://doi.org/10.1186/s12935-022-02605-y, http://www.cancerci.com/.

Shen, M. Y., Liu, T. I., Yu, T. W., Kv, R., Chiang, W. H., Tsai, Y. C., Chen, H. H., Lin, S. C., & Chiu, H. C. (2019). Hierarchically targetable polysaccharide-coated solid lipid nanoparticles as an oral chemo/thermotherapy delivery system for local treatment of colon cancer. *Biomaterials, 197*. https://doi.org/10.1016/j.biomaterials.2019.01.019, http://www.journals.elsevier.com/biomaterials/.

Shrivastava, P., Sharma, R., Gautam, L., Vyas, S., & P. Vyas, S. (2020). Nanoparticles and colon cancer. In *nano drug delivery strategies for the treatment of cancers, Shrivastava et al. (2020) discuss nanoparticles and colon cancer.*. India: Elsevier, 191–223. https://doi.org/10.1016/B978-0-12-819793-6.00009-6, https://www.sciencedirect.com/book/9780128197936.

Singh, P., Waghambare, P., Khan, T. A., & Omri, A. (2022). Colorectal cancer management: Strategies in drug delivery. *Expert Opinion on Drug Delivery*. https://doi.org/10.1080/17425247.2022.2084531, http://www.tandfonline.com/loi/iedd20.

Taylor, C. R. (2018). Introduction to predictive biomarkers: Definitions and characteristics. In *Predictive biomarkers in oncology: Applications in precision medicine*. Springer International Publishing, 3–18. https://link.springer.com/book/10.1007/978-3-319-95228-4, https://doi.org/10.1007/978-3-319-95228-4_1.

Tsakiris, N., Papavasileiou, M., Bozzato, E., Lopes, A., Vigneron, A. M., & Préat, V. (2019). Combinational drug-loaded lipid nanocapsules for the treatment of cancer. *International Journal of Pharmaceutics*. http://www.elsevier.com/locate/ijpharm, https://doi.org/10.1016/j.ijpharm.2019.118588.

Wathoni, N., Nguyen, A. N., Rusdin, A., Umar, A. K., Mohammed, A. F. A., Motoyama, K., Joni, I. M., & Muchtaridi, M. (2020). Enteric-coated strategies in colorectal cancer nanoparticle drug delivery system. *Drug Design, Development and Therapy, 14*. https://www.dovepress.com/getfile.php?fileID=62787, https://doi.org/10.2147/DDDT.S273612.

Won, D. D., Lee, J. I., Lee, I. K., Oh, S. T., Jung, E. S., & Lee, S. H. (2017). The prognostic significance of KRAS and BRAF mutation status in Korean colorectal cancer patients. *BMC Cancer*. https://doi.org/10.1186/s12885-017-3381-7, http://www.biomedcentral.com/bmccancer/.

Xiong, M., Lei, Q., You, X., Gao, T., Song, X., Xia, Y., Ye, T., Zhang, L., Wang, N., & Yu, L. (2017). Mannosylated liposomes improve therapeutic effects of paclitaxel in colon cancer models. *Journal of Microencapsulation*, 513–521. https://doi.org/10.1080/02652048.2017.1339739.

Xu, J., Sun, Y., Yuan, Z., Bao, Y., Li, R., Liu, C., Qiu, Y., Xu, K., Shi, X., Yu, H., Chen, T., Duan, Y., & Yin, P. (2019). Bufalin-Loaded CaP/DPPE-PEG-EGF Nanospheres: Preparation, Cellular Uptake, Distribution, and Anti-Tumor Effects. *Journal of Biomedical Nanotechnology, 15*(2). https://doi.org/10.1166/jbn.2019.2681.

Yassin, Anwer, Mowafy, El-Bagory, Bayomi, & Alsarra (2010). Optimization of 5-flurouracil solid-lipid nanoparticles: A preliminary study to treat colon cancer. *International Journal of Medical Sciences, 7*(6).

You, X., Kang, Y., Hollett, G., Chen, X., Zhao, W., Gu, Z., & Wu, J. (2016). Polymeric nanoparticles for colon cancer therapy: Overview and perspectives. *Journal of Materials Chemistry B, 4*(48). https://doi.org/10.1039/c6tb01925k.

Zheng, Y., Wang, W., Zhao, J., Wu, C., Ye, C., Huang, M., & Wang, S. (2019). Preparation of an injectable, temperature-sensitive chitosan-based hydrogel for combined hyperthermia and chemotherapy of colon cancer. *Carbohydrate Polymers, 222*, 115039. https://doi.org/10.1016/j.carbpol.2019.115039.

CHAPTER

# Lipid nanocarrier-based drug delivery for the treatment of colon cancer

3

**Shikha Baghel Chauhan**[1,2]

[1]*Amity Institute of Pharmacy, Amity University, Noida, Uttar Pradesh, India,*
[2]*Centre of Industrial Pharmacy, Uttar Pradesh, India*

## 3.1 Introduction

### 3.1.1 Overview

Colon cancer is the third most prevalent cancer and ranks fourth in the list of diseases that cause the largest number of cancer deaths in the world (Bray et al., 2018; Morgan et al., 2023; Siegel et al., 2023), with almost 900,000 individuals dead per year due to this illness, which is nearly 10% of all cancer deaths (Bray et al., 2018). Approximately 10%–20% of cases of colorectal cancer (CRC) are known to have a heritable basis (Henrikson et al., 2015). Prolonged standardized lifestyle-related adverse effects like obesity, smoking, alcohol consumption, and overweight also play a significant role (Botteri et al., 2008; Cai et al., 2014; Chan et al., 2011; Kyrgiou et al., 2017; Zhang et al., 2018). Higher rates of recurrence after surgical resection, which is particularly effective in stages I through III and provides an estimated 90% five-year survival probability for CRC at an early stage, further impair the prognosis. However, the survival rates at that level when the disease is at an advanced stage are as low as 12.5% (DeSantis et al., 2014; Favoriti et al., 2016; Janout & Kollárová, 2001; Miller et al., 2019). Postoperative recurrence rates for stage I vary from 0% to 13%, while those for stage III recurrence vary from 28% to 73% (Hellinger & Santiago, 2006).

### 3.1.2 Challenges in conventional drug delivery

Treatments for CRC right now are surgical options, radiotherapy, chemotherapy, targeted therapy, and immunotherapy, but they come with considerable toxicities (Cervantes et al., 2023; Kroschinsky et al., 2017; Punt et al., 2017; Weng et al., 2022). Chemotherapy agents such as 5-FU and capecitabine have serious long-term sequelae that include cardiotoxicity and myelosuppression; important dose reductions must be used due to their effects. Radiotherapy, generally delivered in 25 Gy (short-course) or 45 Gy (long-course), leads to serious late-stage side effects such as fecal incontinence and urogenital dysfunction (Braat et al., 2017; Kye & Cho, 2014; Riaz et al., 2014; Tiwari et al., 2020). These unwanted adverse effects have the potential to cease or even interrupt treatment altogether; hence, an alternative strategy is required (Brami et al., 2016; Chau & Cunningham, 2002; Chintala et al., 2011; Zhang, Chen, et al., 2017).

Nano Drug Delivery Systems for Colon Cancer. DOI: https://doi.org/10.1016/B978-0-443-40443-6.00015-5

### 3.1.3 Need for targeted therapies

Interventions that decrease nonspecific distribution of medication and reduce harmful side effects are urgently needed to enhance the quality of life and outcomes for patients with colon cancer. While many anticancer agents show promise in vitro, their nonspecific distribution in vivo results in adverse effects (Duran et al., 2020; Petrelli et al., 2020; Sastre et al., 1999). Nonspecific biodistribution remains a problem even with attempts to improve medication combinations and dosing intervals (Lund et al., 2016; Madi et al., 2018; Schuurhuizen et al., 2018; Šišić et al., 2015). Colon-specific medication delivery systems, especially those based on nanotechnology, present a viable solution to address these issues (Zhang et al., 2024).

### 3.1.4 Advantages of lipid nanocarriers over other drug delivery systems

Nanoparticles (NPs), especially liposomal NPs (LNPs), improve efficiency and safety by increasing drug stability, solubility, and the circulation duration of the medication (Mitchell et al., 2021; Zhang et al., 2021). LNPs are carrier systems made of ionizable lipids and other lipids that can deliver active components to the site of lesions while sparing healthy tissues. They have low immunogenicity and can pass physiological barriers, which makes them a good fit for colon-targeted therapy where colon cancer inhibition is needed and toxicity caused by drugs in drug-induced enteritis is lessened (Eygeris et al., 2022; Mitchell et al., 2021; Schlich et al., 2021; Shepherd et al., 2021). Doxil, a formulation of DOX in which the drug is enclosed in liposomes, was one of the first approved for use in patients, thus illustrating the impact of new technologies on improving the efficacy and safety of cancer treatment (Alavi & Hamidi, 2019). Cancer nanotechnology holds great promise for drug delivery systems (DDSs) and lipid-based nanoparticles (LBNPs) in particular, with easy access to bioactive drugs, better tumor targeting, and lower toxicity profiles; therefore, it is more effective for clinical practice (Fig. 3.1) (García-Pinel et al., 2019).

## 3.2 Liposomes

Liposomes, known for their biocompatibility and biodegradability, are composed of phospholipids organized in a bilayer. These amphipathic molecules form vesicles in water, enhancing the solubility and stability of loaded anticancer pharmaceuticals (García-Pinel et al., 2019; Yingchoncharoen et al., 2016).

### 3.2.1 Principle

Liposomes can incorporate both lipophilic and hydrophilic drugs into their structure. They are typically incorporated to decrease fluidity, improve hydrophobic drug solubility, and enhance the sustainability of nanoparticles in circulation. Cholesterol-modified liposomes encompass multilaminar vesicles consisting of 0.5–10 nm, large multilamellar vesicles above 100 nm, and small multilamellar vesicles of 10–100 nm in size (García-Pinel et al., 2019; Yingchoncharoen et al., 2016). Liposomes increase circulation duration, enhance drug targeting, and decrease the toxicity of drugs, especially in the case of lipophilic drugs (Das & Huang, 2019). Particles, along with polymers (Østergaard & Moeller, 2010), are well explored in neoplasia treatments, gene transduction (Guan et al., 2017;

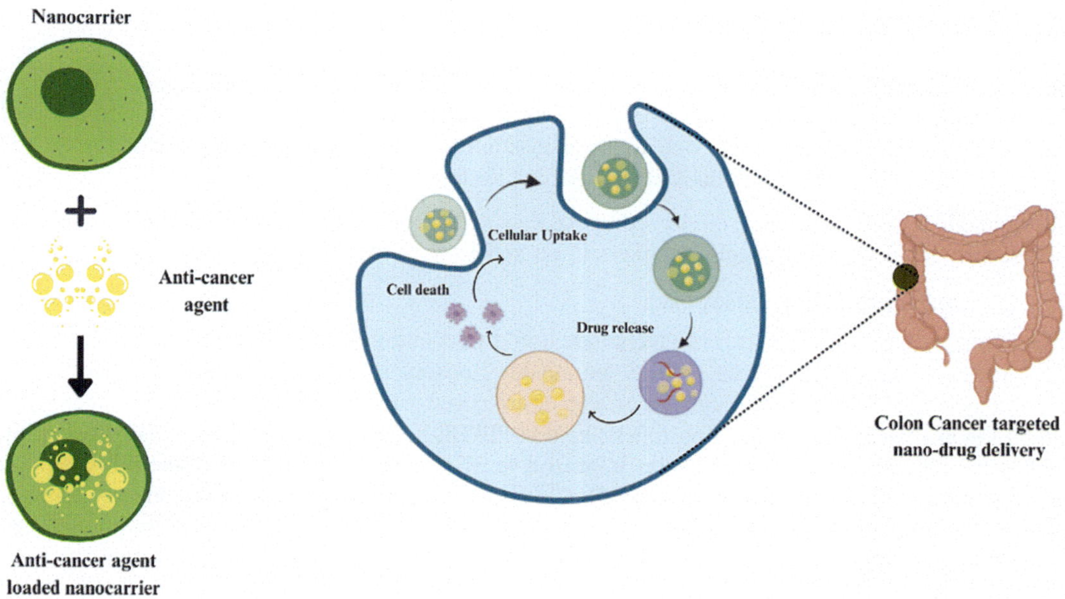


**FIGURE 3.1**

Mechanism of lipid nanocarrier-based drug delivery for the treatment of colon cancer.

Wang, Liu, et al., 2017), immunology treatments (Kwong et al., 2013), and conventional drugs (Huang et al., 2019; Yang & Merlin, 2020; Zhao et al., 2013).

### 3.2.2 Mechanism

Passive targeting and active targeting are two primary methods for nanocarrier vectorization. While active targeting uses structurally modified liposomes with antibodies to selectively detect tumor cells, passive targeting involves liposomes entering tumor cells by molecular migration across the membrane (Gogoi et al., 2016). For these purposes, $Fe_3O_4$ cores are being utilized more often to functionalize nanoparticles, enabling more efficient and regulated administration of anticancer medications (García-Pinel et al., 2019). To allow liposomes of a suitable size to circulate into cancer areas, passive targeting liposomes rely on the porous nature of tumor-associated arteries that contain gaps between endothelial cells. Through high-affinity contacts, this allows the targeted liposomes to aggregate at the tumor location (Kunjachan et al., 2015; Sang et al., 2021).

Lipid-drug conjugates (LDCs) are used in the liposomal prodrug method to create self-organized structures. Drugs are covalently attached to lipids through a linker to form LDCs, which are amphiphilic molecules. These amphiphilic LDCs self-assemble into nanoscale liposomal particles in buffer solutions. Secretory phospholipase A2, among other disease-related enzymes, can break down the covalent bond or connected lipid in LDCs. The particles dissociate and activate upon reaching the target, allowing for accurate medication release for targeted treatment at the intended location (Xing et al., 2019; Yang & Merlin, 2020).

### 3.2.3 Classification of liposomes

#### 3.2.3.1 Conventional liposomes

Natural phospholipids form first-generation liposomes that are biocompatible and have an improved drug-carrying capacity. However, due to their instability, medications are released abruptly and quickly removed by the mononuclear phagocyte system (Akbarzadeh et al., 2013; Allen & Hansen, 1991).

#### 3.2.3.2 Long-circulating liposomes

PEGylation (Polyethylene Glycol) is a technique used by second-generation liposomes, often referred to as long-circulating liposomes (LCLs) or stealth liposomes, to improve stability, circulation duration, and tumor delivery through the effects of enhanced retention and permeability (Chan et al., 2012; Li et al., 2016). PEGylated liposomes like dox/PEG-L have a two- to three-day half-life and up to six times higher tumor concentration (Gabizon et al., 2016). However, PEG modification can hinder cellular uptake, known as the "PEG dilemma." Alternatives like ganglioside or chitosan coatings improve immune evasion and drug delivery efficiency, demonstrating their potential as stable, safe, and effective DDSs (Alomrani et al., 2019).

#### 3.2.3.3 Active-targeting liposomes

New liposome-based DDSs incorporate active-targeting nanotechnologies (Naeem et al., 2020), focusing on CRC pathophysiology. Urotensin-II receptor (UTR)-conjugated liposomes (LipoUTs) (Liang et al., 2016; Zappavigna et al., 2019) and folate receptor-targeted FA-coated liposomes enhance drug selectivity and uptake in CRC cells, improving therapeutic efficiency, particularly in UTR-overexpressing CRC cells (Handali et al., 2018; Moghimipour et al., 2018a).

#### 3.2.3.4 Stimuli-sensitive liposomes

Liposomes can be designed for triggered drug release in response to cancer microenvironmental cues (e.g., pH, reactive oxygen species [ROS]) (Juang et al., 2019; Moghimipour et al., 2018b) or stimuli (e.g., heat, light) (Wang et al., 2011; Zhang et al., 2019). pH-sensitive liposomes using styrene maleic acid (SMA) show efficient 5-fluorouracil delivery to CRC cells, while thermosensitive liposomes, such as ThermoDox, enable controlled doxorubicin release with enhanced tumor targeting during hyperthermia (Gholami & Engel, 2018; Lyon et al., 2018).

#### 3.2.3.5 Cationic liposomes

Cationic liposomes, created by adding positive lipids or cationic polymers, enhance cell binding and gene delivery. First used in EndoTAG-1 for cancer, they form stable lipoplexes (Smyth Templeton, 2002). Despite their potential, challenges include interactions with serum proteins and rapid clearance, mitigated by PEG coating (Chan et al., 2012) in formulations like stable nucleic acid lipid particles (SNALPs) that improve stability and systemic delivery of nucleic acids (Judge et al., 2009; Semple et al., 2010).

### 3.2.4 Effectiveness

A study demonstrated binary mixtures of DPPC/prodrug (10–50 mol% prodrug). Elevated prodrug hydrolysis rates correlate with DPPC's vulnerability to phospholipase A2. PEGylation accelerated the rate of hydrolysis. These outcomes indicate that modifying the components of liposomes may regulate prodrug release (Arouri & Mouritsen, 2012).

### 3.2.5 Applications

Dipalmitoylphosphatidylcholine (DPPC), hydrogenated soy phosphatidylcholine (HSPC), and 1,2-Distearoyl-sn-glycero-3-phosphoethanolamine-poly(ethylene glycol) 2000 (DSPE-PEG2000) are components of a stable liposomal formulation of the lipophilic SN38 prodrug (moeixitecan). Liposomes containing moeixitecan (moe/CLs) substantially enhance the solubility, stability, and sustained release of SN38. Moe/CLs show improved cytotoxicity and targeted HT-29 cell death in contrast to irinotecan. In contrast to both free moeixitecan and irinotecan, moe/CLs show superior therapeutic activity in a colorectal xenograft model (Xing et al., 2019).

## 3.3 Solid lipid nanoparticles

Solid lipid nanoparticles (SLNs) are colloidal lipid molecules with a solid core (Mishra et al., 2018), produced using methods like homogenization or ultra-sonication, and are safer than polymer-based nanoparticles due to nontoxic lipids (Doktorovová et al., 2016; Reis et al., 2016). Food and Drug Administration (FDA)-approved colloidal carriers have a size distribution of 10 nm–0.001 mm, comprising solid lipids that are compact at room temperature and body temperature (Geszke-Moritz & Moritz, 2016). SLNs enhance the regulated release, targeting, and bioavailability of lipophilic medications—many anticancer medicines (Mehnert, 2001). They have high biocompatibility and stability, and do not need organic solvents for administration (Mehnert, 2001; Parhi & Suresh, 2012). SLNs are often covered with polymers such as chitosan (Garciafuentes et al., 2005), PEG (De Campos et al., 2003; Heiati et al., 1997), or polyvinyl alcohol (Pandey et al., 2005) to achieve high biocompatibility and stability, and do not need organic solvents for administration, thus enhancing stability and membrane transport (De Campos et al., 2001; Jani et al., 2011; Vila et al., 2004). Mono-, di-, and triglycerides, stabilized by surfactants (Wong et al., 2007), polysorbate 80, and poloxamer 188, increase tumor targeting (Dong et al., 2012), and PEG coatings promote longevity of circulation and permeability (Yuan et al., 2013). According to a study, SLNs injected with metformin moved and gathered inside the colon tumor, stopping its growth (Ngwuluka et al., 2017).

### 3.3.1 Classification of SLNs

Depending on how the drug is incorporated into the lipid matrix, SLNs can be divided into drug type, active type, and active type drug type.

#### 3.3.1.1 Solid solution SLNs

When it comes to solid solution types, the medication has a strong affinity for lipid molecules and is molecularly distributed throughout the lipid matrix. This type is often made without the use of surfactants or solubilizing agents by employing the cold homogenization process (Zur Mühlen et al., 1998).

#### 3.3.1.2 Drug-enriched core SLNs

With respect to the core-type drug-enriched systems, the formulation consists of a thermally melted matrix containing drug salts for the core, while lipids are retained on the surface. If the temperature of the dispersion is decreased further, the lipids begin to coat the drug, which recrystallizes.

#### 3.3.1.3 Drug-enriched shell SLNs

A solid lipid center forms in the modified shell DDS when the lipid casting approaches the lipid's recrystallization temperature. The medication associates with the SLN's outer liquid shell when the temperature decreases further (Heiati et al., 1997; Müller et al., 2002).

### 3.3.2 Effectiveness

Using suitable polymers, a polymer-lipid hybrid nanoparticle (PLN) complex of verapamil HCl was developed, resulting in 33% repartition in the lipid and 90% entrapment efficiency. Ionic drugs like verapamil HCl and doxorubicin HCl were shown to have an entrapment efficiency of above 80% (Wong et al., 2004).

### 3.3.3 Application

Optimized paclitaxel-loaded solid lipid nanoparticles (OPSLNs) and tagged folic acid oxaliplatin were designed for CRC. The higher cytotoxicity of optimized paclitaxel-loaded solid lipid nanoparticle formulations (OPSLNFs) in HT-29 cells proved that they are efficacious as a novel treatment strategy for CRC (Rajpoot & Jain, 2018).

## 3.4 Nanostructured lipid carriers

Nanostructured lipid carriers (NLCs), which have a wide diameter ranging from 10 to 1000 nm, are the subsequent generations of competent transport mechanisms that were created to overcome the limitations of solid lipid nanoparticles (SLNs), including their limited ability to load drugs and their propensity to release drugs while being stored (Elmowafy & Al-Sanea, 2021; Jain et al., 2021). NLCs are composed of a binary combination of liquid and solid lipids that are solid at room temperature and body temperature, just as SLNs. Additionally, their toxicological potential is minimal.

### 3.4.1 Mechanism

NLCs contain either a solid amorphous form or a less-ordered crystalline framework, along with partly crystallized lipid droplets, to provide excellent efficiency in encapsulation and durability (Tamjidi et al., 2013). Additionally, they offer an alternative for improving the bioavailability, controlled release, and chemical sustainability of useful lipophilic chemicals that are trapped in the lipophilic core (Kumbhar & Pokharkar, 2013; Ucar et al., 2017). NLCs have historically been used as sustained-release medication carriers.

### 3.4.2 Effectiveness

For targeting, a lactobionic acid-functionalized 5-FU-NLC was created, and it showed 34% encapsulation efficiency (EE) (Varshosaz et al., 2012). When tested in vitro in a human hepatocarcinoma (HepG) lineage (Ahn & Park, 2016), entrapment in the NLC reduced the cytotoxicity of 5-FU and enhanced drug uptake.

### 3.4.3 Application

Using rosehip oil as a carrier, alkyl polyglucoside (APG) was incorporated into the NLCs, yielding particles with a narrow size distribution and high EE. APG-NLCs delivered antiangiogenic properties and were also effective against leukemia, lung, and breast cancers. Control Empty NLCs appeared to exhibit selective cell death, which is probably attributed to the active rosehip oil component. This advantageous combination of agents provides a unique strategy to combat cancer (Bonilla-Vidal et al., 2023).

## 3.5 Core-shell lipid nanoparticles

Lipid nanoparticles with a core-shell consist of a shell made of lipid and a core that is nonaqueous, composed of different materials (Campani et al., 2018). Because the core of this design interacts with the drug and supports the lipid layer, it improves particle stability. The nonaqueous core also has the ability to release drugs under controlled conditions (Gao et al., 2013). These core-shell structures have been researched using materials such as hydrogels, micelles, metal nanoparticles, and synthetic polymers (Campani et al., 2018; Sharifabad et al., 2016; Wang et al., 2010).

### 3.5.1 Mechanism

Lipid-based core-shell nanoplatforms provide the biocompatibility and cellular absorption processes of conventional liposomes, but they can also incorporate a variety of medicines and enhance pharmacokinetics. According to a study, metal-core nanoparticles within liposomes had little effect on liposomal characteristics and behaved like pure liposomes when transported by cancer cells through clathrin-mediated endocytosis (Zhu et al., 2018). The polymeric cores increase surface area for controlled release while improving morphological control, size uniformity, and mechanical strength. The lipid shell enhances medication delivery efficiency, circulation time, and biocompatibility (Yang & Merlin, 2020).

### 3.5.2 Effectiveness

In one study, hyaluronic acid (HA)-modified, lipid-coated poly (lactic-co-glycolic acid) (PLGA) nanoparticles loaded with mRIP3-pDNA showed excellent biocompatibility and tumor-targeting efficacy. In the CT26 mouse model of colon cancer, core-shell nanoparticles combined with chloroquine administration achieved more than 80% tumor inhibition. This strategy combines chemotherapy with gene therapy from core-shell nanoparticles to drive tumor growth by activating multiple pathways of coordinated cell death (PCD). It may be effective in the treatment of colon cancer (Hou et al., 2017).

### 3.5.3 Application

The effectiveness of oxaliplatin in colon cancer treatment has been proven to improve with curcumin (CUR), but the drug's low solubility and metabolic instability restrict its clinical application. CUR-encapsulated hyaluronic acid–zein nanoparticles (HZ or Hydrazone-CUR) showed better synergy with OXA and stronger anti-CRC activities than free CUR, particularly in HCT116 and HT29 cell lines. Mechanistic research revealed that HZ-CUR absorption was augmented by OXA-induced CD44 overexpression. Ex vivo and in situ CRC model results confirm HZ-CUR as a potentially effective oral adjuvant to enhance OXA-based chemotherapy (Liu et al., 2022).

## 3.6 Lipid micelles

In aqueous fluids, amphiphilic substances known as lipid micelles develop into spherical forms (Duan et al., 2012). They have been investigated extensively as hydrophobic DDSs. Optimizing their ability to load drugs, increasing their stability in vivo, and reducing their contact with cells while in circulation have been the primary goals of recent studies.

### 3.6.1 Mechanism

PEG–lipid micelles, particularly PEG–DSPE conjugates, have shown promise as DDSs because of their flexibility in grafting as well as their ability to easily encapsulate therapeutics, making them versatile carriers for cancer treatment (Hu, He, et al., 2020). Cationic PEG–lipid micelles for siRNA delivery by modifying mPEG-polycaprolactone (PCL) micelles with 1,2-dioleoyl-3-trimethylammonium-propane (DOTAP) were studied. Loaded with siMcl1 or siBcl-xl, these micelles significantly suppressed colon cancer growth in a mouse model following intra-tumoral injection, demonstrating potential for gene therapy in colon cancer (Lu et al., 2019).

### 3.6.2 Effectiveness

Curcumin-loaded mPEG-PCL micelles, or Cur/FA-Mic, were created in a study to preferentially target CRC cells that overexpress the FA receptor. The size of these micelles' particles was 30.47 nm. Cur/FA-Mic demonstrated pharmacokinetic features that were significantly better than those of free curcumin, including a prolonged half-life (T1/2) and enhanced drug exposure. Research conducted in vivo demonstrated that Cur/FA-Mic was superior to free curcumin or empty micelles in its ability to effectively suppress tumor development, increase apoptosis, and minimize angiogenesis. These results highlight the ability of FA-coated micelles as a powerful technique for the transport of hydrophobic drugs (Hu, He, et al., 2020).

### 3.6.3 Application

Lut-M (LutM-F127) was developed for local CRC treatment. Drug release experiments revealed that LutM-F127 had a slower release rate than Lut-M, converting gel from solution at temperature of body. LutM-F127 has higher therapeutic effects in vivo, including increased tumor apoptosis, proliferative reduction, and angiogenesis inhibition. LutM-F127 has the potential to cure CRC (Hu, Chen, et al., 2020) (Table 3.1, Table 3.2).

## 3.7 Targeting sites for colon cancer

### 3.7.1 Passive targeting

The use of negative technology in nanocarrier systems is quite common because of the unique scientific support and physical properties of the tumor (Fig. 3.2). Large areas of vascular endothelial cells, thick extracellular matrix (ECM), and fluid deficits are characteristic features of tumors. The enhanced permeability and retention (EPR) effect caused by this phenomenon facilitates the passage of nanomedicine through tumor cells (Takakura & Takahashi, 2022).

The fundamental idea underlying passive targeting is this procedure, which alters the physical and chemical properties of the enclosed drugs and further minimizes the toxicity of off-target effects (Fulton & Najahi-Missaoui, 2023).

The anthracycline chemotherapy medication doxorubicin (DOX) comes in liposomal form. The plasma half-life of DOX in humans is extended to 45 hours by the addition of PEG-lipid conjugates (Montanha et al., 2022). In a similar vein, liposomal daunorubicin, which transports daunorubicin (DNX), displays modified distribution and metabolism, resulting in a decrease in systemic toxicity and an increase in tumor accumulation (Crommelin et al., 2020).

The surface modification of LNPs is PEGylation. PEG lipid permeability creates a hydrophilic barrier that prevents binding to plasma proteins, making the healing tumor compatible with the EPR effect and thus allowing it to prolong due to its special properties. The two main obstacles to reaching the target of nanomedicines are high interstitial fluid pressure (IFP) and their rapid elimination by the reticuloendothelial system (RES). The latter limits the penetration of the drug into deep tissue, while the former shortens the half-life of LNPs, preventing the accumulation of the target. Therefore, improving the targeting performance of LNPs requires changes in their structure, size, and surface charge (Mundekkad & Cho, 2022; Wang, Li, et al., 2021).

Addition of chitosan coating is one way to improve the targeting of nanodrugs. This shell has two functions: it prevents the nanodrug from being broken down in the stomach and intestines, and when the drug enters the intestine, some glycohydrolase enzymes digest the glucan of the shell.

**Table 3.1 Advantages and disadvantages of lipid nanocarrier-based drug delivery for the treatment of colon cancer.**

| Lipid-based nanocarriers | Advantages | Disadvantages | References |
|---|---|---|---|
| **Liposomes** | They are easy to formulate, prolong systemic circulation, treat multidrug-resistant tumors, alter pharmacokinetics and biodistribution, reduce cardiotoxicity, and use lipids compatible with parenteral administration for intravenous delivery. | They can face stability challenges, particularly when active-targeting ligands are incorporated into the formulation. These ligands may compromise the overall stability of the liposome structure. | García-Pinel et al. (2019), Yang and Merlin (2020) |
| **Solid lipid nanoparticles (SLNs)** | They target drug-resistant tumors, protect drugs from degradation, and reduce systemic toxicity. | Because of their crystal structure, they are naturally limited in their capacity to load drugs. They are also susceptible to drug ejection, unexpected accumulation, and a greater chance of conformational changes, all of which can have an impact on their stability and functionality. | Mehnert (2001), Mishra et al. (2018) |
| **Nanostructured lipid carriers** | In addition to achieving a desired size of particles for improved drug administration, this approach tackles the problems associated with SLNs, such as poor drug payload, drug ejection, and high-water content. | Using a significant amount of the liquid phase can dilute the solution excessively, and this approach is not ideal for thermo-labile medicines. It also presents a danger of contamination with metals during probe sonication. | Elmowafy and Al-Sanea (2021), Jain et al. (2021) |
| **Core-shell lipid nanoparticles** | They offer fewer adverse reactions, regulated release of drugs, greater cellular absorption, better drug dissolution and equilibrium, and simple drug delivery. | They face challenges, including complex formulation and manufacturing, potential immune responses to lipid components, and limited clinical experience. | Campani et al. (2018), Gao et al. (2013) |
| **Lipid micelles** | They offer prolonged controlled release, better dissolution of insoluble medications, reduced systemic toxicity through targeted delivery, improved cellular uptake via stealth properties, and simpler manufacturing processes for scalability. | They have limited drug loading capacity due to their smaller size, stability issues from aggregation, and potential immune responses from lipid components, which may require strategies for surface modification to mitigate immunogenicity. | Duan et al. (2012) |

**Table 3.2 Efficacy and applications of lipid nanocarrier-based drug delivery for colon cancer treatment.**

| S. No. | Lipid-based nanocarriers | Efficacy | Application | References |
|---|---|---|---|---|
| 1. | **Liposomes** | PEGylation and DPPC/prodrug mixtures promote prodrug hydrolysis, indicating that liposome modification regulates prodrug delivery. | Moe/CLs liposomes outperformed irinotecan in terms of dissolution, equilibrium, prolonged release, cytotoxicity, and cell targeting. | Arouri and Mouritsen (2012), Xing et al. (2019) |
| 2. | **Solid lipid nanoparticles (SLNs)** | Ionic medications demonstrated over 80% retention efficacy, while a polymer-lipid hybrid nanoparticle of verapamil HCl achieved 90%. | Folic acid-containing oxaliplatin solid lipid nanoparticles (SNLs) demonstrated increased cytotoxicity in HT-29 cells, demonstrating their effectiveness in treating colon cancer. | Wong et al. (2004), Rajpoot and Jain (2018) |
| 3. | **Nanostructured lipid carriers (NLCs)** | By achieving 34% encasing, lactobionic acid-functionalized 5-FU-NLC decreased 5-FU cytotoxicity and enhanced drug absorption in HepG2 cells. | Rosehip oil-based APG-NLCs show remarkable encasing, antiangiogenic attributes, and efficiency against breast, lung, and leukemia malignancies. | Varshosaz et al. (2012), Bonilla-Vidal et al. (2023) |
| 4. | **Core-shell lipid nanoparticles** | Integrating chemotherapeutic and gene therapy, HA-modified poly (lactic-co-glycolic acid) nanoparticles containing mRIP3-pDNA exhibited more than 80% tumor suppression in colon cancer. | In HCT116/HT29 cells and colorectal cancer (CRC) models, CUR-entrapped HZ nanoparticles improved the anticolon cancer activities by enhancing oxaliplatin interaction. | Hou et al. (2017), Liu et al. (2022) |
| 5. | **Lipid micelles** | Targeting CRC cells, curcumin-loaded FA-coated micelles (30.47 nm) promoted pharmacokinetics, tumor inhibition, apoptosis, and decreased vasculature. | With its delayed release, enhanced tumor death, decreased growth, and suppressed angiogenesis, LutM-F127 gel may be used in the treatment of CRC. | Hu, He, et al. (2020), Hu, Chen, et al. (2020) |

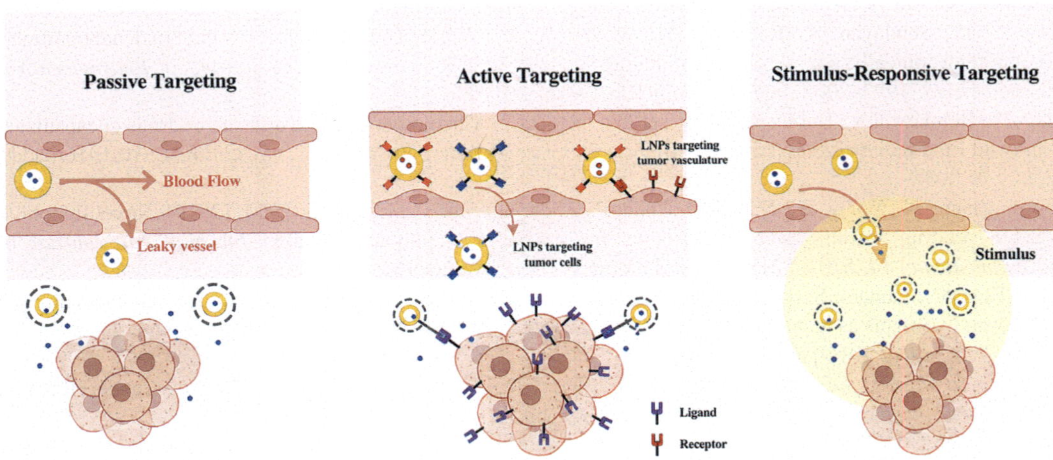


**FIGURE 3.2**

Targeting sites of lipid nanocarrier-based drug delivery for the treatment of colon cancer.

Nanoparticles target tumor cells, and as a result of this damage, folate receptors are overexpressed, exposing folate residues, which can reach targets and facilitate drug release in the intestine. Pharmacodynamic properties (Leonard et al., 2023; Szymańska & Winnicka, 2015).

### 3.7.2 Active targeting

The main goal in nanomedicine is to modify the ligands on the nanoparticle surface. Ligands on the surfaces of nanoparticles facilitate the distribution of drugs to specific cell types, and these ligands tend to bind to receptors on target cells (Jiang et al., 2019; Marques et al., 2020).

For LNP surface modifications, a variety of ligands that fall under the categories of small molecules and macromolecules have been used (Moulahoum et al., 2021). A number of biomarkers unique to CRC are employed in targeted ligand formation (Tiernan et al., 2013).

For example, folate (FA), which is stable, immunogenicity-free, and has a high affinity for folate receptor-α (FR-α), binds naturally to FR-α, which is overexpressed in many cancer types (Shia et al., 2008). FA is, therefore, a well-liked option for nanomedicine targeting. When loaded with kaempferitrin, folate-bound PLGA NPs showed increased cytotoxicity against CRC cells (Govindarasu et al., 2022).

They took advantage of their multipurpose, task-oriented approach, high fidelity, and ability to bind to LNPs without compromising their capabilities. A class of peptides called tumor-homing peptides (THPs) has effects on blood vessels or tumors. They recognize and bind to specific signals or receptors on blood vessels or tumors. In HCT116 colon cancer mice, liposomes functionalized with PIVO-8 (sequence: SNPFSKPYGLTV) and PIVO-24 (sequence: YPHYSLPGSSTL) peptides could promote apoptosis and prevent tumor angiogenesis (Chang et al., 2009).

Even while active targeting greatly improves the specificity and therapeutic efficiency of nanomedicines, additional investigation is still needed to address issues like receptor selection and rapid immune clearance in vivo (Katanasaka et al., 2010; Murase et al., 2010).

### 3.7.3 Stimulus-responsive targeting

A complex delivery technique called stimulus-responsive targeting utilizes the sensitivity of nanoparticles to physiochemical and biological stimuli to deliver drugs precisely to the target site. This strategy entails building a Response System of Stimuli that modifies the structure as well as composition of nanocarriers in order to react to stimuli quickly and accurately. It addresses the issues of the present day, including inadequate targeting, insufficient bioavailability, and delayed medication release (Ahn & Park, 2016; Mi, 2020).

Tumor cell glycolysis produces lactic acid, which lowers pH and causes tissue hypoxia in the tumor microenvironment (TME) of CRC. These traits present major obstacles to the development of successful chemotherapy for CRC because they promote the growth of cancer and increase its resistance to chemotherapy through signaling pathways such as hypoxia-inducible factor (HIF) (Li et al., 2021; Liu et al., 2021; Mu et al., 2023; Tang et al., 2021; Wang et al., 2019).

Tumor targeting is achieved via drug delivery devices that use these polymers; under physiological conditions, they preserve stability, but in acidic tumor environments, they release their payload (Juang et al., 2019).

## 3.8 Formulation of nanostructured lipid carriers

Techniques for making solid nano lipid carriers can also be used to create NLCs due to their ease of expansion, solvent-free nature, and faster manufacturing time than alternative methods. The HPH technique is still one of the most commonly used methods (Singh et al., 2019). The three main subcategories of these preparation methods include approaches using organic solvents, approaches with low energy, and approaches with high energy (Gomaa et al., 2022). Some of the methods are discussed below:

### 3.8.1 High-pressure homogenization

There are two types of high-pressure homogenization (HPH) techniques: (1) Hot high-pressure cooling (HPC) techniques and (2) Cold HPC techniques.

A popular HPH method is to first melt the lipid product, then combine it with the liquid lipid and drug. This dispersion is mixed with an aqueous thermal surfactant solution to form a preemulsion. This homogenization process can be repeated three times at 500 bar pressure and high temperature to produce the desired final product: SLN/NLC. In general, temperature reduces the viscosity of the lipids, reducing PS; however, the degradation potential of the drug or carrier increases. In contrast, in cold HPH, the lipid mixture melts and mixes to form a dispersion of lipid particles. This is then mixed with a cold surfactant solution to form a postulate; this forms the final product SLN/NLC after 1500 to 5–10 cycles of homogenization at room temperature.

Heating lipids over their melting temperatures reduced their viscosity, resulting in smaller particle sizes during the processing stage. There were numerous downsides to the process. Increased temperature was thought to trigger a rapid disintegration of the imbedded medication, as well as unpredictable lipid changeovers (Kasongo et al., 2012).

### 3.8.2 Microemulsion techniques

The lipid nanocarrier is followed by the addition of a drug melted just above the point. Using an auxiliary emulsifier and preheated deionized water creates a clear, thermodynamically stable microemulsion akin to oil-in-water. The resulting mixture is promptly dispersed in ice-cold water using gentle mechanical agitation to generate an NLC dispersion.

### 3.8.3 Inversion of phase temperature technique

This approach uses phase inversion from O/W to W/O emulsion, a cost-effective and solvent-free method for synthesizing LNPs. In Step 1, the lipid, surfactant, and water amounts are optimized before increasing the temperature by 4°C from room temperature to 85°C. To reach the phase inversion zone, the system undergoes three temperature cycles. The addition of cold water (0°C) in Step 2 causes an irreversible shock to the system, resulting in nano capsule formation.

### 3.8.4 The emulsification diffusion of solvent technique

The solvent emulsification-diffusion method (SED) involves saturating a partially water-soluble solvent, such as benzyl alcohol or butyl lactate, with water to achieve thermodynamic equilibrium between the two liquids. The lipid is dissolved in the organic phase (a water-saturated solvent), which is then emulsified with the surfactant-containing aqueous phase. Then, the emulsion is quickly diluted in water and stirred to extract the solvent into the external water phase by diffusion, thus precipitating the lipid and generating the SLNs (Hu et al., 2002).

### 3.8.5 Evaporation solvent injection method

To dissolve lipid, a water-miscible solvent (e.g., acetone, ethanol, methanol, or isopropyl alcohol) or a water-soluble solvent mixture is utilized. The lipid is then swiftly injected into an aqueous surfactant solution and agitated continuously. Excess fat is removed from the resulting dispersion via filtration. This method has the advantages of being simple to use, efficient, and versatile, requiring no specialist equipment (such as a high-pressure homogenizer), and using approved organic solvents (Schubert, 2003).

## 3.9 Delivery routes of lipid-based nanoparticles

Colon-specific delivery enhances CRC treatment by increasing effectiveness, reducing side effects, and enabling long-term low-dosage therapy (Arshad et al., 2020; Naeem et al., 2020).

### 3.9.1 Oral administration

Orally administered CRC treatment is preferred; (Lee et al., 2020) however, the GI tract affects the stability of lipid nanoparticles by pH, transit time, enzymes, and microbiota (Naeem et al., 2020; Zhang, Zhu, et al., 2017). pH-responsive systems, FA-grafted SLNs in Eudragit S100 microbeads (Rajpoot & Jain, 2018, 2019, 2020), release the drugs in the colon (pH more than 7.0) and have a residence time of up to 48 hours (Rajpoot & Jain, 2020).

These dual-responsive systems showed markedly enhanced antitumor activities compared to the controls. Ginger-derived nanoparticles, such as GDNPs, affect colon cells employing multiple response mechanisms since they are plant-derived nanoparticles (Zhang, Viennois, et al., 2016). Despite the fact that colonic exosomes are unstable within the GI tract, these vesicles may have potential value as nanotherapeutics in orally administered CRC treatment (Yang et al., 2020).

### 3.9.2 Injectable administration

In CRC treatment, injectable methods such as IV, intradermal (ID), intramuscular (IM), and subcutaneous (SC) have high systemic bioavailability but are associated with systemic adverse effects (Oussoren et al., 1998). To improve targeting, LBNPs with long circulation are frequently utilized, which combine passive (EPR effect) and active targeting (Fathi & Oyelere, 2016).

A study administered oxaliplatin and anti-TRAIL using chitosan-coated liposomes, which drastically reduced tumor bulk and volume (Tummala et al., 2016). Exosomes, specifically mammalian cell-derived exosomes (MCDEs), provide biocompatibility (Liang et al., 2020). Aspirin-loaded exosomes increased cytotoxicity, apoptosis, and CRC stem cell elimination (Tran et al., 2019).

### 3.9.3 Other delivery routes

Although rectal delivery of nanomedicines for CRC is less common than oral or IV methods, it has benefits such as quick effects, decreased hepatic first-pass metabolism, and less pain (Gulbake, 2016; Xu et al., 2017). Mice treated with lipid-based core-shell nanoparticles containing PEI/p21-saRNA-322 demonstrated anticancer efficacy, as evidenced by 6+ hours of retention in the rectum and CD44-targeted CRC accumulation (Wang, Feng, et al., 2017).

While injectable methods are successful for biologics in all stages, oral or rectal delivery may be beneficial for early-stage CRC to prevent systemic toxicity. Biodegradability, dual targeting of cancer stem cells and metastasized tissue, and systemic toxicity must all be balanced in injectables for advanced CRC (Yang & Merlin, 2020).

### 3.9.4 Applications

A promising approach for targeted administration of medication in colon cancer is provided by nano lipid carriers and with the help of NLCs, the treatment for colon cancer can be revolutionized. These carriers can encapsulate different anticancer drugs, increasing their bioavailability and minimizing

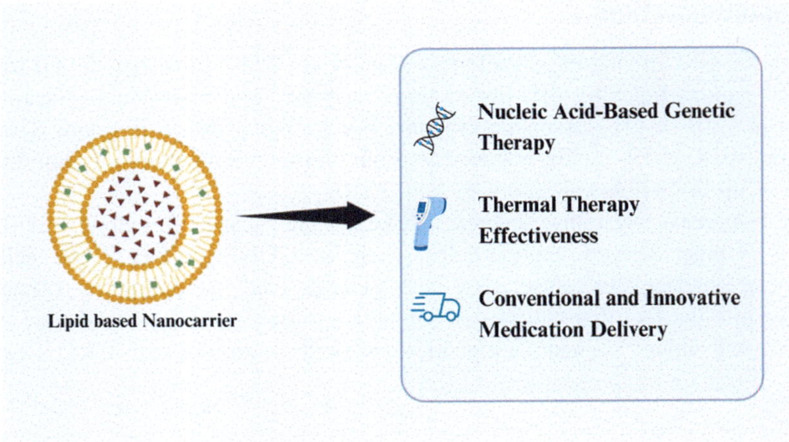


**FIGURE 3.3**

Applications of lipid nanocarrier-based drug delivery for the treatment of colon cancer.

systemic side effects (Fig. 3.3). They are made of biocompatible lipids. Furthermore, because of their special qualities, encapsulated medications can be released gradually, providing longer drug exposure to tissue of the tumor, by off-target toxicity, and optimizing therapeutic advantages.

### *3.9.4.1 Lipid nanoparticle-based nucleic acid-based genetic therapy*

The strategy for treatment highlights its promise in oncological therapies by enabling more accurate and customized antitumor therapy. Based on possible methods of action, gene treatment can be broadly divided into four categories:

Addition or replacement of genes, as demonstrated by mRNA-expressing genes enclosed in LNPs (Agbo et al., 2021).

Control of gene expression using agents such as miRNA, siRNA, and long noncoding RNA (lncRNA) (Ahn and Park, 2016; Hu et al., 2020).

Gene editing with instruments like single-guided RNA (sgRNA) and Cas9 mRNA (Akbarzadeh et al., 2013).

DNA (Al-Asmari et al., 2017; Hou et al., 2021).

A promising treatment option for cancer is gene therapy, which targets harmful genes in a manner that is different from specific sequences (Anguela & High, 2019).

Gene therapy requires a steady, reliable, and safe delivery platform in order to be effective in vivo. This platform needs to protect nucleic acids from deterioration and enable their absorption and subsequent release into cells. LNPs best meet the previously listed specifications (Zhang et al., 2024).

#### 3.9.4.2 Lipid nanoparticles in colorectal cancer therapy: conventional and innovative medication delivery

Lipid nanoparticles have shown great promise as efficient carriers for the targeted administration of a variety of cutting-edge treatments in CRC, providing special benefits over conventional antitumor medications. A study discovered that liposomes carrying scorpion toxin are superior to the toxin's free form in terms of cancer efficacy (Al-Asmari et al., 2017).

As research into colon cancer treatment progresses, the effectiveness of lipid nanoparticles as mediators for various tumor therapies, including lipid nanocarriers, can offer therapeutic benefits that are either superior to or similar to lower pharmaceutical dosages. These are some of the key advantages of using LNPs to deliver medicated treatment to tumor locations (Thi et al., 2021).

5-Fluorouracil (5-Fu), a derivative of uracil that falls under the category of antimetabolic anticancer drugs, is one of the most prevalent chemotherapy treatments for CRC. Its extremely short plasma half-life, significant hazardous side effects, and poor selectivity limit its practical applicability. To overcome these obstacles, Patel et al. encapsulated 5-FU in solid lipid nanoparticles and found that the Caco-2 human CRC cell lines showed a concentration-dependent decrease in cell viability (Patel et al., 2014).

Another popular chemotherapy medication with potent anticancer effects, doxorubicin (DOX), has drawbacks because it is not tumor-specific and has increased cardiotoxicity. In response, Zhang et al. created ginger-derived nanovectors (GDNVs), a type of nanocarrier made from ginger lipids that can effectively load and target the delivery of DOX (Zhang, Xiao, et al., 2016).

### 3.9.5 Limitations

Nano lipid carriers have many abilities for targeted drug delivery, but they face certain limitations, some of them are listed below

- Cytotoxic effects because of the concentration and matrix type are as follows:.

  - Some surfactants' irritating and sensitizing properties,
  - There is still need for improvement in the use and effectiveness of exploited gene delivery systems and medications, including proteins and peptides,
  - Inadequate preclinical and clinical research on these nanoparticles in bone mending (Bunjes et al., 1996).

### 3.9.6 Current advancements

#### 3.9.6.1 New developments in NLCs' capability to administer a variety of treatments

##### 3.9.6.1.1 Phytochemicals

Creanaringenin-loaded NLCs are being developed against non-naringenin-loaded NLCs to combat nonalcoholic fatty liver disease (Hu et al., 2021). NLCs with 1′-acetoxychavicol acetate are being studied for prostate cancer (Subramaniam et al., 2021). Cannabidiol NLCs are administered via the nose to treat neuropathic pain (Matarazzo et al., 2021). The protective ability of curcumin as a depression treatment was enhanced by curcumin-NLCs (Rubab et al., 2021). In vitro tests revealed no toxicity to mitochondria (Yoozbashi et al., 2022).

Bergamot oil NLCs were considered a potential method for treating vitiligo with photothermal therapy (Shaaban et al., 2021).

Since ancient times, active plant ingredients have demonstrated their pharmacological activity against a wide range of disorders; yet, due to their physicochemical characteristics, they exhibit problems with solubility, permeability, and poor bioavailability. To address these phytoconstituents' biopharmaceutical issues, a sophisticated and unique lipid carrier is needed in the formulation. Lipid-based nanocarriers are developed for active ingredients from plants, and they have become innovative medication delivery methods (Chaudhari, Murty, et al., 2021).

Corylin-NLC gel was prepared and demonstrated as a treatment against UV-induced skin aging. It was shown that H. pylori NLC interacts with microbial cells by adhering to cell membranes and causing leakage of cytoplasmic contents (Sharaf et al., 2021). Chrysin is an essential flavonoid found in nature. Its biopharmaceutical performance was markedly enhanced by biotin-conjugated NLCs for oral administration (Sharma et al., 2021).

#### 3.9.6.1.2 Cathartics

The oral bioavailability of perphenazine NLC will increase when it is demonstrated that NLC can overcome the oral limitations of this drug (Saghafi et al., 2021).

The low solubility and instability of hydrochlorothiazide make it unsuitable for oral therapy in children, a problem overcome by conversion to SLN and NLC. NLC was superior to SLN in drug encapsulation; NLC was 80% versus 90% for SLN. The NLC formulation was found to be physically stable (compared to SLN) during six months of storage at 4°C (Mura et al., 2021).

Ciprofloxacin-NLC as a novel strategy to improve ciprofloxacin targeting in noncystic fibrosis bronchiectasis (Almurshedi et al., 2021). Intranasal administration of artesunate NLC holds significant promise as a treatment (Agbo et al., 2021).

In recent years, miltefosine-loaded NLCs have been developed for oral treatment of cutaneous leishmaniasis with the aim of improving safety and reducing hemolytic potential. Mannosylated NLC of clofazimine is a promising tool for safe inhalation drug delivery in the treatment of tuberculosis. A two-fold increase in bioavailability was observed when compared to drug dispersions. Eplerenone-loaded NLCs are effective in the treatment of serous chorioretinopathy when used as an oral target. The optimized drug-loaded NLC showed 134 nm PS, 0.31 polydispersity index (PDI), 76% ± 6.56% w/w EE, and −32.37 mV zeta potential (ZP) (Mura et al., 2021).

Modified-functionalized curcumin NLC has been shown to hold great promise in neurotransmission. The oral bioavailability of nintedanib ethanesulfonate in NLC was increased by more than 26.31-fold compared to suspension, and the oral bioavailability of the drug was enhanced by lymphatic absorption. The use of inhaled itraconazole pegylated NLC in the treatment of pulmonary zoomycosis has also been reported (Shadambikar et al., 2021).

### 3.9.6.2 Antitumor medications

Thermal microemulsion technology was used to deliver sorafenib and ganoderic acid NLCs for the treatment of hepatocellular carcinoma, which exhibited excellent chemoprotective effects compared with ordinary drugs (Wang, Sun, et al., 2021). Nanoencapsulated resveratrol exerts anticancer and apoptotic effects on cell proliferation while protecting against paraquat-induced cytotoxicity (Elgizawy et al., 2021). Imatinib-loaded NLCs were tested in MCF-7 cells, revealing that the drug alone (IC50=52.5 μM) had cytotoxicity 8.75 times lower than the optimal drug–NLC combination (IC50=6 μM) (Makeen et al., 2021). Microneedle-assisted resveratrol NLC for the treatment of localized breast cancer exhibits enhanced internalization and stronger antimicrobial resistance in MDA-MB-231 breast cancer cells compared to the cancer activity of pure resveratrol (Gadag et al., 2021). For nonsmall cell lung cancer, multispecies cancer cell lines have been developed that contain luteinizing hormone-secreting agents, gefitinib, paclitaxel, siRNA targeting the epidermal growth factor (EGF) receptor, and rhodamine for detection (Javed et al., 2022). Metformin administered intraperitoneally to tumor BALB/c mice as NLC ameliorated apoptosis, improved pharmacokinetics, and showed anticancer potential. Folate-conjugated NLC effectively delivered doxorubicin to breast cancer cells (Yu et al., 2021). The taxane derivative cabidetaxel (CAB) was developed using an expert design and multiple parameters to provide a highly lipophilic drug for the treatment of breast cancer using NLC (Chand et al., 2021).

Puerarin-loaded NLC prepared by solvent evaporation showed a better therapeutic response in alcoholic BRL-3A strain with an average size of 159 ± 1.1 nm, EE% 92.16%, DL% 5.75%, and ZP 28.3 mv (Ye et al., 2021). The optimized 5-fluorouracil NLC was developed as an effective cancer treatment with greater bioavailability than that of drugs. When drug-loaded NLCs were administered orally, their bioavailability was higher than that of drug solutions (Borderwala et al., 2021). It was discovered that optimized dual medication-loaded NLCs, including piperine and quercetin, were effective against oral squamous cell cancer (Chaudhari, Gawali, et al., 2021). Chrysin is an essential flavonoid found in nature. Its biopharmaceutical performance was markedly enhanced by biotin-conjugated NLCs for oral administration (Sharma et al., 2021).

## 3.10 Conclusion

Liposomes, SLNs, and NLCs are examples of lipid-based nanocarriers that have demonstrated significant potential in the regional and focused delivery of drugs for chemotherapy in the treatment of colon cancer. They are effective remedies for the shortcomings of conventional chemotherapy because they can distribute both polar and nonpolar medications, shield them from deterioration, and increase their bioavailability while lowering systemic toxicity. Because passive and active-targeting mechanisms enable the improvement of drug concentration at the tumor site, clinical pharmacology of lipid-based systems can be enhanced in this regard. This will increase the efficacy of therapy and decrease its negative effects. Furthermore, the selection and blood retention of these carriers have been enhanced by the introduction of coatings such as PEGylation and ligand–carrier fusion. Even with these advancements, there are still a number of drawbacks, such as the possibility of immunogenic responses, manufacturing scaling issues, and stability issues. Subsequent investigations ought to concentrate on surmounting these obstacles by creating more resilient, economical manufacturing techniques and refining compositions for therapeutic implementation. Additionally, there is a great deal of promise to transform the treatment of colon cancer by combining lipid nanocarriers with other therapeutic approaches, such as gene and immunotherapy. In conclusion, additional research is needed to realize the therapeutic potential of lipid nanocarrier-based systems, which present a promising path for more efficient and targeted colon cancer therapy.

## References

Agbo, C. P., Ugwuanyi, T. C., Ugwuoke, W. I., McConville, C., Attama, A. A., & Ofokansi, K. C. (2021). Intranasal artesunate-loaded nanostructured lipid carriers: A convenient alternative to parenteral formulations for the treatment of severe and cerebral malaria. *Journal of Controlled Release, 334*, 224–236. https://doi.org/10.1016/j.jconrel.2021.04.020, www.elsevier.com/locate/jconrel.

Ahn, H., & Park, J. H. (2016). Liposomal delivery systems for intestinal lymphatic drug transport. *Biomaterials Research, 20*(1). https://doi.org/10.1186/s40824-016-0083-1, biomaterialsres.biomedcentral.com/.

Akbarzadeh, A., Rezaei-Sadabady, R., Davaran, S., Joo, S. W., Zarghami, N., Hanifehpour, Y., Samiei, M., Kouhi, M., & Nejati-Koshki, K. (2013). Liposome: Classification, preparation, and applications. *Nanoscale Research Letters, 8*(1), 102. https://doi.org/10.1186/1556-276X-8-102, http://www.springer.com/materials/nanotechnology/journal/11671.

Al-Asmari, A., Ullah, Z., Al Balowi, A., & Islam, M. (2017). In vitro determination of the efficacy of scorpion venoms as anti-cancer agents against colorectal cancer cells: A nano-liposomal delivery approach. *International Journal of Nanomedicine, 12*, 559–574. https://doi.org/10.2147/ijn.s123514.

Alavi, M., & Hamidi, M. (2019). Passive and active targeting in cancer therapy by liposomes and lipid nanoparticles. *Drug Metabolism and Personalized Therapy, 34*(1). https://doi.org/10.1515/dmpt-2018-0032.

Allen, T. M., & Hansen, C. (1991). Pharmacokinetics of stealth versus conventional liposomes: Effect of dose. *Biochimica et Biophysica Acta (BBA) - Biomembranes, 1068*(2), 133–141. https://doi.org/10.1016/0005-2736(91)90201-i.

Almurshedi, A. S., Aljunaidel, H. A., Alquadeib, B., Aldosari, B. N., Alfagih, I. M., Almarshidy, S. S., Eltahir, E. K. D., & Mohamoud, A. Z. (2021). Development of inhalable nanostructured lipid carriers for ciprofloxacin for noncystic fibrosis bronchiectasis treatment. *International Journal of Nanomedicine, 16*, 2405–2417. https://doi.org/10.2147/IJN.S286896, https://www.dovepress.com/getfile.php?fileID=68148.

Alomrani, A., Badran, M., Harisa, G. I., ALshehry, M., Alhariri, M., Alshamsan, A., & Alkholief, M. (2019). The use of chitosan-coated flexible liposomes as a remarkable carrier to enhance the antitumor efficacy of 5-fluorouracil against colorectal cancer. *Saudi Pharmaceutical Journal, 27*(5), 603–611. https://doi.org/10.1016/j.jsps.2019.02.008, http://www.sciencedirect.com/science/journal/13190164.

Anguela, X. M., & High, K. A. (2019). Entering the modern era of gene therapy. *Annual Review of Medicine, 70*, 273–288. https://doi.org/10.1146/annurev-med-012017-043332, http://www.annualreviews.org/journal/med.

Arouri, A., & Mouritsen, O. G. (2012). Denmark phospholipase A 2-susceptible liposomes of anticancer double lipid-prodrugs. *European Journal of Pharmaceutical Sciences, 45*(4), 408–420. https://doi.org/10.1016/j.ejps.2011.09.013.

Arshad, U., Sutton, P. A., Ashford, M. B., Treacher, K. E., Liptrott, N. J., Rannard, S. P., Goldring, C. E., & Owen, A. (2020). Critical considerations for targeting colorectal liver metastases with nanotechnology. *WIREs Nanomedicine and Nanobiotechnology, 12*(2). https://doi.org/10.1002/wnan.1588.

Bonilla-Vidal, L., Świtalska, M., Espina, M., Wietrzyk, J., García, M. L., Souto, E. B., Gliszczyńska, A., & Sánchez López, E. (2023). Dually active apigenin-loaded nanostructured lipid carriers for cancer treatment. *International Journal of Nanomedicine, 18*, 6979–6997. https://doi.org/10.2147/ijn.s429565.

Borderwala, K., Rathod, S., Yadav, S., Vyas, B., & Shah, P. (2021). Eudragit S-100 surface engineered nanostructured lipid carriers for colon targeting of 5-fluorouracil: Optimization and in vitro and in vivo characterization. *AAPS PharmSciTech, 22*(6). https://doi.org/10.1208/s12249-021-02099-3.

Botteri, E., Iodice, S., Bagnardi, V., Raimondi, S., Lowenfels, A. B., & Maisonneuve, P. (2008). Smoking and colorectal cancer. *JAMA: The Journal of the American Medical Association, 300*, 2765. https://doi.org/10.1001/jama.2008.839.

Braat, M. N. G. J. A., van Erpecum, K. J., Zonnenberg, B. A., van den Bosch, M. A. J., & Lam, M. G. E. H. (2017). Radioembolization-induced liver disease: A systematic review. *European Journal of Gastroenterology & Hepatology, 29*, 144–152. https://doi.org/10.1097/MEG.0000000000000772.

Brami, C., Bao, T., & Deng, G. (2016). Natural products and complementary therapies for chemotherapy-induced peripheral neuropathy: A systematic review. *Critical Reviews in Oncology/Hematology, 98*, 325–334. https://doi.org/10.1016/j.critrevonc.2015.11.014.

Bray, F., Ferlay, J., Soerjomataram, I., Siegel, R. L., Torre, L. A., & Jemal, A. (2018). Global cancer statistics 2018: GLOBOCAN estimates of incidence and mortality worldwide for 36 cancers in 185 countries. *CA Cancer Journal for Clinicians, 68*(6), 394–424. https://doi.org/10.3322/caac.21492, http://onlinelibrary.wiley.com/journal/10.3322/(ISSN)1542-4863.

Bunjes, H., Westesen, K., & Koch, M. H. J. (1996). Crystallization tendency and polymorphic transitions in triglyceride nanoparticles. *International Journal of Pharmaceutics, 129*(1-2), 159–173. https://doi.org/10.1016/0378-5173(95)04286-5, www.elsevier.com/locate/ijpharm.

Cai, S., Li, Y., Ding, Y., Chen, K., & Jin, M. (2014). Alcohol drinking and the risk of colorectal cancer death. *European Journal of Cancer Prevention, 23*, 532–539. https://doi.org/10.1097/CEJ.0000000000000076.

Campani, V., Giarra, S., & De Rosa, G. (2018). Lipid-based core-shell nanoparticles: Evolution and potentialities in drug delivery. *OpenNano, 3*, 5–17. https://doi.org/10.1016/j.onano.2017.12.001.

Cervantes, A., Adam, R., Roselló, S., Arnold, D., Normanno, N., Taïeb, J., Seligmann, J., De Baere, T., Osterlund, P., Yoshino, T., & Martinelli, E. (2023). Metastatic colorectal cancer: ESMO clinical practice guideline for diagnosis, treatment and follow-up. *Annals of Oncology, 34*(1), 10–32. https://doi.org/10.1016/j.annonc.2022.10.003.

Chan, C.-L., Majzoub, R. N., Shirazi, R. S., Ewert, K. K., Chen, Y.-J., Liang, K. S., & Safinya, C. R. (2012). Endosomal escape and transfection efficiency of PEGylated cationic liposome–DNA complexes prepared with an acid-labile PEG-lipid. *Biomaterials, 33*, 4928–4935. https://doi.org/10.1016/j.biomaterials.2012.03.038.

Chan, D. S. M., Lau, R., Aune, D., Vieira, R., Greenwood, D. C., Kampman, E., & Norat, T. (2011). Red and processed meat and colorectal cancer incidence: Meta-analysis of prospective studies. *PLoS One, 6*(6), e20456. https://doi.org/10.1371/journal.pone.0020456, http://www.plosone.org/article/fetchObjectAttachment.action?uri=info%3Adoi%2F10.1371%2Fjournal.pone.0020456&representation=PDF.

Chand, P., Kumar, H., Badduri, N., Gupta, N. V., Bettada, V. G., Madhunapantula, S. V., Kesharwani, S. S., Dey, S., & Jain, V. (2021). Design and evaluation of cabazitaxel loaded NLCs against breast cancer cell lines. *Colloids and Surfaces B: Biointerfaces, 199*, 111535. https://doi.org/10.1016/j.colsurfb.2020.111535.

Chang, D.-K., Chiu, C.-Y., Kuo, S.-Y., Lin, W.-C., Lo, A., Wang, Y.-P., Li, P.-C., & Wu, H.-C. (2009). Antiangiogenic targeting liposomes increase therapeutic efficacy for solid tumors. *Journal of Biological Chemistry, 284*, 12905–12916. https://doi.org/10.1074/jbc.M900280200.

Chau, I., & Cunningham, D. (2002). Adjuvant therapy in colon cancer: Current status and future directions. *Cancer Treatment Reviews, 28*, 223–236. https://doi.org/10.1016/S0305-737200047-6.

Chaudhari, V. S., Gawali, B., Saha, P., Naidu, V. G. M., Murty, U. S., & Banerjee, S. (2021). Quercetin and piperine enriched nanostructured lipid carriers (NLCs) to improve apoptosis in oral squamous cellular carcinoma (FaDu cells) with improved biodistribution profile. *European Journal of Pharmacology, 909*, 174400. https://doi.org/10.1016/j.ejphar.2021.174400.

Chaudhari, V. S., Murty, U. S., & Banerjee, S. (2021). Nanostructured lipid carriers as a strategy for encapsulation of active plant constituents: Formulation and in vitro physicochemical characterizations. *Chemistry and Physics of Lipids, 235*, 105037. https://doi.org/10.1016/j.chemphyslip.2020.105037.

Chintala, L., Vaka, S., Baranda, J., & Williamson, S. K. (2011). Capecitabine versus 5-fluorouracil in colorectal cancer: Where are we now? *Oncology Reviews, 5*(2), 129–140. https://doi.org/10.1007/s12156-011-0074-3.

Crommelin, D. J. A., van Hoogevest, P., & Storm, G. (2020). The role of liposomes in clinical nanomedicine development. What now? Now what? *Journal of Controlled Release, 318*, 256–263. https://doi.org/10.1016/j.jconrel.2019.12.023, www.elsevier.com/locate/jconrel.

Das, M., & Huang, L. (2019). Liposomal nanostructures for drug delivery in gastrointestinal cancers. *Journal of Pharmacology and Experimental Therapeutics, 370*, 647–656. https://doi.org/10.1124/jpet.118.254797.

De Campos, A. M., Sánchez, A., & Alonso, M. J. (2001). Chitosan nanoparticles: A new vehicle for the improvement of the delivery of drugs to the ocular surface. Application to cyclosporin A. *International Journal of Pharmaceutics, 224*, 159–168. https://doi.org/10.1016/S0378-517300760-8.

De Campos, A. M., Sánchez, A., Gref, R., Calvo, P., & Alonso, M. J. (2003). The effect of a PEG versus a chitosan coating on the interaction of drug colloidal carriers with the ocular mucosa. *European Journal of Pharmaceutical Sciences, 20*, 73–81. https://doi.org/10.1016/S0928-098700178-7.

DeSantis, C. E., Lin, C. C., Mariotto, A. B., Siegel, R. L., Stein, K. D., Kramer, J. L., Alteri, R., Robbins, A. S., & Jemal, A. (2014). Cancer treatment and survivorship statistics, 2014. *CA: A Cancer Journal for Clinicians, 64*, 252–271. https://doi.org/10.3322/caac.21235.

Doktorovová, S., Kovačević, A. B., Garcia, M. L., & Souto, E. B. (2016). Preclinical safety of solid lipid nanoparticles and nanostructured lipid carriers: Current evidence from in vitro and in vivo evaluation. *European Journal of Pharmaceutics and Biopharmaceutics, 108*, 235–252. https://doi.org/10.1016/j.ejpb.2016.08.001.

Dong, Y., Ng, W. K., Shen, S., Kim, S., & Tan, R. B. H. (2012). Solid lipid nanoparticles: Continuous and potential large-scale nanoprecipitation production in static mixers. *Colloids and Surfaces B: Biointerfaces, 94*, 68–72. https://doi.org/10.1016/j.colsurfb.2012.01.018.

Duan, X. M., Wang, P., Men, K., Gao, X., Huang, M. J., Gou, M. L., Chen, L. J., Qian, Z. Y., & Wei, Y. Q. (2012). Treating colon cancer with a suicide gene delivered by self-assembled cationic MPEG–PCL micelles. *Nanoscale, 4*(7), 2400. https://doi.org/10.1039/c2nr30079f.

Duran, G., Cruz, R., Simoes, A. R., Barros, F., Giráldez, J. M., Bernárdez, B., Anido, U., Candamio, S., López-López, R., Carracedo, Á., & Lamas, M. J. (2020). Efficacy and toxicity of adjuvant chemotherapy on colorectal cancer patients: How much influence from the genetics? *Journal of Chemotherapy, 32*, 310–322. https://doi.org/10.1080/1120009X.2020.1764281.

Elgizawy, H. A., Ali, A. A., & Hussein, M. A. (2021). Resveratrol: Isolation, and its nanostructured lipid carriers, inhibits cell proliferation, induces cell apoptosis in certain human cell lines carcinoma and exerts protective effect against paraquat-induced hepatotoxicity. *Journal of Medicinal Food, 24*, 89–100. https://doi.org/10.1089/jmf.2019.0286.

Elmowafy, M., & Al-Sanea, M. M. (2021). Nanostructured lipid carriers (NLCs) as drug delivery platform: Advances in formulation and delivery strategies. *Saudi Pharmaceutical Journal, 29*, 999–1012. https://doi.org/10.1016/j.jsps.2021.07.015.

Eygeris, Y., Gupta, M., Kim, J., & Sahay, G. (2022). Chemistry of lipid nanoparticles for RNA delivery. *Accounts of Chemical Research, 55*, 2–12. https://doi.org/10.1021/acs.accounts.1c00544.

Fathi, S., & Oyelere, A. K. (2016). Liposomal drug delivery systems for targeted cancer therapy: Is active targeting the best choice? *Future Medicinal Chemistry, 8*, 2091–2112. https://doi.org/10.4155/fmc-2016-0135.

Favoriti, P., Carbone, G., Greco, M., Pirozzi, F., Pirozzi, R. E. M., & Corcione, F. (2016). Worldwide burden of colorectal cancer: A review. *Updates in Surgery, 68*, 7–11. https://doi.org/10.1007/s13304-016-0359-y.

Fulton, M. D., & Najahi-Missaoui, W. (2023). Liposomes in cancer therapy: How did we start and where are we now. *International Journal of Molecular Sciences, 24*, 6615. https://doi.org/10.3390/ijms24076615.

Gabizon, A. A., Patil, Y., & La-Beck, N. M. (2016). New insights and evolving role of pegylated liposomal doxorubicin in cancer therapy. *Drug Resistance Updates, 29*, 90–106. https://doi.org/10.1016/j.drup.2016.10.003, http://www.elsevier.com/locate/drup.

Gadag, S., Narayan, R., Nayak, A. S., Catalina Ardila, D., Sant, S., Nayak, Y., Garg, S., & Nayak, U. Y. (2021). Development and preclinical evaluation of microneedle-assisted resveratrol loaded nanostructured lipid carriers for localized delivery to breast cancer therapy. *International Journal of Pharmaceutics, 606*. https://doi.org/10.1016/j.ijpharm.2021.120877, www.elsevier.com/locate/ijpharm.

Gao, W., Hu, C.-M. J., Fang, R. H., & Zhang, L. (2013). Liposome-like nanostructures for drug delivery. *Journal of Materials Chemistry B, 1*, 6569. https://doi.org/10.1039/c3tb21238f.

Garciafuentes, M., Torres, D., & Alonso, M. (2005). New surface-modified lipid nanoparticles as delivery vehicles for salmon calcitonin. *International Journal of Pharmaceutics, 296*, 122–132. https://doi.org/10.1016/j.ijpharm.2004.12.030.

García-Pinel, B., Porras-Alcalá, C., Ortega-Rodríguez, A., Sarabia, F., Prados, J., Melguizo, C., & López-Romero, J. M. (2019). Lipid-based nanoparticles: Application and recent advances in cancer treatment. *Nanomaterials, 9*, 638. https://doi.org/10.3390/nano9040638.

Geszke-Moritz, M., & Moritz, M. (2016). Solid lipid nanoparticles as attractive drug vehicles: Composition, properties and therapeutic strategies. *Materials Science and Engineering: C, 68*, 982–994. https://doi.org/10.1016/j.msec.2016.05.119.

Gholami, Y. H., & Engel, A. (2018). Theranostic nanoplatforms for treatment and diagnosis of rectal and colon cancer: A brief review. *Mini-Invasive Surgery, 2018*. https://doi.org/10.20517/2574-1225.2018.44.

Gogoi, M., Kumar, N., & Patra, S. (2016). Multifunctional magnetic liposomes for cancer imaging and therapeutic applications. In *Nanoarchitectonics for Smart Delivery and Drug Targeting*. Elsevier, 743–782 https://doi.org/10.1016/B978-0-323-47347-7.00027-6.

Gomaa, E., Fathi, H. A., Eissa, N. G., & Elsabahy, M. (2022). Methods for preparation of nanostructured lipid carriers. *Methods (San Diego, Calif.), 199*, 3–8. https://doi.org/10.1016/j.ymeth.2021.05.003.

Govindarasu, M., Abirami, P., Alharthi, S. S., Thiruvengadam, M., Rajakumar, G., & Vaiyapuri, M. (2022). Synthesis, physicochemical characterization, and in vitro evaluation of biodegradable PLGA nanoparticles entrapped to folic acid for targeted delivery of kaempferitrin. *Biotechnology and Applied Biochemistry, 69*, 2387–2398. https://doi.org/10.1002/bab.2290.

Guan, J., Sun, J., Sun, F., Lou, B., Zhang, D., Mashayekhi, V., Sadeghi, N., Storm, G., Mastrobattista, E., & He, Z. (2017). Hypoxia-induced tumor cell resistance is overcome by synergistic GAPDH-siRNA and chemotherapy co-delivered by long-circulating and cationic-interior liposomes. *Nanoscale, 9*, 9190–9201. https://doi.org/10.1039/C7NR02663C.

Gulbake, A. (2016). Insight to drug delivery aspects for colorectal cancer. *World Journal of Gastroenterology, 22*, 582. https://doi.org/10.3748/wjg.v22.i2.582.

Handali, S., Moghimipour, E., Rezaei, M., Ramezani, Z., Kouchak, M., Amini, M., Angali, K. A., Saremy, S., & Dorkoosh, F. A. (2018). A novel 5-Fluorouracil targeted delivery to colon cancer using folic acid conjugated liposomes. *Biomedicine & Pharmacotherapy, 108*, 1259–1273. https://doi.org/10.1016/j.biopha.2018.09.128.

Heiati, H., Tawashi, R., Shivers, R. R., & Phillips, N. C. (1997). Solid lipid nanoparticles as drug carriers. I. Incorporation and retention of the lipophilic prodrug 3′-azido-3′-deoxythymidine palmitate. *International Journal of Pharmaceutics, 146*, 123–131. https://doi.org/10.1016/S0378-517304782-5.

Hellinger, M. D., & Santiago, C. A. (2006). Reoperation for recurrent colorectal cancer. *Clinics in Colon and Rectal Surgery, 19*, 228–236. https://doi.org/10.1055/s-2006-956445.

Henrikson, N. B., Webber, E. M., Goddard, K. A., Scrol, A., Piper, M., Williams, M. S., Zallen, D. T., Calonge, N., Ganiats, T. G., Janssens, A. C. J. W., Zauber, A., Lansdorp-Vogelaar, I., van Ballegooijen, M., & Whitlock, E. P. (2015). Family history and the natural history of colorectal cancer: Systematic review. *Genetics in Medicine, 17*, 702–712. https://doi.org/10.1038/gim.2014.188.

Hou, X., Yang, C., Zhang, L., Hu, T., Sun, D., Cao, H., Yang, F., Guo, G., Gong, C., Zhang, X., Tong, A., Li, R., & Zheng, Y. (2017). Killing colon cancer cells through PCD pathways by a novel hyaluronic acid-modified shell-core nanoparticle loaded with RIP3 in combination with chloroquine. *Biomaterials, 124*, 195–210. https://doi.org/10.1016/j.biomaterials.2016.12.032.

Hou, X., Zaks, T., Langer, R., & Dong, Y. (2021). Lipid nanoparticles for mRNA delivery. *Nature Reviews Materials, 6*, 1078–1094. https://doi.org/10.1038/s41578-021-00358-0.

Hu, B., Zhong, L., Weng, Y., Peng, L., Huang, Y., Zhao, Y., & Liang, X.-J. (2020). Therapeutic siRNA: State of the art. *Signal Transduction and Targeted Therapy, 5*, 101. https://doi.org/10.1038/s41392-020-0207-x.

Hu, F. Q., Yuan, H., Zhang, H. H., & Fang, M. (2002). Preparation of solid lipid nanoparticles with clobetasol propionate by a novel solvent diffusion method in aqueous system and physicochemical characterization. *International Journal of Pharmaceutics, 239*, 121–128. https://doi.org/10.1016/S0378-517300081-9.

Hu, R., Liu, S., Shen, W., Chen, C., Cao, Y., Su, Z., Sun, M., & Qi, R. (2021). Study on the inhibitory effects of naringenin-loaded nanostructured lipid carriers against nonalcoholic fatty liver disease. *Journal of Biomedical Nanotechnology, 17*, 942–951. https://doi.org/10.1166/jbn.2021.3077.

Hu, Y., Chen, X., Li, Z., Zheng, S., & Cheng, Y. (2020). Thermosensitive *In Situ* Gel containing luteolin micelles is a promising efficient agent for colorectal cancer peritoneal metastasis treatment. *Journal of Biomedical Nanotechnology, 16*, 54–64. https://doi.org/10.1166/jbn.2020.2870.

Hu, Y., He, Y., Ji, J., Zheng, S., & Cheng, Y. (2020). Tumor targeted curcumin delivery by folate-modified MPEG-PCL self-assembly micelles for colorectal cancer therapy. *International Journal of Nanomedicine, 15*, 1239–1252. https://doi.org/10.2147/IJN.S232777.

Huang, J.-R., Lee, M.-H., Li, W.-S., & Wu, H.-C. (2019). Liposomal irinotecan for treatment of colorectal cancer in a preclinical model. *Cancers, 11*, 281. https://doi.org/10.3390/cancers11030281.

Jain, V., Kumar, H., Chand, P., Jain, S., & S, P. (2021). Lipid-based nanocarriers as drug delivery system and its applications. In *Nanopharmaceutical advanced delivery systems*. Wiley, 1–29. https://doi.org/10.1002/9781119711698.ch1.

Jani, P., Halbert, G. W., Langridge, J., & Florence, A. T. (2011). Nanoparticle uptake by the rat gastrointestinal Mucosa: Quantitation and particle size dependency. *Journal of Pharmacy and Pharmacology, 42*, 821–826. https://doi.org/10.1111/j.2042-7158.1990.tb07033.x.

Janout, V., & Kollárová, H. (2001). Epidemiology of colorectal cancer. *Biomedical Papers, 145*, 5–10. https://doi.org/10.5507/bp.2001.001.

Javed, S., Mangla, B., Almoshari, Y., Sultan, M. H., & Ahsan, W. (2022). Nanostructured lipid carrier system: A compendium of their formulation development approaches, optimization strategies by quality by design, and recent applications in drug delivery. *Nanotechnology Reviews, 11*, 1744–1777. https://doi.org/10.1515/ntrev-2022-0109.

Jiang, Z., Guan, J., Qian, J., & Zhan, C. (2019). Peptide ligand-mediated targeted drug delivery of nanomedicines. *Biomaterials Science, 7*, 461–471. https://doi.org/10.1039/C8BM01340C.

Juang, V., Chang, C., Wang, C., Wang, H., & Lo, Y. (2019). pH-responsive PEG-shedding and targeting peptide-modified nanoparticles for dual-delivery of irinotecan and microRNA to enhance tumor-specific therapy. *Small (Weinheim an der Bergstrasse, Germany), 15*. https://doi.org/10.1002/smll.201903296.

Judge, A. D., Robbins, M., Tavakoli, I., Levi, J., Hu, L., Fronda, A., Ambegia, E., McClintock, K., & MacLachlan, I. (2009). Confirming the RNAi-mediated mechanism of action of siRNA-based cancer therapeutics in mice. *Journal of Clinical Investigation, 119*, 661–673. https://doi.org/10.1172/JCI37515.

Kasongo, K. W., Müller, R. H., & Walker, R. B. (2012). The use of hot and cold high pressure homogenization to enhance the loading capacity and encapsulation efficiency of nanostructured lipid carriers for the hydrophilic antiretroviral drug, didanosine for potential administration to paediatric patients. *Pharmaceutical Development and Technology, 17*, 353–362. https://doi.org/10.3109/10837450.2010.542163.

Katanasaka, Y., Ishii, T., Asai, T., Naitou, H., Maeda, N., Koizumi, F., Miyagawa, S., Ohashi, N., & Oku, N. (2010). Cancer antineovascular therapy with liposome drug delivery systems targeted to BiP/GRP78. *International Journal of Cancer, 127*, 2685–2698. https://doi.org/10.1002/ijc.25276.

Kroschinsky, F., Stölzel, F., von Bonin, S., Beutel, G., Kochanek, M., Kiehl, M., & Schellongowski, P. (2017). New drugs, new toxicities: Severe side effects of modern targeted and immunotherapy of cancer and their management. *Critical Care, 21*, 89. https://doi.org/10.1186/s13054-017-1678-1.

Kumbhar, D. D., & Pokharkar, V. B. (2013). Engineering of a nanostructured lipid carrier for the poorly water-soluble drug, bicalutamide: Physicochemical investigations. *Colloids and Surfaces A: Physicochemical and Engineering Aspects, 416*, 32–42. https://doi.org/10.1016/j.colsurfa.2012.10.031.

Kunjachan, S., Ehling, J., Storm, G., Kiessling, F., & Lammers, T. (2015). Noninvasive imaging of nanomedicines and nanotheranostics: Principles, progress, and prospects. *Chemical Reviews, 115*, 10907–10937. https://doi.org/10.1021/cr500314d.

Kwong, B., Gai, S. A., Elkhader, J., Wittrup, K. D., & Irvine, D. J. (2013). Localized immunotherapy via liposome-anchored anti-CD137 + IL-2 prevents lethal toxicity and elicits local and systemic antitumor immunity. *Cancer Research, 73*, 1547–1558. https://doi.org/10.1158/0008-5472.CAN-12-3343.

Kye, B.-H., & Cho, H.-M. (2014). Overview of radiation therapy for treating rectal cancer. *Annals of Coloproctology, 30*, 165–174. https://doi.org/10.3393/ac.2014.30.4.165.

Kyrgiou, M., Kalliala, I., Markozannes, G., Gunter, M. J., Paraskevaidis, E., Gabra, H., Martin-Hirsch, P., & Tsilidis, K. K. (2017). Adiposity and cancer at major anatomical sites: Umbrella review of the literature. *BMJ (Online), 356*. https://doi.org/10.1136/bmj.j477, http://www.bmj.com/.

Lee, S. H., Bajracharya, R., Min, J. Y., Han, J.-W., Park, B. J., & Han, H.-K. (2020). Strategic approaches for colon targeted drug delivery: An overview of recent advancements. *Pharmaceutics, 12*, 68. https://doi.org/10.3390/pharmaceutics12010068.

Leonard, T. E., Liko, A. F., Gustiananda, M., Putra, A. B. N., Juanssilfero, A. B., & Hartrianti, P. (2023). Thiolated pectin-chitosan composites: Potential mucoadhesive drug delivery system with selective cytotoxicity towards colorectal cancer. *International Journal of Biological Macromolecules, 225*, 1–12. https://doi.org/10.1016/j.ijbiomac.2022.12.012.

Li, X., Zhang, J., Chen, Y., Liang, X., & Luo, X. (2016). The influence of different long-circulating materials on the pharmacokinetics of liposomal vincristine sulfate. *International Journal of Nanomedicine, 11*, 4187–4197. https://doi.org/10.2147/ijn.s109547.

Li, J., Yang, P., Chen, F., Tan, Y., Huang, C., Shen, H., Peng, C., Feng, Y., & Sun, Y. (2021). Hypoxic colorectal cancer-derived extracellular vesicles deliver microRNA-361-3p to facilitate cell proliferation by targeting TRAF3 via the noncanonical NF-κB pathways. *Clinical and Translational Medicine, 11*. https://doi.org/10.1002/ctm2.349.

Liang, G., Zhu, Y., Ali, D. J., Tian, T., Xu, H., Si, K., Sun, B., Chen, B., & Xiao, Z. (2020). Engineered exosomes for targeted co-delivery of miR-21 inhibitor and chemotherapeutics to reverse drug resistance in colon cancer. *Journal of Nanobiotechnology, 18*(1). https://doi.org/10.1186/s12951-019-0563-2, http://www.jnanobiotechnology.com/start.asp.

Liang, X., Luo, M., Wei, X. W., Ma, C. C., Yang, Y. H., Shao, B., Liu, Y. T., Liu, T., Ren, J., Liu, L., He, Z. Y., & Wei, Y. Q. (2016). A folate receptor-targeted lipoplex delivering interleukin-15 gene for colon cancer immunotherapy. *Oncotarget, 7*(32), 52207–52217. https://doi.org/10.18632/oncotarget.10537, http://www.impactjournals.com/oncotarget/index.php?journal=oncotarget&page=article&op=download&path%5B%5D=10537&path%5B%5D=33293.

Liu, C., Zhang, W., Wang, J., Si, T., & Xing, W. (2021). Tumor-associated macrophage-derived transforming growth factor-β promotes colorectal cancer progression through HIF1-TRIB3 signaling. *Cancer Science, 112*, 4198–4207. https://doi.org/10.1111/cas.15101.

Liu, L., Yang, S., Chen, F., & Cheng, K.-W. (2022). Hyaluronic acid–zein core-shell nanoparticles improve the anticancer effect of curcumin alone or in combination with oxaliplatin against colorectal cancer via CD44-

mediated cellular uptake. *Molecules (Basel, Switzerland), 27*, 1498. https://doi.org/10.3390/molecules27051498.

Lu, Y., Zhong, L., Jiang, Z., Pan, H., Zhang, Y., Zhu, G., Bai, L., Tong, R., Shi, J., & Duan, X. (2019). Cationic micelle-based siRNA delivery for efficient colon cancer gene therapy. *Nanoscale Research Letters, 14*, 193. https://doi.org/10.1186/s11671-019-2985-z.

Lund, C. M., Nielsen, D., Dehlendorff, C., Christiansen, A. B., Rønholt, F., Johansen, J. S., & Vistisen, K. K. (2016). Efficacy and toxicity of adjuvant chemotherapy in elderly patients with colorectal cancer: The ACCORE study. *ESMO Open, 1*, e000087. https://doi.org/10.1136/esmoopen-2016-000087.

Lyon, P. C., Gray, M. D., Mannaris, C., Folkes, L. K., Stratford, M., Campo, L., Chung, D. Y. F., Scott, S., Anderson, M., Goldin, R., Carlisle, R., Wu, F., Middleton, M. R., Gleeson, F. V., & Coussios, C. C. (2018). Safety and feasibility of ultrasound-triggered targeted drug delivery of doxorubicin from thermosensitive liposomes in liver tumours (TARDOX): A single-centre, open-label, phase 1 trial. *The Lancet Oncology, 19*, 1027–1039. https://doi.org/10.1016/S1470-204530332-2.

Madi, A., Fisher, D., Maughan, T. S., Colley, J. P., Meade, A. M., Maynard, J., Humphreys, V., Wasan, H., Adams, R. A., Idziaszczyk, S., Harris, R., Kaplan, R. S., & Cheadle, J. P. (2018). Pharmacogenetic analyses of 2183 patients with advanced colorectal cancer; potential role for common dihydropyrimidine dehydrogenase variants in toxicity to chemotherapy. *European Journal of Cancer, 102*, 31–39. https://doi.org/10.1016/j.ejca.2018.07.009.

Makeen, H. A., Mohan, S., Al-Kasim, M. A., Sultan, M. H., Albarraq, A. A., Ahmed, R. A., Alhazmi, H. A., & Alam, M. I. (2021). Preparation, characterization, and anti-cancer activity of nanostructured lipid carriers containing imatinib. *Pharmaceutics, 13*, 1086. https://doi.org/10.3390/pharmaceutics13071086.

Marques, A. C., Costa, P. J., Velho, S., & Amaral, M. H. (2020). Functionalizing nanoparticles with cancer-targeting antibodies: A comparison of strategies. *Journal of Controlled Release, 320*, 180–200. https://doi.org/10.1016/j.jconrel.2020.01.035.

Matarazzo, A. P., Elisei, L. M. S., Carvalho, F. C., Bonfílio, R., Ruela, A. L. M., Galdino, G., & Pereira, G. R. (2021). Mucoadhesive nanostructured lipid carriers as a cannabidiol nasal delivery system for the treatment of neuropathic pain. *European Journal of Pharmaceutical Sciences, 159*, 105698. https://doi.org/10.1016/j.ejps.2020.105698.

Mehnert, W. (2001). Solid lipid nanoparticles Production, characterization and applications. *Advanced Drug Delivery Reviews, 47*, 165–196. https://doi.org/10.1016/S0169-409X00105-3.

Mi, P. (2020). Stimuli-responsive nanocarriers for drug delivery, tumor imaging, therapy and theranostics. *Theranostics, 10*, 4557–4588. https://doi.org/10.7150/thno.38069.

Miller, K. D., Nogueira, L., Mariotto, A. B., Rowland, J. H., Yabroff, K. R., Alfano, C. M., Jemal, A., Kramer, J. L., & Siegel, R. L. (2019). Cancer treatment and survivorship statistics, 2019. *CA: a Cancer Journal for Clinicians, 69*, 363–385. https://doi.org/10.3322/caac.21565.

Mishra, V., Bansal, K. K., Verma, A., Yadav, N., Thakur, S., Sudhakar, K., & Rosenholm, J. M. (2018). Solid lipid nanoparticles: Emerging colloidal nano drug delivery systems. *Pharmaceutics, 10*, 191. https://doi.org/10.3390/pharmaceutics10040191.

Mitchell, M. J., Billingsley, M. M., Haley, R. M., Wechsler, M. E., Peppas, N. A., & Langer, R. (2021). Engineering precision nanoparticles for drug delivery. *Nature Reviews. Drug Discovery, 20*, 101–124. https://doi.org/10.1038/s41573-020-0090-8.

Moghimipour, E., Rezaei, M., Ramezani, Z., Kouchak, M., Amini, M., Angali, K. A., Dorkoosh, F. A., & Handali, S. (2018a). Folic acid-modified liposomal drug delivery strategy for tumor targeting of 5-fluorouracil. *European Journal of Pharmaceutical Sciences, 114*, 166–174. https://doi.org/10.1016/j.ejps.2017.12.011.

Moghimipour, E., Rezaei, M., Ramezani, Z., Kouchak, M., Amini, M., Angali, K. A., Dorkoosh, F. A., & Handali, S. (2018b). Transferrin targeted liposomal 5-fluorouracil induced apoptosis via mitochondria signaling pathway in cancer cells. *Life Sciences, 194*, 104–110. https://doi.org/10.1016/j.lfs.2017.12.026.

Montanha, M. C., Howarth, A., Mohamed, D. A., Loier, E., Main, L., Rösslein, M., Delmaar, C., Prina-Mello, A., & Siccardi, M. (2022). A physiologically based pharmacokinetic model to predict pegylated liposomal doxorubicin disposition in rats and human. *Drug Delivery and Translational Research, 12*, 2178–2186. https://doi.org/10.1007/s13346-022-01175-w.

Morgan, E., Arnold, M., Gini, A., Lorenzoni, V., Cabasag, C. J., Laversanne, M., Vignat, J., Ferlay, J., Murphy, N., & Bray, F. (2023). Global burden of colorectal cancer in 2020 and 2040: Incidence and mortality estimates from GLOBOCAN. *Gut, 72*, 338–344. https://doi.org/10.1136/gutjnl-2022-327736.

Moulahoum, H., Ghorbanizamani, F., Zihnioglu, F., & Timur, S. (2021). Surface biomodification of liposomes and polymersomes for efficient targeted drug delivery. *Bioconjugate Chemistry, 32*, 1491–1502. https://doi.org/10.1021/acs.bioconjchem.1c00285.

Mu, M., Zhang, Q., Li, J., Zhao, C., Li, X., Chen, Z., Sun, X., & Yu, J. (2023). USP51 facilitates colorectal cancer stemness and chemoresistance by forming a positive feed-forward loop with HIF1A. *Cell Death & Differentiation, 30*, 2393–2407. https://doi.org/10.1038/s41418-023-01228-8.

Müller, R. H., Radtke, M., & Wissing, S. A. (2002). Solid lipid nanoparticles (SLN) and nanostructured lipid carriers (NLC) in cosmetic and dermatological preparations. *Advanced Drug Delivery Reviews, 54*, S131–S155. https://doi.org/10.1016/S0169-409X00118-7.

Mundekkad, D., & Cho, W. C. (2022). Nanoparticles in clinical translation for cancer therapy. *International Journal of Molecular Sciences, 23*, 1685. https://doi.org/10.3390/ijms23031685.

Mura, P., Maestrelli, F., D'Ambrosio, M., Luceri, C., & Cirri, M. (2021). Evaluation and comparison of solid lipid nanoparticles (SLNs) and nanostructured lipid carriers (NLCs) as vectors to develop hydrochlorothiazide effective and safe pediatric oral liquid formulations. *Pharmaceutics, 13*, 437. https://doi.org/10.3390/pharmaceutics13040437.

Murase, Y., Asai, T., Katanasaka, Y., Sugiyama, T., Shimizu, K., Maeda, N., & Oku, N. (2010). A novel DDS strategy, "dual-targeting", and its application for antineovascular therapy. *Cancer Letters, 287*, 165–171. https://doi.org/10.1016/j.canlet.2009.06.008.

Naeem, M., Awan, U. A., Subhan, F., Cao, J., Hlaing, S. P., Lee, J., Im, E., Jung, Y., & Yoo, J.-W. (2020). Advances in colon-targeted nano-drug delivery systems: Challenges and solutions. *Archives of Pharmacal Research, 43*, 153–169. https://doi.org/10.1007/s12272-020-01219-0.

Ngwuluka, N. C., Kotak, D. J., & Devarajan, P. V. (2017). Design and Characterization Of Metformin-loaded Solid Lipid Nanoparticles For Colon Cancer. *AAPS PharmSciTech, 18*, 358–368. https://doi.org/10.1208/s12249-016-0505-3.

Østergaard, J., & Moeller, E. H. (2010). Ghrelin–liposome interactions: Characterization of liposomal formulations of an acylated 28-amino acid peptide using CE. *Electrophoresis, 31*, 339–345. https://doi.org/10.1002/elps.200900394.

Oussoren, C., Eling, W. M. C., Crommelin, D. J. A., Storm, G., & Zuidema, J. (1998). The influence of the route of administration and liposome composition on the potential of liposomes to protect tissue against local toxicity of two antitumor drugs. *Biochimica et Biophysica Acta (BBA) - Biomembranes, 1369*, 159–172. https://doi.org/10.1016/S0005-273600221-6.

Pandey, R., Sharma, S., & Khuller, G. K. (2005). Oral solid lipid nanoparticle-based antitubercular chemotherapy. *Tuberculosis, 85*, 415–420. https://doi.org/10.1016/j.tube.2005.08.009.

Parhi, R., & Suresh, P. (2012). Preparation and characterization of solid lipid nanoparticles-A review. *Current Drug Discovery Technologies, 9*, 2–16. https://doi.org/10.2174/157016312799304552.

Patel, M. N., Lakkadwala, S., Majrad, M. S., Injeti, E. R., Gollmer, S. M., Shah, Z. A., Boddu, S. H. S., & Nesamony, J. (2014). Characterization and evaluation of 5-fluorouracil-loaded solid lipid nanoparticles prepared via a temperature-modulated solidification technique. *AAPS PharmSciTech, 15*, 1498–1508. https://doi.org/10.1208/s12249-014-0168-x.

Petrelli, F., Labianca, R., Zaniboni, A., Lonardi, S., Galli, F., Rulli, E., Rosati, G., Corallo, S., Ronzoni, M., Cardellino, G. G., Mattioli, R., Mambrini, A., Ciuffreda, L., Banzi, M., Pusceddu, V., Maiello, E., Zampino,

M., Zagonel, V., Marchetti, P., ... Sobrero, A. (2020). Assessment of duration and effects of 3 vs 6 months of adjuvant chemotherapy in high-risk stage II colorectal cancer. *JAMA Oncology, 6*, 547. https://doi.org/10.1001/jamaoncol.2019.6486.

Punt, C. J. A., Koopman, M., & Vermeulen, L. (2017). From tumour heterogeneity to advances in precision treatment of colorectal cancer. *Nature Reviews Clinical Oncology, 14*, 235–246. https://doi.org/10.1038/nrclinonc.2016.171.

Rajpoot, K., & Jain, S. K. (2018). Colorectal cancer-targeted delivery of oxaliplatin *via* folic acid-grafted solid lipid nanoparticles: Preparation, optimization, and *in vitro* evaluation. *Artificial Cells, Nanomedicine, and Biotechnology, 46*, 1236–1247. https://doi.org/10.1080/21691401.2017.1366338.

Rajpoot, K., & Jain, S. K. (2019). Irinotecan hydrochloride trihydrate loaded folic acid-tailored solid lipid nanoparticles for targeting colorectal cancer: Development, characterization, and *in vitro* cytotoxicity study using HT-29 cells. *Journal of Microencapsulation, 36*, 659–676. https://doi.org/10.1080/02652048.2019.1665723.

Rajpoot, K., & Jain, S. K. (2020). Oral delivery of pH-responsive alginate microbeads incorporating folic acid-grafted solid lipid nanoparticles exhibits enhanced targeting effect against colorectal cancer: A dual-targeted approach. *International Journal of Biological Macromolecules, 151*, 830–844. https://doi.org/10.1016/j.ijbiomac.2020.02.132.

Reis, F., Madureira, A. R., Nunes, S., Campos, D., Fernandes, J., Marques, C., Zuzarte, M., Gullón, B., Rodríguez-Alcalá, L. M., Calhau, C., Sarmento, B., Gomes, A. M., & Pintado, M. M. (2016). Safety profile of solid lipid nanoparticles loaded with rosmarinic acid for oral use: In vitro and animal approaches. *International Journal of Nanomedicine, 11*, 3621–3640. https://doi.org/10.2147/IJN.S104623.

Riaz, A., Awais, R., & Salem, R. (2014). Side Effects of yttrium-90 radioembolization. *Frontiers in Oncology, 4*. https://doi.org/10.3389/fonc.2014.00198.

Rubab, S., Naeem, K., Rana, I., Khan, N., Afridi, M., Ullah, I., Shah, F. A., Sarwar, S., Din, F.ud, Choi, H.-I., Lee, C.-H., Lim, C.-W., Alamro, A. A., Kim, J.-K., & Zeb, A. (2021). Enhanced neuroprotective and antidepressant activity of curcumin-loaded nanostructured lipid carriers in lipopolysaccharide-induced depression and anxiety rat model. *International Journal of Pharmaceutics, 603*, 120670. https://doi.org/10.1016/j.ijpharm.2021.120670.

Saghafi, Z., Mohammadi, M., Mahboobian, M. M., & Derakhshandeh, K. (2021). Preparation, characterization, and *in vivo* evaluation of perphenazine-loaded nanostructured lipid carriers for oral bioavailability improvement. *Drug Development and Industrial Pharmacy, 47*, 509–520. https://doi.org/10.1080/03639045.2021.1892745.

Sang, R., Stratton, B., Engel, A., & Deng, W. (2021). Liposome technologies towards colorectal cancer therapeutics. *Acta Biomaterialia, 127*, 24–40. https://doi.org/10.1016/j.actbio.2021.03.055.

Sastre, J., Navarro, M., Aranda, E., Fonseca, E., Checa, T., Alonso, M. C., Gallen, M., Camps, C., Antón, A., & Díaz-Rubio, E. (1999). Retrospective evaluation of toxicity in three different schedules of adjuvant chemotherapy for patients with resected colorectal cancer. TTD Spanish Cooperative Group. *Oncology Reports*. https://doi.org/10.3892/or.6.6.1421.

Schlich, M., Palomba, R., Costabile, G., Mizrahy, S., Pannuzzo, M., Peer, D., & Decuzzi, P. (2021). Cytosolic delivery of nucleic acids: The case of ionizable lipid nanoparticles. *Bioengineering & Translational Medicine, 6*. https://doi.org/10.1002/btm2.10213.

Schubert, M. (2003). Solvent injection as a new approach for manufacturing lipid nanoparticles – Evaluation of the method and process parameters. *European Journal of Pharmaceutics and Biopharmaceutics, 55*, 125–131. https://doi.org/10.1016/S0939-641100130-3.

Schuurhuizen, C. S., Verheul, H. M., Braamse, A. M., Buffart, L. M., Bloemendal, H. J., Dekker, J., & Konings, I. (2018). The predictive value of cumulative toxicity for quality of life in patients with metastatic colorectal cancer during first-line palliative chemotherapy. *Cancer Management and Research, 10*, 3015–3021. https://doi.org/10.2147/CMAR.S166468.

Semple, S. C., Akinc, A., Chen, J., Sandhu, A. P., Mui, B. L., Cho, C. K., Sah, D. W. Y., Stebbing, D., Crosley, E. J., Yaworski, E., Hafez, I. M., Dorkin, J. R., Qin, J., Lam, K., Rajeev, K. G., Wong, K. F., Jeffs, L. B., Nechev, L., Eisenhardt, M. L., ... Hope, M. J. (2010). Rational design of cationic lipids for siRNA delivery. *Nature Biotechnology, 28*, 172–176. https://doi.org/10.1038/nbt.1602.

Shaaban, M., Nasr, M., Tawfik, A. A., Fadel, M., & Sammour, O. (2021). Bergamot oil as an integral component of nanostructured lipid carriers and a photosensitizer for photodynamic treatment of vitiligo: Characterization and clinical experimentation. *Expert Opinion on Drug Delivery, 18*, 139–150. https://doi.org/10.1080/17425247.2021.1844180.

Shadambikar, G., Marathe, S., Ji, N., Almutairi, M., Bandari, S., Zhang, F., Chougule, M., & Repka, M. (2021). Formulation development of itraconazole PEGylated nano-lipid carriers for pulmonary aspergillosis using hot-melt extrusion technology. *International Journal of Pharmaceutics: X, 3*, 100074. https://doi.org/10.1016/j.ijpx.2021.100074.

Sharaf, M., Arif, M., Khan, S., Abdalla, M., Shabana, S., Chi, Z., & Liu, C. (2021). Co-delivery of hesperidin and clarithromycin in a nanostructured lipid carrier for the eradication of Helicobacter pylori in vitro. *Bioorganic Chemistry, 112*, 104896. https://doi.org/10.1016/j.bioorg.2021.104896.

Sharifabad, M. E., Mercer, T., & Sen, T. (2016). Drug-loaded liposome-capped mesoporous core–shell magnetic nanoparticles for cellular toxicity study. *Nanomedicine: Nanotechnology, Biology, and Medicine*nnm-2016-0248. https://doi.org/10.2217/nnm-2016-0248.

Sharma, T., Katare, O. P., Jain, A., Jain, S., Chaudhari, D., Borges, B., & Singh, B. (2021). QbD-Steered Development of biotin-conjugated nanostructured lipid carriers for oral delivery of chrysin: Role of surface modification for improving biopharmaceutical performance. *Colloids and Surfaces B: Biointerfaces, 197*, 111429. https://doi.org/10.1016/j.colsurfb.2020.111429.

Shepherd, S. J., Warzecha, C. C., Yadavali, S., El-Mayta, R., Alameh, M.-G., Wang, L., Weissman, D., Wilson, J. M., Issadore, D., & Mitchell, M. J. (2021). Scalable mRNA and siRNA lipid nanoparticle production using a parallelized microfluidic device. *Nano Letters, 21*, 5671–5680. https://doi.org/10.1021/acs.nanolett.1c01353.

Shia, J., Klimstra, D. S., Nitzkorski, J. R., Low, P. S., Gonen, M., Landmann, R., Weiser, M. R., Franklin, W. A., Prendergast, F. G., Murphy, L., Tang, L. H., Temple, L., Guillem, J. G., Wong, W. D., & Paty, P. B. (2008). Immunohistochemical expression of folate receptor α in colorectal carcinoma: Patterns and biological significance. *Human Pathology, 39*, 498–505. https://doi.org/10.1016/j.humpath.2007.09.013.

Siegel, R. L., Wagle, N. S., Cercek, A., Smith, R. A., & Jemal, A. (2023). Colorectal cancer statistics, 2023. *CA: A Cancer Journal for Clinicians, 73*, 233–254. https://doi.org/10.3322/caac.21772.

Singh, A., Neupane, Y. R., Mangla, B., & Kohli, K. (2019). Nanostructured lipid carriers for oral bioavailability enhancement of exemestane: Formulation design, in vitro, Ex Vivo, and in vivo studies. *Journal of Pharmaceutical Sciences, 108*, 3382–3395. https://doi.org/10.1016/j.xphs.2019.06.003.

Šišić, I., Pojskić, B., Mekić-Abazović, A., & Kovčin, V. (2015). Therapeutic efficacy and toxicity of bolus application of chemotherapy protocol in the treatment of metastatic colorectal cancer. *Medicinski Glasnik, 13*. https://doi.org/10.17392/802-15.

Smyth Templeton, N. (2002). Cationic liposome-mediated gene delivery *In vivo*. *Bioscience Reports, 22*, 283–295. https://doi.org/10.1023/A:1020142823595.

Subramaniam, B., Arshad, N. M., Malagobadan, S., Misran, M., Nyamathulla, S., Mun, K. S., & Nagoor, N. H. (2021). Development and evaluation of 1′-acetoxychavicol acetate (ACA)-loaded nanostructured lipid carriers for prostate cancer therapy. *Pharmaceutics, 13*, 439. https://doi.org/10.3390/pharmaceutics13040439.

Szymańska, E., & Winnicka, K. (2015). Stability of chitosan—a challenge for pharmaceutical and biomedical applications. *Marine Drugs, 13*, 1819–1846. https://doi.org/10.3390/md13041819.

Takakura, Y., & Takahashi, Y. (2022). Strategies for persistent retention of macromolecules and nanoparticles in the blood circulation. *Journal of Controlled Release, 350*, 486–493. https://doi.org/10.1016/j.jconrel.2022.05.063.

Tamjidi, F., Shahedi, M., Varshosaz, J., & Nasirpour, A. (2013). Nanostructured lipid carriers (NLC): A potential delivery system for bioactive food molecules. *Innovative Food Science & Emerging Technologies, 19*, 29–43. https://doi.org/10.1016/j.ifset.2013.03.002.

Tang, Y., Weng, X., Liu, C., Li, X., & Chen, C. (2021). Hypoxia enhances activity and malignant behaviors of colorectal cancer cells through the STAT3/MicroRNA-19a/PTEN/PI3K/AKT Axis. *Analytical Cellular Pathology, 2021*, 1–19. https://doi.org/10.1155/2021/4132488.

Thi, T. T. H., Suys, E. J. A., Lee, J. S., Nguyen, D. H., Park, K. D., & Truong, N. P. (2021). Lipid-based nanoparticles in the clinic and clinical trials: From cancer nanomedicine to COVID-19 vaccines. *Vaccines, 9*, 359. https://doi.org/10.3390/vaccines9040359.

Tiernan, J. P., Perry, S. L., Verghese, E. T., West, N. P., Yeluri, S., Jayne, D. G., & Hughes, T. A. (2013). Carcinoembryonic antigen is the preferred biomarker for in vivo colorectal cancer targeting. *British Journal of Cancer, 108*, 662–667. https://doi.org/10.1038/bjc.2012.605.

Tiwari, A., Saraf, S., Jain, A., Panda, P. K., Verma, A., & Jain, S. K. (2020). Basics to advances in nanotherapy of colorectal cancer. *Drug Delivery and Translational Research, 10*, 319–338. https://doi.org/10.1007/s13346-019-00680-9.

Tran, P. H. L., Wang, T., Yin, W., Tran, T. T. D., Nguyen, T. N. G., Lee, B.-J., & Duan, W. (2019). Aspirin-loaded nanoexosomes as cancer therapeutics. *International Journal of Pharmaceutics, 572*, 118786. https://doi.org/10.1016/j.ijpharm.2019.118786.

Tummala, S., Gowthamarajan, K., Satish Kumar, M. N., & Wadhwani, A. (2016). Oxaliplatin immuno hybrid nanoparticles for active targeting: An approach for enhanced apoptotic activity and drug delivery to colorectal tumors. *Drug Delivery, 23*, 1773–1787. https://doi.org/10.3109/10717544.2015.1084400.

Ucar, E., Teksoz, S., Ichedef, C., Kilcar, A. Y., Medine, E. I., Ari, K., Parlak, Y., Sayit Bilgin, B. E., & Unak, P. (2017). Synthesis, characterization and radiolabeling of folic acid modified nanostructured lipid carriers as a contrast agent and drug delivery system. *Applied Radiation and Isotopes, 119*, 72–79. https://doi.org/10.1016/j.apradiso.2016.11.002.

Varshosaz, J., Hassanzadeh, F., Sadeghi, H., & Khadem, M. (2012). Galactosylated nanostructured lipid carriers for delivery of 5-FU to hepatocellular carcinoma. *Journal of Liposome Research, 22*, 224–236. https://doi.org/10.3109/08982104.2012.662653.

Vila, A., Gill, H., McCallion, O., & Alonso, M. J. (2004). Transport of PLA-PEG particles across the nasal mucosa: Effect of particle size and PEG coating density. *Journal of Controlled Release, 98*, 231–244. https://doi.org/10.1016/j.jconrel.2004.04.026.

Wang, B., Sun, L., Wen, M., Tan, Y., Almalki, W. H., Katouah, H., Kazmi, I., Afzal, O., Altamimi, A. S. A., Al-Abbasi, F. A., Alrobaian, M., Alharbi, K. S., Alenezi, S. K., Alghaith, A. F., Beg, S., & Rahman, M. (2021). Nano lipidic carriers for codelivery of sorafenib and ganoderic acid for enhanced synergistic antitumor efficacy against hepatocellular carcinoma. *Saudi Pharmaceutical Journal, 29*, 843–856. https://doi.org/10.1016/j.jsps.2021.06.006.

Wang, G., Yu, Y., Wang, Y., Wang, J., Guan, R., Sun, Y., Shi, F., Gao, J., & Fu, X. (2019). Role of SCFAs in gut microbiome and glycolysis for colorectal cancer therapy. *Journal of Cellular Physiology, 234*, 17023–17049. https://doi.org/10.1002/jcp.28436.

Wang, H., Zhao, P., Su, W., Wang, S., Liao, Z., Niu, R., & Chang, J. (2010). PLGA/polymeric liposome for targeted drug and gene co-delivery. *Biomaterials, 31*, 8741–8748. https://doi.org/10.1016/j.biomaterials.2010.07.082.

Wang, J., Li, Y., & Nie, G. (2021). Multifunctional biomolecule nanostructures for cancer therapy. *Nature Reviews Materials, 6*, 766–783. https://doi.org/10.1038/s41578-021-00315-x.

Wang, L., Yang, C., & Wang, J. (2011). Assemble of magnetic nanoparticles into the structure of cisplatin liposome. *Yao Xue Xue Bao=Acta Pharmaceutica Sinica, 46*, 592–598.

Wang, L.-L., Feng, C.-L., Zheng, W.-S., Huang, S., Zhang, W.-X., Wu, H.-N., Zhan, Y., Han, Y.-X., Wu, S., & Jiang, J.-D. (2017). Tumor-selective lipopolyplex encapsulated small active RNA hampers colorectal cancer growth in vitro and in orthotopic murine. *Biomaterials, 141*, 13–28. https://doi.org/10.1016/j.biomaterials.2017.06.029.

Wang, T., Liu, Z., Zhang, Z., Tang, S., Yue, M., Feng, S., Hu, M., Xuan, L., & Chen, Y. (2017). Evaluation of antitumor activity of survivin short interfering RNA delivered by lipid nanoparticles in colon cancer in vitro and in vivo. *Oncology Letters, 14*, 2001–2008. https://doi.org/10.3892/ol.2017.6404.

Weng, J., Li, S., Zhu, Z., Liu, Q., Zhang, R., Yang, Y., & Li, X. (2022). Exploring immunotherapy in colorectal cancer. *Journal of Hematology & Oncology, 15*, 95. https://doi.org/10.1186/s13045-022-01294-4.

Wong, H. L., Bendayan, R., Rauth, A. M., & Wu, X. Y. (2004). Development of solid lipid nanoparticles containing ionically complexed chemotherapeutic drugs and chemosensitizers. *Journal of Pharmaceutical Sciences, 93*, 1993–2008. https://doi.org/10.1002/jps.20100.

Wong, H., Bendayan, R., Rauth, A., Li, Y., & Wu, X. (2007). Chemotherapy with anticancer drugs encapsulated in solid lipid nanoparticles☆ *Advanced Drug Delivery Reviews, 59*, 491–504. https://doi.org/10.1016/j.addr.2007.04.008.

Xing, J., Zhang, X., Wang, Z., Zhang, H., Chen, P., Zhou, G., Sun, C., Gu, N., & Ji, M. (2019). Novel lipophilic SN38 prodrug forming stable liposomes for colorectal carcinoma therapy. *International Journal of Nanomedicine, 14*, 5201–5213. https://doi.org/10.2147/IJN.S204965.

Xu, J., Tam, M., Samaei, S., Lerouge, S., Barralet, J., Stevenson, M. M., & Cerruti, M. (2017). Mucoadhesive chitosan hydrogels as rectal drug delivery vessels to treat ulcerative colitis. *Acta Biomaterialia, 48*, 247–257. https://doi.org/10.1016/j.actbio.2016.10.026.

Yang, C., & Merlin, D. (2020). Lipid-based drug delivery nanoplatforms for colorectal cancer therapy. *Nanomaterials, 10*, 1424. https://doi.org/10.3390/nano10071424.

Yang, C., Zhang, M., Sung, J., Wang, L., Jung, Y., & Merlin, D. (2020). Autologous exosome transfer: A new personalised treatment concept to prevent colitis in a murine model. *Journal of Crohn's and Colitis, 14*, 841–855. https://doi.org/10.1093/ecco-jcc/jjz184.

Ye, Q., Li, J., Li, T., Ruan, J., Wang, H., Wang, F., & Zhang, X. (2021). Development and evaluation of puerarin-loaded controlled release nanostructured lipid carries by central composite design. *Drug Development and Industrial Pharmacy, 47*, 113–125. https://doi.org/10.1080/03639045.2020.1862170.

Yingchoncharoen, P., Kalinowski, D. S., & Richardson, D. R. (2016). Lipid-based drug delivery systems in cancer therapy: What is available and what is yet to come. *Pharmacological Reviews, 68*, 701–787. https://doi.org/10.1124/pr.115.012070.

Yoozbashi, M., Rashidzadeh, H., Kermanian, M., Sadighian, S., Hosseini, M.-J., Kaboli, Z., & Rostamizadeh, K. (2022). Magnetic nanostructured lipid carrier for dual triggered curcumin delivery: Preparation, characterization and toxicity evaluation on isolated rat liver mitochondria. *Journal of Biomaterials Applications, 36*, 1055–1063. https://doi.org/10.1177/08853282211034625.

Yu, G., Ali, Z., Sajjad Khan, A., Ullah, K., Jamshaid, H., Zeb, A., Imran, M., Sarwar, S., Choi, H.-G., & Din, F.ud (2021). Preparation, pharmacokinetics, and antitumor potential of miltefosine-loaded nanostructured lipid carriers. *International Journal of Nanomedicine, 16*, 3255–3273. https://doi.org/10.2147/IJN.S299443.

Yuan, H., Chen, C.-Y., Chai, G., Du, Y.-Z., & Hu, F.-Q. (2013). Improved transport and absorption through gastrointestinal tract by PEGylated solid lipid nanoparticles. *Molecular Pharmaceutics, 10*, 1865–1873. https://doi.org/10.1021/mp300649z.

Zappavigna, S., Abate, M., Cossu, A. M., Lusa, S., Campani, V., Scotti, L., Luce, A., Yousif, A. M., Merlino, F., Grieco, P., De Rosa, G., & Caraglia, M. (2019). Urotensin-II-targeted liposomes as a new drug delivery system towards prostate and colon cancer cells. *Journal of Oncology, 2019*, 1–14. https://doi.org/10.1155/2019/9293560.

Zhang, C., Wang, W., & Zhang, D. (2018). Association between dietary inflammation index and the risk of colorectal cancer: A meta-analysis. *Nutrition and Cancer, 70*, 14–22. https://doi.org/10.1080/01635581.2017.1374418.

Zhang, D., Shah, P. K., Culver, H. R., David, S. N., Stansbury, J. W., Yin, X., & Bowman, C. N. (2019). Photo-responsive liposomes composed of spiropyran-containing triazole-phosphatidylcholine: Investigation of merocyanine-stacking effects on liposome–fiber assembly-transition. *Soft Matter, 15*, 3740–3750. https://doi.org/10.1039/C8SM02181C.

Zhang, J., Ali, K., & Wang, J. (2024). Research advances of lipid nanoparticles in the treatment of colorectal cancer. *International Journal of Nanomedicine, 19*, 6693–6715. https://doi.org/10.2147/IJN.S466490.

Zhang, M., Viennois, E., Prasad, M., Zhang, Y., Wang, L., Zhang, Z., Han, M. K., Xiao, B., Xu, C., Srinivasan, S., & Merlin, D. (2016). Edible ginger-derived nanoparticles: A novel therapeutic approach for the prevention and treatment of inflammatory bowel disease and colitis-associated cancer. *Biomaterials, 101*, 321–340. https://doi.org/10.1016/j.biomaterials.2016.06.018.

Zhang, M., Xiao, B., Wang, H., Han, M. K., Zhang, Z., Viennois, E., Xu, C., & Merlin, D. (2016). Edible ginger-derived nano-lipids loaded with doxorubicin as a novel drug-delivery approach for colon cancer therapy. *Molecular Therapy, 24*, 1783–1796. https://doi.org/10.1038/mt.2016.159.

Zhang, T., Zhu, G., Lu, B., & Peng, Q. (2017). Oral nano-delivery systems for colon targeting therapy. *Pharmaceutical Nanotechnology, 5*. https://doi.org/10.2174/2211738505666170424122722.

Zhang, Y., Chen, Z., & Li, J. (2017). The current status of treatment for colorectal cancer in China. *Medicine, 96*, e8242. https://doi.org/10.1097/MD.0000000000008242.

Zhang, Z., Lu, Y., Qi, J., & Wu, W. (2021). An update on oral drug delivery via intestinal lymphatic transport. *Acta Pharmaceutica Sinica B, 11*, 2449–2468. https://doi.org/10.1016/j.apsb.2020.12.022.

Zhao, C., Feng, Q., Dou, Z., Yuan, W., Sui, C., Zhang, X., Xia, G., Sun, H., & Ma, J. (2013). Local targeted therapy of liver metastasis from colon cancer by galactosylated liposome encapsulated with doxorubicin. *PLoS One, 8*, e73860. https://doi.org/10.1371/journal.pone.0073860.

Zhu, D., Wang, Z., Zong, S., Zhang, Y., Chen, C., Zhang, R., Yun, B., & Cui, Y. (2018). Investigating the intracellular behaviors of liposomal nanohybrids *via* SERS: Insights into the influence of metal nanoparticles. *Theranostics, 8*, 941–954. https://doi.org/10.7150/thno.21173.

Zur Mühlen, A., Schwarz, C., & Mehnert, W. (1998). Solid lipid nanoparticles (SLN) for controlled drug delivery – Drug release and release mechanism. *European Journal of Pharmaceutics and Biopharmaceutics, 45*, 149–155. https://doi.org/10.1016/S0939-641100150-1.

CHAPTER

# 4 Liposomal nanomedicine: applications for drug delivery in cancer therapy

**Manish Kumar[1,2], Debayan Sil[3], Shruti Srivatava[4], Uma Shanker Maurya[5] and Subhendu Chakroborty[6]**

[1]*Department of Pharmaceutics, ISF College of Pharmacy, Moga, Punjab, India,*
[2]*KIET School of Pharmacy, Krishna Institute of Engineering & Technology (KIET), Ghaziabad, Uttar Pradesh, India,*
[3]*Department of Quality Assurance, ISF College of Pharmacy, Moga, Punjab, India,*
[4]*Directorate of Research, Innovation & Consultancy, Chandigarh University, Unnao, Uttar Pradesh, India,*
[5]*Goel Institute of Pharmacy and Sciences, Lucknow, Uttar Pradesh, India,*
[6]*Department of Chemistry, School of Basic Sciences, Chandigarh University, Unnao, Uttar Pradesh, India*

## 4.1 Introduction

Chemotherapy, although a primary approach for cancer treatment, is often limited when using free antitumor drugs due to insufficient sensitivity and specificity. Most currently used anticancer drugs are poorly water-soluble and highly potent, but their disadvantage of low selectivity exerts cytotoxic effects on both tumor and healthy cells. This nonspecific action not only reduces therapeutic efficacy but also results in significant systemic toxicity due to compromised targeted drug accumulation at tumor sites (Rommasi & Esfandiari, 2021). Further, cancer cells are complicated at both genetic and phenotypic levels, resulting in clinical diversity and treatment resistance. Approximately 90% of anticancer drugs have low bioavailability and poor pharmacokinetics. Overall, chemotherapeutic drugs have limited efficacy owing to their rapid metabolism, severe toxic effects, and resistance development. To mitigate these drawbacks, researchers have become increasingly focused on creating selective or targeted cancer drug delivery techniques that can improve the drug bioavailability, increase drug accumulation at the specific tumor sites, minimize side effects on normal tissues, and overall improve the treatment outcomes (Chen et al., 2024; Fulton & Najahi-Missaoui, 2023). Among the several types of nanocarrier systems, including nanoparticles, dendrimers, micelles, nanoconjugates, etc., liposomes have emerged as highly promising nanocarriers for tumor-targeted drug delivery in recent decades. The therapeutic efficacy of liposomes in tumor-targeted drug delivery can be attributed to their distinct physicochemical properties, including a high loading capacity for both hydrophilic and hydrophobic drug molecules, as well as their structural adaptability for surface modification, functionalization, and compositional optimization. Liposomes can be described as biocompatible and biodegradable nanosized carriers with an aqueous core surrounded by lipid bilayers of either natural or synthetic origin (Chen et al., 2024). Liposomes offer multiple advantages, such as enhanced pharmacokinetic profiles of the encapsulated therapeutics, prolonged systemic

Nano Drug Delivery Systems for Colon Cancer. DOI: https://doi.org/10.1016/B978-0-443-40443-6.00006-4

circulation, and passive accumulation at tumor and inflamed sites through the enhanced permeability and retention (EPR) effect. To effectively transport bioactives and other drugs to tumor cells, liposomes must be carefully engineered to overcome physiological and anatomical obstacles (Ortíz et al., 2021). These advancements provide insights into difficulties and potential solutions for tumor targeting with modified liposomes. Advanced strategies involve the conjugation of targeting ligands—such as antibodies, peptides, or proteins—onto the surface of liposomes, enabling selective binding to tumor-specific receptors. This targeted approach enhances cellular uptake and drug accumulation at the tumor site. When combined with stimuli-responsive features, such modifications further increase therapeutic precision while significantly reducing off-target effects and systemic toxicity (Moghimipour & Handali, 2024). This chapter provides an overview of the current advancements in liposome-based drug delivery systems for cancer therapy, emphasizing the design principles, functionalization strategies, and challenges associated with achieving effective tumor targeting and enhanced therapeutic outcomes.

## 4.2 Mechanism of liposome formation

A lot of effort is being put into defining liposomal nanocarriers and reasonably determining their characteristics. Fundamentally, liposomes can be characterized by their formation mechanism, which involves the self-assembly of amphiphilic phospholipids in aqueous environments, primarily driven by hydrophobic interactions among the lipid molecules (Fig. 4.1). Each phospholipid molecule is composed of a hydrophilic (water-attracting) phosphate head and two hydrophobic (water-repelling) fatty acid tails. In an aqueous environment, these molecules spontaneously organize themselves into closed spherical structures to shield their hydrophobic tails from water, creating a hydrophilic core (aqueous cavity) for encapsulating water-soluble drugs and hydrophobic bilayer concentric membranes for incorporating lipid-soluble drugs. The size of liposomal structure can range considerably, from around 10 to 2500 nm. Nonetheless, liposomes are usually designed to be between 50 and 450 nm in size for application in drug delivery since this size range maximizes their cellular absorption, tumor permeability, and circulation duration.

## 4.3 Structure and composition of liposomes

Structurally, liposomes are classified into four types based on size and phospholipid bilayer number: small unilamellar vesicles, large unilamellar vesicles, multilamellar vesicles, and multivesicular vesicles (Giuliano et al., 2021). Unilamellar liposomal structures possess a single phospholipid bilayer, while multilamellar liposomes are characterized by multiple concentric phospholipid bilayers. Multivesicular vesicles create a multilamellar configuration with concentric phospholipid spheres, wherein numerous unilamellar vesicles develop inside larger liposomes. For hydrophilic drugs, encapsulation efficiency typically rises with an increase in vesicle size and declines with the number of surrounding bilayers (Giuliano et al., 2021; Rommasi & Esfandiari, 2021). Furthermore, the number of bilayers and vesicle size has a significant influence on drug-loading capacity and circulation time.

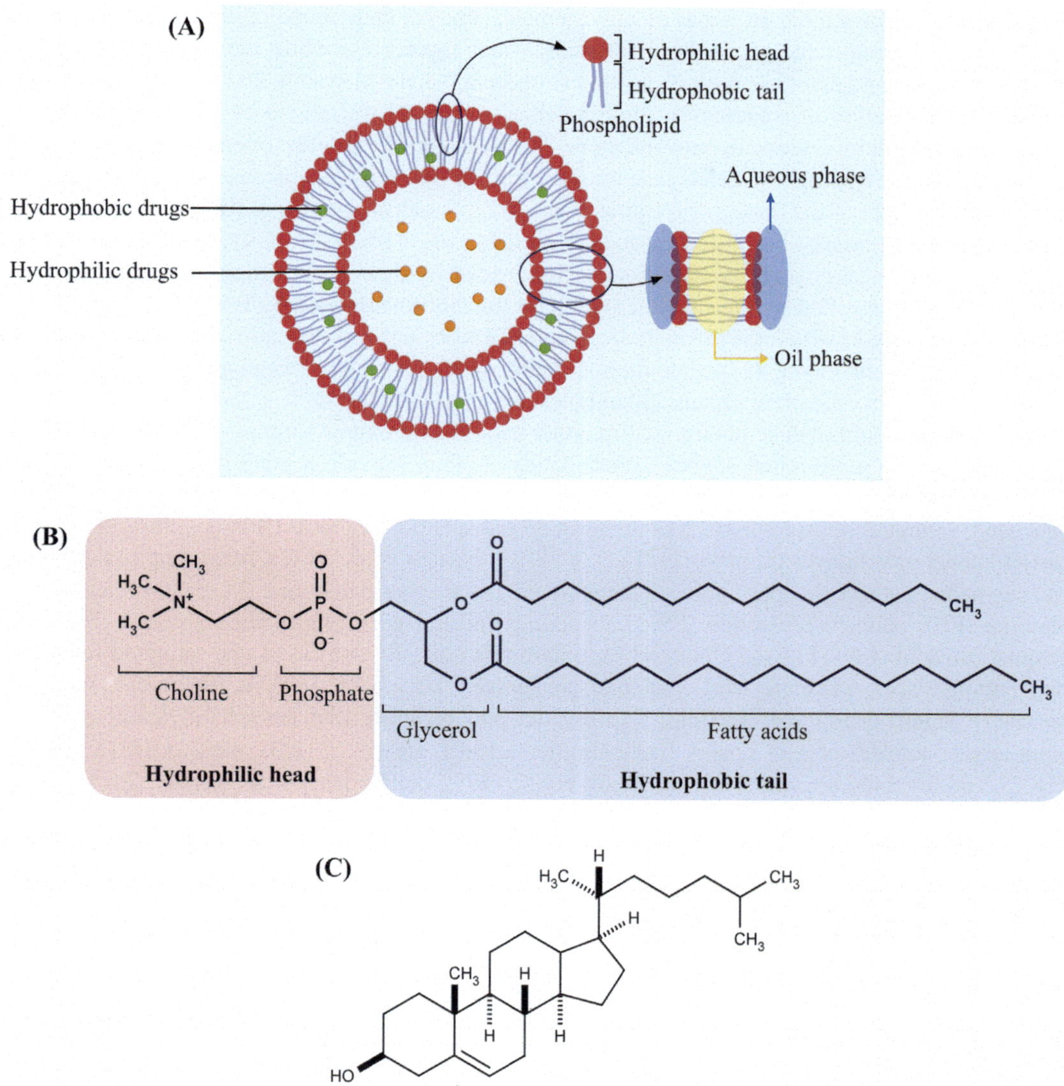


**FIGURE 4.1**

(A) Diagrammatic representation of liposomes and their various components, including hydrophilic core and hydrophobic bilayer, (B) structure of phospholipid, and (C) cholesterol.

## 4.3.1 Phospholipids

Phospholipids, which serve as the fundamental building blocks of liposomes, belong to the class of amphiphilic lipids structurally similar to triglycerides but distinguished by their hydrophilic head (phosphate molecule) and two hydrophobic acyl chains (Fig. 4.1). Phospholipid molecules spontaneously self-assemble in aqueous environments due to their amphiphilic nature. The polar phosphate head groups interact with surrounding water molecules, while the hydrophobic diacyl chains repel water and associate with each other (Fulton & Najahi-Missaoui, 2023). This organization results in the formation of spherical or multilamellar vesicles characterized by a bilayer membrane that develops a lipophilic space in between them. The hydrophobic tails are oriented inward, shielded from water, and the hydrophilic heads face outward toward the external aqueous environment and the internal aqueous core of the liposome (Nsairat et al., 2022; Papahadjopoulos & Kimelberg, 1974). The fundamental characteristics of liposomes, such as particle size, elasticity, rigidity, flexibility, mobility, stability, and electrical charge, are significantly influenced by their phospholipid makeup. Phospholipids can be distinguished by their distinctive phase transition temperature ($T_c$), which refers to the level of temperature at which they change from a gel-like crystallized structure to a liquid crystalline structure. Factors influencing $T_c$ include their aliphatic chain length, polar head group, and the degree of unsaturation in hydrocarbon chains (Sheikholeslami et al., 2022).

These phospholipids can be obtained from either natural or synthetic sources. For example, natural phospholipids, like unsaturated soybean phosphatidylcholine, provide a great permeability but low stability to liposomes, primarily due to the unsaturated properties of the hydrocarbon segment. On the other hand, saturated phospholipids, like dipalmitoyl phosphatidylcholine (DPPC), exhibit robust and nearly impenetrable bilayer frameworks (Li et al., 2019; Nsairat et al., 2022). According to their polar head segments, natural phospholipids are categorized as phosphatidylcholine (PC), phosphatidylethanolamine (PE), phosphatidylserine (PS), phosphatidylinositol (PI), phosphatidylglycerol (PG), and phosphatidic acid (PA) (Fig. 4.2). Natural phospholipids typically consist of one saturated fatty acid (e.g., palmitic acid, margaric acid) and one unsaturated fatty acid, such as oleic acid, which is commonly found in egg yolk lecithin. Egg-derived phospholipids and phosphatidylcholines (PCs) commonly contain fatty acid chains like palmitic (C16:0), stearic (C18:0), oleic (C18:1), linoleic (C18:2), and arachidonic acid (C20:4) (Snook et al., 1999; van Hoogevest & Wendel, 2014).

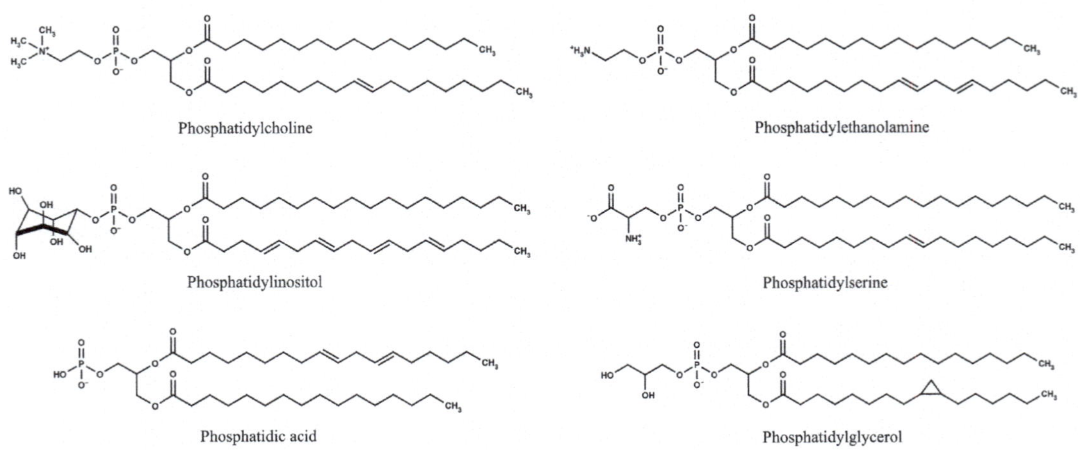


**FIGURE 4.2**

Representative chemical structures of major naturally occurring phospholipids commonly used in liposomal formulations.

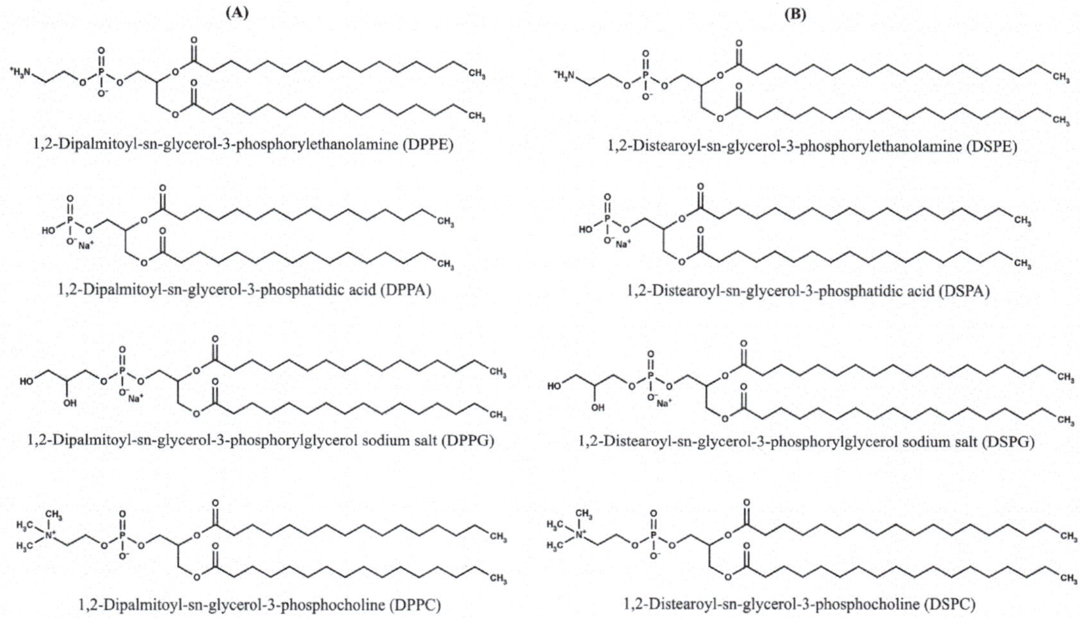


**FIGURE 4.3**

Chemical structures of synthetic saturated phospholipids commonly used in liposomal formulations. (A) Dipalmitoyl-based phospholipid analogs and (B) distearoyl-based phospholipid analogs.

The polar head or nonpolar tail components of natural phospholipids may be chemically altered to create synthetic phospholipids, which enable the production of a large range of distinct and adaptable phospholipids. Although mixed or unsaturated fatty acids in either or both hydrocarbon chains can also be used to create saturated synthetic phospholipids, the majority of them are made from stearic or palmitic fatty acids (Fig. 4.3). These changes provide phospholipids with extensive control over their physical and chemical characteristics, making them critical for modifying liposome features for specified applications (Drescher & van Hoogevest, 2020; van Hoogevest & Wendel, 2014).

## 4.3.2 Steroids

Another very crucial component in liposomal formulations that has a significant potential in controlling the structural integrity is the steroids. Steroids are hydrophobic lipids characterized by a core structure of four fused rings, with various functional groups attached to these rings, contributing to their structural diversity (Fig. 4.1). The most commonly used steroids for liposome formulation are cholesterol, which strengthens liposome structure and stability by incorporating into the phospholipid bilayer (Nsairat et al., 2024). The concentration of cholesterol used in the development of liposomes is a critical parameter that determines the structural integrity of the designed nanocarriers (Ercole et al., 2015). A comprehensive study by various researchers has demonstrated its function as a promising stabilizing agent, with several key effects: (1) promoting the packing density of lipid molecules within the bilayer,

(2) lowering membrane penetrability to both electrolytic and nonelectrolytic compounds, (3) improving vesicular stability by preventing aggregation, (4) modifying membrane fluidity and mobility and therefore promoting rigidity and resistance against mechanical stresses, and (5) potentially increasing drug encapsulation efficiency and drug retention (Ahmed et al., 2019; Krause & Regen, 2014). Cholesterol typically comprises up to 30% of the total lipid content in liposomal formulations, with commonly used lipid-to-cholesterol ratios being 2:1 or 1:1, depending on the desired membrane stability and fluidity (Fulton & Najahi-Missaoui, 2023).

### 4.3.3 Formulation techniques for liposomal systems

Over time, various methods have been employed for liposome preparation, typically involving phospholipid dissolution in organic solvents followed by solvent removal to form liposomes. Drug-loading strategies differ based on the drug's hydrophilicity and the scale of production. Hydrophobic drugs like paclitaxel and docetaxel are efficiently loaded using the lipid film hydration method. Hydrophilic drugs are encapsulated via passive or active (remote) loading; however, passive loading often results in low efficiency due to drug loss into the external aqueous phase (Has & Sunthar, 2020; Lombardo & Kiselev, 2022). Table 4.1 enlists some commonly used liposome preparation techniques.

## 4.4 Liposome-based drug delivery systems

Liposome-based drug delivery approaches incorporate and transport both hydrophilic and hydrophobic pharmaceutical agents using spherical vesicles made of phospholipid bilayers. The distinctive configuration of liposomes provides adaptability for a variety of therapeutic agents by enabling hydrophilic compounds to be contained in the aqueous core and hydrophobic compounds to be integrated within the lipid bilayer (Kim & Jeong, 2021; Sercombe et al., 2015). Liposomes facilitate the optimal biodistribution and targeted drug administration, decrease systemic adverse effects, and increase the stability of incorporated compounds by shielding them from degradation and inactivation. In order to increase efficacy and decrease off-target toxic effects, surface modification using polymers (like polyethylene glycol [PEG]) or targeting ligands (antibodies, peptides, or carbohydrates) might further improve circulation duration and promote site-specific delivery of drugs (Sercombe et al., 2015).

## 4.5 Strategies for drug loading into liposomes

Effective drug loading is a fundamental component of liposome-based drug delivery platforms, with direct implications for therapeutic efficacy, release patterns, and clinical outcomes. The pharmacokinetic and pharmacodynamic characteristics of a particular drug, as well as the stability and encapsulation effectiveness of the liposomal formulation, are all influenced by the intended drug-loading technique. Drugs can be loaded into liposomes using two primary strategies: passive and active (remote) loading (Fulton & Najahi-Missaoui, 2023).

A most prevalent technique for incorporating drugs into liposomes is passive loading, in which the medication is added while the liposome is forming. In the case of hydrophilic compounds, this usually

**Table 4.1 Overview of common liposome preparation methods.**

| Preparation method | Process |
|---|---|
| Thin-film hydration | This method involves dissolving phospholipids in an organic solvent in a round-bottom flask, evaporating the solvent to form a thin lipid film, and then hydrating the film with an aqueous buffer to produce liposomes. The resulting liposomes are typically heterogeneous in size, but this can be improved by sonication or extrusion through polycarbonate membranes (Ahmed et al., 2019; Fulton & Najahi-Missaoui, 2023). |
| Reverse phase evaporation | This method involves creating a water-in-oil emulsion by mixing an aqueous drug solution with a lipid-containing organic phase. The organic solvent is then evaporated under reduced pressure, leading to the formation of liposomes suspended in the aqueous phase. This technique is particularly effective for improving the encapsulation efficiency of both hydrophilic and hydrophobic drugs (Minocha & Kumar, 2022). |
| Ethanol injection method | This method involves dissolving phospholipids in ethanol and rapidly injecting this ethanolic solution into a stirred aqueous phase. As ethanol diffuses into the water, the phospholipids precipitate and self-assemble into bilayer fragments, which then fuse to form liposomal vesicles. After formation, ethanol is removed—typically by evaporation or dialysis—to yield the final liposome suspension (Gouda et al., 2021). |
| Dehydration-rehydration | This method involves first preparing small unilamellar liposomes, which are then lyophilized (freeze-dried) in the presence of lyoprotectants. Upon rehydration with an aqueous solution containing the compound of interest, the liposomes reform and passively encapsulate the compound. This technique is effective for improving drug encapsulation efficiency (Ahmed et al., 2019). |
| Freeze-thaw method | This method involves subjecting preformed liposome suspensions to repeated cycles of freezing (typically at −80°C or in liquid nitrogen) and thawing (at temperatures above the lipid phase transition, such as 40°C–65°C). These cycles disrupt and fuse the lipid bilayers, promoting the encapsulation of solutes, and enhancing entrapment efficiency (Chen et al., 2010; Lian & Ho, 2001). |
| Microfluidic techniques | This method involves injecting a lipid solution through a central channel of a microfluidic device while an aqueous solution flows alongside it. As the aqueous phase replaces the organic solvent, lipids self-assemble into liposomes. By adjusting the flow rates, this method allows precise control over liposome size and uniformity, making it suitable for encapsulating both hydrophilic and hydrophobic drugs (Carugo et al., 2016; Has & Sunthar, 2020). |

consists of encapsulating the drug within the aqueous core by hydrating a dry lipid film using an aqueous drug solution (Fulton & Najahi-Missaoui, 2023; Gubernator, 2011). In the case of lipophilic compounds, the drug is often codissolved in the lipid constituents with organic solvents before film formation, resulting in incorporation into the lipid bilayer. However, passive loading frequently encounters low encapsulation efficiency, particularly for hydrophilic compounds, because they have limited capacity to partition into the developing vesicles. Furthermore, this approach's usefulness may be limited for specific therapeutic agents due to its strong dependence on the drug's solubility, molecular weight, and partitioning properties (Liu et al., 2022; Pande, 2023).

One method for improving the encapsulation efficiency of hydrophilic medications in liposomes is active or remote loading. This technique utilizes transmembrane gradients, including ion or pH gradients, which are incubated with drugs to enable easier drug loading following liposome synthesis. Optimal encapsulation efficiency and minimal drug loss are achieved when the gradient pushes the drug over the membrane, where it is ionized and encapsulated inside the liposome (Nambiar et al., 2024). When uncharged drug molecules approach the acidic or ion-rich aqueous center, they either develop into ion pairs or become protonated after diffusing over the liposomal lipid bilayer. Drugs with suitable pKa values and partition coefficients, which promote successful transmembrane transport and consequent retention, respond particularly well to this strategy (Liu et al., 2022). A well-established example of active loading is the FDA-approved liposomal formulation Doxil, in which doxorubicin is encapsulated using a transmembrane pH gradient created by ammonium sulfate, enabling efficient drug loading and sustained release. This approach results in stable encapsulation, prolonged drug release, and a longer plasma half-life, along with limiting cardiotoxicity (Barenholz, 2012).

## 4.6 Pharmacokinetic and pharmacodynamic profiles of liposomal drug formulations

To comprehend the medicinal and therapeutic potential as well as the clinical application of liposomes as nanocarriers for drug delivery, it is essential to evaluate their pharmacological characteristics. It is necessary to assess liposomal formulations using pharmacokinetic (PK) and pharmacodynamic (PD) research (Allen et al., 2006). Whereas PD studies drug actions and effects at the cellular and molecular level, PK focuses on drug absorption, distribution, metabolism, and excretion. The primary objective of liposomes in cancer treatments is to increase therapeutic effectiveness while inhibiting toxicity to healthy tissues. Pharmacokinetic studies have shown that liposomes prolong systemic circulation, enhance drug bioavailability, and enable sustained drug release, which reduces the need for frequent dosing. Liposomes also exhibit beneficial biodistribution by accumulating more frequently in tumor tissues via the EPR effect, while restricting drug exposure to physiological tissues due to their limited size and permeability. These benefits highlight the great potential of liposomal drug delivery systems over traditional free-drug therapies (Allen & Cullis, 2013; He et al., 2019).

However, in order to improve the solubility, specificity, and sensitivity of liposomal formulations, specific pharmacokinetic adjustments are required, which help overcome chemotherapy-resistant cells, improve half-life and efficacy, and reduce toxicity and the production of undesirable metabolic byproducts. The size and lipid composition of liposomal drugs greatly affect their circulation half-life; however, rapid clearance from the body can limit their therapeutic efficacy, so these formulation strategies are essential for optimizing clinical outcomes. The size and lipid composition of liposomal drugs greatly affect their circulation half-life; however, rapid clearance from the body can limit their therapeutic efficacy, so these formulation strategies are essential for optimizing clinical outcomes (Allen & Cullis, 2013; Sercombe et al., 2015). PEGylation decreases the clearance of liposomes from the bloodstream, while the addition of cholesterol helps modulate membrane fluidity and regulates the rate of drug discharge (Mohamed et al., 2019).

The most significant advantage associated with employing liposomes in drug delivery systems is that they modify the pharmacokinetic profile of incorporated drugs. Plasma proteins, especially opsonins like fibronectin and immunoglobulins, promptly adsorb on the liposomal surface after their delivery. This opsonization shortens the systemic circulation period of the liposomes by identifying them for detection and removal by the mononuclear phagocyte system (MPS), which is mainly involved with Kupffer cells in the liver and phagocytic cells located in bone marrow, spleen, and lymph nodes (Balakrishnan & Gopi, 2025; Lu et al., 2023). Furthermore, the physicochemical properties of liposomes, such as particle size, surface charge, and lipid composition, significantly influence their pharmacokinetics and pharmacodynamics. Firstly, liposome size and clearance rate are directly correlated; the immune system and MPS cells eliminate larger particles with greater efficiency. The net charge of liposome membranes, determined by phospholipids and other components, influences their clearance rate, with increased charge leading to faster elimination. Additionally, membrane lipid composition and structural features like hydrophilic core radius significantly affect liposomal drug pharmacokinetics (Balakrishnan & Gopi, 2025; Paramshetti et al., 2024).

## 4.7 Approaches to enhance tumor targeting using liposomal drug delivery systems

A number of approaches to targeting have been successfully developed to improve liposome accumulation at tumor sites, including active targeting through surface functionalization with ligands and passive targeting through the EPR effect, as shown in Fig. 4.4 Liposomes have grown in significance as promising nanocarrier systems for therapeutic agents, largely due to their unique physicochemical properties (Fulton & Najahi-Missaoui, 2023). Long-circulating liposomes (also known as stealth liposomes) are specifically formulated to evade rapid clearance by the MPS, thereby extending their systemic circulation time and enhancing tumor targeting via the EPR effect by exploiting leaky vasculature and impaired lymphatic drainage. However, many liposomes are cleared by the MPS, primarily in the liver and spleen. To improve stability and circulation time, strategies include cholesterol incorporation to reduce membrane fluidity (e.g., DaunoXome, Myocet, Doxil) and PEGylation, which adds hydrophilic polymers like PEG to form a protective steric barrier, reducing MPS recognition—a phenomenon known as the “stealth effect” (Deodhar & Dash, 2018; Torchilin, 2010).

## 4.8 Passive liposomal targeting (enhanced permeability and retention effect)

First-generation (conventional) liposomes showed limited tumor targeting due to rapid reticular endothelial system (RES) clearance, opsonization by plasma proteins, and drug leakage. Second-generation liposomes improved stability and efficacy through compositional and surface modifications. Cholesterol enhanced membrane rigidity, reducing leakage, while PEGylation provided steric protection, minimized RES uptake, and prolonged circulation time for more effective drug delivery. An effective strategy involves formulating sterically stabilized liposomes with PEG, a hydrophilic polymer that inhibits opsonin binding and prolongs circulation by preventing rapid clearance. Incorporating flexible, hydrophilic, and biocompatible PEG combined with a lipid anchor into liposomes results in a well-hydrated steric barrier that decreases interactions with blood components, extending circulation time and reducing uptake by the liver and spleen. This diminished recognition

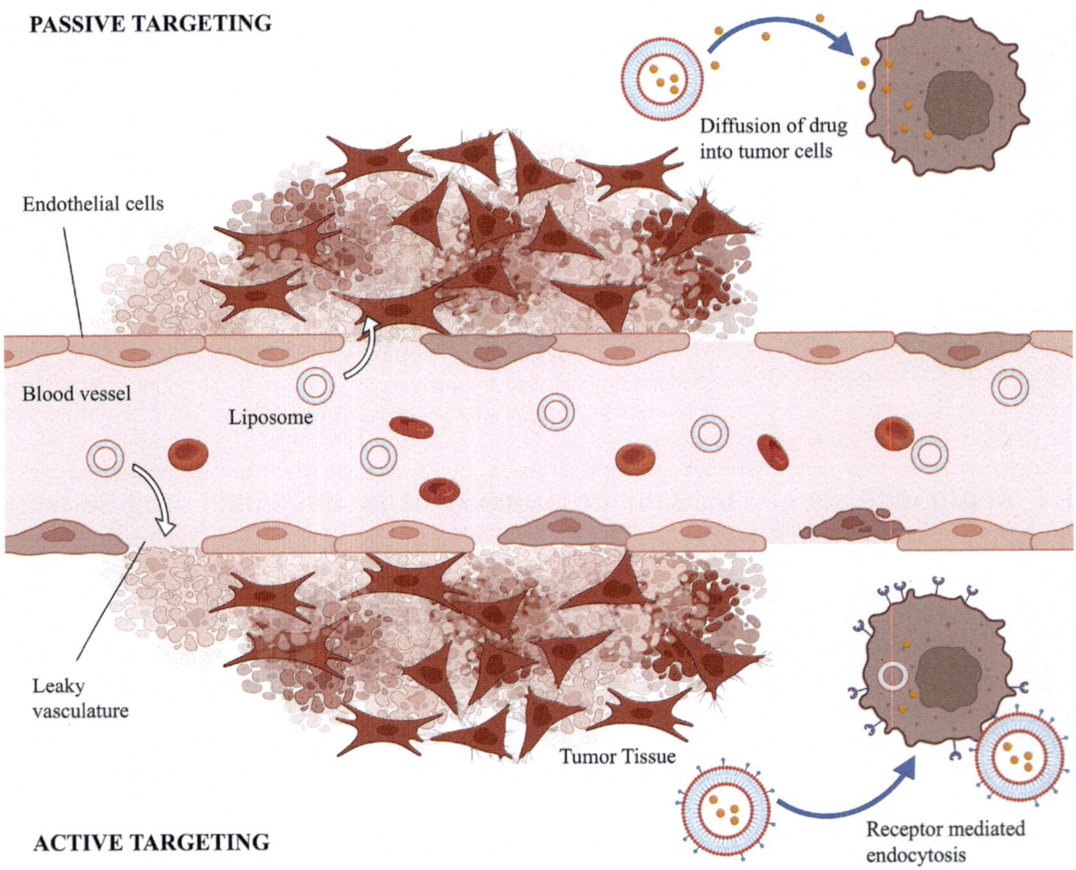


**FIGURE 4.4**

Active and passive targeting of liposomes for targeting tumor cells.

and clearance by the MPS is known as the "stealth effect." The significance of this stealth property is dependent on the molecular weight and surface density of PEG; it has been demonstrated that 5% PEG by weight effectively minimizes protein adsorption and improves systemic circulation.

There are two main PEGylation strategies: preinsertion and postinsertion. Preinsertion involves the utilization of PEG-conjugated phospholipid for developing liposomes. Postinsertion involves the covalent bonding of PEG to the external surface of already developed liposomes. Preinsertion reduces drug loading and the stealth effect, considering PEG chains are designed to orient on both the outer surface and the inner leaflet of the lipid bilayer, whereas postinsertion increases blood circulation time and therapeutic effectiveness in vivo (Fulton & Najahi-Missaoui, 2023).

A promising strategy to prolong the circulation time of liposomes involves coating their surface with superhydrophilic zwitterionic polymers, such as DSPE-poly(carboxybetaine), which form a highly hydrated shield around the vesicle. This modification enhances blood stability and reduces immune recognition, offering an alternative to traditional PEGylation. Two major factors influencing passive targeting of liposomes are the surface charge, which affects interactions with biological barriers, and the accelerated blood clearance (ABC) phenomenon, where repeated administration can lead to rapid removal of liposomes from the bloodstream. Further, charged liposomes (±20 mV) show high affinity for tumor cells and endothelium, but PEGylation can mask this charge, reducing liver uptake and prolonging the circulation duration. Saturated phospholipids and cholesterol do not prevent MPS uptake, but masking surface charge with hydrophilic coatings, such as gangliosides (e.g., monosialoganglioside) and sialic acid derivatives, can reduce clearance—though effects were mainly seen in mice. Several options, like glucuronic acid, reduce MPS absorption. PEG remains the preferred polymer for liposome steric stability and long-term circulation. This technique enhances liposome passive accumulation in highly perfused tissues or organs. Furthermore, PEGylation also promotes the physical stability of liposomes by reducing vesicle coalescence, since PEG chains provide a significant opposition among vesicles that competes with the attracting Van der Waals forces to avoid vesicle aggregation.

The first stealth liposomal formulation approved for cancer therapy was Doxil/Caelyx , which reduced cardiotoxicity and myelotoxicity compared to free doxorubicin and achieved higher tumor drug concentrations using an hydrogenated soybean phosphatidylcholine (HSPC):CL:MPEG2000-DSPE lipid composition. To improve pharmacokinetics, Lipo-Dox was developed by replacing HSPC with distearoylphosphatidylcholine (DSPC), which has a higher phase transition temperature and uniform fatty acid composition. A Phase I study showed that DSPC-based liposomal doxorubicin increased half-life, reduced volume of distribution, and clearance, resulting in better plasma area under cuve (AUC) than Doxil/Caelyx. Although stealth liposomes are clinically useful due to passive tumor accumulation, they lack control over cellular uptake and drug release, limiting efficacy. For instance, SPI-77, a stealth cisplatin liposome similar to Doxil, showed better toxicity but poor patient response in Phase II trials. The third-generation liposomes address this by incorporating ligand-mediated active targeting to enhance biodistribution and drug delivery at tumor sites.

## 4.9 Active liposomal targeting

PEGylation and incorporation of highly saturated phospholipids and cholesterol enhance liposomal pharmacokinetics and stability. However, effective intratumor drug concentration remains uncertain if the released drug is not internalized by targeted tumor cells, limiting its intracellular availability. Passive targeting of liposomes is based on the physiological characteristics of the tumor environment (Gatto et al., 2024). There are two major limitations associated with PEGylation passive targeting of liposomes: (1) limited drug accumulation to the specified tumor cells, that is, less than 5% of the administered drug, plus unwanted distribution of the active drug in the surrounding nontargeted tissues; (2) insufficient vascularization and elevated interstitial fluid pressure, which impede macromolecular transport, making it difficult for nontargeted passive liposomes to enter the tumor core. Active targeting approach uses molecular-based techniques to selectively route liposomes to tumor cells by identifying tumor-specified markers. This is commonly accomplished by conjugating targeting ligands (such as small molecules, RNA aptamers, monoclonal antibodies, vitamins, carbohydrates, antibody fragments, glycolipids, or signal peptides) onto the liposomal surface (Fidan et al., 2024; Sofou & Sgouros, 2008). Active targeting takes into account pathophysiological alterations in tumor microenvironment, for instance, protein overexpression, to allow liposomes to be selectively taken up by cells that exhibit these markers, resulting in improved drug delivery. Active targeting of ligands can be added to improve drug accumulation at targeted tumor sites, including intracellular or vascular targets. Intracellular processing after tumor-specific endocytosis promotes localized drug release, increasing bioavailability and reducing the diffusion of drugs from the tumor cells, both of which lengthen the duration of intratumoral retention (Gatto et al., 2024).

Active targeting can be broadly classified into two categories:

1. Targeting receptors on tumor cells and cancerous tissues: This approach involves conjugating specific molecules to the liposome surface, enabling binding to overexpressed receptors on cancer cells. These receptors are upregulated due to increased metabolic demands associated with rapid tumor cell proliferation. Active targeting utilizes certain molecular alterations to guide liposomes to approach the conjugated surface receptors that are overexpressed on cancer cells, including the epidermal growth factor receptor (EGFR), transferrin receptor (TfR), folate receptor (FR), and others. EGFR (tyrosine kinase receptors) is essential for biological processes, including tissue development and repair. Because of their roles in angiogenesis, cell proliferation, and metastasis, EGFR expression is noticeably higher in cancerous tissues. The FR increases folic acid absorption in tumor cells, whereas the TfR interacts with serum transferrin and thus promotes the internalization of this glycoprotein by endocytosis (Khan et al., 2025; Zahednezhad et al., 2024).
2. Targeting specific components of the tumor microenvironment: By altering the surface of liposomes to target receptors present in the tumor microenvironment, this technique can prevent tumor development, metastasis, treatment resistance, and changes in neovascular endothelial cells. This targeting strategy targets particular receptors found in the tumor microenvironment, including integrins, matrix metalloproteinases (MMPs), vascular endothelial growth factor (VEGF), and vascular cell adhesion molecule (VCAM). These receptors are key areas of intervention for improving the therapeutic efficiency of liposomal drug delivery systems since they are essential for angiogenesis in tumors, invasion, and metastasis (Fidan et al., 2024).

Successful development of targeted liposomes faces challenges such as variable receptor expression on tumor cells, RES uptake, ligand immunogenicity, limited receptor accessibility, low endocytosis efficiency, and the need to optimize drug release and formulation stability. Despite active targeting improving cellular uptake in the tumor's microenvironment, this does not promote tumor-specific accumulation since liposomes remain dependent on passive transportation. Transcytosable nano-based therapy has been recently recognized as a viable option that may penetrate the vascular barrier and improve intratumoral delivery by receptor-mediated, fluid-mediated, or adsorption-mediated transcytosis (Paramshetti et al., 2024).

## 4.10 Recent approaches for colon-targeting drug delivery using liposomes

Recently, numerous studies have been conducted focusing on liposome-based nanosystems for the management and treatment of colorectal cancer, showing notable advancements in preclinical and clinical studies. These include ThermoDox (a temperature (heat)-sensitive liposomal formulation),

LE-SN38 (a liposomal formulation containing SN-38), and CPX-1 (liposomes with a mixture of irinotecan hydrochloride and floxuridine). Phase II clinical studies have demonstrated the potential therapeutic effectiveness and safety of these liposomal formulations in human patients suffering from colon cancer. Apart from these formulations, numerous other liposomal formulations exist in the preclinical phase of development in addition to these clinically tested systems. Some of these include doxorubicin-encapsulated liposomes, liposomal curcumin formulations, and PEGylated long-circulating liposomes loaded with oxaliplatin (PEG-liposomal L-OHP). The primary objectives are to increase tumor targeting, decrease systemic toxicity, and improve drug stability. Furthermore, magnetoliposomes have attracted a lot of attention among the recent advanced technologies because of their multifunctional properties. By combining the benefits of magnetic responsive characteristics along with conventional liposomal drug delivery approach, these devices provide substantial drug-loading capacity, magnetically directed targeting, and treatments that burst release drugs in response to the rise in body temperature (hyperthermia-triggered). Considering this, 5-fluorouracil (5-FU) has been found to have strong anticancer properties, making it a perfect choice for inclusion in magnetoliposomes for colon cancer treatment. In order to enable active targeting, researchers have been working to functionalize such systems with particular specified ligands or moieties—such as PR_b peptides—that identify and attach to receptors overexpressed in cancer cells or tumor vasculature. Fig. 4.5 illustrates the various strategies for functionalizing liposomes to achieve targeted drug delivery, in order to enhance specificity and therapeutic efficacy in cancer treatment.

## 4.11 PEGylated liposomes targeting colon cancer

Early research by several scientists has demonstrated that doxorubicin encapsulated in GM1-modified long-circulating liposomes exhibited enhanced blood retention and tumor accumulation via the EPR effect by evading uptake by the RES. The use of colon-targeted long-circulating liposomes has shown promising results in overcoming the limitations associated with conventional therapies for colon-related pathologies. In inflammatory conditions such as colitis, colon-targeting liposomes exhibited preferential accumulation in inflamed colonic tissues, as evidenced by radiolabeling and γ-scintigraphy studies. A high target-to-nontarget tissue ratio and reduced systemic circulation of liposomes were observed within 24 hours, indicating effective localization at the disease site. This selective accumulation highlights the potential of liposome-based systems to enhance therapeutic efficacy while minimizing off-target effects, making them a valuable strategy for targeted drug delivery in colon cancer management (Shazleen Ibrahim et al., 2024).

To enhance specificity, PEGylated liposomes can be further functionalized with targeting ligands such as the PR_b peptide, which selectively binds to integrin $\alpha5\beta1$—an integrin markedly overexpressed on colon cancer cells. This passive targeting technique increases liposome cellular absorption, resulting in increased cytotoxicity and drug delivery efficiency when compared to nontargeted conventional liposomal formulations. Notably, PR_b-conjugated PEGylated liposomes encapsulating 5-FU showed cytotoxic effects against colon cancer cell lines (CT26.WT, HCT116, RKO) equivalent to the free-drug treatment, while exhibiting higher selectivity and lower systemic toxicity (Marelli et al., 2013). PEGylated liposomes protect incorporated drugs from degradation in

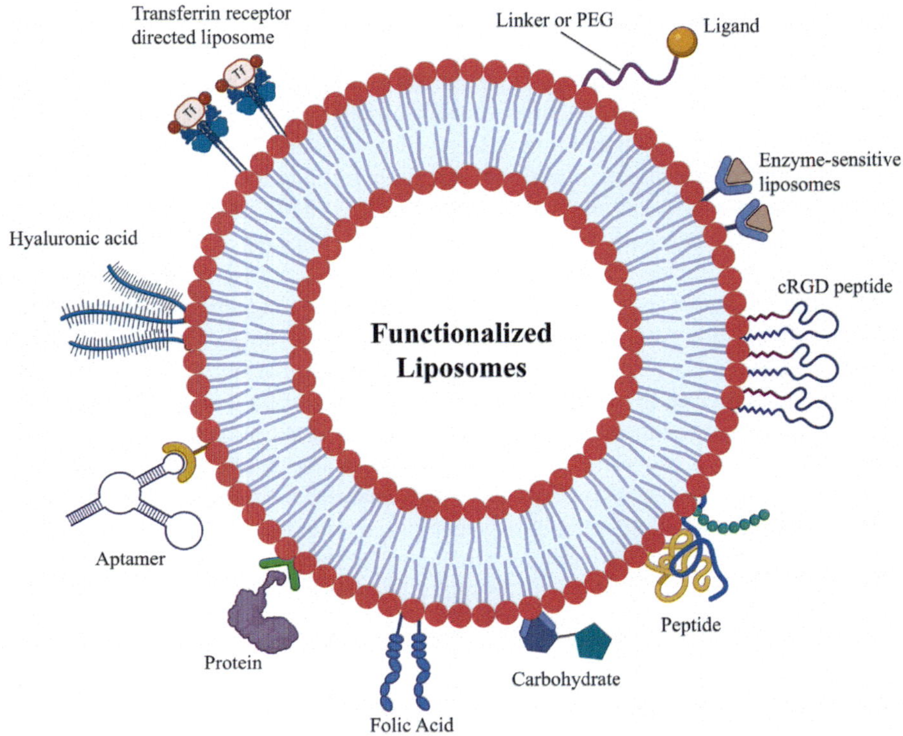


**FIGURE 4.5**

Functionalized liposomes engineered for targeted drug delivery in the management of colon cancer.

harsh gastrointestinal environment, making them ideal for oral or systemic chemotherapy. They emphasize early intervention by demonstrating increased effectiveness in smaller tumors in colon cancer models. Additionally, PEGylated liposomes have the ability to codeliver radionuclides and chemotherapeutic agents, allowing for the combination of imaging for treatment monitoring, chemotherapy, and radiation therapy (Garg et al., 2009).

## 4.12 Folic acid-functionalized liposomes targeting colon cancer

Folate-conjugated liposomes have demonstrated great potential for targeted drug delivery approach for colon cancer therapy because the majority of solid tumor cells, including colorectal cancer cells, overexpress folate receptors on their surfaces. Folic acid, a tiny, water-soluble vitamin B9 (pteroyl-β-glutamate), can be therefore utilized as a targeted ligand for directing liposomes toward the malignant tissues through receptor-mediated endocytosis, thereby enhancing drug absorption and decreasing adverse reactions on healthy tissues (Narmani et al., 2019; Shazleen Ibrahim et al., 2024). Folate-targeted liposomes—such as those incorporating folate-polyethylene glycol-cholesterol hemisuccinate (folate-PEG-CHEMS)—enhance tumor-specific drug delivery and cytotoxicity (including daunorubicin and 5-FU) by exploiting the overexpression of folate receptors (FR) on colon cancer cells. Experiments utilizing cell lines such as Caco-2, HT-29, and HeLa, as well as in vivo models such as CT26-bearing BALB/c mouse model, revealed that folate-targeted 5-FU liposomes demonstrated better therapeutic outcomes than nontargeted formulations or free drugs in terms of antitumor effectiveness and cytotoxicity. Targeted liposomal formulation containing 5-FU showed lower $IC_{50}$, improved cellular uptake, greater reactive oxygen species (ROS) generation, and higher anticancer effectiveness compared to free drug on colorectal cancer cells (Pourmadadi et al., 2023). Additionally, folic acid-functionalized, pH-responsive liposomal formulations encapsulating irinotecan (SpHL-IRN-Fol) demonstrated effective drug release profile in acidic tumor microenvironments, contributing to improved therapeutic efficacy and lower systemic toxicities (Kamal et al., 2024). Similar to this, coloaded liposomes containing pemetrexed (PMX) and docetaxel (DTX) showed pH-dependent release, increased cellular absorption of the drugs, and increased apoptotic rates among CT26 and MC36 colorectal cancer cells. Studies on in vivo models have also demonstrated that liposomes exhibited a significant tumor-inhibiting properties, indicating efficient accumulation at the targeted tumor site (Shazleen Ibrahim et al., 2024).

In another research study, it was demonstrated that liposomes developed using DPPC caused necrosis in HT-29 colon cancer cells. According to these results, the lipid composition used to create liposomes can have a significant impact on both their physical properties and the specific cell-death processes they induce (Zahiri et al., 2021).

## 4.13 Transferrin-functionalized liposomes targeting colon cancer

By taking advantage of the overexpression of TfR1 on cancerous colon epithelial cells, transferrin-conjugated liposomes offer a sophisticated and promising approach for targeted drug delivery in colon cancer management. TfR1 represents a transmembrane glycoprotein that binds to transferrin, a glycopeptide that carries iron throughout the body, to mediate cellular iron absorption (Nogueira-Librelotto et al., 2017). TfR1 is frequently overexpressed at values up to 100 times greater in malignant cells than in healthy tissues, owing to the rapid rate of tumor cell proliferation and iron dependence, making it a desirable molecular target for targeted drug delivery. Transferrin ligand-functionalized liposomes allow receptor-induced endocytosis into colon cancer cells, improving intracellular distribution of incorporated chemotherapeutic drugs while reducing systemic adverse reactions and nonspecific uptake. The effectiveness of this approach has been shown in several studies (Mojarad-Jabali et al., 2022; Tortorella & Karagiannis, 2014). For example, in colon cancer cell lines (such as HCT-8), transferrin-functionalized liposomes loaded with artemisinin or sorafenib demonstrated better cytotoxicity and much greater cellular uptake than their nontargeted formulations. By modulating apoptosis-related indicators, these modified formulations demonstrated improved pro-apoptotic activities through mitochondrial signaling mechanisms and decreased $IC_{50}$ values (Li et al., 2023). Likewise, 5-FU encapsulated in transferrin-targeted liposomes has demonstrated better treatment results, where studies conducted in vitro on HCT-8, HT-29, and Caco-2 cell lines revealed that transferrin-conjugated liposomes enhanced cellular uptake and apoptotic induction more than free drugs or nontargeted liposomes. The impact of composition of lipids on liposome effectiveness was further supported by the improved antitumor activity demonstrated by liposomes produced with phospholipids such as phosphatidylcholine (PC) and DPPC (Riaz et al., 2018).

Transferrin-conjugated liposomes have also been demonstrated in preclinical investigations employing colon tumor-bearing animals (e.g., CT26 in BALB/c) to provide improved tumor suppression and up to approximately 3.7-fold greater tumor accumulation than conventional liposomes, with limited off-target toxicities. Overall, these liposomes promote apoptosis by disrupting iron metabolism and improving drug delivery by targeting TfR1 (Li et al., 2023).

## 4.14 Peptide-functionalized liposomes targeting colon cancer

Integrin α5β1, which is overexpressed in colon cancer and plays a role in tumor growth and metastasis, represents one of the most promising targets for peptide-conjugated liposomes. PEGylated liposomes coupled with PR_b peptide, a fibronectin-mimetic sequence, have the potential to bind to integrin α5β1 on colon cancer cells. Additionally, PR_b functionalized pH-sensitive PEGylated liposomes could have further potential to efficiently release their incorporated payload under acidic endosomal environments, validating the system's response and targeting capability without changing pH sensitivity (Musyanovych & Landfester, 2014; Shazleen Ibrahim et al., 2024).

Additionally, several other peptides exhibit strong affinity for integrins—for instance, RGD peptides (cyclo[Arginine-Glycine-Aspartic acid-D-Tyrosine-Cysteine])—which have led to the development of liposomal formulations incorporating RGD10 lipopeptides, stabilized by a single disulfide bond and encapsulating doxorubicin for targeted cancer therapy. When evaluated in a mouse C26 colon carcinoma model, these liposomes showed superior antitumor activity, primarily because of enhanced targeting of α5-integrin-expressing cancer cells (Cheng & Ji, 2019). The NGR (Asparagine–Glycine–Arginine) peptide has also demonstrated promising therapeutic potential against colon cancer, as it selectively targets the CD13-expressing endothelial cells, such as human umbilical vein endothelial cells (HUVECs) and SLK cells, which are abundantly expressed in the tumor neovasculature. In addition to integrin and CD13 targeting, colon cancer markers such as Obese receptor (ObR) and Mucin1 (Muc1) have also been effectively explored for colon cancer management (Jain & Jain, 2015). Peptides such as the leptin-derived peptide (Lp31) can be conjugated to liposomes to enable efficient binding to leptin receptors ObR, which are overexpressed in colon cancer cells. This targeted approach has demonstrated significantly enhanced cellular uptake in C26 colon cancer cells (Dixit et al., 2024). Similarly, AR13 peptide can be used to target Muc1, which can lead to greater internalization and specific cytotoxicity in C26 colon cancer cells. For gene therapy in metastatic colorectal cancer, iRGD-conjugated liposomes containing pigment epithelium-derived factor (PEDF)-DNA have been formulated in a study. Peptide-conjugated liposomes were proven to be versatile for both drug and gene administration, as demonstrated by the enhanced accumulation of these PEGylated liposomes in integrin-expressing tumor cells, which successfully inhibited angiogenesis and proliferation while triggering death (Cheng & Ji, 2019; Jain & Jain, 2015).

There is another type of peptides researchers are extensively focusing on, and that is cell-penetrating peptides (CPPs)—small peptide sequences capable of ferrying drugs across cellular membranes. In contrast to standard targeting ligands, CPPs have a unique property that enables intracellular drug permeation by bypassing various biological barriers, including endosomal and cellular membranes, which provides the potential to engineer innovative liposomal drug delivery systems for colon cancer therapy (Vale et al., 2020). The most extensively studied CPP is the trans-activator of transcription (TAT) peptide, obtained from the HIV-1 transactivator of transcription protein, which, when functionalized with liposomes, improves cellular uptake via energy-independent translocation approach and additionally evades lysosomal degradation, thus ensuring that the therapeutics reach the cytoplasm more effectively (Bolhassani, 2011).

## 4.15 Aptamer-functionalized liposomes targeting colon cancer

Aptamers can be described as short, artificial single-stranded DNA, RNA, or amino acid sequences that preferentially attach to a wide range of molecular targets, including cell surfaces, proteins, peptides, receptors, and enzymes. Aptamer-conjugated liposomes have demonstrated a wide range of benefits over antibody-based targeting, including increased target selectivity, higher stability, lower molecular size, and reliability of manufacturing and modification (Alshaer et al., 2015; Moosavian & Sahebkar, 2019).

A liposomal formulation encapsulating doxorubicin, coupled with an anti-EpCAM (epithelial cell adhesion molecule) aptamer, has been developed for targeted colon cancer treatment. This aptamer-targeted liposome showed dramatically increased cellular uptake in C26 colon cancer cells, indicating better retention and accumulation of drugs within cancer cells. Furthermore, in vivo investigations showed that aptamer-functionalized liposomes increased intratumoral drug concentration and extended animal life more efficiently compared to Caelyx (marketed liposomal formulation of doxorubicin) (Shazleen Ibrahim et al., 2024; Zhao et al., 2019).

## 4.16 Hyaluronic acid-functionalized liposomes targeting colon cancer

Hyaluronic acid (HA) is a nonsulphated, high molecular weight glycosaminoglycan that is a major component of the extracellular matrix, which consists of N-acetyl-β-D-glucosamine and β-D-glucuronate residues. The HA-binding receptors, majorly including CD44 (particularly CD44 variant isoform v6, CD44v6), along with CD168 and ICAM-1, are overexpressed in colon cancer; therefore targeting CD44 provides a selective and precise method for preventing the progression of colon cancer (Fu et al., 2023; Mattheolabakis et al., 2015). The interaction of HA with CD44v6 has been shown to stimulate many intracellular signaling mechanisms that promote tumor cell proliferation, metastasis, and resistance to conventional cancer treatments. This connection not only encourages tumor growth but also leads to the emergence of chemoresistance, thus rendering it an important target for colon cancer therapy regimens. In addition, experiments have demonstrated that the degradation products of HA, especially oligosaccharide-based products of hyaluronic acid (oHA), might interact with CD44v6. These oHA segments may compete with native HA for CD44v6 binding sites, potentially interrupting the chemoresistance-promoting signaling cascade. Incorporation of oHA with liposomal formulations allows for not only targeted drug delivery but also sensitization of tumor cells to chemotherapy by overcoming resistance mechanisms (Hou et al., 2022; Mattheolabakis et al., 2015).

## 4.17 Onivyde (irinotecan liposomal) in colon cancer

Onivyde (irinotecan liposome injection) is a liposomal formulation of irinotecan, a topoisomerase I inhibitor, developed to enhance the delivery and efficacy of irinotecan in the treatment of solid tumors, including colorectal (colon) cancer. The liposomal encapsulation allows irinotecan to circulate longer in the bloodstream, resulting in higher and more sustained concentrations of the active drug and its metabolite (SN-38) within tumor tissue compared to conventional irinotecan. This improved pharmacokinetic profile translates to greater antitumor activity and reduced systemic toxicity (Milano et al., 2022). Onivyde delivers irinotecan directly to tumor cells, where it is metabolized to SN-38, a potent inhibitor of topoisomerase I. This enzyme is essential for DNA replication and cell division. By inhibiting topoisomerase I, SN-38 induces DNA damage, leading to cancer cell death (Brown & Blair, 2025; Chandran et al., 2017). Preclinical studies have shown that Onivyde demonstrates superior anticancer efficacy over free irinotecan in colorectal cancer cell lines and xenograft models, with enhanced tumor accumulation and reduced tumor progression at equivalent doses. While Onivyde is FDA-approved primarily for metastatic pancreatic cancer (in combination with 5-FU and leucovorin), it is also being investigated in Phase II/III clinical trials for metastatic colorectal cancer and other solid tumors, where early data suggest promising antitumor activity (Chandran et al., 2017).

Onivyde represents a significant advance in the delivery of irinotecan for colon cancer, offering improved tumor targeting, greater efficacy, and a potentially better safety profile compared to conventional formulations. Its use in metastatic colorectal cancer remains under clinical investigation, with early studies supporting its potential as an effective treatment option.

## 4.18 Clinical trials on liposomal-based management approaches for colon cancer

Recent clinical trials have demonstrated that liposomal irinotecan, especially in combination with agents like 5-FU and bevacizumab, shows promising efficacy and tolerability as a second-line treatment for metastatic colon cancer (Table 4.2).

## 4.19 Challenges related to clinical translation of liposomal formulations

Despite significant developments in liposomal nanomedicine for cancer therapy, a number of complications remain to impede its clinical translation and widespread application. The primary concern is the intricacy of large-scale production and rigid quality control, wherein even a minor variation in lipid content, particle size, or surface functionalization can cause batch variability, thereby compromising treatment safety and efficacy. Additionally, in biological systems, liposomes are quickly cleared by the MPS, especially in individuals with weakened immune systems, limiting their accumulation in the circulatory system and at tumor sites. Furthermore, the passive targeting technique based on the EPR effect shows significant variation across various types of tumors and patient demographics, resulting in inconsistent and sometimes inadequate drug delivery effectiveness (Sawant & Torchilin, 2012).

**Table 4.2 Summary of clinical trials investigating liposome-based formulations in the treatment of advanced colon cancer.**

| S. No. | NCT number | Study | Phase | Intervention/treatment |
|---|---|---|---|---|
| 1. | NCT06643793 | Evaluation of safety and efficacy of irinotecan liposome (II) combined with 5-FU, calcium leucovorin, and bevacizumab in the treatment of metastatic colorectal cancer. | Phase 2 | Drug: Irinotecan liposome (ll) + 5-FU/LV + bevacizumab |
| 2. | NCT06443307 | Efficacy and safety of irinotecan-loaded liposomes combined with vermofenib and cetuximab in first-line failure of advanced rat sarcoma viral oncogene homolog (RAS) wild/BRAF-mutated colorectal cancer. | Phase 2 | Irinotecan liposomes combined with vermofenib and cetuximab |
| 3. | NCT00361842 | Effectiveness of CPX-1 (irinotecan HCl: floxuridine) liposome injection in patients suffering from advanced colorectal carcinoma who have already received chemotherapy (oxaliplatin or irinotecan). | Phase 2 | CPX-1 (fixed combination of irinotecan HCl: floxuridine) liposome injection |
| 4. | NCT00311610 | Evaluate the response rate after treatment with SN-38 liposome as a second-line treatment in patients suffering from metastatic colorectal cancer. | Phase 2 | SN-38 liposome IV infusion |
| 5. | NCT01464593 | Evaluate the safety and efficacy of ThermoDox, a thermo-sensitive liposomal doxorubicin formulation, in combination with thermal ablation for hepatic colorectal liver metastases. | Phase 2 | Lyso-thermosensitive liposomal doxorubicin, IV infusion |
| 6. | NCT05969899 | Evaluate the safety and efficacy of irinotecan-incorporated liposomes for first-line therapy of advanced colorectal cancer. | Phase 2 | Liposomal irinotecan-based FOLFIRI combined with bevacizumab |
| 7. | NCT06225622 | Evaluate the safety and efficacy of irinotecan-loaded liposomal injection combined with oxaliplatin +5-FU/LV+ bevacizumab or cetuximab in first-line treatment of advanced metastatic colorectal cancer. | Phase 1 | Irinotecan liposome, oxaliplatin, 5-FU, bevacizumab, cetuximab |

Another major concern is drug leakage, in which encapsulated drugs diffuse out from the liposomal nanostructure over time, resulting in lower drug loading and efficacy before the formulation reaches the patient. Chemical instability of the lipid components or the encapsulated drugs might occur, especially during manufacturing or storage, leading to degradation or denaturation, which further restricts their effectiveness (Allen & Cullis, 2013). Immunogenicity and hypersensitivity responses, though very rare, have been reported with certain liposomal therapies, necessitating close patient monitoring. Regulatory barriers also remain severe, with the approval procedure for complicated nanomedicines requiring comprehensive characterization, repeatability, and long-term safety evidence. Finally, the expensive process of formulation, manufacturing, and quality assurance for these nanoformulations may restrict accessibility, especially in resource-constrained situations. These challenges emphasize the necessity for further research to optimize liposome design, improve targeting efficiency, increase stability, and expedite manufacturing processes in order to fully realize the therapeutic promise of liposomal nanomedicine in cancer therapy (Urbinati et al., 2012).

## 4.20 Conclusion

The implementation of liposomes as nanoformulations for targeted anticancer drug delivery is a significant breakthrough in oncological therapy. Their fundamental biocompatibility, biodegradability, along with their capability to encapsulate both hydrophilic and hydrophobic compounds, makes them promising for increasing treatment efficacy while reducing systemic toxicity. Functionalization with specific ligands such as folic acid, transferrin, aptamers, HA, and cRGD significantly improves cellular uptake, apoptosis, and cytotoxicity in cancer cells—particularly in colon cancer—compared to free drugs and nontargeted liposomes. Despite these promising outcomes, the clinical translation of targeted liposomal systems necessitates rigorous preclinical and clinical evaluation, including pharmacokinetic, biodistribution, and toxicokinetic studies.

## References

Ahmed, K. S., Hussein, S. A., Ali, A. H., Korma, S. A., Lipeng, Q., & Jinghua, C. (2019). Liposome: Composition, characterisation, preparation, and recent innovation in clinical applications. *Journal of Drug Targeting, 27*(7), 742–761. https://doi.org/10.1080/1061186X.2018.1527337, http://www.tandfonline.com/loi/idrt20.

Allen, T. M., & Cullis, P. R. (2013). Liposomal drug delivery systems: From concept to clinical applications. *Advanced Drug Delivery Reviews, 65*(1), 36–48. https://doi.org/10.1016/j.addr.2012.09.037.

Allen, T. M., Cheng WW, K., Hare, J. I., & Laginha, K. M. (2006). Pharmacokinetics and pharmacodynamics of lipidic nano-particles in cancer. *Anti-Cancer Agents in Medicinal Chemistry (Formerly Current Medicinal Chemistry-Anti-Cancer Agents), 6*(6), 513–523. https://doi.org/10.2174/187152006778699121.

Alshaer, W., Hillaireau, H., Vergnaud, J., Ismail, S., & Fattal, E. (2015). Functionalizing liposomes with anti-CD44 aptamer for selective targeting of cancer cells. *Bioconjugate Chemistry, 26*(7), 1307–1313. https://doi.org/10.1021/bc5004313.

Balakrishnan, P., & Gopi, S. (2025). Pharmacokinetic and pharmacodynamic insights of bioactive molecules encapsulated in liposomal formulation. *Current Pharmacology Reports, 11*(1). https://doi.org/10.1007/s40495-025-00406-1.

Barenholz, Y.(C.) (2012). Doxil®—The first FDA-approved nano-drug: Lessons learned. *Journal of Controlled Release, 160*(2), 117–134. https://doi.org/10.1016/j.jconrel.2012.03.020.

Bolhassani, A. (2011). Potential efficacy of cell-penetrating peptides for nucleic acid and drug delivery in cancer. *Biochimica et Biophysica Acta (BBA) - Reviews on Cancer, 1816*(2), 232–246. https://doi.org/10.1016/j.bbcan.2011.07.006.

Brown, M. B., & Blair, H. A. (2025). Liposomal irinotecan: A review as first-line therapy in metastatic pancreatic adenocarcinoma. *Adis, New Zealand Drugs, 85*(2), 255–262. https://doi.org/10.1007/s40265-024-02133-1, https://www.springer.com/journal/40265.

Carugo, D., Bottaro, E., Owen, J., Stride, E., & Nastruzzi, C. (2016). Liposome production by microfluidics: Potential and limiting factors. *Scientific Reports, 6*(1). https://doi.org/10.1038/srep25876.

Chandran, S. P., Natarajan, S. B., Chandraseharan, S., & Mohd Shahimi, M. S. B. (2017). Nano drug delivery strategy of 5-fluorouracil for the treatment of colorectal cancer. *Journal of Cancer Research and Practice, 4*(2), 45–48. https://doi.org/10.1016/j.jcrpr.2017.02.002.

Chen, C., Han, D., Cai, C., & Tang, X. (2010). An overview of liposome lyophilization and its future potential. *Journal of Controlled Release, 142*(3), 299–311. https://doi.org/10.1016/j.jconrel.2009.10.024.

Chen, W., Shi, K., Yu, Y., Yang, P., Bei, Z., Mo, D., Yuan, L., Pan, M., Chen, Y., & Qian, Z. (2024). Drug delivery systems for colorectal cancer chemotherapy. *Chinese Chemical Letters, 35*(2), 109159. https://doi.org/10.1016/j.cclet.2023.109159.

Cheng, Y., & Ji, Y. (2019). RGD-modified polymer and liposome nanovehicles: Recent research progress for drug delivery in cancer therapeutics. *European Journal of Pharmaceutical Sciences, 128*, 8–17. https://doi.org/10.1016/j.ejps.2018.11.023.

Deodhar, S., & Dash, A. K. (2018). Long circulating liposomes: Challenges and opportunities. *Therapeutic Delivery, 9*(12), 857–872. https://doi.org/10.4155/tde-2018-0035, http://www.future-science.com/loi/tde.

Dixit, A.·A., Mandlik, D. S., & Mandlik, S. K. (2024). Functionalised ligand-based nanomaterial drug targeting approaches for colorectal cancer therapy. *Recent Advances in Drug Delivery and Formulation, 18*(3), 170–187. https://doi.org/10.2174/0126673878301004240703073107, https://www.eurekaselect.com/article/141517.

Drescher, S., & van Hoogevest, P. (2020). The phospholipid research center: Current research in phospholipids and their use in drug delivery. *Pharmaceutics, 12*(12), 1235. https://doi.org/10.3390/pharmaceutics12121235.

Ercole, F., Whittaker, M. R., Quinn, J. F., & Davis, T. P. (2015). Cholesterol modified self-assemblies and their application to nanomedicine. *Biomacromolecules, 16*(7), 1886–1914. https://doi.org/10.1021/acs.biomac.5b00550, http://pubs.acs.org/journal/bomaf6.

Fidan, Y., Muçaj, S., Timur, S. S., & Gürsoy, R. N. (2024). Recent advances in liposome-based targeted cancer therapy. *Journal of Liposome Research, 34*(2), 316–334. https://doi.org/10.1080/08982104.2023.2268710.

Fu, C. P., Cai, X. Y., Chen, S. L., Yu, H. W., Fang, Y., Feng, X. C., Zhang, L. M., & Li, C. Y. (2023). Hyaluronic acid-based nanocarriers for anticancer drug delivery. *Polymers, 15*(10). https://doi.org/10.3390/polym15102317, http://www.mdpi.com/journal/polymers.

Fulton, M. D., & Najahi-Missaoui, W. (2023). Liposomes in cancer therapy: How did we start and where are we now. *International Journal of Molecular Sciences, 24*(7). https://doi.org/10.3390/ijms24076615, http://www.mdpi.com/journal/ijms.

Garg, A., Tisdale, A. W., Haidari, E., & Kokkoli, E. (2009). Targeting colon cancer cells using PEGylated liposomes modified with a fibronectin-mimetic peptide. *International Journal of Pharmaceutics, 366*(1-2), 201–210. https://doi.org/10.1016/j.ijpharm.2008.09.016.

Gatto, M. S., Johnson, M. P., & Najahi-Missaoui, W. (2024). Targeted liposomal drug delivery: Overview of the current applications and challenges. *Life, 14*(6). https://doi.org/10.3390/life14060672, www.mdpi.com/journal/life/.

Giuliano, C. B., Cvjetan, N., Ayache, J., & Walde, P. (2021). Multivesicular vesicles: Preparation and applications. *ChemSystemsChem, 3*(2). https://doi.org/10.1002/syst.202000049, https://onlinelibrary.wiley.com/journal/25704206.

Gouda, A., Sakr, O. S., Nasr, M., & Sammour, O. (2021). Ethanol injection technique for liposomes formulation: An insight into development, influencing factors, challenges and applications. *Journal of Drug Delivery Science and Technology, 61*, 102174. https://doi.org/10.1016/j.jddst.2020.102174.

Gubernator, J. (2011). Active methods of drug loading into liposomes: Recent strategies for stable drug entrapment and increased in vivo activity. *Expert Opinion on Drug Delivery, 8*(5), 565–580. https://doi.org/10.1517/17425247.2011.566552.

Has, C., & Sunthar, P. (2020). A comprehensive review on recent preparation techniques of liposomes. *Journal of Liposome Research, 30*(4), 336–365. https://doi.org/10.1080/08982104.2019.1668010.

He, H., Yuan, D., Wu, Y., & Cao, Y. (2019). Pharmacokinetics and pharmacodynamics modeling and simulation systems to support the development and regulation of liposomal drugs. *Pharmaceutics, 11*(3), 110. https://doi.org/10.3390/pharmaceutics11030110.

van Hoogevest, P., & Wendel, A. (2014). The use of natural and synthetic phospholipids as pharmaceutical excipients. *European Journal of Lipid Science and Technology, 116*(9), 1088–1107. https://doi.org/10.1002/ejlt.201400219.

Hou, X., Zhong, D., Chen, H., Gu, Z., Gong, Q., Ma, X., Zhang, H., Zhu, H., & Luo, K. (2022). Recent advances in hyaluronic acid-based nanomedicines: Preparation and application in cancer therapy. *Carbohydrate Polymers, 292*, 119662. https://doi.org/10.1016/j.carbpol.2022.119662.

Jain, A., & Jain, S. K. (2015). Colon targeted liposomal systems (CTLS): Theranostic potential. *Current Molecular Medicine, 15*(7), 621–633. https://doi.org/10.2174/1566524015666150831131320.

Kamal, R., Awasthi, A., Paul, P., Mir, M. S., Singh, S. K., & Dua, K. (2024). Novel drug delivery systems in colorectal cancer: Advances and future prospects. *Pathology - Research and Practice, 262*, 155546. https://doi.org/10.1016/j.prp.2024.155546.

Khan, I. N., Arshad, N., Shaheen, F., Shakoor, R., Hassan, A., & Waqar, M. A. (2025). Recent composition and applications of liposomes in cancer therapy. *International Journal of Polymeric Materials and Polymeric Biomaterials, 74*(8), 720–732. https://doi.org/10.1080/00914037.2024.2368896, www.tandf.co.uk/journals/titles/00914037.asp.

Kim, & Jeong (2021). Liposomes: Biomedical applications. *Chonnam medical journal, 57*(1), 27–35. https://doi.org/10.4068/cmj.2021.57.1.27.

Krause, M. R., & Regen, S. L. (2014). The structural role of cholesterol in cell membranes: From condensed bilayers to lipid rafts. *Accounts of Chemical Research, 47*(12), 3512–3521. https://doi.org/10.1021/ar500260t, http://pubs.acs.org/journal/achre4.

Li, J., Zhang, Z., Zhang, B., Yan, X., & Fan, K. (2023). Transferrin receptor 1 targeted nanomedicine for brain tumor therapy. *Biomaterials Science, 11*(10), 3394–3413. https://doi.org/10.1039/d2bm02152h.

Li, M., Du, C., Guo, N., Teng, Y., Meng, X., Sun, H., Li, S., Yu, P., & Galons, H. (2019). Composition design and medical application of liposomes. *European Journal of Medicinal Chemistry, 164*, 640–653. https://doi.org/10.1016/j.ejmech.2019.01.007.

Lian, T., & Ho, R. J. Y. (2001). Trends and developments in liposome drug delivery systems. *Journal of Pharmaceutical Sciences, 90*(6), 667–680. https://doi.org/10.1002/jps.1023.

Liu, P., Chen, G., & Zhang, J. (2022). A review of liposomes as a drug delivery system: Current status of approved products, regulatory environments, and future perspectives. *Molecules, 27*(4), 1372. https://doi.org/10.3390/molecules27041372.

Lombardo, D., & Kiselev, M. A. (2022). Methods of liposomes preparation: Formation and control factors of versatile nanocarriers for biomedical and nanomedicine application. *Pharmaceutics, 14*(3). https://doi.org/10.3390/pharmaceutics14030543, https://www.mdpi.com/1999-4923/14/3/543/pdf.

Lu, J., Gao, X., Wang, S., He, Y., Ma, X., Zhang, T., & Liu, X. (2023). Advanced strategies to evade the mononuclear phagocyte system clearance of nanomaterials. *Exploration, 3*(1). https://doi.org/10.1002/exp.20220045.

Marelli, U. K., Rechenmacher, F., Sobahi, T. R. A., Mas-Moruno, C., & Kessler, H. (2013). Tumor targeting via integrin ligands. *Frontiers in Oncology, 3*. https://doi.org/10.3389/fonc.2013.00222.

Mattheolabakis, G., Milane, L., Singh, A., & Amiji, M. M. (2015). Hyaluronic acid targeting of CD44 for cancer therapy: From receptor biology to nanomedicine. *Journal of Drug Targeting, 23*(7-8), 605–618. https://doi.org/10.3109/1061186X.2015.1052072.

Milano, G., Innocenti, F., & Minami, H. (2022). Liposomal irinotecan (Onivyde): Exemplifying the benefits of nanotherapeutic drugs. *Cancer Science, 113*(7), 2224–2231. https://doi.org/10.1111/cas.15377.

Minocha, N., & Kumar, V. (2022). Nanostructure system: Liposome – A bioactive carrier in drug delivery systems. *Materials Today: Proceedings, 69*, 614–619. https://doi.org/10.1016/j.matpr.2022.09.494.

Moghimipour, E., & Handali, S. (2024). Functionalized liposomes as a potential drug delivery systems for colon cancer treatment: A systematic review. *International Journal of Biological Macromolecules, 269*(Pt 1), 132023. https://doi.org/10.1016/j.ijbiomac.2024.132023.

Mohamed, M., Abu Lila, A. S., Shimizu, T., Alaaeldin, E., Hussein, A., Sarhan, H. A., Szebeni, J., & Ishida, T. (2019). PEGylated liposomes: Immunological responses. *Science and Technology of Advanced Materials, 20*(1), 710–724. https://doi.org/10.1080/14686996.2019.1627174, http://www.tandfonline.com/loi/tsta20#.V0KNDU1f3cs.

Mojarad-Jabali, S., Mahdinloo, S., Farshbaf, M., Sarfraz, M., Fatahi, Y., Atyabi, F., & Valizadeh, H. (2022). Transferrin receptor-mediated liposomal drug delivery: Recent trends in targeted therapy of cancer. *Expert Opinion on Drug Delivery, 19*(6), 685–705. https://doi.org/10.1080/17425247.2022.2083106.

Moosavian, S. A., & Sahebkar, A. (2019). Aptamer-functionalized liposomes for targeted cancer therapy. *Cancer Letters, 448*, 144–154. https://doi.org/10.1016/j.canlet.2019.01.045, www.elsevier.com/locate/canlet.

Musyanovych, A., & Landfester, K. (2014). Polymer micro- and nanocapsules as biological carriers with multifunctional properties. *Macromolecular Bioscience, 14*(4), 458–477. https://doi.org/10.1002/mabi.201300551.

Nambiar, N. R., Gaur, S., Ramachandran, G., Pandey, R. S., M, S., Nath, L. R., Dutta, T., & Sudheesh, M. S. (2024). Remote loading in liposome: A review of current strategies and recent developments. *Journal of Liposome Research, 34*(4), 658–670. https://doi.org/10.1080/08982104.2024.2315449, http://www.tandfonline.com/loi/ilpr20.

Narmani, A., Rezvani, M., Farhood, B., Darkhor, P., Mohammadnejad, J., Amini, B., Refahi, S., & Abdi Goushbolagh, N. (2019). Folic acid functionalized nanoparticles as pharmaceutical carriers in drug delivery systems. *Drug Development Research, 80*(4), 404–424. https://doi.org/10.1002/ddr.21545.

Nogueira-Librelotto, D. R., Codevilla, C. F., Farooqi, A., & Rolim, C. M. B. (2017). Transferrin-conjugated nanocarriers as active-targeted drug delivery platforms for cancer therapy. *Current Pharmaceutical Design, 23*(3), 454–466. https://doi.org/10.2174/1381612822666161026162347, http://www.ingentaconnect.com/content/ben/cpd.

Nsairat, H., Ibrahim, A. A., Jaber, A. M., Abdelghany, S., Atwan, R., Shalan, N., Abdelnabi, H., Odeh, F., El-Tanani, M., & Alshaer, W. (2024). Liposome bilayer stability: Emphasis on cholesterol and its alternatives. *Journal of Liposome Research, 34*(1), 178–202. https://doi.org/10.1080/08982104.2023.2226216, http://www.tandfonline.com/loi/ilpr20.

Nsairat, H., Khater, D., Sayed, U., Odeh, F., Al Bawab, A., & Alshaer, W. (2022). Liposomes: Structure, composition, types, and clinical applications. *Heliyon, 8*(5), e09394. https://doi.org/10.1016/j.heliyon.2022.e09394.

Ortíz, R., Quiñonero, F., García-Pinel, B., Fuel, M., Mesas, C., Cabeza, L., Melguizo, C., & Prados, J. (2021). Nanomedicine to overcome multidrug resistance mechanisms in colon and pancreatic cancer: Recent progress. *Cancers, 13*(9), 2058. https://doi.org/10.3390/cancers13092058.

Pande, S. (2023). Liposomes for drug delivery: Review of vesicular composition, factors affecting drug release and drug loading in liposomes. *Artificial Cells, Nanomedicine, and Biotechnology, 51*(1), 428–440. https://doi.org/10.1080/21691401.2023.2247036.

Papahadjopoulos, D., & Kimelberg, H. K. (1974). Phospholipid vesicles (liposomes) as models for biological membranes: Their properties and interactions with cholesterol and proteins. *Progress in Surface Science, 4*(C), 141–232. https://doi.org/10.1016/s0079-6816(74)80006-7.

Paramshetti, S., Angolkar, M., Talath, S., Osmani, R. A. M., Spandana, A., Al Fatease, A., Hani, U., Ramesh, K. V. R. N. S., & Singh, E. (2024). Unravelling the in vivo dynamics of liposomes: Insights into biodistribution and cellular membrane interactions. *Life Sciences, 346*, 122616. https://doi.org/10.1016/j.lfs.2024.122616.

Pourmadadi, M., Ghaemi, A., Shamsabadipour, A., Rajabzadeh-Khosroshahi, M., Shaghaghi, M., Rahdar, A., & Pandey, S. (2023). Nanoparticles loaded with Daunorubicin as an advanced tool for cancer therapy. *European Journal of Medicinal Chemistry, 258*, 115547. https://doi.org/10.1016/j.ejmech.2023.115547.

Riaz, M. K., Riaz, M. A., Zhang, X., Lin, C., Wong, K. H., Chen, X., Zhang, G., Lu, A., & Yang, Z. (2018). Surface functionalization and targeting strategies of liposomes in solid tumor therapy: A review. *International Journal of Molecular Sciences, 19*(1). https://doi.org/10.3390/ijms19010195, http://www.mdpi.com/1422-0067/19/1/195/pdf.

Rommasi, F., & Esfandiari, N. (2021). Liposomal nanomedicine: Applications for drug delivery in cancer therapy. *Nanoscale Research Letters, 16*(1). https://doi.org/10.1186/s11671-021-03553-8.

Sawant, R. R., & Torchilin, V. P. (2012). Challenges in development of targeted liposomal therapeutics. *AAPS Journal, 14*(2), 303–315. https://doi.org/10.1208/s12248-012-9330-0.

Sercombe, L., Veerati, T., Moheimani, F., Wu, S. Y., Sood, A. K., & Hua, S. (2015). Advances and challenges of liposome assisted drug delivery. *Frontiers in Pharmacology, 6*. https://doi.org/10.3389/fphar.2015.00286, http://journal.frontiersin.org/article/10.3389/fphar.2015.00286/full.

Shazleen Ibrahim, I., Starlin Chellathurai, M., Mahmood, S., Hakim Azmi, A., Harun, N., Ulul Ilmie Ahmad Nazri, M., Muzamir Mahat, M., & Mohamed Sofian, Z. (2024). Engineered liposomes mediated approach for targeted colorectal cancer drug delivery: A review. *International Journal of Pharmaceutics, 651*, 123735. https://doi.org/10.1016/j.ijpharm.2023.123735.

Sheikholeslami, B., Lam, N. W., Dua, K., & Haghi, M. (2022). Exploring the impact of physicochemical properties of liposomal formulations on their in vivo fate. *Life Sciences, 300*, 120574. https://doi.org/10.1016/j.lfs.2022.120574.

Snook, J. T., Park, S., Williams, G., Tsai, Y.-H., & Lee, N. (1999). Effect of synthetic triglycerides of myristic, palmitic, and stearic acid on serum lipoprotein metabolism. *European Journal of Clinical Nutrition, 53*(8), 597–605. https://doi.org/10.1038/sj.ejcn.1600815.

Sofou, S., & Sgouros, G. (2008). Antibody-targeted liposomes in cancer therapy and imaging. *Expert Opinion on Drug Delivery, 5*(2), 189–204. https://doi.org/10.1517/17425247.5.2.189.

Torchilin, V. P. (2010). Passive and active drug targeting: Drug delivery to tumors as an example. *Handbook of Experimental Pharmacology, 197*, 3–53. https://doi.org/10.1007/978-3-642-00477-3_1.

Tortorella, S., & Karagiannis, T. C. (2014). The significance of transferrin receptors in oncology: The development of functional nano-based drug delivery systems. *Current Drug Delivery, 11*(4), 427–443. https://doi.org/10.2174/1567201810666140106115436, http://www.benthamdirect.org/pages/all_b_bypublication.php.

Urbinati, G., Marsaud, V., & Renoir, J. M. (2012). Anticancer drugs in liposomal nanodevices: A target delivery for a targeted therapy. *Current Topics in Medicinal Chemistry, 12*(15), 1693–1712. https://doi.org/10.2174/156802612803531423.

Vale, N., Duarte, D., Silva, S., Correia, A. S., Costa, B., Gouveia, M. J., & Ferreira, A. (2020). Cell-penetrating peptides in oncologic pharmacotherapy: A review. *Pharmacological Research, 162*, 105231. https://doi.org/10.1016/j.phrs.2020.105231.

Zahednezhad, F., Allahyari, S., Sarfraz, M., Zakeri-Milani, P., Feyzizadeh, M., & Valizadeh, H. (2024). Liposomal drug delivery systems for organ-specific cancer targeting: Early promises, subsequent problems, and recent breakthroughs. *Expert Opinion on Drug Delivery, 21*(9), 1363–1384. https://doi.org/10.1080/17425247.2024.2394611.

Zahiri, M., Taghdisi, S. M., Abnous, K., Ramezani, M., & Alibolandi, M. (2021). Fabrication of versatile targeted lipopolymersomes for improved camptothecin efficacy against colon adenocarcinoma in vitro and in vivo. *Expert Opinion on Drug Delivery, 18*(9), 1309–1322. https://doi.org/10.1080/17425247.2021.1928631, http://www.tandfonline.com/loi/iedd20.

Zhao, Y., Xu, J., Le, V. M., Gong, Q., Li, S., Gao, F., Ni, L., Liu, J., & Liang, X. (2019). EpCAM aptamer-functionalized cationic liposome-based nanoparticles loaded with miR-139-5p for targeted therapy in colorectal cancer. *Molecular Pharmaceutics, 16*(11), 4696–4710. https://doi.org/10.1021/acs.molpharmaceut.9b00867.

CHAPTER

# 5 siRNA–polymer conjugates for the delivery of RNAi therapeutics for the treatment of colon cancer

**Swetha Belavadi, Sampriti Paul, Sunil Kumar Kadiri and Prashant Tiwari**
*Department of Pharmacology, College of Pharmaceutical Sciences, Dayananda Sagar University, Bengaluru, Karnataka, India*

## 5.1 Introduction

### 5.1.1 Overview of colorectal cancer

Colorectal cancer (CRC) is a multifactorial disease that affects the colon (large intestine) or rectum. CRC is ranked as the third common cancer in global burden and the second in cancer-related mortality. The burden of CRC is projected to increase to 3.2 million cases per year and 1.6 million deaths per year by 2040. Rectal cancer constitutes approximately 49.66% of cases, while colon cancer accounts for 49.09%, with 1.25% involving both sites concurrently. Within colon cancers, the sigmoid colon represents the most frequent site of occurrence (55%), followed by the ascending colon (23.3%), transverse colon (8.5%), descending colon (8.1%), cecum (8.0%), and crossing site (2.1%). Globally, males are at a higher risk than females (Duan et al., 2022). CRC is diagnosed globally in individuals over the age of 50, with the median age at diagnosis being around 60–65 years. However, the latest epidemiological studies have reported a rising incidence in younger adults (<50 years), particularly in high-income countries, which has raised concerns about early-onset CRC (Fadlallah et al., 2024).

The incidence of CRC shows marked geographical variation, with the highest rates reported in North America, Europe, Australia, and New Zealand. In contrast, the lowest incidence is observed in Africa, Asia, and South America. Occurrence is characterized by a higher incidence in the economically developed eastern coastal regions and a lower incidence in the less developed western areas.

A temporal trend has been observed in the morbidity due to CRC. High-incidence regions have shown a slowing or even a decline in CRC rates, largely attributable to the implementation of effective screening programs and preventive strategies in high-income countries. In contrast, traditionally low-incidence regions are now experiencing a rising trend, which is believed to be driven by an aging population, lifestyle modifications, environmental changes, and increasing westernization (Duan et al., 2022).

CRC arises from a complex interplay of internal gene modifications and external influences, including lifestyle, diet, inflammation, and the tumor microenvironment. Unhealthy food habits, including high intake of red and processed meat, low fiber, and high-fat diets, increase susceptibility, while obesity, smoking, and alcohol further elevate risk. Chronic inflammation, particularly in

Nano Drug Delivery Systems for Colon Cancer. DOI: https://doi.org/10.1016/B978-0-443-40443-6.00013-1

conditions such as inflammatory bowel disease, creates a pro-tumorigenic environment (World Health Organization: WHO & World Health Organization: WHO, 2023). Additionally, alterations in the gut microbiota and interactions within the tumor microenvironment, including stromal support and immune evasion, promote tumor initiation, growth, and metastasis (Mármol et al., 2017).

#### 5.1.1.1 Current therapeutics in colorectal cancer

Significant progress has been made in the treatment of CRC, especially through the identification of specific protein targets and the development of corresponding therapeutics. These targeted strategies are designed to address the drawbacks of conventional chemotherapy, including its lack of specificity and associated toxicity to healthy cells.

In CRC, several key molecular targets have been identified and are central to therapeutic strategies. Oncogenic drivers include epidermal growth factor receptor (EGFR) and its downstream effectors (RAS, BRAF), with Kirsten rat sarcoma viral oncogene homolog (KRAS) mutations being highly prevalent and BRAF-V600E mutations occurring in 8%–12% of cases. Tumor suppressors such as adenomatous polyposis coli (APC), SMAD4, and p53 are frequently inactivated, leading to dysregulation of Wnt/β-catenin signaling, impaired TGF-β signaling, and loss of cell cycle control. Angiogenesis-related pathways, particularly vascular endothelial growth factor (VEGF)/VEGFR signaling, are critical for tumor growth and metastasis, with multikinase inhibitors like regorafenib targeting VEGFRs, RET, c-KIT, and PDGFR. Immune checkpoints, including PD-1/programmed death-ligand 1 (PD-L1), cytotoxic T-lymphocyte–associated protein-4 (CTLA-4), LAG-3, and TIM-3, are upregulated in the tumor microenvironment, enabling immune escape and serving as targets for immunotherapy. Additional targets include HER2 overexpression in ~5% of metastatic CRCs, COX-2 enzymes, which can be inhibited by nonsteroidal anti-inflammatory drugs (NSAIDs) to reduce adenoma formation, and PDE5, where inhibitors induce apoptosis. Together, these targets provide the foundation for precision medicine approaches in CRC (Makaremi et al., 2021, Cornista et al., 2023).

#### 5.1.1.2 Active therapeutics available against colorectal cancer

Management of CRC is multidisciplinary, incorporating surgery, radiotherapy, chemotherapy, targeted therapies, and immunotherapy, with treatment choices tailored according to the cancer type, disease stage, patient history, and specific tumor characteristics.

Chemotherapy continues to play a central role in the treatment of CRC, with several well-established regimens. Fluoropyrimidines, such as 5-fluorouracil (5-FU) and its oral prodrug capecitabine, act as antimetabolites by inhibiting thymidylate synthase, thereby disrupting DNA synthesis and repair, ultimately leading to apoptosis. Leucovorin (folinic acid) is often coadministered with 5-FU because it stabilizes the 5-FU–thymidylate synthase complex, thereby enhancing cytotoxicity while also reducing adverse effects. Oxaliplatin, a third-generation platinum compound, exerts its effect by forming DNA adducts and crosslinks, which interfere with DNA replication and transcription, resulting in apoptosis. Irinotecan, a camptothecin derivative, inhibits topoisomerase I, an enzyme essential for relieving torsional strain during DNA replication, leading to lethal DNA strand breaks (Jafarlou & Jafarlou, 2025).

Combination protocols include FOLFOX (5-FU, leucovorin, oxaliplatin), FOLFIRI (5-FU, leucovorin, irinotecan), and FOLFOXIRI (5-FU, leucovorin, oxaliplatin, irinotecan), which are selected based on disease burden, patient condition, and therapeutic goals. Additionally, trifluridine-tipiracil (TAS-102), an oral fluoropyrimidine derivative, provides benefits in refractory metastatic CRC. Trifluridine incorporates directly into DNA in place of thymidine, thereby disrupting DNA function, while tipiracil acts as a thymidine phosphorylase inhibitor, prolonging trifluridine's bioavailability and therapeutic effect. Collectively, these regimens exploit distinct yet complementary mechanisms to inhibit DNA synthesis, induce DNA damage, or block repair, thereby reducing tumor burden and prolonging survival in CRC patients (Jafarlou & Jafarlou, 2025).

Targeted therapy in CRC focuses on selectively inhibiting specific molecular pathways that drive tumor growth and progression. Agents directed against the EGFR pathway include cetuximab, a chimeric IgG monoclonal antibody (mAb) approved for metastatic CRC in patients without KRAS mutations, and panitumumab, a fully human mAb with similar indications. Other anti-EGFR antibodies, such as nimotuzumab and necitumumab, have also been investigated, with nimotuzumab demonstrating a favorable safety profile. Targeting the VEGF/VEGFR pathway, bevacizumab (anti-VEGF-A) is widely used to inhibit angiogenesis, while newer agents include fruquintinib, an oral tyrosine kinase inhibitor of all VEGFR subtypes approved in 2023 for refractory metastatic CRC, and aflibercept, a recombinant fusion protein acting as a decoy receptor for VEGFR1 and VEGFR2, approved as second-line therapy. Additional options include ramucirumab, a VEGFR-2–specific antibody, and regorafenib, an oral multitargeted tyrosine kinase inhibitor active against VEGFRs and several other kinases, used in previously treated metastatic CRC. For BRAF-mutant CRC, encorafenib, in combination with cetuximab, received Food and Drug Administration (FDA) approval in 2020 for BRAF-V600E metastatic disease, with binimetinib (MEK inhibitor) and vemurafenib also explored in combination strategies. In HER2-positive metastatic CRC, targeted approaches such as trastuzumab, a recombinant humanized mAb against HER2, combined with tucatinib, an oral HER2-selective tyrosine kinase inhibitor, received accelerated FDA approval in January 2023 for patients refractory to standard chemotherapy. Collectively, these targeted therapies represent precision medicine strategies that improve outcomes in biomarker-defined subgroups of CRC patients (Manzi et al., 2023).

Immunotherapy has emerged as an important treatment modality in CRC, particularly for tumors with high microsatellite instability (MSI-H) or deficient mismatch repair (dMMR), which are characterized by a high mutational burden and increased neoantigen formation that make them more responsive to immune-based approaches. The most widely used agents are immune checkpoint inhibitors (ICIs), which work by blocking inhibitory pathways that normally suppress T-cell activation, thereby restoring antitumor immune responses. Pembrolizumab, a mAb against programmed death receptor-1 (PD-1), prevents the interaction between PD-1 on T cells and its ligands (PD-L1/PD-L2) on tumor or immune cells, enhancing cytotoxic T-cell activity; it is FDA-approved for both first- and second-line treatment of unresectable or metastatic MSI-H/dMMR CRC. Similarly, nivolumab, another anti-PD-1 antibody, has demonstrated efficacy in MSI-H/dMMR metastatic CRC and is also approved in combination with low-dose ipilimumab, a mAb targeting CTLA-4. By inhibiting CTLA-4, ipilimumab promotes early T-cell activation and expansion, while nivolumab sustains effector T-cell activity, together providing synergistic immune activation. Dostarlimab, another anti-PD-1 antibody, received FDA approval in January 2023 for the treatment of dMMR/MSI-H locally advanced rectal cancer, showing remarkable clinical benefit in this subset. Collectively, these ICIs unleash antitumor immunity by counteracting immune evasion mechanisms in CRC, offering durable responses in patients with genetically defined subtypes (Golshani & Zhang, 2020).

In addition to standard treatment, several repurposed drugs and emerging approaches are being investigated for CRC. Sildenafil, a phosphodiesterase-5 (PDE5) inhibitor originally developed for hypertension and erectile dysfunction, has shown potential in inducing apoptosis and enhancing the efficacy of 5-fluorouracil (5-FU). NSAIDs such as celecoxib, which reduces colorectal adenoma occurrence and augments the antitumor activity of cetuximab, and mesalamine, which promotes anchorage-independent CRC cell death through COX inhibition, have also demonstrated therapeutic promise. Chloroquine, an antimalarial drug, can overcome 5-FU resistance by blocking autophagy. Advances in biomarker-driven strategies include Tasin-1 and statins targeting APC mutations, deltarasin and cysmethynil targeting KRAS, vactosertib acting on SMAD4, and nutlin-3 and PRIMA-1MET restoring p53 activity. Immunotherapeutic innovations such as adoptive cell transfer (ACT), employing autologous lymphocytes including tumor-infiltrating lymphocytes (TILs) or CAR-T cells targeting carcinoembryonic antigen (CEA), and cancer vaccines directed against tumor-associated antigens (e.g., CEA, MUC1, HER2, Survivin-2B, SART3, RNF43, TOMM34) using peptide-, messenger RNA (mRNA)-, cell-, or vector-based platforms, are under clinical evaluation. Other novel modalities include nanoparticle-mediated radiotherapy (RT) to improve tumor specificity, proton beam therapy (PBT) and heavy ion therapy (HIT) for precise energy delivery to recurrent tumors, and noncoding RNA-based therapeutics (microRNAs [miRNAs] and small interfering RNAs [siRNAs]) designed to regulate oncogenic gene expression, though none have yet received FDA approval. Finally, oncolytic viral therapies, which exploit viruses to selectively lyse tumor cells while stimulating antitumor immunity, are being explored as potential future options for CRC treatment (Yu et al., 2025).

### 5.1.2 RNA interference therapeutics

RNA interference (RNAi) is a highly conserved biological process in which double-stranded RNA (dsRNA) mediates sequence-specific gene silencing. Serving as a natural defense mechanism, RNAi protects against endogenous parasitic and exogenous pathogenic nucleic acids, while also regulating the expression of protein-coding genes. This phenomenon, also referred to as posttranscriptional gene silencing (PTGS), is primarily driven by siRNAs, which are short dsRNA molecules approximately 21–22 bp in length with a characteristic two-nucleotide 3′ overhang. These siRNAs are either derived endogenously from precursor molecules or introduced exogenously for experimental or therapeutic purposes. By guiding cellular silencing complexes to complementary mRNAs, siRNAs play a pivotal role in targeted posttranscriptional regulation of gene expression (RNA Interference (RNAI), n.d.).

The RNAi pathway mechanizes through two major steps: the processing of trigger RNA and gene silencing via the RNA-induced silencing complex (RISC). In mammalian cells, long dsRNAs or precursor microRNAs (pre-miRNAs) are processed into short interfering RNAs (siRNAs) by the Drosha–DGCR8 complex in the nucleus and the Dicer–TRBP complex in the cytoplasm. The siRNA is incorporated into the RISC, where the guide strand directs RISC to complementary target mRNA, leading to mRNA cleavage or translational repression, thereby silencing gene expression. Collectively, this sequence-specific targeting leads to effective gene silencing. Moreover, RNAi can also be experimentally induced by introducing exogenous dsRNA or constructs expressing short hairpin RNAs (shRNAs), expanding its utility in both research and therapeutic applications (Aagaard & Rossi, 2007).

siRNA plays a crucial role in both general therapeutic development and biomedical research, offering a potent and specific method to inhibit disease targets and potentially transform modern medicine.

In therapeutic development, siRNA therapeutics address a significant challenge by providing a means to rapidly identify specific and potent inhibitors for disease targets across all molecular classes, especially those that have been intractable to traditional small-molecule and protein-based pharmaceutical approaches. The broad potential application of RNAi therapeutics has been demonstrated through numerous proof-of-concept studies in animal models of human diseases, and multiple clinical trials with RNAi therapeutic candidates are currently underway. The primary hurdle for successful drug development, however, remains the identification of effective delivery strategies that can be translated into clinical use.

siRNA has been widely applied in biomedical research as a proof-of-concept therapeutic for diverse diseases and conditions. Its applications include inhibiting infection, inflammation, fibrosis, and addressing metabolic regulation, body weight disorders, and chronic neuropathic pain. siRNA has also shown efficacy in neurodegenerative diseases such as spinocerebellar ataxia, Huntington's disease, amyotrophic lateral sclerosis (ALS), and Alzheimer's disease. In oncology and virology, it has been utilized against HPV-16 E6 oncogene expression, glioblastoma, colon adenocarcinoma, prostate cancer, Ewing's sarcoma, hepatitis B virus replication, and Ebola virus infection (Bumcrot et al., 2006; Kang et al., 2023).

### 5.1.3 Role of small interfering RNA in cancer therapy

siRNA represents a powerful therapeutic tool in oncology due to its capacity for precise, sequence-specific gene silencing. By leveraging the endogenous RNAi pathway, siRNAs degrade target mRNAs, thereby blocking the production of oncogenic proteins that drive tumor progression, angiogenesis, and drug resistance. This mechanism allows for modulation of pathways previously considered "undruggable," positioning siRNA as a cornerstone of emerging precision therapies.

Therapeutic applications of siRNA in cancer: siRNAs can silence a broad spectrum of cancer-associated genes. Antiapoptotic proteins such as Bcl-2, Mcl-1, survivin, and HIF-1α are prime targets, with silencing enhancing apoptosis and sensitizing tumors to chemotherapy. Similarly, inhibition of angiogenic mediators such as VEGF reduces tumor vascularization, while targeting oncogenes, including KRAS, EGFR, HER2, and c-MET, disrupts critical growth and survival pathways. siRNAs have also been designed to suppress immune modulators such as MUC1, thereby enhancing immunogenicity and supporting immunotherapies (Zhang et al., 2023).

Overcoming drug resistance: Resistance to chemotherapy and targeted agents remains a major clinical challenge. siRNA provides a strategy to overcome this by targeting drug efflux pumps (e.g., MDR1/ABCB1, ABCG2), DNA repair enzymes (e.g., PARP1, ERCC1), and oncogenic signaling pathways (e.g., KRAS, PI3K/AKT/mTOR). Inhibiting these mechanisms restores drug sensitivity, improves intracellular drug retention, and enhances tumor cell apoptosis, offering synergistic benefits when combined with standard regimens (Charbe et al., 2020).

Advantages in therapy: The precision of siRNA therapy allows for selective targeting of cancer cells while minimizing toxicity to normal tissues. This specificity enables dose reduction of conventional chemotherapeutics, lowering systemic side effects such as myelosuppression, neurotoxicity, and gastrointestinal damage. For example, VEGF-targeted siRNAs in combination with paclitaxel have been shown to enhance antitumor responses at reduced drug doses (Zhang et al., 2020).

Delivery challenges and advances: Despite its promise, siRNA therapy faces barriers including nuclease degradation, rapid clearance, poor cellular uptake, and potential off-target effects (OTEs). To address these limitations, advanced delivery platforms have been developed. Lipid nanoparticles (LNPs), already used clinically in RNA-based therapies, protect siRNAs and aid in cytoplasmic release. Polymeric nanoparticles, gold nanoparticles (AuNPs), mesoporous silica carriers, and exosomes further enhance tumor targeting. Chemical modifications (e.g., 2′-O-methyl, 2′-fluoro, and phosphorothioate linkages) increase stability and reduce immunogenicity, while ligand-based and pH-sensitive systems ensure tumor-specific release. Efficient endosomal escape mechanisms remain a critical factor in maximizing efficacy (Hoang et al., 2025).

siRNA-based therapeutics offer a novel, versatile, and highly specific approach for cancer treatment, capable of silencing oncogenic drivers, overcoming resistance, and improving therapeutic efficacy with reduced toxicity. While delivery challenges remain, continued innovation in nanotechnology and chemical engineering is paving the way for siRNA to become a key modality in precision oncology.

## 5.2 Small interfering RNA–polymer conjugates: design and synthesis

Polymeric nanocarriers for siRNA delivery
Chemical conjugation strategies
Stimuli-responsive polymers

### 5.2.1 Development of small interfering RNA

siRNA design involves selecting specific sequences and applying chemical modifications to enhance its therapeutic properties, while the steps involved include its synthesis and the subsequent mechanism of action within cells to silence target genes.

Initially, multiple RNA strands are designed and screened through experimental and computational methods to identify sequences capable of effectively targeting and binding to the desired mRNA. Canonical siRNAs typically consist of 21–23 bp double-stranded fragments mimicking Dicer products, while Dicer-substrate siRNAs (25–30 nucleotides) offer enhanced potency and reduced immune stimulation. siRNAs can also be engineered into single-stranded, hairpin, or dumbbell-shaped forms, provided the antisense strand is efficiently incorporated into the RISC (Dong et al., 2019).

To enhance stability and efficacy, chemical modifications are introduced at specific sites: 2′-position modifications (e.g., 2′-O-methyl, 2′-O-methoxyethyl, fluoro, and locked nucleic acids) improve half-life and binding affinity; phosphate backbone modifications (e.g., phosphorothioates, phosphorodithioates) increase nuclease resistance; ribose ring modifications (e.g., morpholino nucleosides, UNAs) reduce OTEs and enhance biostability; and nucleobase modifications (e.g., 2-thiouracil) improve duplex stability. The sense strand can tolerate extensive modifications, whereas the antisense strand requires careful optimization to preserve silencing activity (Chernikov et al., 2023).

Synthesis of siRNA is primarily achieved by automated solid-phase synthesis, which involves repetitive deprotection, coupling, capping, and oxidation steps, yielding high efficiency and purity, followed by cleavage, deprotection, and purification. Alternative methods include rolling circle transcription, generating multimeric siRNAs with self-assembly properties, and chemical cross-linking, where thiol-modified strands form stable multimeric constructs for delivery. Collectively, these design and synthesis strategies enable the production of stable, potent, and therapeutically relevant siRNAs (Beaucage & Reese, 2009; Dong et al., 2019).

#### 5.2.1.1 The need for optimization of small interfering RNA

siRNA conjugates are designed to address the challenges of siRNA delivery, including instability, rapid clearance, and poor cellular uptake. siRNA molecules are inherently unstable and vulnerable to degradation; they require additional support before being used therapeutically. Polymers play a crucial role in siRNA conjugates by shielding them from nucleases, boosting their stability, and prolonging their presence in the bloodstream. Also, they pack the siRNA into nanoparticles, facilitate targeted delivery, manage controlled release based on stimulus-sensitive bonds, and reduce toxicity, overall improving the efficacy of siRNA-based treatments (Patel et al., 2023).

### 5.2.2 Polymeric nanocarriers for small interfering RNA delivery

Polymeric nanocarriers are widely investigated materials that protect siRNA from degradation and facilitate its uptake by target cells for gene silencing therapies. Polymeric nanocarriers represent a significant advancement in siRNA-based therapeutics, addressing key limitations such as instability, poor cellular uptake, and lack of target specificity. siRNA, a potent tool for sequence-specific gene silencing via RNAi, holds promise for treating cancer, genetic disorders, inflammatory diseases, and neurodegenerative diseases, yet its clinical translation is hindered by enzymatic degradation, inefficient delivery across biological barriers, and OTEs (Wang et al., 2021).

Polymeric nanocarriers overcome these challenges by enhancing stability, ensuring efficient encapsulation, enabling targeted and controlled release, and offering biocompatibility and scalability. A variety of systems have been developed, including poly lactic-co-glycolic acid-polyethylene glycol (PLGA-PEG), poly β-amino esters (PBAEs), poly(amido amine) (PAMAM) dendrimers, pH-responsive polymers, and ligand-functionalized carriers, each contributing to improved protection, cellular uptake, and endosomal escape of siRNA. These platforms have demonstrated efficacy in cancer therapy by silencing oncogenic targets, enhancing immunotherapy through checkpoint inhibition, and facilitating imaging-guided delivery for precision medicine.

#### 5.2.2.1 Chitosan

Chitosan is a natural, biodegradable polysaccharide widely explored for siRNA delivery due to its electrostatic bonds and ability to condense negatively charged siRNA into stable nanoparticles. The native chitosan can be modified to trimethyl chitosan (TMC), which improves stability and charge density, carboxymethyl chitosan with pH sensitivity, PEGylated or ligand-modified chitosan for enhanced stability and targeting, and low-molecular-weight chitosan-oligosaccharides for better biocompatibility. Chitosan is composed of D-glucosamine and N-acetyl-D-glucosamine units, which enable strong ionic interactions with siRNA. The release of the loaded siRNA can occur through endosomal acidification, which weakens polymer–siRNA interactions, enzymatic degradation by lysozyme or related enzymes, or through engineered stimuli-responsive linkages such as pH-labile and redox-sensitive bonds. These properties make chitosan a primary carrier of efficient siRNA delivery (Al-Absi et al., 2022).

#### 5.2.2.2 Cyclodextrin

Cyclodextrins (CD) are cyclic oligosaccharides with a hydrophilic outer surface and a hydrophobic cavity. For siRNA delivery, modified CDs are chosen, as native CDs are neutral, and cationic CD polymers, like β-cyclodextrin-containing polycations, along with PEGylated or ligand-functionalized CDs, are commonly used in siRNA delivery to enhance binding capacity, stability, and targeting. Chemically, CDs possess multiple hydroxyl groups that can be linked with amines or guanidinium groups to enable electrostatic condensation of siRNA. Incorporation of polyethylene glycol (PEG) or other polymers improves pharmacokinetics and reduces toxicity. The loaded siRNAs are released through acid environment destabilization, competitive exchange, or cleavage of stimuli-responsive linkers. These properties make cyclodextrin-based polymers versatile, biocompatible carriers capable of protecting siRNA, and, more importantly, CDs have the benefit of versatility (Singh et al., 2019; Mousazadeh et al., 2020).

#### 5.2.2.3 Poly lactic-co-glycolic acid-polyethylene glycol

PLGA-PEG refers to a polymeric conjugation of poly lactic-co-glycolic acid (PLGA) and a hydrophilic polymer, PEG, along with a biodegradable conjugate that enhances siRNA permeability upon the inclusion of amino acids like arginine and lysine to the core. The PEGylation of PLGA enhances the half-life by preventing protein adsorption and opsonization, preventing nuclease degradation, and improving the pharmacokinetic profile by evading the reticuloendothelial system (Kumari et al., 2023; Zhang et al., 2014).

Initially, small molecules conjugated with PLGA-PEG were developed to enhance the potential of PLGA-PEG nanocarriers in improving therapeutic delivery and minimizing adverse effects. Various molecules, such as the F(ab) fragment of anti-PD-L1 mAb, folic acid (FA), curcumin, berberine, and folate (FOL), have been incorporated into PLGA/PEG-based nanocarrier systems, demonstrating remarkable potential to enhance drug delivery, improve tumor specificity, increase cellular uptake, and amplify therapeutic efficacy in colon cancer (Kesharwani et al., 2025).

siRNA targeting critical oncogenic and immune regulatory pathways, including EGFR, B-cell lymphoma 2 (BCL-2), programmed death 1/programmed cell death-ligand-1 (PD-1/PD-L1), polo-like kinase 1 (PLK-1), and survivin, has been successfully delivered using PLGA-PEG nanoparticles. This nanocarrier system has demonstrated remarkable therapeutic potential by enhancing cellular permeability, ensuring high target specificity, and achieving robust gene silencing efficiency, while simultaneously maintaining a favorable safety profile characterized by low cytotoxicity. Collectively, these outcomes highlight the versatility of PLGA-PEG as a delivery platform for siRNA-based therapeutics in cancer treatment (Kwak et al., 2019; Murata et al., 2007; Zhi et al., 2021).

#### 5.2.2.4 Polycationic cross-linked polymers

Polycationic cross-linked polymers are positively charged networks designed to hold and protect siRNA through electrostatic interactions while providing enhanced stability and release. Common backbone polymers include polyethyleneimine (PEI), poly(β-amino esters) (PBAE), poly(L-lysine) (PLL), and PAMAM dendrimers. Their crosslinking chemistry involves redox-based disulfide bonds, pH-labile linkages, biodegradable ester bonds, or photo-based crosslinks. These linkages are tailored for the intracellular release of siRNA (Yuan et al., 2020; Steele et al., 2011).

#### 5.2.2.4.1 Poly(ethylene glycol)-poly(lactic acid)

Poly(ethylene glycol)-poly(lactic acid) (PEG-PLA) is an amphiphilic, biodegradable polymer that has gained considerable attention as a delivery system for siRNA in CRC therapy. The PEG region provides hydrophilicity, improving the loaded siRNA's systemic circulation by reducing opsonization and clearance by the reticuloendothelial system. The polylactic acid (PLA) segment, on the other hand, provides a hydrophobic envelope for siRNA, protecting it from nuclease degradation. This conjugation facilitates stable nanocarriers, enhancing cellular uptake, solubility by almost 5000 times, less toxicity, and controlled release of siRNA (Mundel et al., 2022; Schädlich et al., 2011). The utilization of PEG-PLA loaded siRNA in real-time CRC studies shows improved uptake by endocytosis, inhibited proliferation, migration, and reduced degradation in a synergistic manner (Zhang et al., 2025).

#### 5.2.2.4.2 Poly β-amino ester

PBAE are one of the most efficient polymers for nanocarriers and delivery. PBAE are biodegradable, cationic nanodelivery systems with enhanced specificity. These are highly effective in case of anionic siRNA, as their reversed charge handles the spontaneous nanoparticle formation, endosomal escape, hydrolytic degradability, improves RNA binding, and the hydrolysable backbone aids in selective release, improving the delivery (Dosta et al., 2018; Karlsson et al., 2020). The flexible structure allows for customization, such as end-capping, helper lipids, and peptide conjugation, can optimize the tailored delivery of siRNA to specific cells. siRNA utilizing PBAE enhances the endosomal escape, boosting gene silencing and thus reducing toxicity compared to older polymers like PEI (Chen et al., 2024). PBAE-based siRNA nanoparticles represent a promising, versatile, and nonviral platform for advancing RNAi therapies in clinical settings (Kozielski et al., 2013).

#### 5.2.2.4.3 Poly(amido amine) polymer

PAMAM is also a type of cationic polymer widely used for nucleic acid delivery. It differs from other previously discussed polymers in its structure. PAMAMs are highly branched, dendritic polymers, comprising a central core and repeated synthesis of branching units with abundant terminals. PAMAM is rich in a network of amines, providing a high cationic charge and a sponge-like effect, which aids in easy endosomal release upon stimulation. The loaded siRNA is typically released by endosomal acidification, redox-responsive links, or acid-labile modifications, which destabilize the loaded polymer. These factors facilitate PAMAM as one of the strong and versatile platforms for effective control, release, and gene silencing (Wu et al., 2013; Kheiriabad et al., 2020).

Beyond oncology, they are also being explored for hepatic, cardiovascular, ocular, inflammatory, and neurodegenerative conditions. Despite substantial progress, challenges remain, including toxicity, OTEs, and the need for sustainable production and large-scale clinical translation. Future directions involve the development of novel degradable polymers, integration of green chemistry principles, and the application of artificial intelligence (AI) and nanoinformatics to optimize siRNA design and delivery. With ongoing innovation, polymeric nanocarriers hold strong potential to revolutionize human therapeutics.

### 5.2.3 Chemical conjugation strategies

Chemical conjugation strategies in siRNA delivery are designed to improve its stability, circulation time, cellular uptake, and target specificity, while reducing off-target reactions and toxicity. Naked siRNAs are unstable and prone to clearance and degradation; chemical modifications through the attachment of functional groups, ligands, and polymers will enhance the delivery. Various strategies are available as chemical conjugates to enhance siRNA stability and delivery (Kakkis et al., 2025).

#### 5.2.3.1 Lipid conjugation

The most common concern in the development of siRNA is overcoming the barriers to delivery, as siRNAs are large, charged molecules that lack a decent half-life and circulation time, affecting their permeability. Lipid conjugation represents conjugate-mediated delivery, with promising targeted siRNA delivery. Various options in lipid conjugates provide a broader permutation in choosing and deciding on the appropriate lipid for siRNA. Lipophilic conjugates enhance siRNA pharmacokinetics and facilitate efficacy in a broad range of tissues, including the kidney, liver, lung, adipose tissue, and gastrointestinal tract. Lipid-conjugated siRNAs exhibit prolonged circulation time in vivo, as the lipid moieties display strong binding affinity toward circulating proteins and readily integrate into the plasma membrane, thereby facilitating rapid cellular internalization through endocytosis (Osborn & Khvorova, 2018; Kubo et al., 2020).

#### 5.2.3.2 Peptide conjugation

Peptide-conjugated siRNA employs small peptides as delivery carriers to transport siRNA into the cells, enabling them to exert their gene-silencing effects on selective targets for the therapeutic intervention in CRC. These are derived from the combination of 20 amino acids, as peptides display various three-dimensional conformations, charges, polarities, and hydrophobic or hydrophilic properties, which enhance uptake, target delivery, and endosomal escape. Furthermore, as peptides are naturally abundant biomolecules within living systems, they offer an extensive proteomic reservoir for the identification and design of functional peptides tailored to enhance siRNA-based therapies for CRC (Sciences, 2024; Tai & Gao, 2016).

The addition of specific peptides can enhance the necessary function in siRNA delivery and action. Utilization of cell permeation peptides (CPPs) can enhance the permeability and uptake of siRNAs for an enhanced therapeutic effect. CPPs function by either energy-dependent endocytosis or energy-independent direct translocation (Gori et al., 2023).

Utilization of endosome-disrupting peptides facilitates the destabilization of the endosomal membrane and releases the loaded particles, such as siRNA. EDPs employ amphipathic peptides that are positively charged, bind to the negatively charged membrane, and disrupt it by electrostatic binding (Grau & Wagner, 2024).

Targeting peptides are commonly conjugated with siRNA to enhance the specificity and targetability. These short peptides guide the siRNA to the specific regions and facilitate silencing of genes that have a specific affinity for disease-related biomolecules (Tai & Gao, 2016).

#### 5.2.3.3 Aptamer conjugation

Aptamers are short, single-stranded nucleic acid molecules that fold into unique three-dimensional structures, enabling them to bind specifically and with high affinity to the target cell surface. siRNA conjugated with aptamers serves as targeting ligands that guide siRNA into desired cells and silence the targeted protein. Aptamers are one of the most suitable conjugates for siRNA due to their chemical stability, low immunogenicity, easy synthesis, and high specificity (Sivakumar et al., 2018).

#### 5.2.3.4 Antibody conjugates

In the field of targeted therapy, an advanced method known as antibody-siRNA conjugates has emerged following the steps of antibody-drug conjugates. ARCs specifically aim to overcome the inherent drawbacks of natural siRNA, such as poor stability, excessive negative charge, and lack of uptake by cells. Unlike exogenous lipid-based carriers, ARCs utilize endogenous molecules, offering better safety, permeability, and organ specificity. The activity of ARCs can be potentiated by the addition of CPPs to enhance permeability and promote endosomal escape (Cao et al., 2022).

Conjugation of target-specific antibodies can synergize the therapeutic effect and can be utilized to overcome therapeutic concerns like drug resistance, as seen in CRC (Bäumer et al., 2015).

#### 5.2.3.5 Small-molecule conjugate

Small-molecule conjugation of siRNA involves attaching small molecules like lipids or vitamins (cholesterol, vitamin B, and N-acetylgalactosamine) to improve the stability, cellular uptake, and tissue targeting of siRNA. These conjugates use natural transport pathways to enhance siRNA delivery and prolong the circulation time. Once inside the cell, siRNA is released by the cleaving linkers, and it interferes with the RNA mechanism. Compared to unconjugated siRNA, small ligand conjugates present better efficacy, stability, and lower immunogenic structures (Chernikov et al., 2019).

### 5.2.4 Stimuli-responsive polymers

Stimuli-responsive polymers (SRPs) are advanced components used in designing effective siRNA or their nanocarriers. SRPs are designed to change swelling, aggregation, degradation, charge reversal, or dissociation in response to subtle changes in the biological or external environment. This property allows for demanded siRNA delivery, cellular uptake, and controlled siRNA release. SRPs differ from stimuli-responsive linkers as the latter undergo modification upon entering the cell and have no significant role in the delivery of the siRNA, only in the release. On the other hand, SRPs aid in the specific delivery and launch of the loaded siRNA. SRPs can be chosen based on various stimulation factors seen in the target environment, as mentioned in Table 5.1 (Chernikov et al., 2019; Saw et al., 2019; Zhou et al., 2023; Mi, 2020).

## 5.3 Mechanisms of action

siRNA delivery involves encapsulation and delivery to target cells, cell entry via endocytosis, endosomal escape, and lastly activation and binding to the RISC complex for cleaving mRNA and gene silencing.

**Table 5.1 Stimuli-responsive polymers and their mechanisms of action.**

| Stimuli | Environmental factor | Response mechanism | Examples |
|---|---|---|---|
| **Endogenous stimuli** | | | |
| pH | Low pH in tumor microenvironment (pH 6.5–6.8)Highly acidic endosomal/lysosomal organelles (pH 4–6) | Protonation, leading to nuclear escapeDissociation followed by small interfering RNA (siRNA) release | Tertiary amines, imidazoles, polyhistidine, polyethylenimine, poly(β-amino ester), imine, acetal/ketal, maleic acid amide derivatives |
| Enzyme activity | Overexpression of specific enzymes | Breakdown of enzyme-sensitive linkers, resulting in siRNA release | MMP-2, HLE, Cathepsin B substrates |
| Hypoxia | Hypoxic levels in solid tumors | Reduction of azo or nitro groups by the reductase enzymes | Nitroimidazole, azobenzene |
| Thermal | Elevated temperatures | Breakdown of temperature-sensitive polymersIncorporation of $NH_4HCO_3$ in liposomes generates $CO_2$ at elevated temperatures, resulting in the rupture | Poly(N-isopropylacrylamide), poly(N-inyl isobutyramide) (PAMAM), poly[2-(2-methoxyethoxy)ethyl methacrylate] |
| Reactive oxygen species (ROS) | Elevated ROS levels in the tumor environment | Utilization of ROS-mediated cleavage, sulfur-linked polymersBoric acid linkers react with a $H_2O_2$-rich environmentGSH-mediated linkers | Polyethylene glycol shielded poly(L-lysine), poly(thioketal urethane), thioether polymers |
| Adenosine triphosphate (ATP) | Tumor cells | Polymers are designed to release the siRNA when the ATP concentration is greater than (~10–100 μM) | Phenylboronic acid-based polymers, ATP-triggered self-immolative systems |

## 5.3.1 Cellular uptake and endosomal escape

Utilization of native siRNA is not a suitable choice due to its electronegativity, size, and hydrophilicity, which make it difficult to diffuse through the plasma membrane. However, incorporating siRNA inside carriers resolves its cell permeability. Conjugated siRNA can be internalized by various methods, such as endocytosis and phagocytosis. The cellular uptake of siRNA primarily occurs by endocytosis, where siRNA-loaded carriers such as liposomes, polymers, or conjugates bind to cell surface receptors based on the attached conjugates, which facilitates cell specificity. The bound siRNA is internalized via clathrin-mediated, caveolae-mediated, or macropinocytic pathways. The uptake is often dependent on temperature, ranging from 37°C to 4°C. After internalization, the complexes are transported to endosomes, which mature into late endosomes or lysosomes. The maturation process alters the pH environment, which could lead to the degradation of siRNA. To overcome this natural cellular process, the endosomal escape step needs to be incorporated, which can be achieved by two escape mechanisms: membrane fusion and membrane disruption. Membrane fusion involves the endosomes containing the siRNA fusing with the plasma membrane, releasing the siRNA into the cytoplasm, using enzymes like GTPase, SNAREs, and ESCRT. Alternatively, membrane disruption involves destabilization of siRNA-loaded endosomes by pore-forming cationic peptides or polymers. Suitable delivery systems like nanoparticles prompt endosomal escape by the proton sponge effect, where the NPs cause influx of protons, leading to endosomal swelling and osmotic swelling (Bavishi et al., 2025).

### 5.3.2 Gene silencing in colon cancer cells

The core machinery of siRNA involves gene targeting and silencing, which operates through the RNAi pathway to suppress target oncogene expression. Synthetic siRNA molecules are designed to be complementary to specific mRNA sequences that are overexpressed in colon cancer cells, such as KRAS, VEGF, PD1/PDL1, and BCL2. The process begins with siRNA entering the cytoplasm and being incorporated into the RISC-loading complex (RLC), which subsequently unwinds the siRNA duplex, and the guide strand is incorporated. With the aid of the guided RLC complex, it locates its target mRNA, induces cleavage, and prevents translation into complementary proteins. This results in the downregulation of oncoproteins, leading to inhibition of cancer cell proliferation, apoptosis, reduced angiogenesis, and impaired immune evasion (Zaidi et al., 2023).

### 5.3.3 Targeting tumor microenvironment

Targeting tumor microenvironment (TME) in colon cancer is a dynamic strategy aimed at modulating various cancerous and noncancerous components. The TME consists of complex stromal cells, extracellular matrix, cytokines, growth factors, and gut microbiota, which interact with the cancer cells. Targeting TME via siRNA in colon cancer results in immunomodulation by TGF-β, disrupts cell growth by silencing the VEGF and CXCL12 genes, and stromal and extracellular matrix (ECM) by enhancing specificity and tumor cell vulnerability (Xiang et al., 2025; Pieniądz et al., 2024). The heterogeneity of the TME in CRC provides a wider RNA-interfering therapy, which enhances efficacy and reduces adverse effects.

## 5.4 Innovations in small interfering RNA delivery for colon cancer

### 5.4.1 pH-responsive nanoparticles

The TME of CRC, like many solid tumors, is characterized by a slightly acidic extracellular pH (pH 6.0–7.0) compared to the neutral pH (7.2–7.4) of healthy tissues. Furthermore, intracellular organelles such as endosomes (pH 5–6) and lysosomes (pH 4–5) maintain an acidic environment crucial for their function. This intrinsic pH gradient offers a potent internal stimulus for designing "smart" drug delivery systems (DDSs) that can selectively release therapeutic cargo at the tumor site or inside target cells, thereby enhancing therapeutic efficacy and reducing adverse systemic effects (Ding et al., 2022).

pH-responsive nanoparticles are engineered to undergo specific changes, such as swelling, degradation, or charge reversal, when exposed to these pH fluctuations (Chu et al., 2021). This mechanism is particularly beneficial for siRNA delivery, as it protects fragile siRNA molecules from premature degradation in the bloodstream, enhances cellular uptake, and, crucially, facilitates their endosomal escape into the cytoplasm, where they can execute their gene-silencing function. The "proton sponge effect" is a notable mechanism employed by some pH-responsive systems, where protonable groups within the nanoparticle become protonated in the acidic endosome, leading to an osmotic influx of water and subsequent rupture of the endosomal membrane, thereby releasing the cargo into the cytosol (Vermeulen et al., 2018).

Several types of pH-responsive nanoparticles have been developed for siRNA delivery, with promising applications in colon cancer. For instance, chitosan-based nanoparticles (CS-NPs) are favored due to their biocompatibility and ability to effectively bind negatively charged siRNA (Mu et al., 2021). Guo et al. reported a pH-sensitive DDS using PLGA polymer coated with chitosan for the codelivery of PD-L1 siRNA, IFN-$\gamma$, and glucose oxidase (GOx) to treat colon cancer (Guo et al., 2022). This system, approximately 175 nm in size, demonstrated accelerated release of its cargo at acidic pH (pH 5.5) and achieved significant tumor volume reduction in a mouse model through a combined starvation and immunotherapeutic approach (Eilleia et al., 2018). Another innovative approach by Zhang et al. utilized zinc oxide nanoparticles (ZnO NPs), known for their pH-responsive properties, in combination with chitosan. These cancer cell membrane (CCM)-coated ZnO NPs encapsulating siSurvivin were designed to dissolve in acidic lysosomes upon uptake by tumor cells, releasing siSurvivin and disrupting lysosomal membranes to enhance endosomal escape and gene silencing. This system also synergistically induced cellular apoptosis by elevating oxidative stress levels in tumor cells, leading to significant in vitro and in vivo inhibition of 4T1 tumor growth (Zhang et al., 2025).

Further advancements include amphiphilic carriers based on branched polyethyleneimine (bPEI) modified with hexadecyl chains, designed to encapsulate anticancer drugs and then form nanoconstructs with siRNA. Polyethylenimine, recognized for its proton sponge effect, is often chemically modified to mitigate its charge-associated toxicity (Ito et al., 2004). pH-sensitive LNPs have also shown utility, demonstrating stability at gastric pH (e.g., pH 1.2) but disassembling at lower endosomal pH to release siRNA into the cytosol. These LNPs typically comprise ionizable cationic lipids, cholesterol, phospholipids, and PEG-lipids, facilitating high encapsulation efficiency and stable siRNA binding (Chen et al., 2020). For oral delivery to the colon, a pH-sensitive magnetic nanocomposite hydrogel bead incorporating $Fe_3O_4$ and hydroxyapatite nanoparticles within sodium alginate was developed for 5-Fluorouracil (5-FU). This system exhibited minimal swelling in the acidic gastric environment (pH 1.2) and significant swelling in the alkaline intestinal environment (pH 7.4), enabling targeted drug release in the colon (Foroutan et al., 2022). Moreover, "ultra-pH-sensitive" (UPS) nanotechnology systems maintain their structural integrity during systemic circulation but rapidly dissociate into smaller unimers (<10 nm) in response to even subtle pH changes ($\Delta$pH <0.25) within the acidic TME, enhancing tumor penetration, cellular internalization, and intracellular drug release. These diverse pH-responsive nanoparticle designs collectively enhance siRNA's bioavailability and therapeutic efficacy by addressing challenges like enzymatic degradation, nonspecific distribution, and poor cellular uptake, marking them as powerful tools in advancing colon cancer treatment (Liuqing Yang et al., 2024).

### 5.4.2 Ligand-mediated targeting for colon cancer

siRNAs are attractive for overcoming the limitations of conventional chemotherapy, such as poor clinical outcomes, adverse effects, and multidrug resistance (MDR) (Kumar et al., 2022). However, systemic administration of bare siRNA faces significant hurdles, including rapid enzymatic degradation in biological fluids, renal clearance, and difficulty in crossing negatively charged cell membranes due to electrostatic repulsion and large size (Tabatt et al., 2004). Ligand-mediated targeting, often employing nanocarriers, is a sophisticated strategy to circumvent these issues by enhancing siRNA stability, achieving efficient cellular uptake, and enabling targeted delivery to malignant cells (Gottesman et al., 2002).

Nanocarriers functionalized with specific ligands can selectively bind to receptors or antigens overexpressed on tumor cells, facilitating receptor-mediated endocytosis and increasing intracellular siRNA concentrations. This active targeting approach allows for precise delivery, reducing OTEs and systemic toxicity (Fernández et al., 2018; Zhao et al., 2020). Several ligands have been extensively investigated for targeted siRNA delivery in CRC:

**FA:** Folate receptors (FRs), particularly FR-α, are frequently overexpressed in various human carcinomas, including colon cancer cells, with minimal expression in normal tissues. FA is essential for nucleotide biosynthesis in rapidly proliferating cells (Bajracharya et al., 2022). Nanocarriers conjugated with FA, such as PEGylated polyethyleneimine (PEI) nanoparticles, have been shown to significantly increase the accumulation of PD-L1 siRNA in tumor tissues, leading to effective downregulation of PD-L1 mRNA and protein expression, thereby enhancing immunotherapy (Choi et al., 2023). FA-targeted solid lipid nanoparticles (SLNs) loaded with irinotecan demonstrated improved sustained drug release and high encapsulation efficiency, showing potential for CRC treatment. Advantages of FA as a ligand include its low molecular weight, stability, cost-effectiveness, and low immunogenicity (Rajpoot & Jain, 2020).

**Hyaluronic acid (HA):** HA is a key component of the tumor microenvironment, and its primary receptor, CD44, is frequently overexpressed in CRC cells. This overexpression allows HA-functionalized nanocarriers to selectively target and be internalized by tumor cells. The molecular weight of HA influences its biological effects, with low molecular weight HA (LMW-HA) being associated with angiogenesis, lymphangiogenesis, and invasiveness in CRC. HA-modified chitosan (CS) nanoparticles have been developed for targeted delivery of siRNA to CD44-overexpressing cells, specifically silencing the peptidylprolyl isomerase B (PPIB) gene in CRC cell lines (HT-29, HCT-116), demonstrating enhanced cellular uptake and gene silencing with minimized OTEs. HA-enveloped nanomicelles have also been designed for nose-to-brain siRNA delivery, showing enhanced targeting and uptake in glioma models (Al Jayoush et al., 2022).

**Biotin:** Biotin receptors, specifically sodium-dependent multivitamin transporters (SMVT), are overexpressed on the surface of various cancer cells, including those in the colon. Biotinylated drug vehicles enhance the selectivity and uptake of active agents in tumor cells. Early studies indicated that biotin-labeled doxorubicin-hydroxypropylmethacrylic acid (Dox-HMPA) complexes enhanced efficacy in colon cancer models. Optimal ligand density is crucial for maximal cellular internalization (Jurczyk et al., 2021).

**Peptides and cell membranes:** Peptides offer high specificity and affinity with reduced immunogenicity. For example, KDEL peptide-modified exosomes (KME) have been developed for endoplasmic reticulum (ER)-targeted delivery of PD-L1 siRNA in CRC. This strategy enhanced siRNA uptake and ER-specific intracellular trafficking, leading to effective PD-L1 downregulation on the cell membrane in orthotopic CRC mouse models. Similarly, anti-TAG-72 PEG-immunoliposomes have been developed for targeted gene delivery to TAG-72 overexpressing CRC cells (Wang et al., 2023; Pi et al., 2018). Additionally, CCM-coated nanoparticles, such as chitosan-zinc oxide (CS/ZnO) nanoparticles, can provide homologous targeting effects and immune evasion, protecting siSurvivin from degradation and enhancing its delivery to suppress tumor growth synergistically with ZnO nanoparticle-induced apoptosis (Zhang et al., 2025).

**Codelivery and pH-responsiveness:** Ligand-mediated systems often incorporate codelivery strategies for siRNAs and other therapeutic agents, such as chemotherapeutics or immunomodulatory agents, to achieve synergistic anticancer effects and overcome MDR. siRNAs are capable of silencing various oncogenes or MDR-related genes, including BCL-2, KRAS, PD-L1, AURKA, VEGF, STAT3, CPT1A, Nanog, and secretory leukocyte protease inhibitor (SLPI) (Xiao et al., 2017). Furthermore, the acidic TME is exploited to design pH-responsive delivery systems that release their payload preferentially at tumor sites. For instance, pH-sensitive DDSs codelivering PD-L1 siRNA, IFN-$\gamma$, and GOx have significantly reduced tumor volume in colon cancer mouse models through a combination of starvation and immunotherapy. Oral administration of such targeted nanocarriers is also being explored as a patient-friendly approach for localized diseases like CRC, bypassing invasive procedures (Hou et al., 2022).

### 5.4.3 Combination therapy with chemotherapeutic agents

Single-agent chemotherapy frequently yields suboptimal outcomes in colon cancer due to its inherent limitations, including poor efficacy, systemic adverse effects, and the widespread issue of MDR. To overcome these hurdles, combination therapy, particularly the codelivery of chemotherapeutic drugs and siRNAs, has emerged as a highly promising strategy to achieve superior anticancer activity through synergistic effects (Table 5.2). This approach leverages siRNA's ability to precisely silence genes critical for cancer cell proliferation, metastasis, or drug resistance, thereby enhancing the potency of conventional chemotherapeutic agents and circumventing resistance mechanisms (Kamran et al., 2022).

## 5.5 Challenges in small interfering RNA delivery

Effective delivery of siRNA remains a primary objective for a successful application in cancer therapy due to several biological barriers. siRNAs are inherently unstable in biological fluids, as they are degraded by nucleases. Their large size and negatively charged nature hinder their uptake and delivery. Additionally, targeting specific tissues while evading clearance mechanisms and immune detection limits clinical translation.

### 5.5.1 Biological barriers (mucosal, enzymatic, and cellular)

The therapeutic delivery of siRNA is hindered by several extracellular barriers that vary with the route of administration. Systemically delivered siRNAs are rapidly degraded by serum endonucleases, cleared through renal filtration due to their small size, and recognized by immune receptors such as

**Table 5.2** **siRNA-based combination strategies for colon cancer using chemotherapeutic drug codelivery.**

| Combination therapy name | Nanocarriers used | Target pathway/gene | Outcome | Type of delivery system | References |
|---|---|---|---|---|---|
| DOX + Aspirin + BCL-2 siRNA | Amphiphilic carrier | BCL-2 (antiapoptotic gene) | Enhanced apoptosis, G2/M cell cycle arrest, reduced mitochondrial membrane potential in HCT-116 cells | Synergistic apoptosis induction | Ray (2021) |
| DOX + BCL-2 siRNA | Gold nanoparticles | BCL-2 | Superior apoptosis compared to free DOX | Gene-chemo codelivery | Xue et al. (2021) |
| DOX + BCL-2 siRNA | pH-responsive chitosan-based vector | BCL-2 | 88% tumor inhibition in vivo | Prodrug-based gene-chemo codelivery | Kumar et al. (2022) |
| CEL + PD-L1 siRNA | ER-targeted exosomes | PD-L1 (immune checkpoint) | Induced immunogenic cell death and immune modulation | Chemoimmunotherapy | |
| PD-L1 siRNA + IFN-γ + glucose oxidase | pH-sensitive nanocarrier | PD-L1 | PD-L1 downregulation, T-cell activation, tumor volume reduction | Starvation and immunotherapeutic delivery | Guo et al. (2022) |
| OXA + CPT1A siRNA | Nanoparticle-based system | CPT1A | Reversal of oxaliplatin resistance | Gene-chemo codelivery | Pallathadka et al. (2023) |
| OXA + ASPN siRNA | Nanoparticles | Wnt/β-catenin pathway | Overcomes chemoresistance | Synergistic modulation of signaling pathway | Huang, Zhou, et al. (2022) |
| 5-FU + Nanog siRNA | Not specified | Nanog oncogene | Increased sensitivity of colorectal cancer (CRC) cells to 5-FU | Cotargeting therapy | Pallathadka et al. (2023) |
| 5-FU + PIK3CA + KRAS siRNAs | Not specified | PIK3CA and KRAS | Enhanced therapeutic response in mutation-positive CRC cells | Triple gene-chemo codelivery | |
| DTX + RelA siRNA | Folate-targeted PEGylated β-cyclodextrin | NF-κB (RelA) | Enhanced DTX-induced apoptosis slowed CRC progression in vivo | Targeted gene-chemo delivery | de la Torre et al. (2022) |
| Cisplatin + SLPI siRNA | Not specified | SLPI-PUMA-mediated apoptosis | Enhanced chemosensitivity in CRC cells | Gene-chemo synergy | Nugteren and Samsom (2021) |

*Abbreviation:* DOX, *Doxorubicin;* BCL-2, *B-cell lymphoma 2;* siRNA, *small interfering RNA;* CEL, *Celecoxib;* PD-L1, *programmed death-ligand 1;* IFN-γ, *interferon gamma;* OXA, *Oxaliplatin;* CPT1A, *carnitine palmitoyltransferase 1A;* ASPN, *Asporin;* 5-FU, *5-Fluorouracil;* PIK3CA, *phosphatidylinositol 3-kinase catalytic subunit alpha;* KRAS, *Kirsten rat sarcoma viral oncogene homolog;* DTX, *Docetaxel;* RelA, *NF-κB p65 subunit;* SLPI, *secretory leukocyte protease inhibitor.*

TLRs and RLRs, leading to cytokine release and potential toxicity. Their strong negative charge also promotes opsonization and clearance by the reticuloendothelial system, while nonspecific uptake by healthy cells can cause OTEs. Locally delivered siRNAs face additional challenges, including the negatively charged mucus layer that limits diffusion and the tight epithelial junctions that restrict paracellular transport. Moreover, the inherent anionic and hydrophilic nature of siRNA prevents its passive diffusion across the hydrophobic cell membrane (Zaidi et al., 2023).

Mucosal RNAi therapeutics administered locally encounter significant challenges due to the complex nature of mucosal barriers. These barriers include the mucosal gel layer and densely packed epithelial cells, which collectively impede particle diffusion and facilitate rapid clearance. The mucus itself is a negatively charged, hydrated protein gel that restricts the movement of anionic siRNA molecules through electrostatic interactions. Moreover, the continuous replenishment and movement in the intestines further hinder siRNA penetration. The characteristics of the mucosal membrane, such as its thickness, pH, and microenvironment, vary across different intestinal regions, influencing drug transport. Additionally, the presence of tight epithelial junctions limits siRNA permeability, while the electrochemical properties and inherent hydrophilicity of siRNA further restrict its translocation across the mucosal barrier.

However, rectal administration of siRNA is much more favorable, as the pH is much higher than that of the stomach, and the colon also presents a low enzymatic degradation of the siRNA. Thus, this route is a promising route of administration in the case of CRC (González et al., 2012).

Enzymatic degradation represents a major barrier in siRNA delivery, as unmodified siRNA is highly prone to endonuclease reactions in both extracellular and intracellular environments. In the case of systemic delivery, extracellular RNases cleave naked siRNA, resulting in a short half-life of just a few minutes. Even after cellular uptake, intracellular RNases and lysosomal degradation can further degrade it before it reaches the RISC. To overcome these enzymatic barriers, siRNAs are often chemically modified or encapsulated within protective delivery systems. Lipid-based nanoparticles, and enzyme-responsive conjugates can overcome this challenge.

The therapeutic application of siRNA faces multiple intracellular barriers that restrict its clinical success, even after improvements in stability and immune evasion. Its inherent properties—large size and strong negative charge—prevent efficient membrane penetration, necessitating advanced carrier systems for cellular uptake. Once internalized, siRNA is highly susceptible to enzymatic degradation by RNases and phosphatases, reducing its intracellular stability. A major challenge is endosomal and lysosomal entrapment, as only about 1%–2% of internalized siRNA typically escape into the cytoplasm, where RISC-mediated gene silencing occurs. Moreover, even when active, siRNA can induce off-target gene silencing due to partial sequence complementarity and may trigger immune activation through receptors like Toll-like receptor 3 (TLR3), further limiting its therapeutic precision and safety (Hu et al., 2020).

### 5.5.2 Stability and off-target effects

The therapeutic success of siRNA is fundamentally challenged by issues related to its stability and the risk of OTEs. Regarding stability, unmodified or naked siRNA is inherently unstable in the systemic circulation, being highly susceptible to degradation by endonucleases in the serum and plasma. Due to their small size (approximately 7 nm or 13 kDa molecular weight) and low molecular weight, siRNAs are subject to swift elimination through the renal system by glomerular filtration, resulting in a severely diminished half-life often ranging from a few minutes to a maximum of a few hours (Kumari et al., 2023).

Conversely, OTEs represent a major therapeutic drawback, where the siRNA inadvertently silences nontargeted genes due to partial sequence homology—sometimes as few as 6–9 nucleotides in the "seed region"—with unintended mRNA targets. This nonspecific gene silencing can result in toxicity or lead to undesirable interferon responses. OTEs are influenced by improper sequence design, misuse of guide and passenger strands by the RISC, and the oversaturation of the RNAi mechanism. To address both stability and specificity simultaneously, modifications are critical, often involving chemical alterations and substitutions on the ribose sugar backbone, or the use of sophisticated delivery systems such as nanoparticles and conjugation to enhance stability, reduce immune recognition, and improve target specificity (Sajid et al., 2020; Paul et al., 2022).

### 5.5.3 Scalability and regulatory hurdles

The commercialization of siRNA drugs faces significant hurdles concerning scalability and production, largely due to limitations in traditional synthesis methods. Historically, siRNA synthesis relied on phosphoramidite chemistry, which is a complex process involving a four-step cycle: deprotection, coupling, capping, and oxidation. This chemical synthesis method is no longer adequate for the demands of large-scale production because it is associated with high risks, significant consumption of raw materials, minimal product yield, and poor economic profitability, which collectively act as barriers to the widespread development of siRNA drugs. To address these issues, innovation in manufacturing is shifting toward enzymatic synthesis of oligonucleotides, which has been shown to improve these problems. This newer approach simplifies the traditional four-step chemical process into a two-step extension and deblocking process, which reduces the consumption of reagents. Furthermore, the enzymatic method is an aqueous-based process that avoids the extensive use of organic solvents and the generation of harmful by-products, thereby enhancing the overall atom economy and enabling high-yield and high-purity siRNA production needed for commercial success (Xiao et al., 2024).

Regulatory hurdles in siRNA delivery stem from the novelty and complexity of RNA-based therapeutics and their delivery systems. Unlike traditional small-molecule drugs, siRNA formulations often combine a biologically active oligonucleotide with a nanocarrier, making them hybrid products that require evaluation of both components. Regulatory agencies such as the FDA and the European Medicines Agency (EMA) demand extensive data on safety, efficacy, immunogenicity, and biodistribution. A key challenge lies in demonstrating consistent manufacturing quality, stability, and reproducibility, as even minor variations in formulation can alter pharmacokinetics or trigger immune responses. Additionally, long-term toxicity, off-target gene silencing, and immune activation risks must be thoroughly assessed in preclinical and clinical studies (Tables 5.3 and. 5.4). Standardized guidelines for siRNA therapeutics are still evolving, and differences in global regulatory frameworks further complicate approval pathways. Therefore, addressing these hurdles requires early regulatory engagement, comprehensive characterization of delivery systems, and rigorous validation of manufacturing and safety data to ensure clinical and commercial viability (Field, 2024; Vimal, 2025).

**Table 5.3 Preclinical studies of small interfering RNA (siRNA) delivery systems for therapeutic applications in colon cancer.**

| S. No. | Drug/target | Delivery platform | Dose/formulation | Model | Outcomes | References |
|---|---|---|---|---|---|---|
| 1. | PLK1 siRNA | Cholesterol-enriched exosomes (Chol/MEs) | siPLK1 via Chol/MEs | HCT116 colorectal cancer (CRC) cells, xenograft mouse models | Increased G2/M phase arrest and apoptosis; tumor growth suppression; bioluminescence reduced to 0.05-fold of control; superior to conventional transfection | Zhang et al. (2025) |
| 2. | PLK1 siRNA | Chol/MEs (oral delivery) | siPLK1 orally administered | CRC models | Significant tumor growth inhibition; noninvasive platform | |
| 3. | β-catenin siRNA | Mannose-functionalized multiwalled carbon nanotubes | Targeted siRNA delivery | Caco-2 CRC cells | 50% β-catenin knockdown; 32% reduction in viability; synergy with 5-FU leading to ~80% viability reduction | Ghareeb et al. (2021) |
| 4. | LSD1 siRNA | Not specified | siRNA targeting LSD1 | CRC cells (in vitro) | Suppressed proliferation; migration; invasion | |
| 5. | AMIGO2-targeted siRNA (implied) | Not specified | Medications targeting AMIGO2 | Liver metastasis models | Reduced liver metastasis by inhibiting tumor adherence | Tanio et al. (2021) |
| 6. | YTHDF1-related siRNA (implied) | Not specified | Targeting YTHDF1-m6A-ARHGEF2 axis | Metastatic CRC models | Therapeutic potential in metastatic CRC due to overexpression | Wang et al. (2022) |
| 7. | B7-H3 (CD276) siRNA | Not specified | Targeting immune checkpoint B7-H3 | CRC models | Inhibited angiogenesis via NF-κB pathway | Wang et al. (2020) |
| 8. | Survivin siRNA | Folate-displaying extracellular vesicles (EVs) | Folic acid/EV/siSurvivin | Patient-derived CRC xenograft (nude mice) | Significantly reduced tumor size and weight compared to controls; effective in vivo delivery and tumor suppression | |
| 9. | PFK1 siRNA | Not specified | Targeting PFK1 | In vitro and in vivo | Enhanced radiosensitivity by inhibiting glycolysis | Kubo and Seyama (2021) |
| 10. | ASPN siRNA | Polyethene glycol (PEG)-PAMAM-OXA nanoparticle | Codelivering oxaliplatin and ASPN siRNA | CRC models | Overcame oxaliplatin resistance; suppressed Wnt pathway | Huang, Sharma, et al. (2022) |
| 11. | Nanog siRNA | Not specified | Targeting Nanog | CRC cells | Improved 5-FU sensitivity; reduced proliferation; induced apoptosis | Cuciniello et al. (2021) |

**Table 5.4 Clinical trials exploring small interfering RNA (siRNA)-based therapeutics for colon cancer management.**

| Study IDs | Therapeutic name | Indications | Target gene | Route of administration | Phase |
|---|---|---|---|---|---|
| NCT03087591 | APN401 (siRNA-transfected PBMC) | Recurrent or metastatic pancreatic cancer, colorectal cancer (CRC), or other solid tumors | Not specified (enzyme-blocking mechanism) | Intravenous infusion | Phase I |
| NCT04089631 | Capecitabine | Colon cancer stage II (adjuvant therapy) | Circulating tumor DNA (ctDNA)-guided therapy | Oral | Phase III |
| NCT02110563 | DCR-MYC | Solid tumors, multiple myeloma, or lymphoma | MYC oncogene | Intravenous | Phase I |
| NCT01158079 | ALN-VSP02 | Solid tumors | Not specified (antiangiogenic activity) | Intravenous | Phase I |
| NCT01591356 | EphA2-targeting DOPC-encapsulated siRNA | Advanced or recurrent solid tumors | Ephrin type-A receptor 2 | Intravenous | Phase I |
| NCT05174169 | mFOLFOX6, CAPOX, mFOLFIRINOX | Stage III colon cancer (adjuvant chemotherapy) | ctDNA-guided therapy | Intravenous/oral combination | Phase II/III |
| NCT04622423 | Advanced therapy medicinal products-investigational | Pancreatic ductal adenocarcinoma, CRC, liver metastasis | Tumor-associated antigens, immune pathways | To be determined | Observational |
| NCT04264702 | SIGNATERA ctDNA test (diagnostic) | Stage I to IV CRC | ctDNA biomarkers | Diagnostic test (blood-based) | Observational |
| NCT05037149 | STP707 | Solid tumors | Not specified | Intravenous | Phase I |
| NCT02060188 | Nivolumab (alone or in combination with Ipilimumab, Cobimetinib, Daratumumab, BMS-986016) | Microsatellite unstable/stable CRC, mismatch repair proficient/deficient CRC | PD-1 pathway, CTLA-4 (when combined with ipilimumab) | Intravenous | Phase II |

# 5.6 Applications in colon cancer therapy

## 5.6.1 Targeted gene silencing

In colon cancer therapy, siRNA/RNAi holds significant promise by targeting various genes crucial for tumor survival and drug resistance:

**E2F1:** The E2F1 transcription factor is a key regulator of the cell cycle, proliferation, differentiation, apoptosis, metastasis, and chemoresistance in CRC. Studies using the HT-29 CRC model demonstrated that siRNA targeting E2F1 (siE2F1), delivered via gene electrotransfer (GET), significantly reduced E2F1 expression at both mRNA and protein levels, leading to decreased cell survival, delayed tumor growth, reduced proliferation, and increased necrosis in tumors. GET proved to be an efficient delivery system for siE2F1 in vivo, overcoming limitations of other methods, such as nanoliposomes, which can suffer from immunogenicity and poor stability.

**Livin:** Livin, an inhibitor of apoptosis protein (IAP), is overexpressed in various cancers, including CRC, and protects cells from programmed cell death. Research in HCT116 colon cancer cell lines showed that siRNA-mediated downregulation of Livin dramatically decreased its expression, suppressed cellular proliferation, and increased the rate of apoptosis in vitro. In a BALB/c nude mice xenograft model, Livin-siRNA injection significantly reduced tumor size without causing significant body weight changes or toxic reactions in vital organs such as the kidney, liver, and brain. These findings suggest Livin as an effective target for CRC treatment using siRNA (Oh et al., 2011).

**Survivin:** As a member of the IAP family, survivin (BIRC5) is frequently overexpressed in many human malignancies, including CRC, but is negligible in normal adult cells. Its overexpression correlates with cancer progression, recurrence, chemoresistance, and poor prognosis. Silencing survivin expression reduces cancer cell proliferation, enhances apoptosis, and sensitizes tumors to chemotherapy. For instance, a study demonstrated that siRNA targeting survivin induced apoptosis and inhibited tumor growth in SW480 colorectal cells in xenograft models. The differential expression makes survivin a selective target, potentially reducing side effects on healthy cells (Liu et al., 2016).

**EGFR:** EGFR is a receptor tyrosine kinase that promotes proliferation, survival, and migration of cancer cells, and its overexpression or mutation is prevalent in CRC. siRNA targeting EGFR disrupts these oncogenic pathways, inhibiting tumor cell proliferation and survival. Combining EGFR-targeting siRNA with conventional therapies can significantly improve anticancer activity, especially in drug-resistant cases (Carolina Sousa et al., 2024).

**PD-L1:** PD-L1 expression on cancer cells can suppress T-cell activity, enabling immune evasion. In CRC, exosomes loaded with PD-L1 and CTLA-4 siRNAs were shown to knock down gene expression, inhibit cancer cell proliferation, and suppress tumor growth in vivo by activating antitumor immune responses. This dual silencing strategy effectively repressed immune escape (Ubanako et al., 2023).

**Drug resistance genes (e.g., accelerated blood clearance [ABC] transporters, CPT1A):** siRNA can also target genes responsible for drug resistance, such as ABCB1 (MDR1) and ABCG2, which pump chemotherapeutic drugs out of cancer cells. Silencing these transporters restores intracellular drug accumulation and enhances the efficacy of agents like doxorubicin and paclitaxel. Additionally, CPT1A, an enzyme involved in fatty acid oxidation (FAO), is overexpressed in colon cancer and correlates with oxaliplatin resistance. iRGD-modified exosomes encapsulating CPT1A siRNA effectively reversed oxaliplatin resistance and inhibited tumor growth in colon cancer by suppressing FAO (Lin et al., 2021).

### 5.6.2 Theranostic applications of siRNA/RNAi in colon cancer therapy

Theranostics represents an innovative strategy in medical management, seamlessly integrating diagnostic and therapeutic functionalities within a single platform. In the context of CRC, theranostic nanomedicine offers a revolutionary approach to overcome limitations of conventional treatments, such as off-target toxicity, drug resistance, and inadequate diagnostic precision. The unique physicochemical properties of nanoparticles, including their small size, high surface-to-volume ratio, and tunable surface chemistry, enable them to simultaneously carry imaging agents and therapeutic molecules, such as siRNA or other gene-silencing oligonucleotides (Zhang et al., 2025). This dual capability allows for real-time monitoring of drug delivery and therapeutic response while precisely targeting cancer cells.

#### *5.6.2.1 Nanocarrier-mediated siRNA/RNAi theranostics in colorectal cancer*

Nanotechnology provides crucial delivery systems for siRNA/RNAi in theranostic applications. Liposomes, spherical vesicles composed of phospholipid bilayers, are particularly effective nanocarriers due to their biocompatibility, ability to encapsulate both hydrophilic and hydrophobic substances, and amenability to surface functionalization (Azimizonuzi et al., 2022). These features make liposomes and other nanoparticles ideal for delivering gene-silencing agents to tumor sites while incorporating diagnostic markers.

Specific examples of siRNA/RNAi theranostics in CRC demonstrate this potential:

**Epithelial cell adhesion molecule (EpCAM) silencing with hybrid liposomes:** Hybrid liposomes (HLs) have been developed to encapsulate a poly(L-lysine)-siRNA complex targeting the EpCAM, a protein often overexpressed in epithelial cancers. In a SCID mouse model, these HLs demonstrated a significant reduction in tumor volume (therapeutic effect). The study highlighted the efficient absorption of these HLs, showcasing the diagnostic potential of tracking their delivery and accumulation to EpCAM-positive cells, thus embodying a "nano-in-nano" concept where the therapeutic siRNA is delivered via a complex nanocarrier system (Bhavsar et al., 2017).

**EGFR-targeted nanohybrid cerasomes with immunotherapy:** An innovative theranostic platform for CRC involved EGFR-targeted liposomal nanohybrid cerasomes. These cerasomes were loaded with porphyrin, an imaging agent, and further functionalized with magnetic resonance imaging (MRI) contrast agent DOTA-Gd and near infrared (NIR) dye IRDye800CW, along with cetuximab, an antibody targeting EGFR. This system was designed for in vivo tumor identification and photodynamic therapy (PDT), while also incorporating PD-L1 immunotherapy as an adjuvant. The active targeting achieved by the EGFR antibody enhanced the MRI signal intensity at the tumor site, allowing for precise tumor localization and monitoring of therapeutic efficacy, which included robust inhibition of tumor development without systemic toxicity. Although not explicitly stated as PD-L1 siRNA, this multimodal approach integrates an immunotherapy that can be mediated by gene silencing, alongside sophisticated imaging (Rampado et al., 2020).

**General gene delivery for theranostics:** Nanosized liposomes are clinically validated for delivering a variety of therapeutic and imaging agents, including gene/nucleic acids such as siRNA, to solid tumors. This enables image-guided therapies and combined therapeutic applications. These liposomal systems can be engineered to integrate imaging probes for various modalities, like optical imaging, MRI, positron emission tomography (PET), and single-photon emission computed tomography (SPECT) (Lee & Im, 2019).

### 5.6.3 Preclinical and clinical studies

Tables 5.3 and 5.4

## 5.7 Future perspectives

### 5.7.1 Emerging nanocarrier technologies

Future perspectives for emerging nanocarrier technologies utilizing siRNA in CRC therapy emphasize the advancement of precision and targeted medicine. Nanoparticle-mediated siRNA delivery offers a promising approach to overcome key limitations of traditional siRNA therapeutics, including enzymatic degradation, poor cellular uptake, and OTEs. By silencing oncogenes and MDR-related genes, siRNA-loaded nanoparticles present a genetic means to counter chemoresistance. Advanced delivery platforms—such as lipid-based, polymeric, and viral vectors, as well as codelivery systems like lipopolyplexes—have demonstrated improved tumor accumulation, pH-responsive siRNA release, and synergistic effects with chemotherapeutics. Nevertheless, challenges including immunogenicity, toxicity, limited stability, and suboptimal tumor localization remain. Continued optimization of these nanocarrier systems is essential to facilitate clinical translation and establish siRNA-based nanotherapies as a transformative strategy in CRC treatment (Singh et al., 2023; Aghamiri et al., 2019; Udappusamy et al., 2025).

### 5.7.2 Integration with AI and 3D printing

Current reviews on AI applications in CRC focus mainly on prediction, classification, and prognosis in therapies such as chemotherapy, immunotherapy, and targeted therapy. Although AI, particularly machine learning (ML) and deep learning (DL), effectively analyzes molecular and clinical data for personalized treatment, no direct application to siRNA therapy in CRC has been reported. The most relevant use involves ML models, such as LASSO regression, integrating immune-related long noncoding RNA (lncRNA) signatures to predict chemotherapy response, highlighting AI's potential in individualized treatment assessment but not yet in siRNA optimization (Yang et al., 2024).

The integration of 3D-printed polymer systems represents a promising strategy for personalized, localized, and sustained delivery of siRNA therapeutics. Extrusion-based 3D printing enables the fabrication of core–shell hydrogel implants designed to overcome siRNA's poor pharmacokinetics and rapid clearance. Typically, a thermoresponsive shell composed of agarose, Pluronic F127, and gelatin provides a tissue-mimicking diffusion barrier, while the siRNA-loaded PF127 core preserves RNA integrity. The system allows precise customization of implant geometry and controlled siRNA release—sustained for over 30 days—by adjusting printing parameters. Although primarily proposed for postsurgical local therapy in soft tissue cancers, specific applications in CRC remain to be explored (Mahmoud et al., 2023).

### 5.7.3 Personalized medicine approaches

Personalized siRNA therapy in CRC focuses on silencing oncogenes based on patient-specific mutations. The KRAS mutation, present in about 40% of CRC cases, causes resistance to anti-EGFR therapies. siRNAs targeting mutant KRAS, particularly the G12D variant, have shown tumor growth inhibition in preclinical and early clinical studies. Additionally, combination approaches targeting multiple pathways, such as PI3K/AKT/mTOR and RAS/ERK, using multiple siRNAs have demonstrated superior efficacy compared to single-gene targeting. Modified siRNA can address tumor-specific drug resistance mechanisms by silencing MDR genes like MDR1 (ABCB1), which encode efflux pumps. Knockdown of genes such as β-catenin or MUC13 has been shown to enhance the sensitivity of CRC cells to chemotherapeutics like 5-fluorouracil (5-FU), offering improved treatment outcomes.

Effective siRNA therapy requires targeted delivery systems to ensure stability and specificity. Ligand-targeted nanoparticles, functionalized with antibodies or folate, enhance tumor selectivity while minimizing off-target toxicity. Stimuli-responsive nanoparticles release siRNA in tumor-specific conditions, such as acidic pH. Exosomes, natural nanovesicles, are also being explored for siRNA delivery, with KRAS siRNA-loaded exosomes showing promising tumor-targeting efficacy in preclinical models.

Major challenges include limited delivery efficiency, endosomal entrapment, OTEs, and immunogenicity. Successful implementation of personalized siRNA therapy requires robust biomarker identification and comprehensive molecular profiling to guide target selection. Continued advancements in delivery platforms and precision diagnostics are essential for translating siRNA-based personalized medicine into effective CRC treatment (Xie & Wang, 2022; Sousa & Videira, 2025).

## References

Aagaard, L., & Rossi, J. J. (2007). RNAi therapeutics: Principles, prospects and challenges. *Advanced Drug Delivery Reviews, 59*(2–3), 75–86. https://doi.org/10.1016/j.addr.2007.03.005.

Aghamiri, S., Jafarpour, A., Malekshahi, Z. V., Gomari, M. M., & Negahdari, B. (2019). Targeting siRNA in colorectal cancer therapy: Nanotechnology comes into view. *Journal of Cellular Physiology, 234*(9), 14818–14827. https://doi.org/10.1002/jcp.28281.

Al Jayoush, A. R., Haider, M., Khan, S. A., & Hussain, Z. (2022). Hyaluronic acid-functionalized nanomedicines for CD44-receptors-mediated targeted cancer therapy: A review of selective targetability and biodistribution to tumor microenvironment. *International Journal of Biological Macromolecules, 224*, 1414–1432. https://doi.org/10.1016/j.ijbiomac.2022.10.237.

Al-Absi, M. Y., Caprifico, A. E., & Calabrese, G. (2022). Chitosan and its structural modifications for siRNA delivery. *Advanced Pharmaceutical Bulletin, 13*(2), 275–282. https://doi.org/10.34172/apb.2023.030.

Azimizonuzi, H., Ghayourvahdat, A., Ahmed, M. H., Kareem, R. A., Zrzor, A. J., Mansoor, A. S., Athab, Z. H., & Kalavi, S. (2022). A state-of-the-art review of the recent advances of theranostic liposome hybrid nanoparticles in cancer treatment and diagnosis. *Journal of Drug Delivery Science and Technology, 75*, 103663. https://doi.org/10.1016/j.jddst.2022.103663.

Bajracharya, R., Song, J. G., Patil, B. R., et al. (2022). Functional ligands for improving anticancer drug therapy: Current status and applications to drug delivery systems. *Drug Delivery, 29*(1), 1959–1970. https://doi.org/10.1080/10717544.2022.2089296.

Bäumer, S., Bäumer, N., Appel, N., Terheyden, L., Fremerey, J., Schelhaas, S., Wardelmann, E., Buchholz, F., Berdel, W. E., & Müller-Tidow, C. (2015). Antibody-Mediated delivery of Anti–KRAS-siRNA in vivo overcomes therapy resistance in colon cancer. *Clinical Cancer Research, 21*(6), 1383–1394. https://doi.org/10.1158/1078-0432.ccr-13-2017.

Bavishi, L., Butani, S. B., & Patel, S. S. (2025). Small interfering RNA (SiRNA) based therapies for cancer treatment: A special focus on cervical cancer. *PubMed*178198. https://doi.org/10.1016/j.ejphar.2025.178198.

Beaucage, S. L., & Reese, C. B. (2009). Recent advances in the chemical synthesis of RNA. *Current Protocols in Nucleic Acid Chemistry, 38*(1). https://doi.org/10.1002/0471142700.nc0216s38.

Bhavsar, D., Subramanian, K., Sethuraman, S., & Krishnan, U. M. (2017). Nano-in-nano' hybrid liposomes increase target specificity and gene silencing efficiency in breast cancer induced SCID mice. *European Journal of Pharmaceutics and Biopharmaceutics: Official Journal of Arbeitsgemeinschaft fur Pharmazeutische Verfahrenstechnik e.V, 119*, 96–106. https://doi.org/10.1016/j.ejpb.2017.06.009.

Bumcrot, D., Manoharan, M., Koteliansky, V., & Sah, D. W. Y. (2006). RNAi therapeutics: A potential new class of pharmaceutical drugs. *Nature Chemical Biology, 2*(12), 711–719. https://doi.org/10.1038/nchembio839.

Cao, W., Li, R., Pei, X., Chai, M., Sun, L., Huang, Y., Wang, J., Barth, S., Yu, F., & He, H. (2022). Antibody–siRNA conjugates (ARC): Emerging siRNA drug formulation. *Medicine in Drug Discovery, 15*, 100128. https://doi.org/10.1016/j.medidd.2022.100128.

Charbe, N. B., Amnerkar, N. D., Ramesh, B., Tambuwala, M. M., Bakshi, H. A., Aljabali, A. A., Khadse, S. C., Satheeshkumar, R., Satija, S., Metha, M., Chellappan, D. K., Shrivastava, G., Gupta, G., Negi, P., Dua, K., & Zacconi, F. C. (2020). Small interfering RNA for cancer treatment: Overcoming hurdles in delivery. *Acta Pharmaceutica Sinica B, 10*(11), 2075–2109. https://doi.org/10.1016/j.apsb.2020.10.005.

Chen, Y., Yang, J., Fu, S., & Wu, J. (2020). Gold nanoparticles as radiosensitizers in cancer radiotherapy. *International Journal of Nanomedicine, 15*, 9407–9430. https://doi.org/10.2147/IJN.S272902.

Chen, B., Ren, Q., Jiang, P., Wu, Q., Shuai, Q., & Yan, Y. (2024). Combinatorial synthesis of alkyl chain-capped poly(β-Amino Ester)s for effective siRNA delivery. *Macromolecular Bioscience*. https://doi.org/10.1002/mabi.202400168.

Chernikov, I. V., Ponomareva, U. A., & Chernolovskaya, E. L. (2023). Structural modifications of sIRNA improve its performance in vivo. *International Journal of Molecular Sciences, 24*(2), 956. https://doi.org/10.3390/ijms24020956.

Chernikov, I. V., Vlassov, V. V., & Chernolovskaya, E. L. (2019). Current development of siRNA bioconjugates: From research to the clinic. *Frontiers in Pharmacology, 10*. https://doi.org/10.3389/fphar.2019.00444.

Choi, Y., Seok, S. H., Yoon, H. Y., et al. (2023). Advancing cancer immunotherapy through siRNA-based gene silencing for immune checkpoint blockade. *Biomaterials, 297*, 122109. https://doi.org/10.1016/j.biomaterials.2023.122109.

Chu, S., Shi, X., Tian, Y., & Gao, F. (2021). pH-Responsive polymer nanomaterials for tumor therapy. *Polymers (Basel), 13*(12), 1944. https://doi.org/10.3390/polym13121944.

Cornista, A. M., Giolito, M. V., Baker, K., Hazime, H., Dufait, I., Datta, J., Khumukcham, S. S., De Ridder, M., Roper, J., Abreu, M. T., Breckpot, K., & Van Der Jeught, K. (2023). Colorectal cancer immunotherapy: State of the art and future directions. *Gastro Hep Advances, 2*(8), 1103–1119. https://doi.org/10.1016/j.gastha.2023.09.007.

Cuciniello, R., Filosa, S., & Crispi, S. (2021). Novel approaches in cancer treatment: Preclinical and clinical development of small non-coding RNA therapeutics. *Cancers (Basel), 13*(15), 3736. https://doi.org/10.3390/cancers13153736.

de la Torre, C., Játiva, P., Posadas, I., et al. (2022). A β-Cyclodextrin-based nanoparticle with very high transfection efficiency unveils siRNA-activated TLR3 responses in human prostate cancer cells. *Pharmaceutics, 14*, 2424. https://doi.org/10.3390/pharmaceutics14112424.

Ding, H., Tan, P., Fu, S., Tian, X., Zhang, H., Ma, X., Gu, Z., & Luo, K. (2022). Preparation and application of pH-responsive drug delivery systems. *Journal of Controlled Release: Official Journal of the Controlled Release Society, 349*, 791–808. https://doi.org/10.1016/j.jconrel.2022.07.015.

Dong, Y., Siegwart, D. J., & Anderson, D. G. (2019). Strategies, design, and chemistry in siRNA delivery systems. *Advanced Drug Delivery Reviews, 144*, 133–147. https://doi.org/10.1016/j.addr.2019.05.004.

Dosta, P., Ramos, V., & Borrós, S. (2018). Stable and efficient generation of poly(β-amino ester)s for RNAi delivery. *Molecular Systems Design & Engineering, 3*(4), 677–689. https://doi.org/10.1039/c8me00006a.

Duan, B., Zhao, Y., Bai, J., Wang, J., Duan, X., Luo, X., Zhang, R., Pu, Y., Kou, M., Lei, J., & Yang, S. (2022). *Colorectal Cancer: An overview.* Exon Publications eBooks, 1–12. https://doi.org/10.36255/exon-publications-gastrointestinal-cancers-colorectal-cancer.

Eilleia, S. Y., Soliman, M. E., Mansour, S., & SG, A. (2018). Novel technique of insulin loading into porous carriers for oral delivery. *Asian J Pharm Sci, 13*(4), 297–309. https://doi.org/10.1016/j.ajps.2017.07.002.

Fadlallah, H., Masri, J. E., Fakhereddine, H., Youssef, J., Chemaly, C., Doughan, S., & Abou-Kheir, W. (2024). Colorectal cancer: Recent advances in management and treatment. *World Journal of Clinical Oncology, 15*(9), 1136–1156. https://doi.org/10.5306/wjco.v15.i9.1136.

Fernández, M., Javaid, F., & Chudasama, V. (2018). Advances in targeting the folate receptor in the treatment/imaging of cancers. *Chemistry Sciences, 9*, 790–810. https://doi.org/10.1039/c7sc04004k.

Field, P. (2024, October 11). Small interfering RNA: regulatory challenges and therapeutic potential. Regulatory Rapporteur. https://www.regulatoryrapporteur.org/atmp/small-interfering-rna-regulatory-challenges-and-therapeutic-potential/754.article

Foroutan, R., Mohammadzadeh, A., Javanbakht, S., Mohammadi, R., & Ghorbani, M. (2022). Alginate/magnetic hydroxyapatite bio-nanocomposite hydrogel bead as a pH-responsive oral drug carrier for potential colon cancer therapy. *Carbohydrate Polymers, 293*, 119728. https://doi.org/10.1016/j.carbpol.2022.119728.

Ghareeb, A., Ghazal, N., Assali, M., et al. (2021). Novel siRNA delivery system targeting β-catenin in colorectal cancer Cells: Synergistic effect with 5-fluorouracil. *Biomedicine & Pharmacotherapy=Biomedecine & Pharmacotherapie, 138*, 111436. https://doi.org/10.1016/j.biopha.2021.111436.

Golshani, G., & Zhang, Y. (2020). Advances in immunotherapy for colorectal cancer: A review. *Therapeutic Advances in Gastroenterology, 13*. https://doi.org/10.1177/1756284820917527.

González, B. B., Nielsen, E. B., Thomsen, T. B., & Howard, K. A. (2012). Mucosal delivery of RNAI therapeutics. *In Advances in Delivery Science and Technology*, 97–125. https://doi.org/10.1007/978-1-4614-4744-3_5.

Gori, A., Lodigiani, G., Colombarolli, S. G., Bergamaschi, G., & Vitali, A. (2023). Cell penetrating peptides: Classification, mechanisms, methods of study, and applications. *ChemMedChem, 18*(17). https://doi.org/10.1002/cmdc.202300236.

Gottesman, M. M., Fojo, T., & Bates, S. E. (2002). Multidrug resistance in cancer: Role of ATP-dependent transporters. *Nature Reviews Cancer, 2*, 48–58. https://doi.org/10.1038/nrc706.

Grau, M., & Wagner, E. (2024). Strategies and mechanisms for endosomal escape of therapeutic nucleic acids. *Current Opinion in Chemical Biology, 81*, 102506. https://doi.org/10.1016/j.cbpa.2024.102506.

Guo, K., Liu, Y., Ding, M., Sun, Q., & Shubhra, Q. T. H. (2022). Enhanced drug release from a pH-responsive nanocarrier can augment colon cancer treatment by blocking PD-L1 checkpoint and consuming tumor glucose. *Biomaterials, 290*, 121827. https://doi.org/10.1016/j.biomaterials.2022.121827.

Hoang, C. N. M., Nguyen, S. H., & Tran, M. T. (2025). Nanoparticles in cancer therapy: Strategies to penetrate and modulate the tumor microenvironment – A review. *Smart Materials in Medicine*. https://doi.org/10.1016/j.smaim.2025.07.004.

Hou, Y., Jin, J., Duan, H., et al. (2022). Targeted therapeutic effects of oral inulin-modified double-layered nanoparticles containing chemotherapeutics on orthotopic colon cancer. *Biomaterials, 283*, 121440. https://doi.org/10.1016/j.biomaterials.2022.121440.

Hu, B., Zhong, L., Weng, Y., Peng, L., Huang, Y., Zhao, Y., & Liang, X. (2020). Therapeutic siRNA: State of the art. *Signal Transduction and Targeted Therapy, 5*(1). https://doi.org/10.1038/s41392-020-0207-x.

Huang, C. Z., Zhou, Y., Tong, Q. S., et al. (2022). Precision medicine-guided co-delivery of ASPN siRNA and oxaliplatin by nanoparticles to overcome chemoresistance of colorectal cancer. *Biomaterials, 290*, 121827. https://doi.org/10.1016/j.biomaterials.2022.121827.

Huang, C., Sharma, A., Thakur, R., et al. (2022). Asporin, an extracellular matrix protein, is a beneficial regulator of cardiac remodeling. *Matrix Biology: Journal of the International Society for Matrix Biology, 110*, 40–59. https://doi.org/10.1016/j.matbio.2022.04.009.

Ito, T., Sun, L., Bevan, M. A., & Crooks, R. M. (2004). Comparison of nanoparticle size and electrophoretic mobility measurements using a carbon-nanotube-based coulter counter, dynamic light scattering, transmission electron microscopy, and phase analysis light scattering. *Langmuir: The ACS Journal of Surfaces and Colloids, 20*, 6940–6945. https://doi.org/10.1021/la049524t.

Jafarlou, V., & Jafarlou, M. (2025). Current approaches, challenges, and future perspectives in colorectal cancer therapeutics and integrated care. *waocp.com*. https://doi.org/10.31557/apjcc.2025.10.3.861-869.

Jurczyk, M., Jelonek, K., Musiał-Kulik, M., et al. (2021). Single- versus dual-targeted nanoparticles with folic acid and biotin for anticancer drug delivery. *Pharmaceutics, 13*(3), 326. https://doi.org/10.3390/pharmaceutics13030326.

Kakkis, A., Sun, L., Xie, Z., Qaddourah, M., Dawson, P. E., Sanchez, B. B., Orzolek, B. J., Brunel, F. M., Stock, J. R., Kolakowski, R. V., Korboukh, I., & Leon, J. C. (2025). Synthesis of chemically diverse siRNA-Lipid conjugates enabled by reversible adsorption to solid support (RASS). *Chembiochem: A European Journal of Chemical Biology*. https://doi.org/10.1002/cbic.202500169.

Kamran, S., Seyedrezazadeh, E., Shanehbandi, D., et al. (2022). Combination therapy with KRAS and P38α siRNA suppresses colorectal cancer growth and development in SW480 cell line. *J Gastrointest Cancer, 53*, 597–604. https://doi.org/10.1007/s12029-021-00648-8.

Kang, H., Ga, Y. J., Kim, S. H., Cho, Y. H., Kim, J. W., Kim, C., & Yeh, J. (2023). Small interfering RNA (siRNA)-based therapeutic applications against viruses: Principles, potential, and challenges. *Journal of Biomedical Science, 30*(1). https://doi.org/10.1186/s12929-023-00981-9.

Karlsson, J., Rhodes, K. R., Green, J. J., & Tzeng, S. Y. (2020). Poly(beta-amino ester)s as gene delivery vehicles: Challenges and opportunities. *Expert Opinion on Drug Delivery, 17*(10), 1395–1410. https://doi.org/10.1080/17425247.2020.1796628.

Kesharwani, P., Kumar, V., Goh, K. W., Gupta, G., Alsayari, A., Wahab, S., & Sahebkar, A. (2025). PEGylated PLGA nanoparticles: Unlocking advanced strategies for cancer therapy. *Molecular Cancer, 24*(1). https://doi.org/10.1186/s12943-025-02410-x.

Kheiriabad, S., Ghaffari, M., Dolatabadi, J. E. N., & Hamblin, M. R. (2020). PAMAM dendrimers as a delivery system for small interfering RNA. *In Methods in molecular biology*, 91–106. https://doi.org/10.1007/978-1-0716-0290-4_5.

Kozielski, K. L., Tzeng, S. Y., & Green, J. J. (2013). A bioreducible linear poly(β-amino ester) for siRNA delivery. *Chemical Communications, 49*(46), 5319. https://doi.org/10.1039/c3cc40718g.

Kubo, T., & Seyama, T. (2021). Preclinical development of siRNA conjugates to target tumor antigens. *Drug Discovery Today, 26*(11), 2584–2593. https://doi.org/10.1016/j.drudis.2021.06.005.

Kubo, T., Nishimura, Y., Sato, Y., Yanagihara, K., & Seyama, T. (2020). Sixteen different types of lipid-conjugated SIRNAs containing saturated and unsaturated fatty acids and exhibiting enhanced RNAI potency. *ACS Chemical Biology, 16*(1), 150–164. https://doi.org/10.1021/acschembio.0c00847.

Kumar, K., Rani, V., Mishra, M., & Chawla, R. (2022). New paradigm in combination therapy of siRNA with chemotherapeutic drugs for effective cancer therapy. *European Journal of Medicinal Chemistry, 239*, 114528. https://doi.org/10.1016/j.ejmech.2022.114528.

Kumari, A., Kaur, A., & Aggarwal, G. (2023). The emerging potential of siRNA nanotherapeutics in treatment of arthritis. *Asian Journal of Pharmaceutical Sciences, 18*(5), 100845. https://doi.org/10.1016/j.ajps.2023.100845.

Kwak, S. Y., Lee, S., Han, H. D., Chang, S., Kim, K., & Ahn, H. J. (2019). PLGA nanoparticles codelivering siRNAs against programmed cell death protein-1 and Its ligand gene for suppression of colon tumor growth. *Molecular Pharmaceutics, 16*(12), 4940–4953. https://doi.org/10.1021/acs.molpharmaceut.9b00826.

Lee, W., & Im, H. J. (2019). Theranostics based on liposome: Looking back and forward. *Nuclear Medicine and Molecular Imaging, 53*(4), 242–246. https://doi.org/10.1007/s13139-019-00551-y.

Lin, D., Zhang, H., Liu, R., Deng, T., Ning, T., Bai, M., et al. (2021). iRGD-modified exosomes effectively deliver CPT1A siRNA to colon cancer cells, reversing oxaliplatin resistance by regulating fatty acid oxidation. *Molecular Oncology, 15*(12), 3430–3446. https://doi.org/10.1002/1878-0261.13052.

Liu, H. Y., Yu, X., Liu, H., Wu, D., & She, J. X. (2016). Co-targeting EGFR and survivin with a bivalent aptamer-dual siRNA chimera effectively suppresses prostate cancer. *Scientific Reports, 6*, 30346. https://doi.org/10.1038/srep30346.

Mahmoud, D. B., Wölk, C., & Schulz-Siegmund, M. (2023). Fabrication of 3D printed, Core-and-Shell implants as controlled release systems for local siRNA delivery. *Advanced Healthcare Materials, 12*(31). https://doi.org/10.1002/adhm.202301643.

Makaremi, S., Asadzadeh, Z., Hemmat, N., Baghbanzadeh, A., Sgambato, A., Ghorbaninezhad, F., Safarpour, H., Argentiero, A., Brunetti, O., Bernardini, R., Silvestris, N., & Baradaran, B. (2021). Immune checkpoint inhibitors in colorectal cancer: Challenges and future prospects. *Biomedicines, 9*(9), 1075. https://doi.org/10.3390/biomedicines9091075.

Manzi, J., Hoff, C. O., Ferreira, R., Pimentel, A., Datta, J., Livingstone, A. S., Vianna, R., & Abreu, P. (2023). Targeted therapies in colorectal cancer: Recent advances in biomarkers, landmark trials, and future perspectives. *Cancers, 15*(11), 3023. https://doi.org/10.3390/cancers15113023.

Mármol, I., Sánchez-De-Diego, C., Dieste, A. P., Cerrada, E., & Yoldi, M. R. (2017). Colorectal carcinoma: A general overview and future perspectives in colorectal cancer. *International Journal of Molecular Sciences, 18*(1), 197. https://doi.org/10.3390/ijms18010197.

Mi, P. (2020). Stimuli-responsive nanocarriers for drug delivery, tumor imaging, therapy and theranostics. *Theranostics, 10*(10), 4557–4588. https://doi.org/10.7150/thno.38069.

Mousazadeh, H., Pilehvar-Soltanahmadi, Y., Dadashpour, M., & Zarghami, N. (2020). Cyclodextrin based natural nanostructured carbohydrate polymers as effective non-viral siRNA delivery systems for cancer gene therapy. *Journal of Controlled Release, 330*, 1046–1070. https://doi.org/10.1016/j.jconrel.2020.11.011.

Mu, M., Liang, X., Chuan, D., et al. (2021). Chitosan coated pH-responsive metal-polyphenol delivery platform for melanoma chemotherapy. *Carbohydrate Polymer, 264*, 118000. https://doi.org/10.1016/j.carbpol.2021.118000.

Mundel, R., Thakur, T., & Chatterjee, M. (2022). Emerging uses of PLA–PEG copolymer in cancer drug delivery. *3 Biotechnology, 12*(2). https://doi.org/10.1007/s13205-021-03105-y.

Murata, N., Takashima, Y., Toyoshima, K., Yamamoto, M., & Okada, H. (2007). Anti-tumor effects of anti-VEGF siRNA encapsulated with PLGA microspheres in mice. *Journal of Controlled Release, 126*(3), 246–254. https://doi.org/10.1016/j.jconrel.2007.11.017.

Nugteren, S., & Samsom, J. N. (2021). Secretory leukocyte protease inhibitor (SLPI) in mucosal tissues: Protects against inflammation, but promotes cancer. *Cytokine & Growth Factor Reviews, 59*, 22–35. https://doi.org/10.1016/j.cytogfr.2021.01.003.

Oh, B. Y., Lee, R. A., & Kim, K. H. (2011). siRNA targeting Livin decreases tumor in a xenograft model for colon cancer. *World Journal of Gastroenterology: WJG, 17*(20), 2563–2571. https://doi.org/10.3748/wjg.v17.i20.2563.

Osborn, M. F., & Khvorova, A. (2018). Improving siRNA deliveryin vivothrough lipid conjugation. *Nucleic Acid Therapeutics, 28*(3), 128–136. https://doi.org/10.1089/nat.2018.0725.

Pallathadka, H., Hsu, C. Y., Saleh, R. O., et al. (2023). Specific small interfering RNAs (siRNAs) for targeting the metastasis, immune responses, and drug resistance of colorectal cancer cells (CRC). *Pathology, Research and Practice, 246*, 154562. https://doi.org/10.1016/j.prp.2023.154562.

Patel, S. S., Hoogenboezem, E. N., Yu, F., DeJulius, C. R., Fletcher, R. B., Sorets, A. G., Cherry, F. K., Lo, J. H., Bezold, M. G., Francini, N., D'Arcy, R., Brasuell, J. E., Cook, R. S., & Duvall, C. L. (2023). Core polymer optimization of ternary siRNA nanoparticles enhances in vivo safety, pharmacokinetics, and tumor gene silencing. *Biomaterials, 297*, 122098. https://doi.org/10.1016/j.biomaterials.2023.122098.

Paul, A., Muralidharan, A., Biswas, A., Kamath, B. V., Joseph, A., & Alex, A. T. (2022). siRNA therapeutics and its challenges: Recent advances in effective delivery for cancer therapy. *OpenNano, 7*, 100063. https://doi.org/10.1016/j.onano.2022.100063.

Pi, F., Binzel, D. W., Lee, T. J., et al. (2018). Nanoparticle orientation to control RNA loading and ligand display on extracellular vesicles for cancer regression. *Nature Nanotechnology, 13*(1), 82–89. https://doi.org/10.1038/s41565-017-0012-z.

Pieniądz, P., Pięt, M., & Paduch, R. (2024). Characteristics of the colorectal cancer microenvironment—Role in cancer progression and therapeutic possibilities. *Applied Sciences, 14*(7), 2930. https://doi.org/10.3390/app14072930.

Rajpoot, K., & Jain, S. K. (2020). Oral delivery of pH-responsive alginate microbeads incorporating folic acid-grafted solid lipid nanoparticles exhibits enhanced targeting effect against colorectal cancer: A dual-targeted approach. *International Journal of Biological Macromolecules, 151*, 830–844. https://doi.org/10.1016/j.ijbiomac.2020.02.127.

Rampado, R., Crotti, S., Caliceti, P., Pucciarelli, S., & Agostini, M. (2020). Nanovectors design for theranostic applications in colorectal cancer. *Journal of Clinical Medicine, 9*(1), 33. https://doi.org/10.3390/jcm9010033.

Ray, L. (2021). Synergistic anticancer activity by co-delivered nanosized dual therapeutic agents and siRNA in colon cancer. *Drug Delivery and Translational Research, 11*(4), 1463–1481. https://doi.org/10.1007/s13346-020-00841-2.

RNA interference (RNAI). (n.d.). https://www.ncbi.nlm.nih.gov/probe/docs/techrnai/

Sajid, M. I., Moazzam, M., Kato, S., Cho, K. Y., & Tiwari, R. K. (2020). Overcoming barriers for siRNA therapeutics: From bench to bedside. *Pharmaceuticals, 13*(10), 294. https://doi.org/10.3390/ph13100294.

Saw, P. E., Yao, H., Lin, C., Tao, W., Farokhzad, O. C., & Xu, X. (2019). StiMuli-responsive polymer–prodrug hybrid nanoplatform for multistage siRNA delivery and combination cancer therapy. *Nano Letters, 19*(9), 5967–5974. https://doi.org/10.1021/acs.nanolett.9b01660.

Schädlich, A., Caysa, H., Mueller, T., Tenambergen, F., Rose, C., Göpferich, A., Kuntsche, J., & Mäder, K. (2011). Tumor accumulation of NIR fluorescent PEG–PLA nanoparticles: Impact of particle size and human xenograft tumor model. *ACS Nano, 5*(11), 8710–8720. https://doi.org/10.1021/nn2026353.

Sciences, B. (2024, August 22). Peptide-siRNA Conjugates. BOC Sciences. https://www.bocsci.com/peptide-sirna-conjugates.html

Singh, R. P., Hidalgo, T., Cazade, P., Darcy, R., Cronin, M. F., Dorin, I., O'Driscoll, C. M., & Thompson, D. (2019). Self-Assembled cationic B-Cyclodextrin nanostructures for siRNA delivery. *Molecular Pharmaceutics, 16*(3), 1358–1366. https://doi.org/10.1021/acs.molpharmaceut.8b01307.

Singh, S. K., Solanki, R., & Patel, S. (2023). Nano pharmaceutical delivery in combating colorectal cancer. *Medicine in Drug Discovery, 21*, 100173. https://doi.org/10.1016/j.medidd.2023.100173.

Sivakumar, P., Kim, S., Kang, H. C., & Shim, M. S. (2018). Targeted siRNA delivery using aptamer-siRNA chimeras and aptamer-conjugated nanoparticles. *Wiley Interdisciplinary Reviews Nanomedicine and Nanobiotechnology, 11*(3). https://doi.org/10.1002/wnan.1543.

Sousa, C., & Videira, M. (2025). Dual approaches in oncology: The promise of siRNA and chemotherapy combinations in cancer therapies. *Onco, 5*(1), 2. https://doi.org/10.3390/onco5010002.

Steele, T. W., Zhao, X., Tarcha, P., & Kissel, T. (2011). Factors influencing polycation/siRNA colloidal stability toward aerosol lung delivery. *European Journal of Pharmaceutics and Biopharmaceutics, 80*(1), 14–24. https://doi.org/10.1016/j.ejpb.2011.08.008.

Tabatt, K., Kneuer, C., Sameti, M., et al. (2004). Transfection with different colloidal systems: Comparison of solid lipid nano-particles and liposomes. *Journal of Controlled Release: Official Journal of the Controlled Release Society, 97*(2), 321–332. https://doi.org/10.1016/j.jconrel.2004.02.029.

Tai, W., & Gao, X. (2016). Functional peptides for siRNA delivery. *Advanced Drug Delivery Reviews, 110–111*, 157–168. https://doi.org/10.1016/j.addr.2016.08.004.

Tanio, A., Saito, H., Amisaki, M., et al. (2021). AMIGO2 as a novel indicator of liver metastasis in patients with colorectal cancer. *Oncology Letters, 21*(2), 158. https://doi.org/10.3892/ol.2021.12419.

Ubanako, P., Mirza, S., Ruff, P., & Penny, C. (2023). Exosome-mediated delivery of siRNA molecules in cancer therapy: Triumphs and challenges. *Pharmaceutics, 15*(7), 1982. https://doi.org/10.3390/pharmaceutics15071982.

Udappusamy, V., Thangam, S. N., Ramasamy, N. K., Ganeshan, S., Rajan, D. K., & Moovendhan, M. (2025). Nanoparticle-mediated delivery of SIRNA: Targeting gene silencing in cancer therapy. *Journal of Drug Delivery Science and Technology*106973. https://doi.org/10.1016/j.jddst.2025.106973.

Vermeulen, J. M., De Smedt, S. C., Remaut, K., & Braeckmans, K. (2018). The proton sponge hypothesis: Fable or fact? *European Journal of Pharmaceutics and Biopharmaceutics: Official Journal of Arbeitsgemeinschaft fur Pharmazeutische Verfahrenstechnik e.V, 129*, 184–190. https://doi.org/10.1016/j.ejpb.2018.05.034.

Vimal, J. (2025, August 6). Small interfering RNA (siRNA) Therapeutics | DelveInsight. DelveInsight Business Research. https://www.delveinsight.com/blog/small-interfering-rna-sirna-therapeutics-competitive-landscape-technology-and-pipeline-insights-2015-a-delveinsights-report

Wang, J., Zhang, Z., Zhuo, Y., et al. (2023). Endoplasmic reticulum-targeted delivery of celastrol and PD-L1 siRNA for reinforcing immunogenic cell death and potentiating cancer immunotherapy. *Biomaterials, 295*, 122045. https://doi.org/10.1016/j.biomaterials.2023.122045.

Wang, R., Ma, Y., Zhan, S., et al. (2020). B7–H3 promotes colorectal cancer angiogenesis through activating the NF-κB pathway to induce VEGFA expression. *Cell Death & Disease, 11*(1), 55. https://doi.org/10.1038/s41419-020-2252-3.

Wang, S., Gao, S., Zeng, Y., et al. (2022). N6-methyladenosine reader YTHDF1 promotes ARHGEF2 translation and RhoA signaling in colorectal cancer. *Gastroenterology, 162*(4), 1183–1196. https://doi.org/10.1053/j.gastro.2021.12.244.

Wang, H., Zhang, S., Lv, J., & Cheng, Y. (2021). Design of polymers for siRNA delivery: Recent progress and challenges. *View, 2*(3). https://doi.org/10.1002/viw.20200026.

World Health Organization: WHO & World Health Organization: WHO. (2023, July 11). Colorectal cancer. https://www.who.int/news-room/fact-sheets/detail/colorectal-cancer

Wu, J., Huang, W., & He, Z. (2013). Dendrimers as carriers for siRNA delivery and gene silencing: A review. *The Scientific World JOURNAL, 2013*(1). https://doi.org/10.1155/2013/630654.

Xiang, J., Wang, J., Xiao, H., Huang, C., Wu, C., Zhang, L., Qian, C., & Xiang, D. (2025). Targeting tumor-associated macrophages in colon cancer: Mechanisms and therapeutic strategies. *Frontiers in Immunology, 16*. https://doi.org/10.3389/fimmu.2025.1573917.

Xiao, B., Ma, L., & Merlin, D. (2017). Nanoparticle-mediated co-delivery of chemotherapeutic agent and siRNA for combination cancer therapy. *Expert Opinion on Drug Delivery, 14*(1), 65–73. https://doi.org/10.1080/17425247.2016.1205583.

Xiao, B., Wang, S., Pan, Y., Zhi, W., Gu, C., Guo, T., Zhai, J., Li, C., Chen, Y. Q., & Wang, R. (2024). Development, opportunities and challenges of SiRNA Nucleic acid Drugs. *Molecular Therapy—Nucleic Acids, 36*(1), 102437. https://doi.org/10.1016/j.omtn.2024.102437.

Xie, J., & Wang, S. (2022). Small interfering RNA in colorectal cancer liver metastasis therapy. *Technology in Cancer Research & Treatment, 21*. https://doi.org/10.1177/15330338221103318.

Xue, R., Pan, Y., Xia, L., & Li, J. (2021). Non-viral vectors combined delivery of siRNA and anti-cancer drugs to reverse tumor multidrug resistance. *Acta Biomaterialia, 128*, 94–110. https://doi.org/10.1016/j.actbio.2021.04.028.

Yang, J., Huang, J., Han, D., & Ma, X. (2024). Artificial intelligence applications in the treatment of colorectal Cancer: A narrative review. *Clinical Medicine Insights Oncology, 18*. https://doi.org/10.1177/11795549231220320.

Yu, X., Wu, H., Wu, B., Zhi, S., Lan, Y., Chen, W., & Liu, J. (2025). Advances and challenges in drug repurposing in precision therapeutics of colorectal cancer. *World Journal of Gastrointestinal Oncology, 17*(7). https://doi.org/10.4251/wjgo.v17.i7.107681.

Yuan, Z., Guo, X., Wei, M., Xu, Y., Fang, Z., Feng, Y., & Yuan, W. (2020). Novel fluorinated polycationic delivery of anti-VEGF siRNA for tumor therapy. *NPG Asia Materials, 12*(1). https://doi.org/10.1038/s41427-020-0216-9.

Zaidi, S. S. A., Fatima, F., Zaidi, S.a A., Zhou, D., Deng, W., & Liu, S. (2023). Engineering siRNA therapeutics: Challenges and strategies. *Journal of Nanobiotechnology, 21*(1). https://doi.org/10.1186/s12951-023-02147-z.

Zhang, C., Zhao, Y., Zhang, E., Jiang, M., Zhi, D., Chen, H., Cui, S., Zhen, Y., Cui, J., & Zhang, S. (2020). Co-delivery of paclitaxel and anti-VEGF siRNA by tripeptide lipid nanoparticle to enhance the anti-tumor activity for lung cancer therapy. *Drug Delivery, 27*(1), 1397–1411. https://doi.org/10.1080/10717544.2020.1827085.

Zhang, J., Chen, B., Gan, C., Sun, H., Zhang, J., & Feng, L. (2023). A comprehensive review of small interfering RNAs (SiRNAs): Mechanism, therapeutic targets, and delivery strategies for cancer therapy. *International Journal of Nanomedicine, 18*, 7605–7635. https://doi.org/10.2147/ijn.s436038.

Zhang, K., Tang, X., Zhang, J., Lu, W., Lin, X., Zhang, Y., Tian, B., Yang, H., & He, H. (2014). PEG–PLGA copolymers: Their structure and structure-influenced drug delivery applications. *Journal of Controlled Release, 183*, 77–86. https://doi.org/10.1016/j.jconrel.2014.03.026.

Zhang, Z., Wang, H., Zhang, Y., Wang, J., An, Z., Du, Z., Yu, J., & Gong, Y. (2025). Preparation of co-loaded HHLA2 and EGFR siRNA nanoparticles and its delivery in colorectal cancer cells. *Journal of Drug Delivery Science and Technology*107185. https://doi.org/10.1016/j.jddst.2025.107185.

Zhao, Z., Ukidve, A., Kim, J., & Mitragotri, S. (2020). Targeting strategies for tissue-specific drug delivery. *Cell, 181*(1), 151–167. https://doi.org/10.1016/j.cell.2020.02.001.

Zhi, K., Raji, B., Nookala, A. R., Khan, M. M., Nguyen, X. H., Sakshi, S., Pourmotabbed, T., Yallapu, M. M., Kochat, H., Tadrous, E., Pernell, S., & Kumar, S. (2021). PLGA nanoparticle-based formulations to cross the blood–brain barrier for drug delivery: From R&D to cGMP. *Pharmaceutics, 13*(4), 500. https://doi.org/10.3390/pharmaceutics13040500.

Zhou, H., Chen, D. S., Hu, C. J., Hong, X., Shi, J., & Xiao, Y. (2023). Stimuli-Responsive Nanotechnology for RNA Delivery. *Advanced Science, 10*(36). https://doi.org/10.1002/advs.202303597.

CHAPTER

# 6 Carbon-based quantum dots and nanotubes for colon cancer targeting and drug delivery

**Madiha Ghazanfar[1,*], Abdul Basit[1,*], Muneeb Ullah[2], Jin-Wook Yoo[2] and Muhammad Naeem[1]**

[1]*Department of Biological Sciences, National University of Medical Sciences (NUMS), Rawalpindi, Punjab, Pakistan,*

[2]*Department of Manufacturing Pharmacy, College of Pharmacy, Pusan National University, Busan, Republic of Korea*

## 6.1 Introduction

Colon cancer (CC) is the third most common malignant tumor worldwide, which continues to pose formidable challenges due to its late diagnostics, highly diverse complex pathophysiology, silent progression, and limited efficacy of conventional therapies (Cappell, 2008). According to American Cancer Society, in 2025 approximately 154,270 new cases of CC are estimated to be diagnosed with around 52,900 projected deaths expected to occur in the United States (Siegel et al., 2025; Douaiher et al., 2017). Initially, neoplastic tubular adenomas in the colon form benign pedunculated polyps projecting into the lumen. Over time, these adenomas develop increasingly disorganized villous architecture and exhibit dysplastic changes at the cellular level (Ray, 2022). Early detection improves CC prognosis, but stages of CC are often asymptomatic; as the disease progresses, symptoms may include hematochezia, intestinal obstruction, constipation, palpable abdominal mass, unexplained weight loss, low iron levels, and fatigue (Mork et al., 2015; Benson et al., 2021).

Despite advancements in conventional diagnostic methods, including radiological techniques, computed tomography (CT), magnetic resonance (MR), colonography, and laboratory markers, these methods are less sensitive. The most widely studied marker is carcinoembryonic antigen (CEA), but this antigen has a low predictive value for asymptomatic patients (Mork et al., 2015; Kuipers et al., 2015). Although there is progress in therapeutic modalities, resistance still remains a significant challenge, often leading to treatment failure, poor progression-free survival, and cancer recurrence. Surgery is the primary and mainstay effective treatment for CC. However, despite advances in surgical techniques, recurrence is a significant limitation factor impacting long-term survival (Ryuk et al., 2014). Adjuvant to surgery, chemotherapy is a cornerstone treatment for CC. However, one of its major limitations is the development of resistance; in addition, resistance to chemotherapy poses significant toxicity challenges. Prevalent toxicities such as gastrointestinal (GI) toxicity, nephrotoxicity, cardiotoxicity, and secondary malignancies pose long-term complications (Oh et al., 2025; Agranovich & Berthelet, 2000). After surgery, radiotherapy (RT) and immunotherapy (IT) are

[*]Contribute equally to this work.

Nano Drug Delivery Systems for Colon Cancer. DOI: https://doi.org/10.1016/B978-0-443-40443-6.00010-6

considered standard preoperative treatments for CC. However, RT and IT have limited application in colonic cancer cells because of undefined targets and the fact that they have been shown to make precise radiation and immunotherapy delivery challenging without affecting surrounding organs that are sensitive to radiation damage (Le et al., 2017).

Nanotechnology has emerged as a revolutionary field in cancer research by addressing the shortcomings of conventional therapies. One of the key advantages of nanotechnology is the enhanced permeability and retention (EPR) effect that promotes nanoparticle accumulation in tumor tissues due to leaky vasculature, facilitating precise targeting of tumor cells, improved pharmacokinetics, and reduction in systemic toxicity and drug resistance (Goel et al., 2025). Furthermore, multifunctional theranostic nanoparticles combine immunomodulation with imaging and drug delivery, offering real-time monitoring of therapeutic responses. The emergence of two versatile classes of carbon-based nanomaterials, such as carbon quantum dots (CQDs) and carbon nanotubes (CNTs), has gained significant attention due to their unique structural, chemical, electronic, and biological properties. CQDs are typically nanosized, spherical in shape, and possess excellent photoluminescence properties, making them exceptionally biocompatible, eco-friendly, and easy to surface functionalize with minimal side effects (Shi et al., 2022). Sumio Iijima discovered CNTs in 1991, which are cylindrical nanomolecules that exhibit high aspect ratios composed of graphitic carbon. Like CQDs, CNTs are photodynamic, nontoxic, and are used to directly destroy cancer cells (Pechnikova et al., 2025; Aftab et al., 2025).

CQDs are a class of zero-dimensional nanoparticles with remarkable structural versatility and flexibility, which allows precise tuning by doping with diverse functional groups such as carboxyl, hydroxyl, epoxy, and amino groups on surface, enhancing their use in bioimaging and drug delivery (Carethers & Jung, 2015). Whereas CNTs are one-dimensional rolling graphene sheets, which exist as structures with variable lengths and diameters (single-walled carbon nanotubes [SWCNTs] or multi-walled carbon nanotubes [MWCNTs]). These structurally adaptable properties of CQDs and CNTs make nanotechnology more valuable for versatile applications like bioimaging, biosensing, reinforcing biomaterials, serving as drug carriers, or acting as scaffolds in tissue engineering (Ray, 2022). Unlike conventional therapies, CQDs and CNTs are generally less toxic, have high drug-loading capacity, and offer dual functionality—diagnosis and therapy (theranostics)—leading to more efficient nanomedicine strategies (Li et al., 2025). This book chapter aims to provide an in-depth understanding of carbon-based quantum dots and nanotubes applied for precise CC targeting and drug delivery, ultimately contributing to more effective, safer, and patient-specific therapeutic strategies for CC.

## 6.2 Colon cancer overview

CC is one of the most common cancers worldwide. Most CC cases occur in individuals aged 50 and older (Cappell, 2008; Douaiher et al., 2017). Early CC detection combines history, physical exam, colonoscopy with biopsy, imaging (CT, magnetic resonance imaging [MRI]), and blood markers (CEA, CA19-9); treatment varies by stage from surgery to targeted therapies and minimally invasive techniques (Benson et al., 2021; Ullah et al., 2023). CC recurrence occurs in about 15%–30% of patients within five years after surgery, with most recurrences peaking around 11 months and influenced by tumor stage, invasion, surgical complications, and biomarkers like CEA and CA19-9; treatment resistance and toxicities further limit outcomes (Ryu et al., 2023). Nanotechnology has emerged as a revolutionary field in cancer research by addressing the shortcomings of conventional therapies. By using of nanometer range (1–100 nm) carbon nanoparticles, CQDs and CNTs with unique physico-chemical properties will increase therapeutic efficacy, minimize systemic toxicity, enhance bioavailability, solubility, and sustain targeted drug release (Piktel et al., 2016; Aftab et al., 2025; Alifah et al., 2024). Beyond targeted drug delivery, both CQDs and CNTs possess significant potential in CC diagnosis and therapy monitoring, revolutionizing the era of therapeutics.

## 6.3 Targeting strategies of carbon quantum dots and carbon nanotubes for colon cancer

Targeting strategies enhance the specificity and accumulation of carbon-based nanocarriers in CC (Banerjee et al., 2017). Active targeting involves decorating CNTs with ligands (e.g., folic acid [FA], antibodies, peptides) that recognize overexpressed receptors on CC cells (Krasteva & Georgieva, 2022; Aftab et al., 2025). Stimuli-responsive systems can be engineered to release drugs in response to tumor-specific conditions such as acidic pH, elevated glutathione, or enzymatic activity, thereby minimizing premature release and systemic toxicity (Ding et al., 2016). CQDs can be conjugated with chemotherapeutic agents via covalent bonds (amide, ester) or noncovalent interactions (electrostatic $\pi$-$\pi$ stacking) (Pareek et al., 2025; Hasan et al., 2023). Additionally, their surfaces can be modified with polyethylene glycol (PEG), targeting ligands, or stimuli-responsive linkers to improve pharmacokinetics and tumor specificity (Siafaka et al., 2016; Safdar et al., 2025, 2025). CQDs functionalized with FA and loaded with 5-FU have demonstrated enhanced selective uptake by folate receptor-overexpressing CC cells, improving therapeutic efficacy and reducing systemic side effects (Sarkar et al., 2024).

Combining CQDs or CNTs with multiple therapeutic agents, or integrating them with other nanomaterials can further enhance treatment efficacy (Gaur et al., 2021; Ullah, Wahab, et al., 2023). Codelivery of chemotherapeutic drugs and gene silencing agents (e.g., small interfering RNA [siRNA]) on CNTs or CQDs platforms can simultaneously inhibit tumor growth and overcome drug resistance (Carvalho et al., 2021; Ahmed et al., 2025).

## 6.4 Passive targeting mechanism

Passive targeting is one of the most fundamental and widely utilized strategies in nanomedicine for CC (Mohamed et al., 2025) as illustrated in Fig. 6.1. It relies on the inherent pathophysiology of the tumor tissues, such as leaky vasculature, abnormal lymphatic drainage, acidic microenvironment, and elevated metabolic activity, to achieve preferential accumulation of therapeutic nanoparticles without the need for specific ligands (Fernandes et al., 2018). CQDs and CNTs offer passive targeting as an efficient first-line approach to enhance drug delivery to colon tumors by leveraging their unique size, shape, surface charge properties, and circulation behavior (Pareek et al., 2025). EPR effect is the cornerstone of passive tumor targeting (Rama et al., 2025). Rapidly proliferating tumor cells stimulate

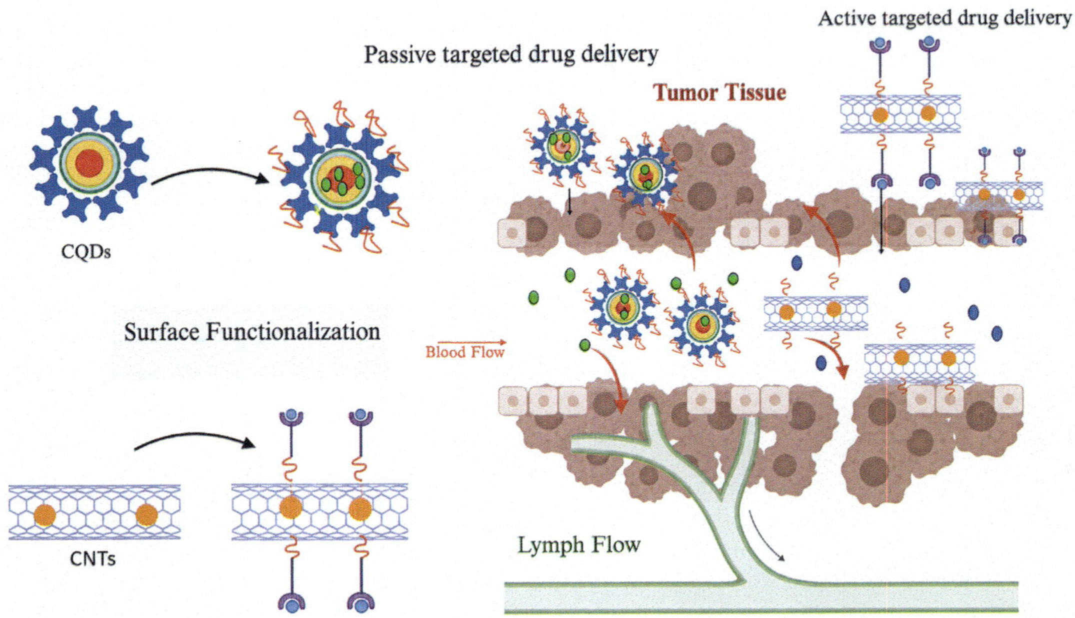


**FIGURE 6.1**

Active and passive targeted drug delivery using carbon quantum dots (*CQDs*) and carbon nanotubes (*CNTs*).

the formation of new blood vessels (angiogenesis) to sustain their growth (Cao, 2005; Ullah et al., 2021). However, these new formed vessels are often irregular, disorganized, and leaky due to poorly aligned endothelial cells, defective basement membranes, and wide fenestrations (100–800 nm) in the vascular wall (Gifre-Renom & Jones, 2021). At the same time, tumors lack an efficient lymphatic drainage system, resulting in the retention of macromolecules and nanoparticles within the tumor interstitium (Thomas & Schudel, 2015; Zeb et al., 2024). Nanocarriers designed within an optimal size range can thus preferentially accumulate in tumor tissues compared to normal tissues (Wang et al., 2019).

CQDs are typically 10 nm and may be rapidly cleared by renal filtration if used as free particles (Soo Choi et al., 2007). CQDs can be incorporated into larger carriers or conjugated to macromolecules (20–100 nm) to effectively exploit the EPR effect (Privitera et al., 2016; Ullah et al., 2024). CNTs, due to their high aspect ratio, can be engineered with controlled lengths and functional coatings to fall within this ideal size range, enhancing passive tumor uptake (Saleem, Wang, & Chen, 2018; Aftab et al., 2025). Hydrophilic coatings such as PEG or zwitterionic polymer systems are often applied to CQDs and CNTs to prevent rapid clearance by the phagocyte system (mononuclear phagocyte system [MPS]), extend blood circulation time, and improve the probability of tumor accumulation (Pechnikova et al., 2025; Safdar et al., 2024). The physiochemical properties of carbon-based nanocarriers significantly influence their passive targeting efficiency (Sajjadi et al., 2021). The rod-like geometry of CNTs affects margination and endothelial interaction differently from spherical nanoparticles, which can alter tumor penetration and retention (Chaudhuri et al., 2010).

Passive targeting alone is often insufficient due to intertumoral heterogeneity in vascular permeability, so it is commonly combined with active targeting and stimuli-responsive release strategies to improve selective therapeutic index (Parra-Nieto et al., 2024). Neutral, slightly negative surfaces are preferable for minimizing nonspecific protein adsorption and clearance (Lichtenberg et al., 2019). However, moderate positive charges can improve cellular internalization once the particles reach the tumor site (Wang et al., 2016).

## 6.5 Active targeting mechanism

Passive targeting exploits the natural physiology of tumors; active targeting introduces an additional layer of selectivity and precision by functionalizing nanocarriers with ligands that specifically recognize and bind to overexpressed receptors or biomarkers on the cancer cell or tumor-associated vasculature (Attia et al., 2019; Hasan et al., 2025). This strategy significantly enhances the cellular uptake, intratumoral accumulation, and therapeutic efficacy of nanocarriers while minimizing off-targeting effects (Wang et al., 2016). In context of CQDs and CNTs, active targeting transforms these versatile carbon nanomaterials into precision delivery vehicles capable of receptor-mediated endocytosis, controlled intracellular trafficking, and site-specific drug release in CC (Pardo et al., 2018).

## 6.6 Drug delivery system for colon cancer

In recent years, CQDs and CNTs have gained significant attention for their effectiveness in CC as shown in Fig. 6.2, as well as in both in vitro and in vivo bioimaging, gene delivery, and therapeutic applications (Du et al., 2019; Amin et al., 2024; Ullah et al., 2025). This depicts a wide range of advanced nanoparticle-based delivery systems designed for targeted therapy in CC. To enhance disease detection and targeted therapy, fluorescent probes such as FA, hyaluronic acid (HA), and Arginine–Glycine–Aspartic acid (RGD)-peptide have been successfully conjugated with CQDs (Li et al., 2025; Nawaz et al., 2025).

Polyethyleneimine (PEI)-passivated carbon dots were synthesized via microwave-assisted pyrolysis of glycerol, with PEI enhancing photoluminescence and facilitating DNA condensation for gene transfection. The pyrolysis duration significantly affected DNA condensation and cytotoxicity, with optimized CD-PEI composites showing reduced toxicity and enhanced plasmid DNA expression in COS-7 and HepG2 cells (Demirci et al., 2025; Ullah et al., 2025). According to Biswal and Bhattia (2021), CQDs enter cells through multiple mechanisms such as phagocytosis, endocytosis, and micropinocytosis. Endocytic pathway include clathrin and caveolae-mediated uptake, both essential for internalization of nanoparticles. Using specific inhibitors such as glucose, filipin III, DMA, and CPZ, researchers demonstrate that clathrin and caveolae-mediated pathways dominate in CD-based DNA and siRNA internalization, confirming the efficiency of endocytic routes. CNTs have been successfully used to deliver anticancer drugs such as doxorubicin (DOX), 5-fluorouracil (5-FU), and irinotecan directly to CC (Zarghami Dehaghani et al., 2021).

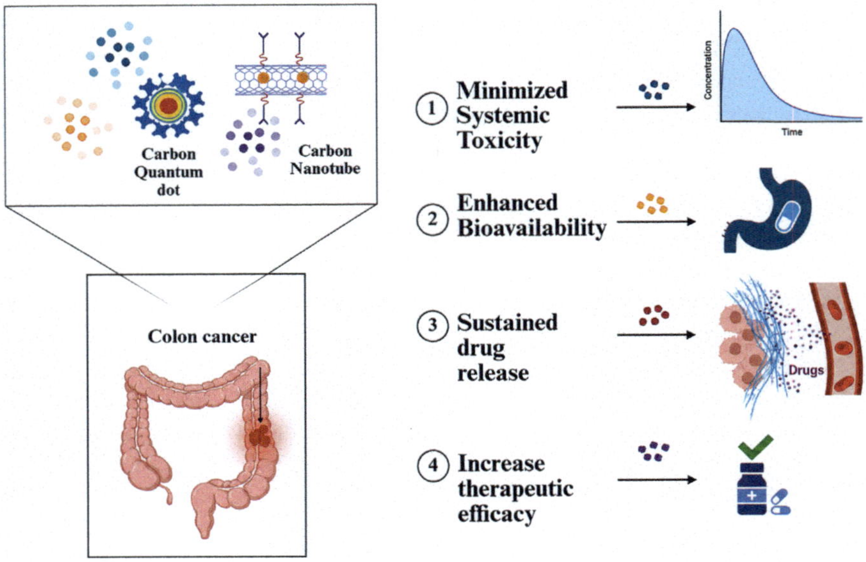


**FIGURE 6.2**

Drug delivery approaches used in the targeted therapy of colon cancer.

A PEGylated MWCNT system conjugated with FA and loaded with DOX achieved enhanced cytotoxicity and tumor suppression in CC models while reducing off-targeting effects.

A selective list of the drug delivery vehicles is shown in Table 6.1.

## 6.7 Challenges

Before a material is classified as a CQD or CNT-based nanocarrier, it must undergo rigorous characterization, which includes verification of nanoscale carbon domains, surface functional groups, and crystalline structure. Translating laboratory results into clinical practice will require standardizing synthesis procedures and guaranteeing reproducibility. Second, a thorough understanding of the mechanisms underlying cellular interactions is necessary. The logical development of next-generation nanotherapeutics will be accelerated by thorough investigations into the molecular mechanisms of absorption, intracellular trafficking, and drug release, as well as immunological interactions and tumor microenvironment regulation. Investigating the effects of ligand density, surface charge, and nanoparticle geometry on biodistribution, receptor engagement, and therapeutic response in vivo is part of this. Third, the major priorities must continue to be safety and long-term biocompatibility. Concerns regarding possible toxicity, bioaccumulation, and inflammatory reactions are continually raised by CNTs, even though CQDs are typically thought to be safe and biodegradable. Many of these problems can be avoided using surface functionalization and creative coating techniques, but in order to meet regulatory standards and foster clinical confidence, thorough in vivo investigations, long-term toxicity evaluations, and standardized safety procedures are necessary.

**Table 6.1 Representative examples of carbon quantum dots (*CQDs*) and carbon nanotubes (*CNTs*)-based drug delivery systems for colon cancer therapy.**

| Carbon nanomaterials | Drug loaded | Ligand attached | Cell line | Characterization methods | Drug loading/ efficacy | References |
|---|---|---|---|---|---|---|
| CQDs | Doxorubicin (DOX) | Folic acid (FA) | HCT116 | Ultraviolet–visible spectroscopy (UV-Vis), fourier transform infrared spectroscopy (FTIR), transmission electron microscopy (TEM), fluorescence, zeta potential | 82.3% | Kazeminava et al. (2024) |
| | 5-Fluorouracil (5-FU) | Hyaluronic acid | HT-29 | FTIR, nuclear magnetic resonance spectroscopy (NMR), TEM, UV-Vis, confocal microscopy | High cellular uptake | Udaipuria and Bhattacharya (2025) |
| | SN-38 (irinotecan active metabolite) | Aptamer AS1411 | SW620 | TEM, UV-Vis, x-ray photoelectron spectroscopy (XPS), fluorescence | Enhanced cytotoxicity | Ramesh, Ahlawat, and Srinivas (2010), Nawaz et al. (2025) |
| | Curcumin | FA | Caco-2 | TEM, dynamic light scattering (DLS), fluorescence, FTIR | 76% | Narayana et al. (2024) |
| CNTs | Doxorubicin | FA | HT-29 | UV-Vis, TEM, zeta, FTIR | 145% | Yusefi (2022) |
| | 5-Fluorouracil | Hyaluronic acid | HCT116 | SEM, TEM, Raman, confocal microscopy | Enhanced drug release | Bayda (2017) |
| | Oxaliplatin | Anti-epithelial cell adhesion molecule antibody | SW480 | FTIR, TEM, UV-Vis, flow cytometry | Targeted delivery, increased apoptosis | Wu et al. (2013) |
| | Camptothecin | Transferrin | LoVo | Atomic force microscopy (AFM), TEM, zeta, fluorescence | Higher cytotoxicity versus free drug | Zhang et al. (2023) |
| | Doxorubicin | Aptamer AS1411 | Caco-2 | UV-Vis, SEM, confocal microscopy | 2.3× uptake compared to free DOX | Taghavi et al. (2017) |

## 6.8 Conclusion and future perspective

Both CQDs and CNTs have been recently developed as extremely versatile and efficient nanoplatforms for CC therapy. Crucially, due to their potential of combining passive tumor accumulation, receptor-mediated active targeting, and stimuli response together, the potential to enhance therapy effectiveness with reduced systemic toxicity and off-targets effects is demonstrated. Future research should focus on patient-derived tumor models, organoids, and sophisticated preclinical systems that more accurately represent the complexity and heterogeneity of human colorectal cancer, going beyond in vitro and small-animal models. Including CQDs and CNTs in combination therapies, such as with gene editing, immunotherapy, or molecularly targeted medications, may improve therapy outcomes even further. Furthermore, the creation of theranostic systems that combine therapeutic monitoring, controlled medication release, and real-time imaging will open the door to genuinely customized CC treatment. In conclusion, CQDs and CNTs are positioned as transformative agents in oncology because of their inherent abundance, affordability, and functional diversity, as well as the constant potential of light, surface chemistry, and tailored molecular design.

## Acknowledgment

The author would like to acknowledge National University of Medical Sciences, Rawalpindi, Pakistan.

## Conflict of interest

The author declares that there is no conflict of interest related to this book chapter.

## References

Aftab, M., Ikram, S., Ullah, M., Khan, S. U., Wahab, A., & Naeem, M. (2025). Advancement of 3D bioprinting towards 4d bioprinting for sustained drug delivery and tissue engineering from biopolymers. *Journal of Manufacturing and Materials Processing, 9*(8), 285. https://doi.org/10.3390/jmmp9080285.

Aftab, M., Ikram, S., Ullah, M., Khan, N., Naeem, M., Khan, M. A., Bakhtiyor o'g'li, R. B., Qizi, K. S. S., Erkinjon Ugli, O. O., Abdurasulovna, B. M., & Qizi, O. K. A. (2025). Recent trends and future directions in 3d printing of biocompatible polymers. *Journal of Manufacturing and Materials Processing, 9*(4), 129. https://doi.org/10.3390/jmmp9040129.

Aftab, M., Mehreen, S., Ullah, M., Ikram, S., & Naeem, M. (2025). Recent advancements in inorganic based nanomaterials for wound healing; challenges and future opportunities. *Journal of Cluster Science, 36*(6). https://doi.org/10.1007/s10876-025-02932-w.

Agranovich, A., & Berthelet, E. J. B. C. M. J. (2000). *Radiotherapy for colorectal cancer, 42*(3), 139–141.

Ahmed, Z., Ullah, M., Zeshan, D., Khan, S. U., Ali, F., & Wahab, A. (2025). Exploring the tumor microenvironment in solid cancer: From biology to therapy. *Methods in Cell Biology, 198*, 359–385. https://doi.org/10.1016/bs.mcb.2025.02.020, https://www.sciencedirect.com/science/bookseries/0091679X.

Alifah, N., Palungan, J., Ardayanti, K., Ullah, M., Nurkhasanah, A. N., Mustopa, A. Z., Lallo, S., Agustina, R., Yoo, J.-W., & Hasan, N. (2024). Development of clindamycin-releasing polyvinyl alcohol hydrogel with self-healing property for the effective treatment of biofilm-infected wounds. *Gels, 10*(7), 482. https://doi.org/10.3390/gels10070482.

Amin, Z., et al. (2024). Fundamentals of nanotechnology. In *Revolutionizing agriculture: A comprehensive exploration of agri-nanotechnology*. Springer, 1–16.

Attia, M. F., Anton, N., Wallyn, J., Omran, Z., & Vandamme, T. F. (2019). An overview of active and passive targeting strategies to improve the nanocarriers efficiency to tumour sites. *Journal of Pharmacy and Pharmacology, 71*(8), 1185–1198. https://doi.org/10.1111/jphp.13098, http://onlinelibrary.wiley.com/journal/10.1111/(ISSN)2042-7158.

Banerjee, A., Pathak, S., Subramanium, V. D., Dharanivasan, G., Murugesan, R., & Verma, R. S. (2017). Strategies for targeted drug delivery in treatment of colon cancer: Current trends and future perspectives. *Drug Discovery Today, 22*(8), 1224–1232. https://doi.org/10.1016/j.drudis.2017.05.006.

Bayda, S. (2017). *Synthesis of fluorescent carbon nanoparticles (CNPs) and their applications in drug delivery.*

Benson, A. B., Venook, A. P., Al-Hawary, M. M., Arain, M. A., Chen, Y.-J., Ciombor, K. K., Cohen, S., Cooper, H. S., Deming, D., Farkas, L., Garrido-Laguna, I., Grem, J. L., Gunn, A., Hecht, J. R., Hoffe, S., Hubbard, J., Hunt, S., Johung, K. L., Kirilcuk, N., Krishnamurthi, S., Messersmith, W. A., Meyerhardt, J., Miller, E. D., Mulcahy, M. F., Nurkin, S., Overman, M. J., Parikh, A., Patel, H., Pedersen, K., Saltz, L., Schneider, C., Shibata, D., Skibber, J. M., Sofocleous, C. T., Stoffel, E. M., Stotsky-Himelfarb, E., Willett, C. G., Gregory, K. M., & Gurski, L. A. (2021). Colon cancer, version 2.2021, NCCN clinical practice guidelines in oncology. *Journal of the National Comprehensive Cancer Network, 19*(3), 329–359. https://doi.org/10.6004/jnccn.2021.0012.

Cao, Y. (2005). Tumor angiogenesis and therapy. *Biomedicine & Pharmacotherapy, 59*(2), S340. https://doi.org/10.1016/s0753-3322(05)80070-8.

Cappell, M. S. (2008). Pathophysiology, clinical presentation, and management of colon cancer. *Gastroenterology Clinics of North America, 37*(1), 1–24. https://doi.org/10.1016/j.gtc.2007.12.002.

Carethers, J. M., & Jung, B. H. (2015). Genetics and genetic biomarkers in sporadic colorectal cancer. *Gastroenterology, 149*(5), 1177. https://doi.org/10.1053/j.gastro.2015.06.047, http://www.journals.elsevier.com/gastroenterology/.

Carvalho, B. G., Vit, F. F., Carvalho, H. F., Han, S. W., & de la Torre, L. G. (2021). Recent advances in co-delivery nanosystems for synergistic action in cancer treatment. *Journal of Materials Chemistry B, 9*(5), 1208–1237. https://doi.org/10.1039/d0tb02168g, http://pubs.rsc.org/en/journals/journal/tb.

Chaudhuri, P., Harfouche, R., Soni, S., Hentschel, D. M., & Sengupta, S. (2010). Shape effect of carbon nanovectors on angiogenesis. *ACS Nano, 4*(1), 574–582. https://doi.org/10.1021/nn901465h.

Choi, H. S., Liu, W., Misra, P., Tanaka, E., Zimmer, J. P., Itty Ipe, B., Bawendi, M. G., & Frangioni, J. V. (2007). Renal clearance of quantum dots. *Nature Biotechnology, 25*(10), 1165–1170. https://doi.org/10.1038/nbt1340.

Demirci, S., Suner, S. S., Sahiner, M., Akcali, A., Guven, O., & Sahiner, N. (2025). A comparative study of nitrogen doped carbon dots prepared from linear polyethyleneimine (L-PEI) and branched polyethyleneimine (B-PEI): Thermal, optical, biocompatibility, sensor, antibacterial, and light-induced antibacterial activity. *Journal of Fluorescence*. https://doi.org/10.1007/s10895-025-04494-6.

Ding, C., Tong, L., Feng, J., & Fu, J. (2016). Recent advances in stimuli-responsive release function drug delivery systems for tumor treatment. *Molecules (Basel, Switzerland), 21*(12), 1715. https://doi.org/10.3390/molecules21121715.

Douaiher, J., Ravipati, A., Grams, B., Chowdhury, S., Alatise, O., & Are, C. (2017). Colorectal cancer—global burden, trends, and geographical variations. *Journal of Surgical Oncology, 115*(5), 619–630. https://doi.org/10.1002/jso.24578.

Du, J., Xu, N., Fan, J., Sun, W., & Peng, X. (2019). Carbon dots for in vivo bioimaging and theranostics. *Small (Weinheim an der Bergstrasse, Germany), 15*(32). https://doi.org/10.1002/smll.201805087.

Fernandes, C., Suares, D., & Yergeri, M. C. (2018). Tumor microenvironment targeted nanotherapy. *Frontiers in Pharmacology, 9*. https://doi.org/10.3389/fphar.2018.01230, http://www.frontiersin.org/Pharmacology.

Gaur, M., Misra, C., Yadav, A. B., Swaroop, S., Maolmhuaidh, F.Ó., Bechelany, M., & Barhoum, A. (2021). Biomedical applications of carbon nanomaterials: fullerenes, quantum dots, nanotubes, nanofibers, and graphene. *Materials, 14*(20), 5978. https://doi.org/10.3390/ma14205978.

Gifre-Renom, L., & Jones, E. A. V. (2021). Vessel enlargement in development and pathophysiology. *Frontiers in Physiology, 12*. https://doi.org/10.3389/fphys.2021.639645, http://www.frontiersin.org/Physiology/archive/.

Goel, A., Saraswat, I., Sharma, S., & Joshi, R. (2025). Harnessing nanotechnology and gene editing for cancer therapy: Asynergistic approach to precision medicine. *Current Gene Therapy, 25*. https://doi.org/10.2174/0115665232367601250717090618.

Hasan, N., Jiafu, C., Mustopa, A. Z., Himawan, A., Umami, R. N., Ullah, M., Wathoni, N., & Yoo, J.-W. (2023). Recent advancements of nitric oxide-releasing hydrogels for wound dressing applications. *Journal of Pharmaceutical Investigation, 53*(6), 781–801. https://doi.org/10.1007/s40005-023-00636-9.

Hasan, N., Palungan, J., & Ullah, M. (2025). Gene editing techniques in cancer research. *Methods in Cell Biology, 198*, 287–312. https://doi.org/10.1016/bs.mcb.2025.03.002, https://www.sciencedirect.com/science/bookseries/0091679X.

Kazeminava, F., Javanbakht, S., Latifi, Z., Rasoulzadehzali, M., Abbaszadeh, M., Alimohammadzadeh, B., Mahdipour, M., Fattahi, A., Hamishehkar, H., Adibag, Z., & Nouri, M. (2024). Ultrasound-assisted encapsulating folic acid-based carbon quantum dots within breast cancer cell-derived exosomes as a co-receptors-mediated anticancer nanocarrier for enhanced breast cancer therapy. *Scientific Reports, 14*(1). https://doi.org/10.1038/s41598-024-67934-6.

Krasteva, N., & Georgieva, M. (2022). Promising therapeutic strategies for colorectal cancer treatment based on nanomaterials. *Pharmaceutics, 14*(6), 1213. https://doi.org/10.3390/pharmaceutics14061213.

Kuipers, E. J., Grady, W. M., Lieberman, D., Seufferlein, T., Sung, J. J., Boelens, P. G., Van De Velde, C. J. H., & Watanabe, T. (2015). Colorectal cancer. *Nature Reviews Disease Primers, 1*. https://doi.org/10.1038/nrdp.2015.65, http://www.nature.com/nrdp/.

Le, D. T., Durham, J. N., Smith, K. N., Wang, H., Bartlett, B. R., Aulakh, L. K., Lu, S., Kemberling, H., Wilt, C., Luber, B. S., Wong, F., Azad, N. S., Rucki, A. A., Laheru, D., Donehower, R., Zaheer, A., Fisher, G. A., Crocenzi, T. S., Lee, J. J., Greten, T. F., Duffy, A. G., Ciombor, K. K., Eyring, A. D., Lam, B. H., Joe, A., Kang, S. P., Holdhoff, M., Danilova, L., Cope, L., Meyer, C., Zhou, S., Goldberg, R. M., Armstrong, D. K., Bever, K. M., Fader, A. N., Taube, J., Housseau, F., Spetzler, D., Xiao, N., Pardoll, D. M., Papadopoulos, N., Kinzler, K. W., Eshleman, J. R., Vogelstein, B., Anders, R. A., & Diaz, L. A. (2017). Mismatch repair deficiency predicts response of solid tumors to PD-1 blockade. *Science (New York, N.Y.), 357*(6349), 409–413. https://doi.org/10.1126/science.aan6733, http://science.sciencemag.org/content/sci/357/6349/409.full.pdf.

Lichtenberg, J. Y., Ling, Y., & Kim, S. (2019). Non-specific adsorption reduction methods in biosensing. *Sensors, 19*(11), 2488. https://doi.org/10.3390/s19112488.

Li, M., Huang, Y., Shen, C., Wang, Y., Lin, Y.'ai, Wang, Z., Chen, N., & Luo, Y. (2025). Application of quantum dots in cancer diagnosis and treatment: Advances and perspectives. *Nano Research, 18*(1), 94907163. https://doi.org/10.26599/nr.2025.94907163.

Mohamed, R. G. A., Ali, S. M., Ahmed, I. S., Rawas-Qalaji, M., & Hussain, Z. (2025). Next-generation nanocarriers for colorectal cancer: Passive, active, and stimuli-responsive strategies for precision therapy. *Emirates Biomaterials Science, 13*(20), 5626–5664. https://doi.org/10.1039/d5bm01176k, http://pubs.rsc.org/en/journals/journal/bm.

Mork, M. E., You, Y. N., Ying, J., Bannon, S. A., Lynch, P. M., Rodriguez-Bigas, M. A., & Vilar, E. (2015). High prevalence of hereditary cancer syndromes in adolescents and young adults with colorectal cancer.

*Journal of Clinical Oncology, 33*(31), 3544–3549. https://doi.org/10.1200/JCO.2015.61.4503, http://jco.ascopubs.org/content/33/31/3544.full.pdf.

Narayana, S., Gowda, B. H. J., Hani, U., Shimu, S. S., Paul, K., Das, A., Ashique, S., Ahmed, M. G., Tarighat, M. A., & Abdi, G. (2024). Inorganic nanoparticle-based treatment approaches for colorectal cancer: Recent advancements and challenges. *Journal of Nanobiotechnology, 22*(1). https://doi.org/10.1186/s12951-024-02701-3.

Nawaz, K., Iqr, a, Ullah, M., Yoo, J.-W., Aiman, U., Ghazanfar, M., & Naeem, M. (2025). Role of nutrition in the management of inflammatory bowel disease. *Recent Progress in Nutrition, 05*(01), 1–76. https://doi.org/10.21926/rpn.2501002.

Nawaz, K., Ullah, M., Yoo, J. W., Wahab, A., Hasan, N., & Naeem, M. (2025). *Tissue engineering for wound healing: Recent advancements and opportunities nanotechnology in wound healing: Innovations, applications, and future perspectives*. CRC Press, Russian Federation CRC Press, Russian Federation, 149–167. https://doi.org/10.1201/9781003605966-7, https://www.taylorfrancis.com/books/edit/10.1201/9781003605966/nanotechnology-wound-healing-mohammed-gulzar-ahmed-sumel-ashique-arshad-farid-gokhan-zengin.

Oh, J. M., Kim, S., Tsung, C., Kent, E., Jain, A., Ruff, S. M., & Zhang, H. (2025). Comprehensive review of the resistance mechanisms of colorectal cancer classified by therapy type. *Frontiers in Immunology, 16*. https://doi.org/10.3389/fimmu.2025.1571731, https://www.frontiersin.org/journals/immunology#.

Pardo, J., Peng, Z., & Leblanc, R. M. (2018). Cancer targeting and drug delivery using carbon-based quantum dots and nanotubes. *Molecules (Basel, Switzerland), 23*(2). https://doi.org/10.3390/molecules23020378, http://www.mdpi.com/1420-3049/23/2/378/pdf.

Pareek, A., Kumar, D., Pareek, A., & Gupta, M. M. (2025). Advancing cancer therapy with quantum dots and other nanostructures: A review of drug delivery innovations, applications, and challenges. *Cancers, 17*(5), 878. https://doi.org/10.3390/cancers17050878.

Parra-Nieto, J., de Carcer, I. A., García del Cid, M. A., Jimenez-Falcao, S., Gónzalez-Larre, J., & Baeza, A. (2024). Stimuli-responsive nanocarriers as active enhancers of antitumoral immunotherapy. *Advanced Materials Interfaces, 11*(30). https://doi.org/10.1002/admi.202400343.

Pechnikova, N. A., Domvri, K., Porpodis, K., Istomina, M. S., Iaremenko, A. V., & Yaremenko, A. V. (2025). Carbon quantum dots in biomedical applications: Advances, challenges, and future prospects. *Aggregate, 6*(3). https://doi.org/10.1002/agt2.707, https://onlinelibrary.wiley.com/journal/26924560.

Piktel, E., Niemirowicz, K., Wątek, M., Wollny, T., Deptuła, P., & Bucki, R. (2016). Recent insights in nanotechnology-based drugs and formulations designed for effective anti-cancer therapy. *Journal of Nanobiotechnology, 14*(1). https://doi.org/10.1186/s12951-016-0193-x.

Privitera, A., Righetto, M., Mosconi, D., Lorandi, F., Isse, A. A., Moretto, A., Bozio, R., Ferrante, C., & Franco, L. (2016). Boosting carbon quantum dots/fullerene electron transfer via surface group engineering. *Physical Chemistry Chemical Physics, 18*(45), 31286–31295. https://doi.org/10.1039/c6cp05981c.

Rama, E., May, J.-N., Rix, A., Lammers, T., & Kiessling, F. (2025). Image-guided strategies to improve drug delivery to tumors beyond using the enhanced permeability and retention (EPR) effect. *Biochemical and Biophysical Research Communications, 778*, 152346. https://doi.org/10.1016/j.bbrc.2025.152346.

Ramesh, M., Ahlawat, P., & Srinivas, N. R. (2010). Irinotecan and its active metabolite, SN-38: Review of bioanalytical methods and recent update from clinical pharmacology perspectives. *Biomedical Chromatography, 24*(1), 104–123. https://doi.org/10.1002/bmc.1345.

Ray, K. (2022). Mapping the malignant transformation from polyps to CRC. *Nature Reviews Gastroenterology & Hepatology, 19*(8), 488. https://doi.org/10.1038/s41575-022-00660-4.

Ryuk, J. P., Choi, G.-S., Park, J. S., Kim, H. J., Park, S. Y., Yoon, G. S., Jun, S. H., & Kwon, Y. C. (2014). Predictive factors and the prognosis of recurrence of colorectal cancer within 2 years after curative resection. *Annals of Surgical Treatment and Research, 86*(3), 143. https://doi.org/10.4174/astr.2014.86.3.143.

Ryu, H. S., Kim, J., Park, Y. R., Cho, E. H., Choo, J. M., Kim, J.-S., Baek, S.-J., & Kwak, J.-M. (2023). Recurrence patterns and risk factors after curative resection for colorectal cancer: insights for postoperative surveillance strategies. *Cancers, 15*(24), 5791. https://doi.org/10.3390/cancers15245791.

Safdar, M., Amin, Z., Ullah, M., Wahab, A., Hasan, N., & Naeem, M. (2025). Cancer stem cell analysis and targeting. *Methods in Cell Biology, 198*, 251–271. https://doi.org/10.1016/bs.mcb.2025.02.017, https://www.sciencedirect.com/science/bookseries/0091679X.

Safdar, M., Naeem, M., Aftab, M., Ullah, M., & Wahab, A. (2025). *Genetic engineering of fungi fungal biotechnology: advancements in drug discovery, healthcare, and environmental solutions*. Pakistan: CRC Press, Pakistan CRC Press, 36–44. https://doi.org/10.1201/9781003594840-4, https://www.routledge.com/Fungal-Biotechnology-Advancements-in-Drug-Discovery-Healthcare-and-Environmental-Solutions/Farid-Preet/p/book/9781032976839.

Safdar, M., Ullah, M., Hamayun, S., Wahab, A., Ullah Khan, S., Abdikakhorovich, S. A., Ul Haq, Z., Mehreen, A., Naeem, M., Mustopa, A. Z., & Hasan, N. (2024). Microbiome miracles and their pioneering advances and future frontiers in cardiovascular disease. *Current Problems in Cardiology, 49*(9), 102686. https://doi.org/10.1016/j.cpcardiol.2024.102686.

Sajjadi, M., Nasrollahzadeh, M., Jaleh, B., Soufi, G. J., & Iravani, S. (2021). Carbon-based nanomaterials for targeted cancer nanotherapy: Recent trends and future prospects. *Journal of Drug Targeting, 29*(7), 716–741. https://doi.org/10.1080/1061186x.2021.1886301.

Saleem, J., Wang, L., & Chen, C. (2018). Carbon-based nanomaterials for cancer therapy via targeting tumor microenvironment. *Advanced Healthcare Materials, 7*(20). https://doi.org/10.1002/adhm.201800525.

Sarkar, S., Kiren, S., & Gmeiner, W. H. (2024). Review of prodrug and nanodelivery strategies to improve the treatment of colorectal cancer with fluoropyrimidine drugs. *Pharmaceutics, 16*(6). https://doi.org/10.3390/pharmaceutics16060734, http://www.mdpi.com/journal/pharmaceutics.

Shi, W., Han, Q., Wu, J., Ji, C., Zhou, Y., Li, S., Gao, L., Leblanc, R. M., & Peng, Z. (2022). Synthesis mechanisms, structural models, and photothermal therapy applications of top-down carbon dots from carbon powder, graphite, graphene, and carbon nanotubes. *International Journal of Molecular Sciences, 23*(3), 1456. https://doi.org/10.3390/ijms23031456.

Siafaka, P., Üstündağ Okur, N., Karavas, E., & Bikiaris, D. (2016). Surface modified multifunctional and stimuli responsive nanoparticles for drug targeting: Current status and uses. *International Journal of Molecular Sciences, 17*(9), 1440. https://doi.org/10.3390/ijms17091440.

Siegel, R. L., Kratzer, T. B., Giaquinto, A. N., Sung, H., & Jemal, A. (2025). Cancer statistics, 2025. *Ca-A Cancer Journal for Clinicians, 75*(1), 10–45. https://doi.org/10.3322/caac.21871, http://onlinelibrary.wiley.com/journal/10.3322/(ISSN)1542-4863.

Taghavi, S., Nia, A. H., Abnous, K., & Ramezani, M. (2017). Polyethylenimine-functionalized carbon nanotubes tagged with AS1411 aptamer for combination gene and drug delivery into human gastric cancer cells. *International Journal of Pharmaceutics, 516*(1-2), 301–312. https://doi.org/10.1016/j.ijpharm.2016.11.027.

Thomas, S. N., & Schudel, A. (2015). Overcoming transport barriers for interstitial-, lymphatic-, and lymph node-targeted drug delivery. *Current Opinion in Chemical Engineering, 7*, 65–74. https://doi.org/10.1016/j.coche.2014.11.003, http://www.elsevier.com/wps/find/journaldescription.cws_home/725837/description#description.

Udaipuria, N., & Bhattacharya, S. (2025). Novel carbohydrate polymer-based systems for precise drug delivery in colon cancer: Improving treatment effectiveness with intelligent biodegradable materials. *Biopolymers, 116*(1). https://doi.org/10.1002/bip.23632.

Ullah, M., Awan, U. A., Ali, H., Wahab, A., Ullah Khan, S., Naeem, M., Ruslin, M., Mustopa, A. Z., & Hasan, N. (2025). Carbon dots: New rising stars in the carbon family for diagnosis and biomedical applications. *Journal of Nanotheranostics, 6*(1), 1. https://doi.org/10.3390/jnt6010001.

Ullah, M., Awan, U. A., Muhaymin, A., Naeem, M., Yoo, J.-W., Mehreen, A., Safdar, A., Hasan, N., Haider, A., & Ud-Din, F. (2025). Cancer nanomedicine: Smart arsenal in the war against cancer. *Inorganic Chemistry Communications, 174*, 114030. https://doi.org/10.1016/j.inoche.2025.114030.

Ullah, M., Lee, J., Hasan, N., Hakim, M. L., Kwak, D., Kim, H., Lee, E., Ahn, J., Mun, B., Lee, E. H., Jung, Y., & Yoo, J.-W. (2024). Clindamycin-loaded polyhydroxyalkanoate nanoparticles for the treatment of methicillin-resistant staphylococcus aureus-infected wounds. *Pharmaceutics, 16*(10), 1315. https://doi.org/10.3390/pharmaceutics16101315.

Ullah, M., Wahab, A., Khan, D., Saeed, S., Ullah Khan, S., Ullah, N., & Saleh, T. A. (2021). Modified gold and polymeric gold nanostructures: Toxicology and biomedical applications. *Colloid and Interface Science Communications, 42*, 100412. https://doi.org/10.1016/j.colcom.2021.100412.

Ullah, M., Wahab, A., Ullah Khan, S., Zaman, U., Rehman, K.ur, Hamayun, S., Naeem, M., Ali, H., Riaz, T., Saeed, S., Alsuhaibani, A. M., & Refat, M. S. (2023). Stent as a novel technology for coronary artery disease and their clinical manifestation. *Current Problems in Cardiology, 48*(1), 101415. https://doi.org/10.1016/j.cpcardiol.2022.101415.

Ullah, M., Waleed, A., Hamayun, S., Shaukat, A., Safdar, M., Hussain, M., Waleed, T., Akber, K., Ali, H., Amjad, M., & Ahmed, I. (2023). Nanotechnology and biomedical devices used as a novel tool in biosensing and bioimaging of disease. *Journal of Women Medical and Dental College, 1*(4). https://doi.org/10.56600/jwmdc.v1i4.55.

Wang, N., Cheng, X., Li, N., Wang, H., & Chen, H. (2019). Nanocarriers and their loading strategies. *Advanced Healthcare Materials, 8*(6). https://doi.org/10.1002/adhm.201801002.

Wang, S., Huang, P., & Chen, X. Y. (2016). Hierarchical targeting strategy for enhanced tumor tissue accumulation/retention and cellular internalization. *Advanced Materials, 28*(34), 7340–7364. https://doi.org/10.1002/adma.201601498, http://onlinelibrary.wiley.com/journal/10.1002/(ISSN)1521-4095.

Wu, L., Man, C., Wang, H., Lu, X., Ma, Q., Cai, Y., & Ma, W. (2013). PEGylated multi-walled carbon nanotubes for encapsulation and sustained release of oxaliplatin. *Pharmaceutical Research, 30*(2), 412–423. https://doi.org/10.1007/s11095-012-0883-5.

Yusefi, M. (2022). *Polysaccharide-based magnetic nanocomposites as drug carriers for potential colorectal cancer treatment.*

Zarghami Dehaghani, M., Yousefi, F., Sajadi, S. M., Munir, M. T., Abida, O., Habibzadeh, S., Hamed Mashhadzadeh, A., Rabiee, N., Mostafavi, E., & Reza Saeb, M. (2021). Theoretical encapsulation of fluorouracil (5-FU) anti-cancer chemotherapy drug into carbon nanotubes (CNT) and boron nitride nanotubes (BNNT). *Molecules (Basel, Switzerland), 26*(16), 4920. https://doi.org/10.3390/molecules26164920.

Zeb, F., Mehreen, A., Naqeeb, H., Ullah, M., Waleed, A., Awan, U. A., Haider, A., & Naeem, M. (2024). Nutrition and dietary intervention in cancer: Gaps, challenges, and future perspectives. *Springer Science and Business Media LLC, 191*, 281–307. https://doi.org/10.1007/978-3-031-55622-7_11.

Zhang, L., Wang, C., Jiang, Y., Zhang, S., Ye, D., & Liu, L. (2023). Drug delivery mechanism of doxorubicin and camptothecin on single-walled carbon nanotubes by DFT study. *Applied Surface Science, 614*, 156242. https://doi.org/10.1016/j.apsusc.2022.156242.

CHAPTER

# Recent trends of extracellular vesicles for therapeutic intervention of colon cancer

7

**Sunil Choudhary[1], Pitchaikannu Venkatraman[2], Sanheeta Chakrabarty[3], Lalit Mohan Aggarwal[1] and Shreyans Jain[3]**

[1]*Department of Radiotherapy and Radiation Medicine, Institute of Medical Sciences, Banaras Hindu University, Varanasi, Uttar Pradesh, India,*

[2]*Department of Neurosurgery and Gamma Knife Unit, Neurosciences Centre, All India Institute of Medical Sciences, New Delhi, India,*

[3]*Department of Pharmaceutical Engineering & Technology, Indian Institute of Technology (BHU), Varanasi, Uttar Pradesh, India*

## 7.1 Introduction

One of the most prevalent and deadly cancers in the world is colorectal cancer (CRC), which encompasses both colon and rectal cancer. Globally, CRC ranks second in cancer-related mortality and third in incidence, according to GLOBOCAN 2022 estimates (Giusti et al., 2023; Imran & Shahid, 2021; Tamanna et al., 2025). A sizable percentage of patients receive their diagnoses at an advanced stage, which reduces the efficacy of traditional therapies and limits the curative potential of surgery (Bray et al., 2021, 2024). Surgical resection is usually followed by chemotherapy (such as 5-fluorouracil, oxaliplatin, and irinotecan) and, in certain situations, targeted therapies like anti-EGFR or antivascular endothelial growth factor (VEGF) agents (Folprecht et al., 2022; Shaham et al., 2025). However, a number of drawbacks, such as acquired drug resistance, off-target toxicity, and the incapacity to effectively enter the tumor microenvironment (TME), frequently impede these strategies (Aldea et al., 2021). Personalized treatment is further complicated by interpatient variability and tumor biology heterogeneity (De Palma & Hanahan, 2012; Molinari and Marisi, 2018). These difficulties underscore the pressing need for novel, focused delivery methods that can improve treatment specificity and overcome drug resistance while posing the least amount of systemic toxicity (Cao et al., 2023; Lv et al., 2022; Najafi et al., 2023; Rui et al., 2023).

In recent years, a growing body of research has underscored the importance of tumor-stroma communication, immune modulation, and molecular crosstalk in driving CRC progression, metastasis, and therapeutic resistance (Shakhpazyan et al., 2023). Conventional therapeutics often fail to modulate these intricate signaling networks, highlighting the need for advanced delivery systems capable of interacting with or even reprogramming the TME (Goenka et al., 2023). Against this backdrop, extracellular vesicles (EVs) have garnered considerable attention for their ability to mediate intercellular communication and deliver functional biomolecules across biological barriers. Their endogenous origin allows them to carry complex biological information in a physiologically

Nano Drug Delivery Systems for Colon Cancer. DOI: https://doi.org/10.1016/B978-0-443-40443-6.00023-4

compatible manner, making them ideal carriers for precision-based therapies, including RNA drugs, gene-editing tools, immunomodulatory agents, and combination regimens (Aloi et al., 2024; Chang et al., 2025; Ciardiello et al., 2016; Xu et al., 2018).

EVs have emerged as highly promising candidates for therapeutic delivery in cancer biology due to their unique combination of natural origin, biological compatibility, and functional versatility (Xu et al., 2018; Zhang, Zang et al., 2019). Unlike synthetic nanocarriers, EVs are inherently produced by almost all cell types and therefore possess a native nanoscale architecture optimized for intercellular communication (van Niel et al., 2022). Their lipid bilayer membrane offers exceptional biocompatibility and stability in physiological environments, while simultaneously protecting encapsulated therapeutic cargos, such as microRNAs (miRNAs), siRNAs, mRNAs, proteins, or small-molecule drugs from enzymatic degradation and rapid clearance (Zare et al., 2022; Zhang et al., 2023). This intrinsic protection significantly enhances the bioavailability and functional half-life of therapeutic agents. In addition, EVs display surface proteins, adhesion molecules, and receptor–ligand signatures inherited from their parent cells, which enable inherent targeting abilities and selective uptake by specific cell populations, including tumor and stromal cells in the CRC microenvironment. Such natural tropism provides a distinct advantage over conventional delivery systems that often require complex chemical modifications to achieve similar specificity. Furthermore, EVs can cross biological barriers, circulate without eliciting strong immune responses, and integrate seamlessly with cellular pathways, enabling efficient intracellular delivery. Collectively, these attributes position EVs as biologically intelligent carriers with significant potential to transform therapeutic intervention strategies for colon cancer (Elliott & He, 2021; Gutiérrez-Vázquez et al., 2013; Verma & Arora, 2025; Yong et al., 2022).

Taken together, these developments highlight EVs as a rapidly advancing and highly promising platform for improving therapeutic outcomes in CRC. In the past three to five years, notable progress in EV engineering, targeted delivery, and functional cargo modulation has strengthened their potential to overcome major limitations of conventional therapies, while emerging clinical studies further support their safety and translational relevance. Therefore, this review aims to summarize the recent trends in utilizing EVs for therapeutic intervention in colon cancer, focusing on key advances, evolving applications, and progress toward clinical translation. This will provide a clear understanding of how EV-based strategies are shaping the future landscape of CRC treatment.

### 7.1.1 Extracellular vesicles

EVs are membrane-bound nanoscale particles actively secreted by nearly all cell types under both physiological and pathological conditions, and they have emerged as essential mediators of intercellular communication in multicellular systems (Liu & Wang, 2023; van Niel et al., 2022). These vesicles encapsulate a rich and diverse repertoire of bioactive molecules, including proteins, lipids, DNA fragments, mRNAs, miRNAs, long noncoding RNAs (lncRNAs), and metabolites, thereby functioning as natural carriers capable of transferring functional cargo between local and distant cells (Ahmadi et al., 2024; Gurunathan, Kang, Qasim, Khan & Kim, 2021; Mokhtari et al., 2025; Samadi et al., 2025). Based on their biogenesis and size, EVs are broadly classified into three major categories: exosomes (30–150 nm), which originate from the inward budding of endosomal membranes to form multivesicular bodies (MVBs) that subsequently fuse with the plasma membrane to release their contents; microvesicles (100–1000 nm), which are shed directly from the outward budding of the plasma membrane; and apoptotic bodies (>1000 nm), which arise from membrane blebbing during programmed cell death (Gurunathan, Kang & Kim, 2021; Heijnen et al., 1999; Nunez et al., 2010). This heterogeneity endows EVs with a wide range of biological functions, enabling them to influence processes such as immune modulation, angiogenesis, metastasis, and drug resistance (Yang et al., 2022). Importantly, EVs are abundantly present in easily accessible biological fluids such as blood, urine, saliva, breast milk, cerebrospinal fluid, and ascitic fluid, making them valuable candidates for liquid biopsy–based diagnostics. Their molecular composition often mirrors the physiological or pathological state of their parental cells, allowing EVs to serve as disease-specific biomarkers for early detection, prognosis, and monitoring of therapeutic response. Additionally, their intrinsic biocompatibility, stability in circulation, and ability to protect their cargo from enzymatic degradation have positioned EVs as promising therapeutic vehicles in cancer and other diseases. Together, these characteristics underscore the immense potential of EVs as versatile tools for both diagnostic and therapeutic applications in modern biomedical research (Khoei et al., 2020).

### 7.1.2 Significance of extracellular vesicles in tumor progression, metastasis, and immune modulation

EVs play a multifaceted role in shaping the malignant behavior of colon cancer by orchestrating tumor progression, metastatic dissemination, and immune modulation. Tumor-derived EVs (TDEVs) act as potent mediators of communication between cancer cells and the surrounding TME, carrying a rich cargo of oncogenic proteins, lipids, DNA fragments, mRNAs, and noncoding RNAs that collectively reprogram neighboring and distant cells (Zheng et al., 2019). One of their critical functions is the promotion of angiogenesis: TDEVs deliver pro-angiogenic factors such as VEGF, angiopoietin, and miR-21 to endothelial cells, enhancing vascular proliferation and permeability, which in turn ensures sustained nutrient supply and tumor expansion. EVs are also key drivers of metastatic progression; by transporting integrins, matrix metalloproteinases (MMPs), and pro-invasive miRNAs, they remodel extracellular matrices, modulate stromal fibroblasts, and prepare premetastatic niches in distant organs such as the liver and lungs. This preconditioning creates a receptive microenvironment that supports colonization and growth of disseminated tumor cells (Giusti et al., 2023; Li et al., 2024). In addition to structural and metabolic reprogramming, EVs profoundly influence immune dynamics within the tumor milieu. They carry immunosuppressive molecules like PD-L1, TGF-β, and Fas Ligand (FasL), which inhibit dendritic cell maturation, induce apoptosis or dysfunction of cytotoxic T cells, and promote the expansion of regulatory T cells, collectively facilitating immune escape and enabling tumor survival (Yue et al., 2023). Moreover, EV-mediated transfer of specific miRNAs and drug-efflux proteins contributes to chemoresistance by altering drug transport, promoting epithelial-mesenchymal transition (EMT), and activating survival signaling pathways in cancer cells (Yang et al., 2022). Through these diverse mechanisms, EVs act as powerful regulators of tumor aggressiveness, making them not only fundamental to understanding cancer biology but also highly promising targets for therapeutic intervention. Their natural ability to transfer bioactive molecules with specificity and stability highlights their potential to be engineered as drug-delivery vehicles or to be inhibited to suppress pathways that fuel tumor progression (Chavda et al., 2022).

## 7.2 Extracellular vesicles in colon cancer biology

### 7.2.1 Role of tumor-derived extracellular vesicles in colorectal cancer initiation and progression

The development and progression of CRC are significantly mediated by TDEVs. The bioactive cargo proteins, lipids, mRNAs, and noncoding RNAs found in these EVs, which are secreted in large quantities by CRC cells, alter the behavior of recipient cells (Rizk et al., 2022). By delivering oncogenic miRNAs (such as miR-21 and miR-1246) and inhibiting tumor suppressor genes such as p53 or PTEN, TDEVs can rewire healthy colon epithelial cells during the early phases of carcinogenesis. Increased cellular survival, proliferation, and genomic instability are the results of this molecular reprogramming (Harding et al., 1983; Harding et al., 2013; Pan & Johnstone, 1983; Salimi et al., 2020; Zhang, Higginbotham et al., 2019).

Moreover, TDEVs have the ability to stimulate autocrine and paracrine signaling loops within the tumor mass, which fosters the growth of cancer cells. For example, it has been demonstrated that exosomes derived from CRC that are enriched with components of the Wnt/β-catenin pathway promote the self-renewal and proliferation of cancer stem cells (CSCs), which, in turn, contributes to tumor heterogeneity and treatment resistance (Angius et al., 2021).

Beyond early carcinogenic events, TDEVs actively modulate stromal and immune components to create a supportive niche for tumor expansion. Fibroblasts exposed to CRC-derived EVs undergo activation into cancer-associated fibroblasts (CAFs), which secrete cytokines, extracellular matrix components, and growth factors that facilitate tumor invasion (Rahmati et al., 2024). Similarly, endothelial cells internalize angiogenic EV cargo such as VEGF, E-cadherin fragments, and pro-angiogenic miRNAs, driving angiogenesis and enhancing vascular remodeling to fulfill the metabolic needs of the growing tumor. TDEVs also contribute to therapy failure by transporting multidrug resistance proteins, antiapoptotic factors, and miRNAs such as miR-196b or miR-92a, thereby reprogramming cancer cells to evade chemotherapeutic stress (Maleki et al., 2023; Siveen et al., 2019).

Furthermore, accumulating evidence suggests that TDEVs play a pivotal role in shaping the premetastatic niche in distant organs, particularly the liver, the most common metastatic site for CRC. Through integrin patterns (such as αvβ5), TDEVs selectively home to specific tissues, where they trigger inflammatory signaling, recruit immunosuppressive cells, and modify the extracellular matrix to support metastatic colonization (Brown & Marshall, 2019; Pang et al., 2023). Collectively, these diverse actions highlight TDEVs as master regulators of CRC initiation and progression, underscoring their diagnostic value and therapeutic potential in intercepting tumor evolution.

### 7.2.2 Extracellular vesicle–mediated crosstalk in the tumor microenvironment

The TME in CRC is a highly dynamic ecosystem composed of cancer cells, stromal fibroblasts, endothelial cells, infiltrating immune cells, and extracellular matrix components, all of which collectively influence tumor behavior. EVs function as critical mediators of communication among these compartments by transporting regulatory molecules that remodel the TME in favor of tumor progression. EVs released by CAFs play a central role in enhancing metabolic reprogramming within CRC cells (Li et al., 2021). By delivering growth-promoting factors, glycolytic enzymes, and miRNAs such as miR-21 and miR-155, CAF-derived EVs elevate glycolytic flux, boost adenosine triphosphate (ATP) production, and drive an aggressive, pro-survival metabolic phenotype in tumor cells (Lv et al., 2022). These vesicles also promote EMT, thereby increasing invasiveness and metastatic potential (Gopal et al., 2017).

Simultaneously, TDEVs exert profound immunomodulatory effects within the TME. By transferring immunosuppressive molecules such as miR-146a, PD-L1, and TGF-β, TDEVs impair cytotoxic T lymphocyte (CTL) activation, inhibit dendritic cell maturation, and expand regulatory T-cell (Treg) populations, collectively dampening antitumor immunity (Najaflou et al., 2022). TDEVs also drive macrophage plasticity, with miR-934 and miR-1246 directing macrophage polarization toward the M2 phenotype, an immunosuppressive, tumor-promoting state characterized by enhanced secretion of IL-10, VEGF, and matrix-remodeling enzymes (Harding et al., 1983; Harding et al., 2013; Pan & Johnstone, 1983). Moreover, interactions between EVs and endothelial cells contribute to neovascularization, as EV-carried VEGF, miR-25-3p, and angiopoietic proteins promote vascular permeability and new blood vessel formation.

Beyond these well-established interactions, EVs also influence other stromal components of the CRC microenvironment. For instance, EV-mediated transfer of MMPs to fibroblasts and endothelial cells enhances extracellular matrix degradation, facilitating tumor invasion and metastatic dissemination (Nawaz et al., 2018). Additionally, EVs derived from hypoxic CRC cells carry hypoxia-inducible factors (HIFs) and metabolic regulators that further condition neighboring stromal cells to support survival under low-oxygen conditions (Zonneveld et al., 2019). These hypoxia-modulated EVs promote angiogenesis, suppress immune surveillance, and stimulate CSC enrichment. EVs also participate in reprogramming the tumor vasculature by modulating endothelial tight junction proteins such as ZO-1 and occludin, thereby increasing vascular leakiness. This enhanced permeability not only supports nutrient delivery but also facilitates tumor cell intravasation. Furthermore, interactions between EVs and mesenchymal stem cells (MSCs) promote their differentiation into tumor-associated fibroblasts, expanding the pro-tumorigenic stromal population.

### 7.2.3 Impact on angiogenesis, epithelial-mesenchymal transition, and metastasis

EVs play a crucial role in regulating important cancer hallmarks like metastasis, angiogenesis, and EMT. By delivering VEGF, angiopoietins, and pro-angiogenic miRNAs (such as miR-210 and miR-21) to endothelial cells, TDEVs stimulate angiogenesis and neovascularization, which support tumor growth (Olejarz et al., 2020). Regarding EMT, TDEVs help epithelial CRC cells change into a mesenchymal phenotype, which is characterized by an upregulation of vimentin, N-cadherin, and Snail, and a loss of E-cadherin. This change in phenotype makes CRC cells more migratory and invasive. A key role in this transformation is played by EVs enriched with EMT-inducing factors like TGF-β, ZEB1 mRNA, and members of the miR-200 family (Colombo et al., 2013; Hessvik & Llorente, 2018; McCullough et al., 2018; Zhang et al., 2021).

Furthermore, EV-mediated remodeling of distant organ microenvironments to create premetastatic niches promotes metastasis. By guiding bone marrow-derived cells to specific organs like the liver and lungs, CRC-derived EVs can prime the stromal bed for colonization. Organotropic metastasis has been linked to exosomal integrins, specifically α6β4 and αvβ5, which direct EVs and metastatic cells to particular tissues (Guo et al., 2023).

Building upon these functions, recent studies reveal that EV-mediated angiogenesis not only enhances nutrient supply to expanding tumors but also facilitates intravasation of circulating tumor cells (CTCs). EVs enriched with miR-25-3p, miR-221/222, and hypoxia-induced proteins can disrupt endothelial tight junctions, increasing vascular permeability and enabling tumor cells to enter the bloodstream more efficiently (Kikuchi et al., 2019). Hypoxic CRC cells, in particular, release EVs containing HIF-1α-regulated miRNAs that further stimulate endothelial sprouting and stabilize pathological neovessels. In parallel, EV-driven EMT contributes directly to metastasis by enhancing extracellular matrix degradation. TDEVs transporting MMP-2, MMP-9, urokinase-type plasminogen activator (uPA), and miRNAs that regulate cytoskeletal rearrangement accelerate basement membrane breakdown and stromal invasion (Piperigkou et al., 2021). These EMT-enhanced cells become highly motile, allowing for more efficient dissemination. Moreover, EVs profoundly influence metastatic organotropism. CRC-derived EVs enriched with specific integrins and chemokine receptors educate resident stromal cells in the liver and lungs, inducing pro-inflammatory cytokines (IL-6, TNF-α), fibronectin deposition, and recruitment of myeloid-derived suppressor cells (MDSCs) (He et al., 2023; Rahmati et al., 2024). This results in a preconditioned microenvironment that favors adhesion, survival, and outgrowth of metastatic CRC cells. Additionally, EV-associated lncRNAs, such as lncRNA-UCA1 and lncRNA-MALAT1, have been shown to regulate colonization efficiency by modulating endothelial activation, fibroblast remodeling, and immune evasion (Hermann et al., 2022).

## 7.3 Extracellular vesicles as diagnostic and prognostic biomarkers

Cell-derived membrane particles known as EVs are found in nearly all bodily fluids and carry molecular cargo that represents the physiological or pathological condition of the cells that produced them (Zheng et al., 2019). EVs are becoming more well-known in the field of colon cancer as potent, noninvasive biomarkers for prognosis, early detection, and tracking the course of the disease and the effectiveness of treatment (Fig. 7.1). Their remarkable stability in circulation, combined with the selective enrichment of tumor-specific proteins, nucleic acids, and metabolites, makes EVs exceptionally well-suited for liquid biopsy applications, enabling clinicians to detect molecular changes in real time and monitor disease progression or treatment response through a simple, minimally invasive sample (Bordhan, 2024; Mishra et al., 2022).

Vesicles as noninvasive diagnostic and prognostic tools in colon cancer. EVs isolated from blood, urine, or stool carry tumor-associated cargo, including miRNAs (miR-21, miR-92a), lncRNAs (HOTAIR, MALAT1), and proteins (carcinoembryonic antigen [CEA], epithelial cell adhesion molecule [EpCAM]) that enable early diagnosis, disease staging, prognostic assessment, and real-time monitoring of treatment response and chemoresistance.

### 7.3.1 Liquid biopsy potential

Traditional diagnostic approaches for CRC, such as colonoscopy and tissue biopsy, are invasive, costly, and often limited in their ability to detect early-stage disease or monitor tumor evolution over time. In contrast, liquid biopsy has emerged as a minimally invasive, repeatable, and patient-friendly alternative that enables real-time molecular profiling using easily accessible biological fluids such as blood, urine,

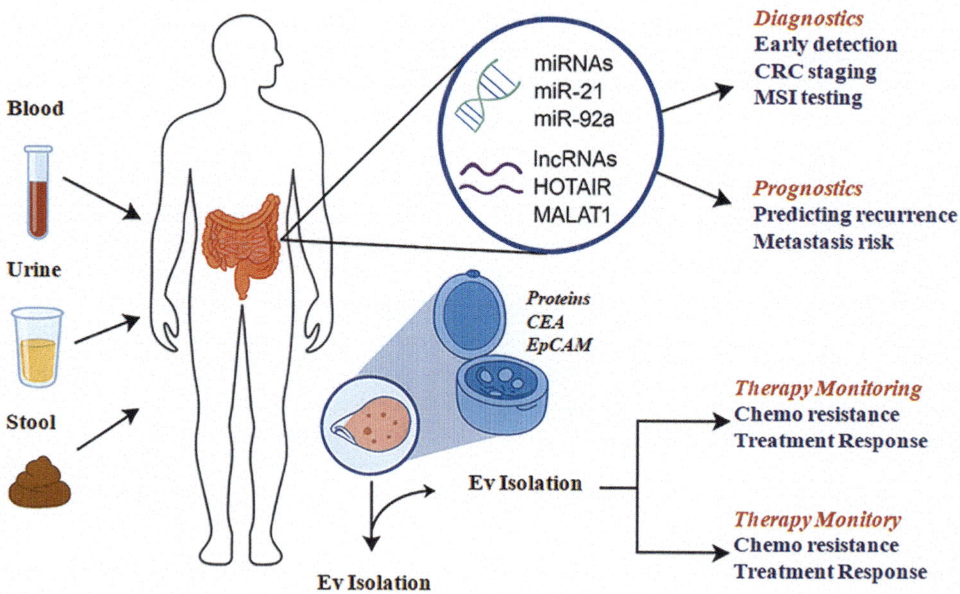


**FIGURE 7.1**

Extracellular vesicles as noninvasive diagnostic and prognostic tools in colon cancer.

and stool (Hessvik & Llorente, 2018). Among the various circulating components, EVs have gained substantial attention due to their remarkable stability, enrichment of tumor-specific cargo, and ability to mirror the molecular landscape of the originating tumor (Möller & Lobb, 2020; Xu et al., 2018).

EVs isolated from different bodily fluids serve as highly informative reservoirs of diagnostic and prognostic markers (Boukouris & Mathivanan, 2015; Verma et al., 2015). Blood-derived EVs contain circulating tumor-specific nucleic acids, proteins, and metabolites that correlate strongly with tumor burden, molecular subtype, metastatic potential, and overall disease progression (Aibaidula, 2023; Bertero et al., 2019; Verma et al., 2015). Urine EVs offer a completely noninvasive sampling route and are increasingly being explored for multiorgan cancer detection (Pevzner et al., 2025; Salar et al., 2025). Their ability to capture systemic tumor signatures makes them promising indicators of disseminated or recurrent disease (Poudineh et al., 2018; Swanton et al., 2024). Stool-derived EVs are particularly valuable in CRC detection because they originate directly from exfoliated epithelial and tumor cells lining the colon (Maani, 2024). This provides a highly enriched, tumor-proximal source of biomarkers, making stool EVs powerful tools for early screening, risk stratification, and posttreatment surveillance.

## 7.3.2 Tumor-specific cargo

EVs carry a rich repertoire of biomolecules, including miRNAs, lncRNAs, mRNAs, DNA fragments, and proteins that mirror the molecular state of their parent cells (Abramowicz & Story, 2020; Paramanantham et al., 2022). In CRC, the composition of this cargo undergoes significant alterations, making EVs highly informative indicators of tumor biology (Zhang, Shen et al., 2025). Numerous studies have demonstrated that EVs derived from CRC patients exhibit elevated levels of oncogenic miRNAs such as miR-21, miR-92a, and miR-1246, each of which contributes to tumor cell proliferation, invasion, and chemoresistance (Desmond et al., 2019; Francavilla et al., 2020; Rahmati et al., 2024). Similarly, CRC-derived EVs are enriched with oncogenic lncRNAs, including HOTAIR, MALAT1, and CCAT1, that modulate key signaling cascades such as the PI3K/Akt and Wnt/β-catenin pathways, thereby promoting tumorigenesis and metastatic progression (Kotelevets & Chastre, 2023; Rahmati et al., 2024). Beyond RNA species, TDEVs also encapsulate proteins of diagnostic and prognostic relevance. Markers such as EpCAM, CEA, ALG-2-Interacting Protein X (ALIX), and mutant variants of Kirsten Rat Sarcoma Viral Oncogene Homolog (KRAS) can be detected within circulating EVs, offering a molecular fingerprint of underlying malignant transformation (Lawrence & Shah, 2024). Since EVs protect their cargo from enzymatic degradation, these molecules are recoverable in a stable form from various biological fluids, enhancing their utility in noninvasive biomarker discovery (Zakari et al., 2024). Importantly, the simultaneous detection of multiple tumor-associated molecules within EVs holds promise for multiplexed diagnostic platforms, which may significantly improve the sensitivity, specificity, and predictive accuracy of CRC detection and monitoring.

### 7.3.3 Clinical relevance and ongoing trials

The growing clinical interest in EVs as minimally invasive biomarkers has accelerated translational research in CRC, positioning EV-based diagnostics as a promising complement or alternative to conventional screening methods (Zhang, Shen et al., 2025). Several ongoing clinical trials are actively evaluating the diagnostic, prognostic, and therapeutic monitoring potential of EVs in CRC. For instance, plasma-derived EVs are being investigated for early, noninvasive detection of CRC (NCT04777717), while another study (NCT05354141) is examining EV-associated miRNA signatures as predictors of recurrence risk and treatment responsiveness. In parallel, stool EV assays are being compared to traditional fecal occult blood testing for early-stage CRC detection (NCT05423826), reflecting a shift toward leveraging EVs directly shed from colorectal lesions for enhanced sensitivity and specificity. EV molecular signatures are increasingly recognized for their potential in precision oncology. Emerging evidence suggests that EV cargo correlates with tumor stage, metastatic potential, and molecular subtypes, including microsatellite instability (MSI) and mismatch repair deficiency, which are the two crucial determinants of immunotherapy responsiveness. EV-derived DNA, RNA, and protein content show high concordance with tumor tissue profiles, enabling real-time molecular monitoring without the need for repeated invasive biopsies (Liang et al., 2021; Yang et al., 2022). Additionally, EV alterations during chemotherapy or immunotherapy can offer early indications of therapeutic efficacy or resistance, thus supporting longitudinal disease management and adaptive treatment strategies (O'Neill et al., 2019).

As analytical platforms for EV isolation, purification, and characterization continue to advance, their integration into clinical workflows becomes increasingly feasible. Techniques such as microfluidics, immunoaffinity capture, and high-throughput sequencing are enhancing the precision and scalability of EV-based assays. Collectively, these developments highlight the strong translational momentum behind EV-based diagnostics and underscore their emerging role as versatile biomarkers capable of informing early detection, prognostication, and personalized therapeutic decision-making in CRC (Chen et al., 2023; Cheng et al., 2021; Lawrence & Shah, 2024; Yan et al., 2021; Youssef et al., 2025; Zhang, Huang et al., 2025; Zhao et al., 2021).

## 7.4 Extracellular vesicle–based therapeutic strategies in colon cancer

EVs have emerged as versatile platforms for innovative therapeutic strategies in CRC, offering advantages in targeted delivery, immunomodulation, and microenvironmental reprogramming (Biondi et al., 2024; Prakash et al., 2023; Rahmati et al., 2024). As summarized in Table 7.1, one of the most widely explored approaches involves the use of EVs as drug carriers, where chemotherapeutics such as 5-fluorouracil and doxorubicin are encapsulated to enhance drug stability, improve tumor-specific uptake, and reduce systemic toxicity. Similarly, EV-mediated RNA delivery enables transport of regulatory miRNAs, siRNAs, and lncRNAs to modulate oncogenic or tumor-suppressive pathways with remarkable gene-specific precision (Singh et al., n.d.). However, challenges remain regarding RNA stability and immune clearance without appropriate modification. Engineered EVs represent another promising strategy, in which vesicles are surface-functionalized with ligands, antibodies, or peptides to improve targeting efficiency (Hu et al., 2025; Johnson et al., 2023). Further, these advantages are counterbalanced by the complexity of large-scale manufacturing and regulatory constraints (Hu et al., 2025).

In addition, EV-based immunotherapies leverage vesicles carrying tumor-associated antigens to stimulate antitumor immune responses or aid in cancer vaccine development, although careful optimization of antigen loading is required to prevent immune tolerance (Bar et al., 2024). EVs derived from immune cells, such as dendritic cells, Natural Killer (NK) cells, and macrophages, offer natural immunostimulatory functions and may synergize with existing immunotherapies, but issues of scalability and preservation of functional integrity limit their translational readiness (Fernández-Delgado et al., 2020; Schioppa et al., 2024). Lastly, EV-mediated TME modulation is gaining clinical interest, as EVs can deliver bioactive molecules to stromal or vascular components of the TME, promoting vascular normalization and reducing metastatic potential (Tao & Guo, 2020). Yet, the inherent complexity and heterogeneity of the TME introduce unpredictability into therapeutic responses (Junttila & De Sauvage, 2013; Zhang et al., 2022). Overall, the diverse mechanisms and translational potential detailed in Table 7.1 underscore the growing importance of EV-based therapeutics as a multifaceted and rapidly advancing frontier in colon cancer treatment.

### 7.4.1 Natural extracellular vesicles as drug carriers

Compared to synthetic nanoparticles, natural EVs have a number of benefits, such as increased stability in circulation, decreased immunogenicity, and better biocompatibility (Table 7.1). They can pass through biological barriers and have the innate ability to home in on tumor tissues (Ran et al., 2022).

**Table 7.1 Extracellular vesicle (*EV*)–based therapeutic strategies under investigation for colorectal cancer, summarizing their mechanisms of action, key therapeutic advantages, and current limitations.**

| Therapeutic strategy | Mechanism of action | Key advantages | Limitations/ challenges |
|---|---|---|---|
| **EVs as drug carriers** | Loading chemotherapeutic agents (e.g., 5-FU, doxorubicin) into EVs for targeted delivery to tumor cells | Increased drug stability, reduced off-target toxicity, enhanced uptake by cancer cells | Limited large-scale production, variable drug-loading efficiency |
| **EV-mediated RNA delivery** | Transport of miRNA, siRNA, or long non-coding RNAs to regulate oncogenic or tumor-suppressor pathways | Gene-specific targeting, ability to modulate tumor microenvironment (TME) | Potential immune clearance, RNA instability without modification |
| **Engineered EVs** | Surface modification (e.g., ligands, antibodies, peptides) for improved targeting | Enhanced tumor specificity, reduced systemic exposure | Complex manufacturing, regulatory hurdles |
| **EV-based immunotherapy** | Use of EVs carrying tumor antigens to stimulate antitumor immune response | Potential for cancer vaccines, induction of long-term immunity | Risk of immune tolerance, optimization of antigen loading required |
| **EVs from immune cells** | Harnessing EVs from dendritic cells, NK cells, or macrophages for anticancer effects | Natural immune activation, potential synergy with checkpoint inhibitors | Scalability issues, maintaining functional integrity |
| **EVs for TME modulation** | Delivering agents to alter stromal or vascular components of TME | Can normalize vasculature, reduce metastasis potential | Complex TME interactions may limit predictability |

Growing evidence supports the utility of EVs for encapsulating a wide range of chemotherapeutic agents. Doxorubicin-loaded EVs, for instance, have shown enhanced intracellular delivery and improved apoptotic effects in CRC models compared to the free drug (Bagheri et al., 2020). Similarly, encapsulation of 5-fluorouracil (5-FU) into EVs significantly improves its therapeutic index by protecting the drug from premature degradation, prolonging circulation time, and increasing tumor-specific uptake (Fader et al., 2009; Gross et al., 2012; Gurung et al., 2021; Hessvik et al., 2023). Notably, MSC-derived EVs loaded with 5-FU have demonstrated potent cytotoxic effects against CRC cells while largely sparing healthy colonocytes, underscoring the selective delivery potential of EV-based formulations (Chen et al., 2020; Riihimäki et al., 2016). This selective cytotoxicity is attributed to the inherent tropism of MSC-EVs toward inflamed or tumor-associated tissues, as well as the controlled release of the drug cargo within malignant cells (Goh et al., 2017).

Natural EVs can also transport hydrophobic drugs, multidrug combinations, or synergistic therapeutic agents, thereby offering a versatile platform for personalized oncology (Chen et al., 2022). Their ability to shield unstable compounds from enzymatic degradation further enhances pharmacokinetics and maximizes intracellular bioavailability. As research advances, EVs are increasingly recognized as biologically intelligent carriers capable of overcoming major limitations associated with traditional chemotherapy, including dose-limiting toxicity, poor tumor penetration, and multidrug resistance (Wu et al., 2025).

### 7.4.2 Engineered extracellular vesicles

The therapeutic potential of EVs has been enhanced through bioengineering techniques. Targeting specificity to CRC cells can be improved by surface functionalization using ligands (such as folate or RGD peptides) or antibodies (such as anti-EGFR) (Dixit et al., 2024; Janani et al., 2022).

Genetic engineering of EV-producing cells offers another powerful strategy to load bioactive molecules into EVs in a controlled and sustained manner (Villata et al., 2020). Donor cells can be modified to express therapeutic cargo, such as siRNAs, shRNAs, mRNAs, or CRISPR/Cas9 components, allowing EVs to deliver gene-silencing tools directly to tumor cells (Song et al., 2024). This approach has promised to target oncogenic drivers in CRC, including KRAS, B-Raf proto-oncogene (BRAF), and components of the Wnt/β-catenin pathway, enabling modulation of molecular aberrations that are otherwise difficult to treat with conventional drugs (Du et al., 2022; Engstrand et al., 2018).

In addition to genetic and surface modifications, recent advances have led to the development of hybrid EVs, created by fusing natural EV membranes with synthetic liposomal structures (Rodríguez & Vader, 2022; Sun et al., 2023). These biomimetic hybrids combine the intrinsic biocompatibility and targeting advantages of EVs with the scalable production and controlled drug-loading efficiency of synthetic nanocarriers (Sundaram & Aryal, 2025). Such hybrid systems offer enhanced stability, higher loading capacity, and tunable pharmacokinetics, representing a new class of intelligent delivery platforms for CRC therapy (Mohamed et al., 2025). Collectively, engineered EVs provide a versatile and customizable therapeutic modality capable of addressing multiple challenges in CRC treatment, from improving drug targeting and reducing systemic toxicity to delivering genetic tools that reprogram tumor biology. These innovations position engineered EVs at the forefront of next-generation precision nanomedicine.

### 7.4.3 Immunomodulatory extracellular vesicles

EVs also hold substantial promise as modulators of antitumor immunity, offering new opportunities to enhance immunotherapeutic strategies in CRC (Bandini et al., 2024). EVs derived from immune cells possess inherent immunological functions that can be exploited for therapeutic benefit (Buzas, 2023). Among the most widely studied are dendritic cell–derived EVs, often referred to as dexosomes (Luo et al., 2023). These vesicles retain the antigen-presenting machinery of their parent cells, including MHC class I and II molecules, costimulatory proteins, and tumor-associated antigens (Buzas, 2023; Matsuzaka & Yashiro, 2022). As a result, dexosomes can stimulate T-cell priming and activation, making them a robust platform for the development of cell-free cancer vaccines (Nikfarjam et al., 2020). Several preclinical studies have reported that dexosome-based vaccination enhances CTL responses and delays tumor growth in CRC models, underscoring their translational value.

In addition to dendritic cell–derived vesicles, EVs released from effector immune cells such as $CD8^+$ T cells and natural killer (Riihimäki et al., 2016) cells exhibit intrinsic antitumor activity. These vesicles carry perforin, granzymes, and immune-stimulatory cytokines capable of inducing apoptosis in CRC cells (McNamara, 2019; Pérez-Domínguez et al., 2025). Importantly, immune cell–derived EVs remain functionally active even under conditions where direct immune cell infiltration into the tumor is limited, providing a unique advantage in the context of the immunosuppressive TME. Such vesicles can also counteract CRC-associated immunosuppression by interfering with TDEV-mediated inhibition of immune cell function (Fidler & Kripke, 2015).

TDEVs can also be reprogrammed to enhance antitumor immunity (Pérez-Domínguez et al., 2025). Engineering strategies have enabled the incorporation of immune-activating agents into TDEVs, transforming them from mediators of immune evasion into vehicles that promote antitumor responses (Chen et al., 2021). For instance, exosomes loaded with stimulator of interferon genes (STING) agonists have demonstrated the ability to trigger potent innate immune signaling, resulting in enhanced type I interferon production and improved recognition of CRC cells by antigen-presenting cells. Similarly, EVs engineered to carry checkpoint inhibitors, such as anti-PD-L1 peptides or siRNAs targeting immunosuppressive molecules, have shown encouraging outcomes in preclinical CRC models by restoring T-cell activity and reducing tumor burden (Prakash et al., 2023; Zhang, Gong et al., 2025).

Together, these advances highlight the multifaceted potential of EVs to strengthen immunotherapeutic strategies in CRC. Whether harnessed directly from immune cells or engineered to deliver immunomodulatory agents, EV-based approaches are shaping a new generation of immunotherapies capable of overcoming the profound immune resistance characteristic of advanced CRC. Continued optimization, standardization, and clinical evaluation of immunomodulatory EVs will be essential to translate these promising findings into tangible clinical benefits.

## 7.5 Recent advancements and clinical applications

Recent advancements in EV engineering have substantially broadened their diagnostic and therapeutic relevance in CRC, as depicted in Fig. 7.2 One of the most impactful developments is the design of hybrid EV-based delivery platforms that integrate natural EV membranes with synthetic nanosystems, including liposomes, polymeric nanoparticles, and gold nanocarriers. These hybrid constructs combine the intrinsic biocompatibility, immune evasion, and tumor-targeting properties of EVs with the enhanced stability, drug-loading efficiency, and controllable release characteristics of synthetic nanomaterials, thereby overcoming key limitations of conventional drug-delivery systems (Bakir et al., 2020; Becker et al., 2016; Pastushenko & Blanpain, 2019).

Another promising direction involves stimuli-responsive ("smart") EV formulations engineered to release therapeutic cargo in response to tumor-specific microenvironmental cues. Hallmarks of the colorectal TME, such as acidic pH, elevated redox potential, and increased enzymatic activity, are exploited to trigger site-specific drug release. Such precision-activated systems reduce off-target toxicity, improve intratumoral drug retention, and enable more effective therapeutic outcomes with reduced systemic burden. Growing preclinical evidence supports these innovations. EVs loaded with

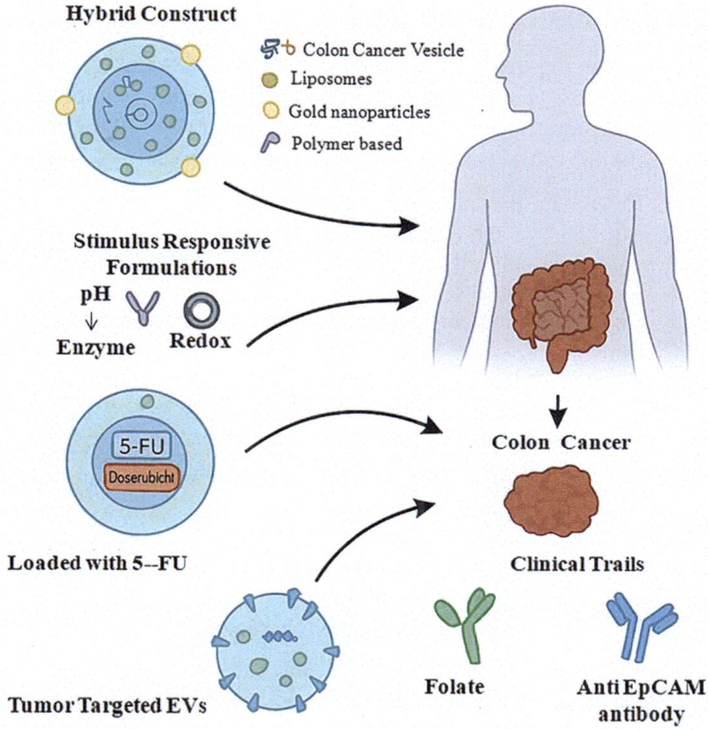


**FIGURE 7.2**

Recent advancements and clinical applications of extracellular vesicle (*EV*)–based therapeutics in colorectal cancer (*CRC*). Schematic overview illustrating emerging EV-based strategies, including hybrid EV-liposome constructs, stimulus-responsive formulations (pH, enzyme, redox), and EVs loaded with chemotherapeutic agents such as 5-fluorouracil (*5-FU*). Tumor-targeted EVs engineered with ligands like folate and antiepithelial cell adhesion molecule (*EpCAM*) antibodies, along with ongoing clinical investigations, highlight the growing translational potential of EV-mediated therapies in CRC.

chemotherapeutics such as doxorubicin or 5-fluorouracil (5-FU) demonstrate superior tumor inhibition and markedly reduced systemic toxicity compared with free-drug administration in murine models (Samadi et al., 2025). Furthermore, surface-functionalized EVs decorated with targeting ligands, including folate, Arginine–Glycine–Aspartic Acid (RGD) peptides, and anti-EpCAM antibodies, exhibit enhanced uptake by CRC cells, reinforcing the potential for personalized and tumor-specific therapeutic delivery (Ribatti et al., 2020).

Clinical translation is also progressing. Ongoing clinical trials are investigating EV-based liquid biopsy approaches, diagnostic biomarkers, and therapeutic formulations for CRC, reflecting growing confidence in EVs as a next-generation biomedical tool. Collectively, these advancements, illustrated in Fig. 7.2, highlight EVs' emerging role in precision oncology and underscore their potential to transform both therapeutic strategies and clinical management in CRC.

## 7.6 Challenges and future directions

Despite significant progress in the field of EV-based therapeutics, several critical challenges continue to impede their widespread clinical application in colon cancer (Zhang, Shen et al., 2025). One of the most persistent obstacles is the efficient and scalable production of high-quality EVs. Conventional isolation techniques, including ultracentrifugation, precipitation-based methods, and size-exclusion chromatography, are labor-intensive, offer limited purity, and lack scalability for industrial or clinical-grade manufacturing. Developing automated, closed-system platforms capable of producing standardized, high-yield EV preparations remains a top priority for successful clinical translation (de Almeida Fuzeta et al., 2019; Shaon et al., 2025).

Equally important are the challenges surrounding EV characterization and standardization. Although the International Society for Extracellular Vesicles (ISEV) has established minimal information for studies of extracellular vesicles (MISEV) guidelines, variability in their implementation persists across laboratories. Robust and harmonized analytical protocols are needed to evaluate EV size distribution, zeta potential, surface marker expression, cargo composition, and biological activity (Vaiaki & Falasca, 2024). Such standardization is essential for improving reproducibility, ensuring batch-to-batch consistency, and navigating the regulatory pathways required for therapeutic approval.

Safety and regulatory concerns also present considerable hurdles. As biologically derived nanoparticles with complex molecular profiles, EVs pose potential risks related to immunogenicity, unpredictable biodistribution, and off-target interactions (Stawarska et al., 2024). Comprehensive safety assessments, along with the establishment of dedicated regulatory frameworks tailored to EV-based therapeutics, will be crucial for mitigating these risks. Regulatory agencies will need to adapt existing guidelines or develop new ones that address the intricacies of EV composition, bioactivity, and manufacturing (Verma & Arora, 2025).

Looking ahead, the future of EV-based therapy in colon cancer lies in precision engineering and integrative technologies (Arrè et al., 2024; Lawrence & Shah, 2024; Youssef et al., 2025). The personalization of EV cargo and surface ligands based on patient-specific tumor signatures may enable highly selective and effective targeting strategies. The incorporation of multiomics approaches, such as transcriptomics, proteomics, and metabolomics, combined with AI-driven data analytics, is expected to refine EV design and accelerate the identification of optimal therapeutic candidates. Additionally, synergistic strategies that pair EVs with immunotherapies, RNA-based modalities, or gene-editing systems (e.g., CRISPR/Cas9) hold promise for next-generation treatment paradigms (Du et al., 2022; Song et al., 2024). These combined approaches may enhance tumor immune modulation, overcome drug resistance, and improve the overall therapeutic index.

## 7.7 Conclusion

CRC remains one of the leading causes of cancer mortality worldwide, driven by its complex pathogenesis, late-stage diagnosis, and limited efficacy of conventional therapeutic options. Over the past decade, the emergence of EVs has transformed our understanding of tumor biology and redefined approaches for cancer diagnosis, prognosis, and targeted therapy. EVs, owing to their biocompatibility, molecular diversity, inherent targeting capacity, and ability to cross physiological barriers, occupy a unique position at the intersection of tumor communication, molecular regulation, and therapeutic intervention. This review highlights the established and emerging roles of EVs in CRC progression, the TME, metastatic dissemination, and their translational potential as diagnostic and therapeutic tools.

Accumulating evidence demonstrates that TDEVs contribute fundamentally to CRC initiation and progression by transferring oncogenic miRNAs, lncRNAs, proteins, and metabolic regulators that reprogram recipient cells. These vesicles orchestrate multiple hallmarks of cancer: they enhance proliferation, sustain metabolic alterations, drive EMT, remodel the extracellular matrix, and establish premetastatic niches in organs such as the liver and lungs. The selective enrichment of pro-tumorigenic molecules within EVs underscores their pivotal role as molecular messengers capable of shaping the malignant phenotype and coordinating interactions between cancer cells and stromal counterparts. Furthermore, their involvement in angiogenesis and immune suppression demonstrates that EVs are not only passive carriers but also active modulators of the tumor ecosystem. Their ability to impair dendritic cell maturation, suppress cytotoxic T-cell activity, and promote regulatory T-cell expansion highlights their contribution to creating an immune-privileged niche for tumor survival.

Beyond their mechanistic significance in cancer biology, EVs have emerged as exceptionally promising biomarkers for early diagnosis, prognostication, and monitoring of treatment response. Liquid biopsy platforms utilizing EVs extracted from blood, urine, or stool offer a minimally invasive, repeatable, and high-fidelity alternative to traditional biopsy. EV-derived miRNAs (such as miR-21, miR-92a, and miR-1246), lncRNAs (including HOTAIR, MALAT1, and CCAT1), and tumor-associated proteins (CEA, EpCAM, and ALIX) demonstrate strong diagnostic and prognostic relevance. Their stability in circulation and protection from enzymatic degradation further enhance their utility for clinical assays. Emerging clinical trials are already evaluating EV signatures for early CRC detection, prediction of recurrence, and real-time assessment of therapeutic efficacy. With advances in microfluidics, sequencing technologies, and high-purity EV isolation methods, the translation of EV-based diagnostics into clinical practice appears increasingly attainable.

In parallel, EV-based therapeutic strategies represent one of the most rapidly evolving frontiers in CRC treatment. Natural EVs offer distinct advantages over synthetic nanocarriers, including reduced immunogenicity, improved tissue penetration, and enhanced biocompatibility. Their ability to encapsulate chemotherapeutics, nucleic acids, gene-editing tools, and immunomodulators paves the way for precision delivery to cancer cells with minimal systemic toxicity. Drug-loaded EVs have shown superior uptake and enhanced cytotoxic effects in CRC models, while EV-mediated delivery of miRNAs or siRNAs offers opportunities to silence oncogenic pathways or restore tumor-suppressor activity. Engineered EVs, functionalized with targeting ligands, antibodies, or peptides, further refine tumor tropism and minimize off-target exposure. Similarly, EV-based immunotherapies, including EV vaccines and immune cell–derived EVs, hold potential to amplify antitumor immunity, synergize with checkpoint inhibitors, and overcome immune resistance. EV-mediated reprogramming of the TME also offers innovative avenues to normalize vasculature, inhibit metastasis, and reverse stromal-driven therapy resistance. Despite these promising developments, several challenges remain before EV-based diagnostics and therapeutics can be fully integrated into routine clinical care. Standardization of EV isolation, characterization, and quantification remains a major bottleneck, as variability across platforms and protocols can limit reproducibility and cross-study comparisons. Large-scale production of therapeutic EVs, ensuring cargo-loading efficiency, and maintaining structural integrity during storage and delivery also present technical and regulatory hurdles.

Furthermore, the heterogeneity of EV populations and the complex biodistribution patterns in vivo warrant deeper investigation to optimize safety and efficacy. Addressing these challenges will require sustained collaborative efforts integrating molecular biology, engineering, clinical oncology, and regulatory science.

Nevertheless, the rapid advancements in EV research spanning fundamental biology, biomarker discovery, nanomedicine, and clinical translation reflect the immense potential of EVs to reshape the future landscape of CRC management. EVs offer a uniquely powerful platform capable of bridging diagnostic precision with targeted therapeutic intervention. Their dual utility, enabling both the detection of TD molecular signatures and the delivery of personalized therapeutic agents, aligns seamlessly with the goals of modern precision oncology. As technologies evolve and clinical trials progress, EV-based strategies may contribute to earlier detection of CRC, improved accuracy in disease stratification, enhanced therapeutic responsiveness, and reduced toxicity.

In conclusion, EVs represent a transformative paradigm in CRC research. Their roles in tumor progression, microenvironmental remodeling, and metastasis deepen our understanding of CRC pathophysiology, while their promise as biomarkers and therapeutic carriers opens new avenues for clinical innovation. Continued advancements in EV engineering, isolation technology, and clinical validation will likely accelerate their transition from experimental systems to mainstream oncology. With their remarkable biological versatility and translational potential, EVs are poised to become integral components of next-generation cancer diagnostics and therapeutics, offering renewed hope for improving outcomes in patients with CRC.

## References

Abramowicz, A., & Story, M. D. (2020). The long and short of it: The emerging roles of non-coding RNA in small extracellular vesicles. *Cancers, 12*(6), 1445.

Ahmadi, M., Abbasi, R., & Rezaie, J. (2024). Tumor immune escape: Extracellular vesicles roles and therapeutics application. *Cell Communication and Signaling, 22*(1), 9.

Aibaidula, A. (2023). *Plasma extracellular vesicles as biomarkers for glioblastoma: Characterization and molecular phenotype*. College of Medicine-Mayo Clinic.

Aldea, M., Andre, F., Marabelle, A., Dogan, S., Barlesi, F., & Soria, J.-C. (2021). Overcoming resistance to tumor-targeted and immune-targeted therapies. *Cancer discovery, 11*(4), 874–899.

de Almeida Fuzeta, M., de Matos Branco, A. D., Fernandes-Platzgummer, A., da Silva, C. L., & Cabral, J. M. (2019). Addressing the manufacturing challenges of cell-based therapies. In *Current applications of pharmaceutical biotechnology*. Springer, 225–278.

Aloi, N., Drago, G., Ruggieri, S., Cibella, F., Colombo, P., & Longo, V. (2024). Extracellular vesicles and immunity: At the crossroads of cell communication. *International journal of molecular sciences, 25*(2), 1205.

Angius, A., Scanu, A. M., Arru, C., Muroni, M. R., Rallo, V., Deiana, G., Ninniri, M. C., Carru, C., Porcu, A., & Pira, G. (2021). Portrait of cancer stem cells on colorectal cancer: Molecular biomarkers, signaling pathways and miRNAome. *International Journal of Molecular Sciences, 22*(4), 1603.

Arrè, V., Mastrogiacomo, R., Balestra, F., Serino, G., Viti, F., Rizzi, F., Curri, M. L., Giannelli, G., Depalo, N., & Scavo, M. P. (2024). Unveiling the potential of extracellular vesicles as biomarkers and therapeutic nanotools for gastrointestinal diseases. *Pharmaceutics, 16*(4), 567.

Bagheri, E., Abnous, K., Farzad, S. A., Taghdisi, S. M., Ramezani, M., & Alibolandi, M. (2020). Targeted doxorubicin-loaded mesenchymal stem cells-derived exosomes as a versatile platform for fighting against colorectal cancer. *Life Sciences, 261*, 118369.

Bakir, B., Chiarella, A. M., Pitarresi, J. R., & Rustgi, A. K. (2020). EMT, MET, plasticity, and tumor metastasis. *Trends in Cell Biology, 30*(10), 764–776.

Bandini, S., Ulivi, P., & Rossi, T. (2024). Extracellular vesicles, circulating tumor cells, and immune checkpoint inhibitors: Hints and promises. *Cells, 13*(4), 337.

Bar, O., Porgador, A., & Cooks, T. (2024). Exploring the potential of the convergence between extracellular vesicles and CAR technology as a novel immunotherapy approach. *Journal of Extracellular Biology, 3*(9), e70011.

Becker, A., Thakur, B. K., Weiss, J. M., Kim, H. S., Peinado, H., & Lyden, D. (2016). Extracellular vesicles in cancer: Cell-to-cell mediators of metastasis. *Cancer Cell, 30*(6), 836–848.

Bertero, L., Siravegna, G., Rudà, R., Soffietti, R., Bardelli, A., & Cassoni, P. (2019). Peering through a keyhole: Liquid biopsy in primary and metastatic central nervous system tumours. *Neuropathology and Applied Neurobiology, 45*(7), 655–670.

Biondi, A., Vacante, M., Catania, R., & Sangiorgio, G. (2024). Extracellular vesicles and immune system function: Exploring novel approaches to colorectal cancer immunotherapy. *Biomedicines, 12*(7), 1473.

Bordhan, P. (2024). Targeting cancer cell-derived small extracellular vesicles (sEVs) to develop nanoparticle-integrated detection systems.

Boukouris, S., & Mathivanan, S. (2015). Exosomes in bodily fluids are a highly stable resource of disease biomarkers. *PROTEOMICS–Clinical Applications, 9*(3-4), 358–367.

Bray, F., Laversanne, M., Sung, H., Ferlay, J., Siegel, R. L., Soerjomataram, I., & Jemal, A. (2024). Global cancer statistics 2022: GLOBOCAN estimates of incidence and mortality worldwide for 36 cancers in 185 countries. *CA: A Cancer Journal for Clinicians, 74*(3), 229–263.

Bray, F., Laversanne, M., Weiderpass, E., & Soerjomataram, I. (2021). The ever-increasing importance of cancer as a leading cause of premature death worldwide. *Cancer, 127*(16), 3029–3030.

Brown, N. F., & Marshall, J. F. (2019). Integrin-mediated TGFβ activation modulates the tumour microenvironment. *Cancers, 11*(9), 1221.

Buzas, E. I. (2023). The roles of extracellular vesicles in the immune system. *Nature Reviews. Immunology, 23*(4), 236–250.

Cao, Y., Xu, P., Shen, Y., Wu, W., Chen, M., Wang, F., Zhu, Y., Yan, F., Gu, W., & Lin, Y. (2023). Exosomes and cancer immunotherapy: A review of recent cancer research. *Frontiers in Oncology, 12*, 1118101.

Chang, S., Chang, Y., Li, F., Xu, Z., Han, X., Liu, Y., Li, H., Hu, S., & Tang, T. (2025). Global, regional, national burden of colorectal cancer from 1990 to 2021, with projections of incidence to 2050: A systematic analysis of the global burden of disease study 2021. *Frontiers in Oncology, 15*, 1597847.

Chavda, V. P., Patel, A. B., Mistry, K. J., Suthar, S. F., Wu, Z.-X., Chen, Z.-S., & Hou, K. (2022). Nano-drug delivery systems entrapping natural bioactive compounds for cancer: Recent progress and future challenges. *Frontiers in oncology, 12*, 867655.

Chen, J., Tan, Q., Yang, Z., & Jin, Y. (2022). Engineered extracellular vesicles: Potentials in cancer combination therapy. *Journal of Nanobiotechnology, 20*(1), 132.

Chen, J., Zhang, H., Zhou, L., Hu, Y., Li, M., He, Y., & Li, Y. (2021). Enhancing the efficacy of tumor vaccines based on immune evasion mechanisms. *Frontiers in Oncology, 10*, 584367.

Chen, J., Zheng, Y., Wang, H., Zhang, D., Zhao, L., Yu, D., Lin, Z., & Zhang, T. (2020). Cause of death among patients with colorectal cancer: A population-based study in the United States. *Aging (Albany NY), 12*(22), 22927.

Chen, M., Lin, S., Zhou, C., Cui, D., Haick, H., & Tang, N. (2023). From conventional to microfluidic: Progress in extracellular vesicle separation and individual characterization. *Advanced Healthcare Materials, 12*(8), 2202437.

Cheng, S., Li, Y., Yan, H., Wen, Y., Zhou, X., Friedman, L., & Zeng, Y. (2021). Advances in microfluidic extracellular vesicle analysis for cancer diagnostics. *Lab on a Chip, 21*(17), 3219–3243.

Ciardiello, C., Cavallini, L., Spinelli, C., Yang, J., Reis-Sobreiro, M., De Candia, P., Minciacchi, V. R., & Di Vizio, D. (2016). Focus on extracellular vesicles: New frontiers of cell-to-cell communication in cancer. *International journal of molecular sciences, 17*(2), 175.

Colombo, M., Moita, C., Van Niel, G., Kowal, J., Vigneron, J., Benaroch, P., Manel, N., Moita, L. F., Théry, C., & Raposo, G. (2013). Analysis of ESCRT functions in exosome biogenesis, composition and secretion highlights the heterogeneity of extracellular vesicles. *Journal of Cell Science, 126*(24), 5553–5565.

Desmond, B. J., Dennett, E. R., & Danielson, K. M. (2019). Circulating extracellular vesicle microRNA as diagnostic biomarkers in early colorectal cancer—A review. *Cancers, 12*(1), 52.

Dixit, A. A., Mandlik, D. S., & Mandlik, S. K. (2024). Functionalised ligand-based nanomaterial drug targeting approaches for colorectal cancer therapy. *Recent Advances in Drug Delivery and Formulation: Formerly Recent Patents on Drug Delivery & Formulation, 18*(3), 170–187.

Du, R., Wang, C., Zhu, L., & Yang, Y. (2022). Extracellular vesicles as delivery vehicles for therapeutic nucleic acids in cancer gene therapy: Progress and challenges. *Pharmaceutics, 14*(10), 2236.

Elliott, R. O., & He, M. (2021). Unlocking the power of exosomes for crossing biological barriers in drug delivery. *Pharmaceutics, 13*(1), 122.

Engstrand, J., Nilsson, H., Strömberg, C., Jonas, E., & Freedman, J. (2018). Colorectal cancer liver metastases–a population-based study on incidence, management and survival. *BMC Cancer, 18*(1), 78.

Fader, C. M., Sánchez, D. G., Mestre, M. B., & Colombo, M. I. (2009). TI-VAMP/VAMP7 and VAMP3/cellubrevin: Two v-SNARE proteins involved in specific steps of the autophagy/multivesicular body pathways. *Biochimica et Biophysica Acta (BBA)-Molecular Cell Research, 1793*(12), 1901–1916.

Fernández-Delgado, I., Calzada-Fraile, D., & Sánchez-Madrid, F. (2020). Immune regulation by dendritic cell extracellular vesicles in cancer immunotherapy and vaccines. *Cancers, 12*(12), 3558.

Fidler, I. J., & Kripke, M. L. (2015). The challenge of targeting metastasis. *Cancer and Metastasis Reviews, 34*(4), 635–641.

Folprecht, G., Martinelli, E., Mazard, T., Modest, D. P., Tsuji, A., Esser, R., Cremolini, C., & Falcone, A. (2022). Triplet chemotherapy in combination with anti-EGFR agents for the treatment of metastatic colorectal cancer: Current evidence, advances, and future perspectives. *Cancer Treatment Reviews, 102*, 102301.

Francavilla, A., Turoczi, S., Tarallo, S., Vodicka, P., Pardini, B., & Naccarati, A. (2020). Exosomal microRNAs and other non-coding RNAs as colorectal cancer biomarkers: A review. *Mutagenesis, 35*(3), 243–260.

Giusti, I., Poppa, G., Di Fazio, G., D'Ascenzo, S., & Dolo, V. (2023). Metastatic dissemination: Role of tumor-derived extracellular vesicles and their use as clinical biomarkers. *International journal of molecular sciences, 24*(11), 9590.

Goenka, A., Khan, F., Verma, B., Sinha, P., Dmello, C. C., Jogalekar, M. P., Gangadaran, P., & Ahn, B. C. (2023). Tumor microenvironment signaling and therapeutics in cancer progression. *Cancer Communications, 43*(5), 525–561.

Goh, W. J., Lee, C. K., Zou, S., Woon, E. C., Czarny, B., & Pastorin, G. (2017). Doxorubicin-loaded cell-derived nanovesicles: An alternative targeted approach for anti-tumor therapy. *International Journal of Nanomedicine,* 2759–2767.

Gopal, S. K., Greening, D. W., Rai, A., Chen, M., Xu, R., Shafiq, A., Mathias, R. A., Zhu, H.-J., & Simpson, R. J. (2017). Extracellular vesicles: Their role in cancer biology and epithelial–mesenchymal transition. *Biochemical Journal, 474*(1), 21–45.

Gross, J. C., Chaudhary, V., Bartscherer, K., & Boutros, M. (2012). Active Wnt proteins are secreted on exosomes. *Nature Cell Biology, 14*(10), 1036–1045.

Guo, S., Huang, J., Li, G., Chen, W., Li, Z., & Lei, J. (2023). The role of extracellular vesicles in circulating tumor cell-mediated distant metastasis. *Molecular Cancer, 22*(1), 193.

Gurunathan, S., Kang, M.-H., Qasim, M., Khan, K., & Kim, J.-H. (2021). Biogenesis, membrane trafficking, functions, and next generation nanotherapeutics medicine of extracellular vesicles. *International Journal of Nanomedicine,* 3357–3383.

Gurunathan, S., Kang, M.-H., & Kim, J.-H. (2021). A comprehensive review on factors influences biogenesis, functions, therapeutic and clinical implications of exosomes. *International Journal of Nanomedicine,* 1281–1312.

Gurung, S., Perocheau, D., Touramanidou, L., & Baruteau, J. (2021). The exosome journey: From biogenesis to uptake and intracellular signalling. *Cell Communication and Signaling, 19*(1), 47.

Gutiérrez-Vázquez, C., Villarroya-Beltri, C., Mittelbrunn, M., & Sánchez-Madrid, F. (2013). Transfer of extracellular vesicles during immune cell-cell interactions. *Immunological Reviews, 251*(1), 125–142.

Harding, C., Heuser, J., & Stahl, P. (1983). Receptor-mediated endocytosis of transferrin and recycling of the transferrin receptor in rat reticulocytes. *The Journal of Cell Biology, 97*(2), 329–339.

Harding, C. V., Heuser, J. E., & Stahl, P. D. (2013). Exosomes: Looking back three decades and into the future. *The Journal of Cell Biology, 200*(4), 367.

He, K., Wang, Z., Luo, M., Li, B., Ding, N., Li, L., He, B., Wang, H., Cao, J., & Huang, C. (2023). Metastasis organotropism in colorectal cancer: Advancing toward innovative therapies. *Journal of Translational Medicine, 21*(1), 612.

Heijnen, H. F., Schiel, A. E., Fijnheer, R., Geuze, H. J., & Sixma, J. J. (1999). Activated platelets release two types of membrane vesicles: Microvesicles by surface shedding and exosomes derived from exocytosis of multivesicular bodies and □-granules. *Blood, The Journal of the American Society of Hematology, 94*(11), 3791–3799.

Hermann, D. M., Xin, W., Bähr, M., Giebel, B., & Doeppner, T. R. (2022). Emerging roles of extracellular vesicle-associated non-coding RNAs in hypoxia: Insights from cancer, myocardial infarction and ischemic stroke. *Theranostics, 12*(13), 5776.

Hessvik, N. P., & Llorente, A. (2018). Current knowledge on exosome biogenesis and release. *Cellular and Molecular Life Sciences, 75*(2), 193–208.

Hessvik, N. P., Sagini, K., Romero, S., Ramirez-Garrastacho, M., Rodriguez, M., Tutturen, A. E. V., Kvalvaag, A., Stang, E., Brech, A., & Sandvig, K. (2023). siRNA screening reveals that SNAP29 contributes to exosome release. *Cellular and Molecular Life Sciences, 80*(7), 177.

Hu, M., Han, Y., Zhang, X., Tian, S., Shang, Z., Yuan, Z., & He, L. (2025). Extracellular vesicles for targeted drug delivery: Advances in surface modification strategies and therapeutic applications. *Journal of Translational Medicine, 23*(1), 1028.

Imran, M. A., & Shahid, H. (2021). A review-colorectal cancer, Prevalence, along with screening, diagnosis, and novel therapies. *International Journal of Sciences: Basic and Applied Research (IJSBAR).*

Janani, B., Vijayakumar, M., Priya, K., Kim, J. H., Prabakaran, D., Shahid, M., Al-Ghamdi, S., Alsaidan, M., Othman Bahakim, N., & Hassan Abdelzaher, M. (2022). EGFR-Based targeted therapy for colorectal cancer—Promises and challenges. *Vaccines, 10*(4), 499.

Johnson, V., Vasu, S., Kumar, U. S., & Kumar, M. (2023). Surface-engineered extracellular vesicles in cancer immunotherapy. *Cancers, 15*(10), 2838.

Junttila, M. R., & De Sauvage, F. J. (2013). Influence of tumour micro-environment heterogeneity on therapeutic response. *Nature, 501*(7467), 346–354.

Khoei, S. G., Dermani, F. K., Malih, S., Fayazi, N., & Sheykhhasan, M. (2020). The use of mesenchymal stem cells and their derived extracellular vesicles in cardiovascular disease treatment. *Current Stem Cell Research & Therapy, 15*(7), 623–638.

Kikuchi, S., Yoshioka, Y., Prieto-Vila, M., & Ochiya, T. (2019). Involvement of extracellular vesicles in vascular-related functions in cancer progression and metastasis. *International journal of molecular sciences, 20*(10), 2584.

Kotelevets, L., & Chastre, E. (2023). Extracellular vesicles in colorectal cancer: From tumor growth and metastasis to biomarkers and nanomedications. *Cancers, 15*(4), 1107.

Lawrence, S. R., & Shah, K. M. (2024). Prospects and current challenges of extracellular vesicle-based biomarkers in cancer. *Biology, 13*(9), 694.

Li, Y., Zheng, Y., Tan, X., Du, Y., Wei, Y., & Liu, S. (2024). Extracellular vesicle-mediated pre-metastatic niche formation via altering host microenvironments. *Frontiers in Immunology, 15*, 1367373.

Li, Z., Sun, C., & Qin, Z. (2021). Metabolic reprogramming of cancer-associated fibroblasts and its effect on cancer cell reprogramming. *Theranostics, 11*(17), 8322.

Liang, Y., Lehrich, B. M., Zheng, S., & Lu, M. (2021). Emerging methods in biomarker identification for extracellular vesicle-based liquid biopsy. *Journal of Extracellular Vesicles, 10*(7), e12090.

Liu, Y.-J., & Wang, C. (2023). A review of the regulatory mechanisms of extracellular vesicles-mediated intercellular communication. *Cell Communication and Signaling, 21*(1), 77.

Luo, S., Chen, J., Xu, F., Chen, H., Li, Y., & Li, W. (2023). Dendritic cell-derived exosomes in cancer immunotherapy. *Pharmaceutics, 15*(8), 2070.

Lv, B., Wang, Y., Ma, D., Cheng, W., Liu, J., Yong, T., Chen, H., & Wang, C. (2022). Immunotherapy: Reshape the tumor immune microenvironment. *Frontiers in Immunology, 13*, 844142.

Maani, R. (2024). Interaction and functional impact of colorectal cancer extracellular vesicles on *Escherichia coli*. Sheffield Hallam University.

Maleki, M., Golchin, A., Javadi, S., Khelghati, N., Morovat, P., Asemi, Z., Alemi, F., Vaghari-Tabari, M., Yousefi, B., & Majidinia, M. (2023). Role of exosomal miRNA in chemotherapy resistance of Colorectal cancer: A systematic review. *Chemical Biology & Drug Design, 101*(5), 1096–1112.

Matsuzaka, Y., & Yashiro, R. (2022). Regulation of extracellular vesicle-mediated immune responses against antigen-specific presentation. *Vaccines, 10*(10), 1691.

McCullough, J., Frost, A., & Sundquist, W. I. (2018). Structures, functions, and dynamics of ESCRT-III/Vps4 membrane remodeling and fission complexes. *Annual Review of Cell and Developmental Biology, 34*(1), 85–109.

McNamara, A. M. (2019). Regulation of anti-tumor immunity in colorectal cancer by interaction of exosome immune complexes with Fc gamma receptors.

Mishra, V., Singh, A., Chen, X., Rosenberg, A. J., Pearson, A. T., Zhavoronkov, A., Savage, P. A., Lingen, M. W., Agrawal, N., & Izumchenko, E. (2022). Application of liquid biopsy as multi-functional biomarkers in head and neck cancer. *British Journal of Cancer, 126*(3), 361–370.

Mohamed, R. G. A., Ali, S. M., Ahmed, I. S., Rawas-Qalaji, M., & Hussain, Z. (2025). Next-generation nanocarriers for colorectal cancer: Passive, active, and stimuli-responsive strategies for precision therapy. *Biomaterials Science, 13*(20), 5626–5664.

Mokhtari, K., Sheykhhasan, M., Shahnazari, M., Ahmadieh-Yazdi, A., Shokrollah, N., Samadi, P., & Yang, P. (2025). Extracellular vesicles in reproductive medicines. In *Extracellular vesicles for therapeutic and diagnostic applications*. Elsevier, 243–281.

Molinari, C., Marisi, G., Passardi, A., Matteucci, L., De Maio, G., & Ulivi, P. (2018). Heterogeneity in colorectal cancer: A challenge for personalized medicine? *International Journal of Molecular Sciences, 19*(12), 3733.

Möller, A., & Lobb, R. J. (2020). The evolving translational potential of small extracellular vesicles in cancer. *Nature Reviews. Cancer, 20*(12), 697–709.

Najafi, S., Majidpoor, J., & Mortezaee, K. (2023). Extracellular vesicle–based drug delivery in cancer immunotherapy. *Drug Delivery and Translational Research, 13*(11), 2790–2806.

Najaflou, M., Shahgolzari, M., Khosroushahi, A. Y., & Fiering, S. (2022). Tumor-derived extracellular vesicles in cancer immunoediting and their potential as oncoimmunotherapeutics. *Cancers, 15*(1), 82.

Nawaz, M., Shah, N., Zanetti, B. R., Maugeri, M., Silvestre, R. N., Fatima, F., Neder, L., & Valadi, H. (2018). Extracellular vesicles and matrix remodeling enzymes: The emerging roles in extracellular matrix remodeling, progression of diseases and tissue repair. *Cells, 7*(10), 167.

van Niel, G., Carter, D. R., Clayton, A., Lambert, D. W., Raposo, G., & Vader, P. (2022). Challenges and directions in studying cell–cell communication by extracellular vesicles. *Nature Reviews. Molecular Cell Biology, 23*(5), 369–382.

Nikfarjam, S., Rezaie, J., Kashanchi, F., & Jafari, R. (2020). Dexosomes as a cell-free vaccine for cancer immunotherapy. *Journal of Experimental & Clinical Cancer Research, 39*(1), 258.

Nunez, R., Sancho-Martinez, S., Novoa, J., & Lopez-Hernandez, F.: (2010). Apoptotic volume decrease as a geometric determinant for cell dismantling into apoptotic bodies. *Cell Death & Differentiation, 17*(11), 1665–1671.

Olejarz, W., Kubiak-Tomaszewska, G., Chrzanowska, A., & Lorenc, T. (2020). Exosomes in angiogenesis and anti-angiogenic therapy in cancers. *International Journal of Molecular Sciences, 21*(16), 5840.

O'Neill, C. P., Gilligan, K. E., & Dwyer, R. M. (2019). Role of extracellular vesicles (EVs) in cell stress response and resistance to cancer therapy. *Cancers, 11*(2), 136.

De Palma, M., & Hanahan, D. (2012). The biology of personalized cancer medicine: Facing individual complexities underlying hallmark capabilities. *Molecular Oncology, 6*(2), 111–127.

Pan, B.-T., & Johnstone, R. M. (1983). Fate of the transferrin receptor during maturation of sheep reticulocytes in vitro: Selective externalization of the receptor. *Cell, 33*(3), 967–978.

Pang, X., He, X., Qiu, Z., Zhang, H., Xie, R., Liu, Z., Gu, Y., Zhao, N., Xiang, Q., & Cui, Y. (2023). Targeting integrin pathways: Mechanisms and advances in therapy. *Signal Transduction and Targeted Therapy, 8*(1), 1.

Paramanantham, A., Asfiya, R., Das, S., McCully, G., & Srivastava, A. (2022). Extracellular vesicle (EVs) associated non-coding RNAs in lung cancer and therapeutics. *International Journal of Molecular Sciences, 23*(21), 13637.

Pastushenko, I., & Blanpain, C. (2019). EMT transition states during tumor progression and metastasis. *Trends in Cell Biology, 29*(3), 212–226.

Pevzner, I. B., Andrianova, N. V., Lomakina, A. K., Cherkesova, K. S., Semenchenko, E. D., & Plotnikov, E. Y. (2025). Organ-specific extracellular vesicles in the treatment of ischemic acute organ injury: Mechanisms, successes, and prospects. *International Journal of Molecular Sciences, 26*(19), 9709.

Piperigkou, Z., Kyriakopoulou, K., Koutsakis, C., Mastronikolis, S., & Karamanos, N. K. (2021). Key matrix remodeling enzymes: Functions and targeting in cancer. *Cancers, 13*(6), 1441.

Poudineh, M., Sargent, E. H., Pantel, K., & Kelley, S. O. (2018). Profiling circulating tumour cells and other biomarkers of invasive cancers. *Nature Biomedical Engineering, 2*(2), 72–84.

Prakash, A., Gates, T., Zhao, X., Wangmo, D., & Subramanian, S. (2023). Tumor-derived extracellular vesicles in the colorectal cancer immune environment and immunotherapy. *Pharmacology & Therapeutics, 241*, 108332.

Pérez-Domínguez, F., Quezada-Monrás, C., Cárcamo, L., Muñoz, J. P., & Carrillo-Beltrán, D. (2025). Oncolytic viruses as a novel therapeutic approach for colorectal cancer: Mechanisms, current advances, and future directions. *Cancers, 17*(11), 1854.

Rahmati, S., Moeinafshar, A., & Rezaei, N. (2024). The multifaceted role of extracellular vesicles (EVs) in colorectal cancer: Metastasis, immune suppression, therapy resistance, and autophagy crosstalk. *Journal of Translational Medicine, 22*(1), 452.

Ran, G. H., Lin, Y. Q., Tian, L., Zhang, T., Yan, D. M., Yu, J. H., & Deng, Y. C. (2022). Natural killer cell homing and trafficking in tissues and tumors: From biology to application. *Signal Transduction and Targeted Therapy, 7*(1), 205.

Ribatti, D., Tamma, R., & Annese, T. (2020). Epithelial-mesenchymal transition in cancer: A historical overview. *Translational Oncology, 13*(6), 100773.

Riihimäki, M., Hemminki, A., Sundquist, J., & Hemminki, K. (2016). Patterns of metastasis in colon and rectal cancer. *Scientific Reports, 6*(1), 1–9.

Rizk, N. I., Abulsoud, A. I., Kamal, M. M., Kassem, D. H., & Hamdy, N. M. (2022). Exosomal-long non-coding RNAs journey in colorectal cancer: Evil and goodness faces of key players. *Life Sciences, 292*, 120325.

Rodríguez, D. A., & Vader, P. (2022). Extracellular vesicle-based hybrid systems for advanced drug delivery. *Pharmaceutics, 14*(2), 267.

Rui, R., Zhou, L., & He, S. (2023). Cancer immunotherapies: Advances and bottlenecks. *Frontiers in Immunology, 14*, 1212476.

Salar, M. Z., Mustafa, G., Hassan, T., Rafiq, I., & Tariq, M. N. (2025). Emerging biomarkers for cancer diagnosis: Paving the way for precision oncology. *Kashmir Journal of Science, 4*(2).

Salimi, L., Akbari, A., Jabbari, N., Mojarad, B., Vahhabi, A., Szafert, S., Kalashani, S. A., Soraya, H., Nawaz, M., & Rezaie, J. (2020). Synergies in exosomes and autophagy pathways for cellular homeostasis and metastasis of tumor cells. *Cell & Bioscience, 10*(1), 64.

Samadi, P., Sheykhhasan, M., Mokhtari, K., Yang, P., Maghool, F., & Kalhor, N. (2025). Extracellular vesicles: Unlocking therapeutic potential in regenerative medicine. In *Extracellular vesicles for therapeutic and diagnostic applications*. Elsevier, 397–435.

Schioppa, T., Gaudenzi, C., Zucchi, G., Piserà, A., Vahidi, Y., Tiberio, L., Sozzani, S., Del Prete, A., Bosisio, D., & Salvi, V. (2024). Extracellular vesicles at the crossroad between cancer progression and immunotherapy: Focus on dendritic cells. *Journal of Translational Medicine, 22*(1), 691.

Shaham, S. H., Vij, P., & Tripathi, M. K. (2025). Advances in targeted and chemotherapeutic strategies for colorectal cancer: Current insights and future directions. *Biomedicines, 13*(3), 642.

Shakhpazyan, N., Mikhaleva, L., Bedzhanyan, A., Gioeva, Z., Sadykhov, N., Mikhalev, A., Atiakshin, D., Buchwalow, I., Tiemann, M., & Orekhov, A. (2023). Cellular and molecular mechanisms of the tumor stroma in colorectal cancer: Insights into disease progression and therapeutic targets. *Biomedicines, 11*(9), 2361.

Shaon, M. A., Farhana, F. Z., Sikder, R., Shrestha, K., Haque, F., Tantu, M. T., Bin Manjur, O. H., Aktar, S., Koo, K. M., & Ross, A. G. (2025). Preeclampsia diagnostics and therapeutics: Advances, challenges, and prospects in nanoscale exosome-based clinical translation. *ACS Nano Medicine*.

Singh, N., Guha, L., Kumar, H. (n.d.). Extracellular vesicle.

Siveen, K. S., Raza, A., Ahmed, E. I., Khan, A. Q., Prabhu, K. S., Kuttikrishnan, S., Mateo, J. M., Zayed, H., Rasul, K., & Azizi, F. (2019). The role of extracellular vesicles as modulators of the tumor microenvironment, metastasis and drug resistance in colorectal cancer. *Cancers, 11*(6), 746.

Song, Z., Tao, Y., Liu, Y., & Li, J. (2024). Advances in delivery systems for CRISPR/Cas-mediated cancer treatment: A focus on viral vectors and extracellular vesicles. *Frontiers in Immunology, 15*, 1444437.

Stawarska, A., Bamburowicz-Klimkowska, M., Runden-Pran, E., Dusinska, M., Cimpan, M. R., Rios-Mondragon, I., & Grudzinski, I. P. (2024). Extracellular vesicles as next-generation diagnostics and advanced therapy medicinal products. *International Journal of Molecular Sciences, 25*(12), 6533.

Sun, M., Yang, J., Fan, Y., Zhang, Y., Sun, J., Hu, M., Sun, K., & Zhang, J. (2023). Beyond extracellular vesicles: Hybrid membrane nanovesicles as emerging advanced tools for biomedical applications. *Advanced Science, 10*(32), 2303617.

Sundaram, V., & Aryal, S. (2025). Emerging biomimetic drug delivery nanoparticles inspired by extracellular vesicles. *Wiley interdisciplinary reviews: Nanomedicine and nanobiotechnology, 17*(4), e70025.

Swanton, C., Bernard, E., Abbosh, C., André, F., Auwerx, J., Balmain, A., Bar-Sagi, D., Bernards, R., Bullman, S., & DeGregori, J. (2024). Embracing cancer complexity: Hallmarks of systemic disease. *Cell, 187*(7), 1589–1616.

Tamanna, S., Jannat, T., & Abbas, A. M. (2025). Introduction to colorectal cancer. In *Colorectal Cancer*. CRC Press, 1–12.

Tao, S.-C., & Guo, S.-C. (2020). Role of extracellular vesicles in tumour microenvironment. *Cell Communication and Signaling, 18*(1), 163.

Vaiaki, E. M., & Falasca, M. (2024). Comparative analysis of the minimal information for studies of extracellular vesicles guidelines: Advancements and implications for extracellular vesicle research. In *Seminars in cancer biology, 2024*, Elsevier, 12–24.

Verma, M., Lam, T. K., Hebert, E., & Divi, R. L. (2015). Extracellular vesicles: Potential applications in cancer diagnosis, prognosis, and epidemiology. *BMC Clinical Pathology, 15*(1), 6.

Verma, N., & Arora, S. (2025). Navigating the global regulatory landscape for exosome-based therapeutics: Challenges, strategies, and future directions. *Pharmaceutics, 17*(8), 990.

Villata, S., Canta, M., & Cauda, V. (2020). EVs and bioengineering: From cellular products to engineered nanomachines. *International Journal of Molecular Sciences, 21*(17), 6048.

Wu, J., Jin, Z., Fu, T., Qian, Y., Bian, X., Zhang, X., & Zhang, J. (2025). Extracellular vesicle-based drug delivery systems in cancer therapy. *International Journal of Molecular Sciences, 26*(10), 4835.

Xu, R., Rai, A., Chen, M., Suwakulsiri, W., Greening, D. W., & Simpson, R. J. (2018). Extracellular vesicles in cancer—implications for future improvements in cancer care. *Nature Reviews Clinical Oncology, 15*(10), 617–638.

Yan, H., Li, Y., Cheng, S., & Zeng, Y. (2021). Advances in analytical technologies for extracellular vesicles. *Analytical Chemistry, 93*(11), 4739–4774.

Yang, Q., Xu, J., Gu, J., Shi, H., Zhang, J., Zhang, J., Chen, Z. S., Fang, X., Zhu, T., & Zhang, X. (2022). Extracellular vesicles in cancer drug resistance: Roles, mechanisms, and implications. *Advanced Science, 9*(34), 2201609.

Yong, T., Wei, Z., Gan, L., & Yang, X. (2022). Extracellular-vesicle-based drug delivery systems for enhanced antitumor therapies through modulating the cancer-immunity cycle. *Advanced Materials, 34*(52), 2201054.

Youssef, E., Palmer, D., Fletcher, B., & Vaughn, R. (2025). Exosomes in precision oncology and beyond: From bench to bedside in diagnostics and therapeutics. *Cancers, 17*(6), 940.

Yue, M., Hu, S., Sun, H., Tuo, B., Jia, B., Chen, C., Wang, W., Liu, J., Liu, Y., & Sun, Z. (2023). Extracellular vesicles remodel tumor environment for cancer immunotherapy. *Molecular Cancer, 22*(1), 203.

Zakari, S., Niels, N. K., Olagunju, G. V., Nnaji, P. C., Ogunniyi, O., Tebamifor, M., Israel, E. N., Atawodi, S. E., & Ogunlana, O. O. (2024). Emerging biomarkers for non-invasive diagnosis and treatment of cancer: A systematic review. *Frontiers in oncology, 14*, 1405267.

Zare, M., Pemmada, R., Madhavan, M., Shailaja, A., Ramakrishna, S., Kandiyil, S. P., Donahue, J. M., & Thomas, V. (2022). Encapsulation of miRNA and siRNA into nanomaterials for cancer therapeutics. *Pharmaceutics, 14*(8), 1620.

Zhang, A., Miao, K., Sun, H., & Deng, C.-X. (2022). Tumor heterogeneity reshapes the tumor microenvironment to influence drug resistance. *International Journal of Biological Sciences, 18*(7), 3019.

Zhang, L., Shen, H., & Zhang, N. (2025). Advances in extracellular vesicle (EV) biomarkers for precision diagnosis and therapeutic in colorectal cancer. *Frontiers in Oncology, 15*, 1581015.

Zhang, G., Huang, X., Liu, S., Xu, Y., Wang, N., Yang, C., & Zhu, Z. (2025). Demystifying EV heterogeneity: emerging microfluidic technologies for isolation and multiplexed profiling of extracellular vesicles. *Lab on a Chip*.

Zhang, H., Gong, L., Yu, L., Xian, C., Ma, Z., Wang, X., & Xia, R. (2025). Emerging roles of non-coding RNA derived from extracellular vesicles in regulating PD-1/PD-L1 pathway: Insights into cancer immunotherapy and clinical applications. *Cancer Cell International, 25*(1), 188.

Zhang, J., Chen, B., Gan, C., Sun, H., Zhang, J., & Feng, L. (2023). A comprehensive review of small interfering RNAs (siRNAs): Mechanism, therapeutic targets, and delivery strategies for cancer therapy. *International Journal of Nanomedicine*, 7605–7635.

Zhang, M., Zang, X., Wang, M., Li, Z., Qiao, M., Hu, H., & Chen, D. (2019). Exosome-based nanocarriers as bio-inspired and versatile vehicles for drug delivery: Recent advances and challenges. *Journal of Materials Chemistry B, 7*(15), 2421–2433.

Zhang, Q., Higginbotham, J. N., Jeppesen, D. K., Yang, Y.-P., Li, W., McKinley, E. T., Graves-Deal, R., Ping, J., Britain, C. M., & Dorsett, K. A. (2019). Transfer of functional cargo in exomeres. *Cell Reports, 27*(3), 940–954 e946.

Zhang, Q., Jeppesen, D. K., Higginbotham, J. N., Graves-Deal, R., Trinh, V. Q., Ramirez, M. A., Sohn, Y., Neininger, A. C., Taneja, N., & McKinley, E. T. (2021). Supermeres are functional extracellular nanoparticles replete with disease biomarkers and therapeutic targets. *Nature Cell Biology, 23*(12), 1240–1254.

Zhao, Z., Wijerathne, H., Godwin, A. K., & Soper, S. A. (2021). Isolation and analysis methods of extracellular vesicles (EVs). *Extracellular Vesicles and Circulating Nucleic Acids, 2*(1), 80.

Zheng, J., Tan, J., Miao, Y.-Y., & Zhang, Q. (2019). Extracellular vesicles degradation pathway based autophagy lysosome pathway. *American Journal of Translational Research, 11*(3), 1170.

Zonneveld, M. I., Keulers, T. G., & Rouschop, K. M. (2019). Extracellular vesicles as transmitters of hypoxia tolerance in solid cancers. *Cancers, 11*(2), 154.

CHAPTER

# 8 Mitochondria-targeted drug delivery systems for the effective treatment of colon cancer

**Manali Rupareliya[1], Jyotsna Vitore[2], Ajinkya Aher[3], Mansi Patil[4], Melroy D'Sa[5], Dhanashree Guram[5], Yash Jadhav[5] and Prabhanjan Giram[6]**

[1]*Department of Pharmaceutics, Shobhaben Pratapbhai Patel School of Pharmacy and Technology Management, SVKM's NMIMS, Mumbai, Maharashtra, India,*
[2]*Department of Pharmaceutics, National Institute of Pharmaceutical Education and Research, Ahmedabad, Gujarat, India,*
[3]*Department of Pharmacology, Arnold and Marie Schwartz College of Pharmacy and Health Sciences, Long Island University, Brooklyn, NY, United States,*
[4]*Department of Pharmaceutical Chemistry, TSSM's Jayawant Institute of Pharmaceutical Sciences and Research Bavdhan, Pune, Maharashtra, India,*
[5]*Department of Pharmaceutics, OES's Oriental College of Pharmacy, Navi-Mumbai, Maharashtra, India,*
[6]*Department of Pharmaceutical Sciences, University at Buffalo, The State University of New York, Buffalo, NY, United States*

## 8.1 Introduction

### 8.1.1 Colon cancer

Cancer is a life-threatening condition that occurs due to uncontrolled proliferation of tumor cells, leading to cancer-related mortalities worldwide. Colonic cancer is also known as bowel cancer or colorectal cancer (CRC), which is the third most common cancer, characterized by colonic cell proliferation and metastasis due to dysfunction of mitochondrial fusion and fission mechanism and mutations in mitochondrial DNA (mtDNA) (Wu et al., 2024). Around 1.93 million patients with colon cancer are found globally, resulting in approximately 9,00,000 deaths annually, and the rate of mortality is exceeding year by year (Qiu et al., 2025). Individuals in older age group, that is, above 50 years, are most likely to be diagnosed with colonic cancer, in which men are more susceptible than women (Boyle & Leon, 2002).

Research findings indicate that the dysfunction of mitochondrial dynamics is a primary cause of initiation of development of colonic cell tumors (Wu et al., 2024). Genetic mutations, inflammation, hormonal factors, imbalance of gut microflora, and other physiological stimuli initiate the formation of tumor cells, while other modifiable risk factors such as obesity, smoking, and unhealthy diets also contribute to progression of tumor cells by affecting the normal metabolic functioning of cells (Al Kamzari & Constantinou, 2025). As the progression of tumor cells initiates, there is occurrence of blood in fecal matter, leading to iron deficiency anemia. The severity of disease increases with the

Nano Drug Delivery Systems for Colon Cancer. DOI: https://doi.org/10.1016/B978-0-443-40443-6.00019-2

symptoms of polyp formation (abnormal growth of tissue in the colon), unusual bowel movements, abdominal cramps and pain, weight loss, and exhaustion (Al Kamzari & Constantinou, 2025).

The molecular mechanism of transition from normal colon epithelial cells to abnormal cell proliferation occurs through alterations in various pathways (Li et al., 2025). Adenomatous polyposis coli (APC) is the important tumor suppressor gene responsible for the development of dysplastic abnormal cells. Mutations in APC promote the development of polyps, which subsequently carry mutations in the Kristen rat sarcoma viral oncogene (KRAS) and B-Raf proto-oncogene (BRAF) genes, leading to formation of dysplastic polyps and metastasis. Further development of malignant tumors occurs through mutations in the Tumor Protein 53 (TP53) tumor suppressor gene (Al Kamzari & Constantinou, 2025). Apart from this pathway, mitochondrial metabolic reprogramming has been a core marker for the colon cancer progression (Qiu et al., 2025). Mitochondria play a key role in energy production through the Krebs cycle and electron transport chain (ETC) reactions. Mitochondria consist of mitochondrial DNA (mtDNA) copies enclosed in the nucleus, which are involved in mitochondrial biogenesis. Mitochondrial biogenesis is the process in which the mtDNA replication occurs to fulfill the need of energy. The replication or transcription of mtDNA is governed by Mitochondrial transcriptional factor-A (TFAM), whose dysregulation results in altered mitochondrial biogenesis followed by carcinogenesis (Lin et al., 2017).

The development of colon cancer proceeds in four stages. At initial stages I and II, due to mutational changes, healthy colon epithelial cells distribute and split up, regulating the normal body function. In stage III, cancerous cells proliferate and develop into tumor cells, which spread to the neighboring lymph nodes. Continuous abnormal proliferation results in metastasis and cancer. If the condition is diagnosed at an early stage of development, it can be easy to control the severe growth of the disease (Alrushaid et al., 2023). Determination of tumor biomarkers, laboratory tests, endoscopy, biopsy, and genetic testing all contribute to diagnostic techniques for cancer detection (Al Kamzari & Constantinou, 2025). Other diagnostic areas involve the determination of mitochondrial dynamics, number of mtDNA copies, and intracellular reactive oxygen species (ROS) levels, while multitarget fecal immunochemical test and stool DNA test facilitate the mutations available that improve the imaging techniques and help in detection of cancer (Al Kamzari and Constantinou, 2025).

Nowadays, artificial intelligence (AI)-driven emerging technologies in cancer treatment often include multiple therapies, but these can have limitations and side effects, which may increase the risk of other health issues (Liu et al., 2025). Conventional treatments for colonic cancer include surgery, radiotherapy, chemotherapeutic agents such as folinic acid, oxaliplatin, 5-fluorouracil, and immunotherapy. The major drawback of conventional therapy is that it acts on cancer cells as well as healthy cells, while targeted therapies focus mainly on suppression of cancer cells without affecting normal cells. In targeted therapies, mitochondrial targeting possesses a greater potential in the treatment of colon cancer (Al Kamzari & Constantinou, 2025) (Fig. 8.1).

### 8.1.2 Mitochondria in colon cancer

Mitochondria are the double-membrane-bound organelles present in the cytoplasm of eukaryotic cells, which play significant role in energy production, oxidoreductive mechanisms, and regulation of apoptosis. The need for adenosine triphosphate (ATP) for survival of cells is accomplished by mitochondria through the tricarboxylic acid (TCA) cycle, electron transport chain (ETC), and oxidative phosphorylation (OXPHOS) system. Mitochondria possess their own double-stranded

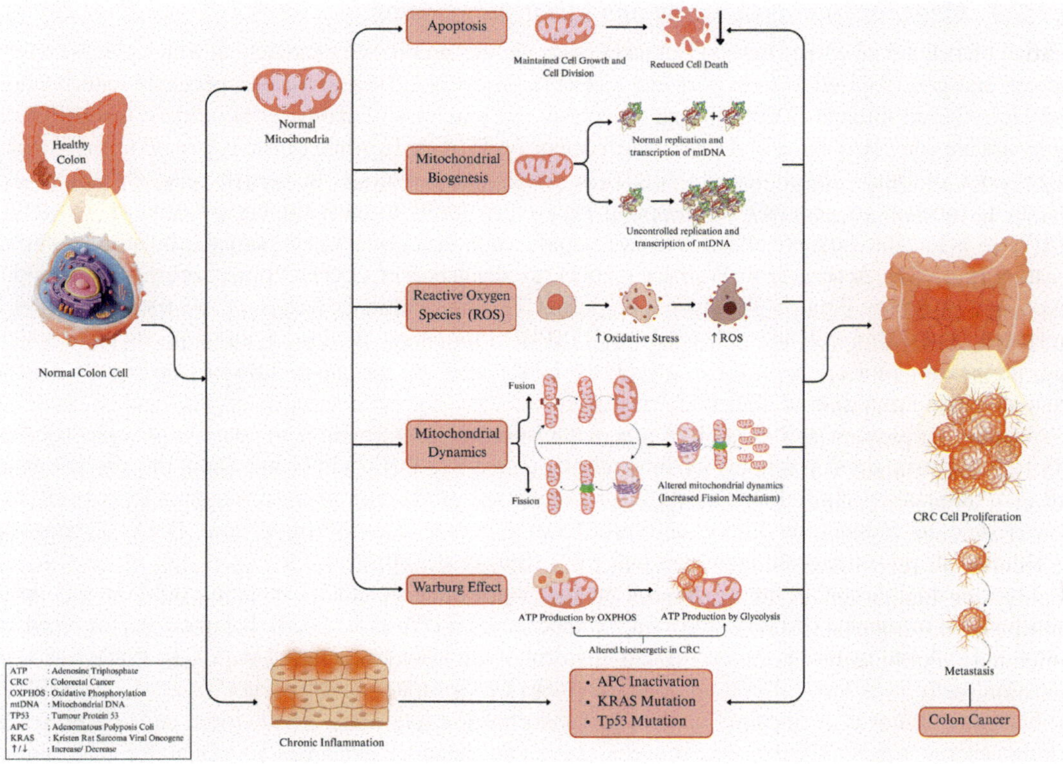


**FIGURE 8.1**

Pathogenesis of colon cancer. The diagrammatic presentation depicts the transformation of healthy colon cell to a cancerous cell. Altered mitochondrial functions, such as apoptosis, uncontrolled mtDNA function, reactive oxygen species (ROS), fission-fusion, and the Warburg effect, along with chronic colonic inflammation associated with activation and inhibition of various factors, play an important role in proliferation and metastasis of CRC.

circular DNA known as mitochondrial DNA (mtDNA). This mtDNA codes for OXPHOS system complexes in ATP synthesis (Luo et al., 2020). Mitochondrial biogenesis is responsible for multiplication of mtDNA copies, as higher number of mtDNA copies is an indication of greater mitochondrial function. Mitochondrial transcriptional factor-A (TFAM) and PGC-1α govern the replication and transcription of mtDNA in the nucleoids of the normal cells (Lin et al., 2017). It also regulates levels of reactive oxidative species (ROS) and mitochondrial OXPHOS (Liu et al., 2025). Mitochondria also possess dynamic properties such as fusion, fission, and metabolic functions that support several cellular processes, cell division, and metabolism of amino acids and fatty acids. Normal mechanisms of these functions and their role in occurrence of colon cancer are discussed further (Wu et al., 2024).

#### 8.1.2.1 Mitochondrial dysfunction and tumor progression

Normal mitochondrial energy metabolism takes place through anaerobic respiration, in which glucose enters the cell and gets oxidized to form pyruvate molecules in cytosol. These pyruvates penetrate mitochondria and are oxidized through OXPHOS and glycolysis (Księżakowska-Łakoma et al., 2014). In OXPHOS, pyruvates are converted into acetyl CoA, which enters the TCA cycle, while lactate is formed from pyruvate in presence of lactate dehydrogenase (LDH) in process of glycolysis. In normal cells, OXPHOS and glycolysis function cooperatively by adapting the oxygen levels to meet the energy demand. However, OXPHOS is found to be more effective in energy production than glycolysis in normal cellular mechanisms (Aisu et al., 2024). According to Warburg's effect, dysregulation of OXPHOS mechanism stimulates the production of energy through glycolytic mechanism. This is the distinct feature of tumors in which they prefer the enhanced glycolytic mechanism over OXPHOS for energy production that supports proliferation and metastasis. Hence, the aerobic glycolysis is elevated in malignant tumors, leading to cancer (Księżakowska-Łakoma et al., 2014).

Aerobic respiration and OXPHOS are major sources for formation of reactive oxygen species (ROS). Oxidative stress promotes the intracellular damage to DNA and brings about the alterations in mtDNA (Księżakowska-Łakoma et al., 2014). Increased production of mitochondrial ROS (mtROS) is observed to induce the tumor cell proliferation either due to irreversible DNA damage or stimulation of pro-oncogenic signaling pathways (Sun et al., 2018).

Mitochondrial fusion is the process of merging outer mitochondrial membrane (OMM) and inner mitochondrial membrane (IMM) to form single mitochondrion (Wu et al., 2024). It encourages the repair of damaged DNA and maintains the internal environment of mitochondria (Zhao et al., 2024). While fission is associated with breakdown of one mitochondrion into two separate mitochondria (Wu et al., 2024). This facilitates the removal of impaired or damaged mitochondria, promotes cell division, and adapts to the cellular energy needs. The balance between these mitochondrial dynamics is crucial for maintaining morphology and functions of mitochondria and effective OXPHOS. Dysfunction in either fusion or fission mechanism may account for tumorigenesis and cancer development (Wu et al., 2024; Zhao et al., 2024).

Mitochondrial biogenesis is primarily governed by mtDNA present in mitochondria. Increase or decrease in number of mtDNA copies results in different types of cancers. For example, mtDNA is found to decrease in gastric cancer, breast cancer, and hepatic carcinoma. In colon cancer, mtDNA copies are significantly increased due to mutations or impairment in mtDNA, which enhance the OXPHOS and promote tumor cell survival and metastasis (Sun et al., 2018). It maintains the balance between cell death and cell growth by regulating apoptotic pathways in normal tissue. Dysregulation of apoptotic pathways leads to reduced cell death and enhanced abnormal cell proliferation (Księżakowska-Łakoma et al., 2014).

Alteration in mitochondrial biogenesis, increased mitochondrial fission, apoptotic resistance, oxidative stress, and OXPHOS imbalance underlie the mitochondrial dysfunctions and play a significant role in colon cancer development and metastasis; this also increases the chemotherapy resistance in tumor cells (Wu et al., 2024; Zhao et al., 2024).

Table 8.1 summarizes the mitochondrial functions in normal colon cells compared with those in colon cancer involved in rapid colonic cancer cell growth and proliferation.

#### 8.1.2.2 Mitochondria as a therapeutic agent

Since mitochondria are a key factor in cancer development and tumorigenesis due to mitochondrial metabolism and dysfunctions, they can also become a promising therapeutic target for treatment of cancer.

**Table 8.1 Mitochondrial functions in normal cells versus in colon cancer cells.**

| S. No. | Mitochondrial functions | | References |
|---|---|---|---|
| | In normal cell | In colon cancer cells | |
| 1. | Production of energy (adenosine triphosphate) through oxidative phosphorylation pathway | Rapid energy production through glycolysis pathway (Warburg effect) | Aisu et al. (2024) |
| 2. | Maintaining balance between oxidoreductive species | Increased production of mtROS | Sun et al. (2018); Zhao et al. (2024) |
| 3. | Regulation of cell differentiation and apoptosis | Dysregulation of apoptotic pathways | Księżakowska-Łakoma et al. (2014) |
| 4. | Maintaining balance between mitochondrial dynamics, that is, fusion and fission | Enhanced fission mechanism | Wu et al. (2024); Zhao et al. (2024) |
| 5. | Controlling replication and transcription of mtDNA copies | Mutations in mtDNA and abnormal increase in mtDNA copies | Luo et al. (2020); Sun et al. (2018) |

Recently, researchers have increased their focus on developing the mitochondria-targeted therapeutic strategies for enhanced effects of chemotherapy (Zhao et al., 2024). Reversal of hyperglycolytic state of tumor cells may induce cell death and develop a selective therapeutic approach (Fulda et al., 2010). Aerobic glycolysis can be suppressed by inhibition of LDH; this would reduce the energy supply to tumor cells and affect their survival (Fulda et al., 2010). Induction of apoptosis helps to regulate the cell death mechanism, which suppresses the cell proliferation and enhances tumor cell death (Księżakowska-Łakoma et al., 2014). Increased mitochondrial fission causes mitochondrial reprogramming, promoting proliferation of colonic cancer cells and developing chemoresistance, while excessive fission enhances apoptotic function of mitochondria. This highlights the need for effective therapies targeting mitochondrial dynamics for treatment of colon cancer (Wu et al., 2024). Damaged mitochondria support tumor cells in surviving and growing rapidly. Mitophagy is the process of abolishing damaged mitochondria; hence, activation of mitophagy induces clearance of dysfunctional mitochondria, thereby improving metabolic reprogramming and increasing sensitivity of tumors to chemotherapy (Zhao et al., 2024).

In the following sections, various therapeutic mitochondrial targets are discussed in detail, followed by their mechanisms (Luo et al., 2020).

## 8.2 Mitochondria: targeted therapy

### 8.2.1 Mitochondria targeting

Mitochondria play an important role in regulation of cellular metabolic activities, energy production, that is, ATP production, cell signaling, apoptosis, and biosynthesis of macromolecules, which are essential for maintaining the normal functioning of the cells. Mitochondria are structurally and functionally dynamic cell organelles, and their role is vital in the eukaryotic cells. Cancerous cells alter these basic mitochondrial functions to serve their increased bioenergetic needs for cell survival, growth, and proliferation (Ghosh et al., 2020). It is known that cancerous cells at all stages of oncogenesis, that is, tumor progression, proliferation, and transformation, are under the influence of mitochondrial metabolism (Ghosh et al., 2020). Often, cancer cells reprogram energy pathways to promote glycolysis, and the use of OXPHOS is minimized; this is known as Warburg effect (Wu et al., 2024). Mitochondria-targeted cancer therapy alters the metabolic pathways essential for abnormal cell proliferation.

Research in oncology has driven attention of cancer therapy to target specific cell organelles such as endoplasmic reticulum, mitochondria, and lysosomes (Chen et al., 2024). Cell organelles are responsible for specific cellular functions and are closely related to development of cancerous cells and various other diseases (Chen et al., 2024). Targeted organelle therapies reduce therapy-related side effects and potentiate precise killing of cancer cells.

Surgery, radiotherapy, and chemotherapeutic agents are implemented for the treatment of colon cancer. 5-Fluorouracil (5-FU) and folinic acid are considered the gold standard, which are used in combination with oxaliplatin for effective CRC treatment (Neitzel et al., 2020). Recent studies have found that altered mitochondrial function is greatly associated with the development of resistance against chemotherapeutic agents in colon/CRC (Zhao et al., 2024). OXPHOS plays a major role in this activity of mitochondria. The proliferation and metastasis of colon cancer cells are driven by various mitochondrial dysfunctions. Mitochondria-targeting therapies specifically attack biomolecules and receptors that contribute to chemotherapy resistance in CRC. Various possible mitochondrial targets in CRC are summarized in Table 8.2.

Table 8.2 summarizes the potential mitochondrial targets, such as OXPHOS, TCA cycle, PGC1-α, dynamin-related protein 1 (DRP1), and ZNF746/parking interacting substrate (PARIS), involved in effective treatment of CRC and the role of these targets in CRC cells.

### 8.2.2 Mechanisms of action

Mitochondria, being a major organelle involved in cellular metabolic pathways, represent a promising approach/target for effective treatment of cancerous cells (Chen et al., 2024). Selective and targeted

**Table 8.2 Specific mitochondrial targets in colorectal cancer cells.**

| S. No. | Mitochondrial targets | Roles in colon cancer | References |
|---|---|---|---|
| 1. | Oxidative phosphorylation (OXPHOS) | Increased energy production | Pang and Wu (2025); Wu et al. (2024); Zhao et al. (2022) |
| 2. | Tricarboxylic acid cycle | Energy production | Neitzel et al. (2020) |
| 3. | PGC1-α | Increase OXPHOS and alter mitochondrial dynamics | Zhao et al. (2024) |
| 4. | Dynamin-related protein 1 (DRP1) and ZNF746/parking-interacting substrate | Increase mitochondrial fission and fusion | Xing et al. (2023); Zhao et al. (2024) |

therapy in colon cancer via mitochondria is possible through specific targets. Recent studies and research on colon cancer have supported the beneficial effects of these targets in altering the growth of CRC. The following section briefly explains how these mitochondrial targets have emerged in CRC, their role in meeting the altered (increased) bioenergetic need, mechanism to target them, and the subsequent actions produced.

#### *8.2.2.1 Targeting oxidative phosphorylation in colorectal cancer*

A distinct feature of cancer cells is metabolic programming to meet their active energy needs for rapid growth, proliferation, and cell survival under stressful conditions. Normal colonic cells rely on energy-rich OXPHOS for ATP production. On the other hand, cancerous colonic cells prefer glycolysis over OXPHOS for rapid energy production. This phenomenon is termed the Warburg effect (Pang & Wu, 2025). It is driven by tumor microenvironment (TME) to enhance CRC cell proliferation. Glycolysis becomes a primary source of energy production in colon cancer cells; this emphasizes targeting glycolysis pathway through inhibition of various enzymes involved in it. Glycolytic enzymes, namely hexokinase, phosphofructokinase, pyruvate kinase, and LDH, are therapeutic targets in effective CRC treatment strategies (Pang & Wu, 2025). The allopathic treatment regimen involved in glycolysis inhibition gradually faces resistance developed by the CRC cells. This condition arises as CRC cells drastically shift energy production from glycolysis to OXPHOS.

Various signaling pathways are involved in promoting OXPHOS activity and are associated with the development of resistance to cancer therapy. CRC cells adapt to OXPHOS to uphold their bioenergetic needs and can thus be targeted by OXPHOS inhibitors (Zhao et al., 2022). OXPHOS inhibitors selectively inhibit and downregulate the overexpression of OXPHOS genes, subsequently suppressing OXPHOS activity. This prevents the development of drug resistance to the conventional anticancer therapies and sensitizes resistant cancer cells to the conventional treatment regimen (Zhao et al., 2022).

#### *8.2.2.2 Targeting tricarboxylic acid cycle in colorectal cancer*

TCA cycle, or Krebs cycle, or citric acid cycle, is the second and important pathway involved in energy production required for essential cellular functions. It is mainly responsible for oxidation of acetyl-CoA to produce NADH2 and FADH2 molecules, which ultimately enter into OXPHOS in the mitochondrial matrix. TCA amplifies the energy yield per glucose molecule through generation of these biomolecules. Isocitrate dehydrogenase, pyruvate dehydrogenase kinase, α-ketoglutarate dehydrogenase (KGDH), and pyruvate dehydrogenase complex (PDC) are the major enzymes involved in running TCA cycle (Neitzel et al., 2020). These enzymes can be viewed as potential targets for disrupting TCA cycle and ultimately the energy hub in CRC. Promising agents are under clinical investigation; one of these is α-lipoic acid (LA) derivative CPI-613 (Neitzel et al., 2020).

LA is an endogenous antioxidant triggering cell death in tumor cells and is also associated with potentiating p53 activity. CPI-613, a LA derivative, selectively targets PDC and KGDH in TCA cycle. CPI-613 inhibits multienzyme complexes that perform oxidative decarboxylation of pyruvate to acetyl-CoA and α-KG to succinyl-CoA (Neitzel et al., 2020). The molecular mechanisms of PDC and KGDH inhibition are distinct. Thus selectively targeting the multienzyme complexes of TCA cycle significantly reduces fueling of OXPHOS and would hamper the bioenergetic-derived metabolic programming in CRC cells.

#### 8.2.2.3 Targeting altered mitochondrial dynamics in colorectal cancer

Mitochondria represent a dynamic network associated with changes in size and distribution; these dynamics are governed by two processes, namely fission and fusion (Xing et al., 2023). Repair of damaged mitochondria, regulation of internal mitochondrial environment, and maintenance of the integrity of mitochondrial DNA (mtDNA) are the roles of fusion process (Zhao et al., 2024). These are regulated through a series of complex processes. Dynamin-related protein 1 (DRP1) is a GTPase protein acting as a master regulator in mitochondrial dynamics. It plays a crucial role in mitochondrial fission. DRP1 overexpression is related to promoting BRAF V600E driven colon cancer progression and proliferation (Padder et al., 2021). This makes DRP1 a potential target in CRC treatment.

Another protein greatly associated with dysfunctions and dysregulation of mitochondrial dynamics is Zinc Finger Protein 746 (ZNF746), also known as PARIS. Studies on CRC cells have found that the overexpression of ZNF746/PARIS leads to inhibition of proteins like Mitofusin-1 (MFN1), Mitofusin-2 (MFN2), and peroxisome proliferator-activated receptor γ-coactivator 1-α (PGC1-α) (Zhao et al., 2024). Inhibition of these proteins alters equilibrium between fission and fusion, thus significantly reducing mitochondrial activity. In addition, impaired mitochondrial dynamics also promote resistance against chemotherapeutic agents (e.g., 5-FU). Further studies suggest that selective inhibition of ZNF746 protein expression reduces resistance to 5-FU in CRC cells (Zhao et al., 2024). Thus, it highlights the importance of ZNF746 in mediating chemotherapeutic resistance and its potential in mitochondria-targeted therapies (Fig. 8.2).

## 8.3 Mitochondria-targeted drug delivery strategies

Mitochondria play a fundamental role in energy production, apoptosis, reactive oxygen species (ROS) generation, and metabolic regulation. Dysfunctional mitochondria have been implicated in various diseases, such as cancer (e.g., breast cancer, colon cancer), neurodegenerative disorders (e.g., Parkinson's and Alzheimer's disease), cardiovascular diseases, and metabolic disorders. As a result, mitochondria-targeted drug delivery strategies are emerging as therapeutic interventions.

Conventional drug delivery methods often fail to reach mitochondria, probably due to various barriers (plasma membrane, cytoplasmic degradation, and inefficient mitochondrial uptake). To address such challenges, strategies have been developed to enhance drug accumulation in mitochondria. This chapter deals with approaches such as mitochondria-targeting ligands, nanocarrier-based approaches, and prodrug systems designed for mitochondrial activation (Murphy & Smith, 2007; Yousif et al., 2009).

### 8.3.1 Mitochondria-targeting ligands

Mitochondria-targeting ligands can be used for selective drug accumulation in mitochondria, which can be achieved by exploiting their unique biochemical properties, such as high negative membrane

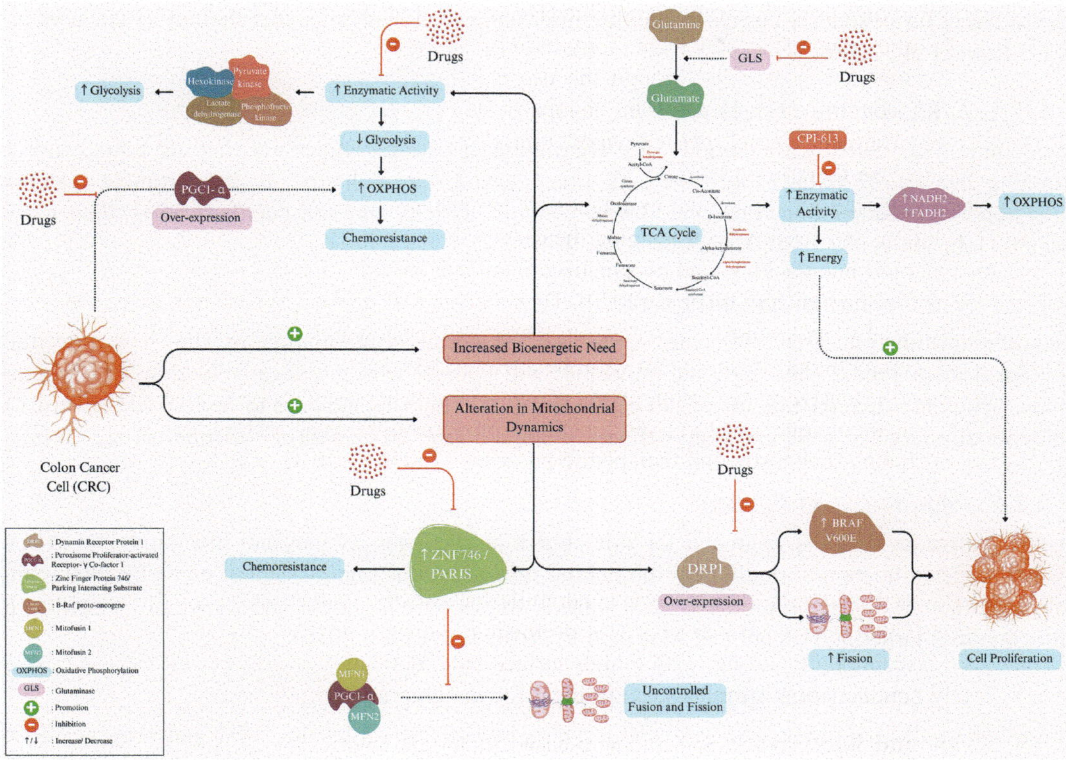


**FIGURE 8.2**

The schematic flow represents various possible mitochondrial targets, how these targets have emerged and the mechanisms for mitochondria-targeted therapy.

potential (−180 mV) of the mitochondrial membrane and specialized mitochondrial transporters (Smith et al., 1999).

#### *8.3.1.1 Cation-based ligands*

Mitochondria maintain an electrochemical gradient across the IMM and the matrix, such that the matrix is negatively charged. This allows positively charged or cationic molecules to accumulate within mitochondria due to membrane potential-driven uptake (Smith et al., 1999).

##### 8.3.1.1.1 Triphenylphosphonium conjugates

One of the widely used mitochondria-targeting moieties is TPP; it is lipophilic and has a delocalized positive charge. TPP-conjugated molecules can cross plasma membrane and accumulate in mitochondria via the mitochondrial membrane potential-driven mechanism. TPP-conjugated drugs, such as MitoQ (mitochondria-targeted antioxidant) and MitoTam (mitochondria-targeted tamoxifen derivative), have shown promising results. TPP-conjugated nanocarriers have been developed that ensure direct mitochondria targeting, thereby improving treatment in cancer therapy (Biswas et al., 2012; Ross et al., 2006).

##### 8.3.1.1.2 Rhodamine derivatives

Rhodamine dyes (such as rhodamine 123) are a class of cationic molecules that are mitochondria-targeting ligands. They have been used for imaging and drug delivery, as they promote selective mitochondrial accumulation. However, due to potential toxicity and nonspecific binding, their use is limited (Johnson et al., 1980; Lampidis et al., 1983).

##### 8.3.1.1.3 Guanidinium-based ligands

Guanidinium-functionalized molecules show mitochondrial accumulation as they can penetrate biological membranes. These cationic molecules will interact with the negatively charged phospholipids, which are the constituents of the cell membrane. These ligands are to be looked forward for mitochondria-targeted gene and drug delivery (Ahn and Lee, 2018; Sibrian-Vazquez et al., 2008).

#### *8.3.1.2 Peptide-based ligands*

Peptides offer an efficient approach for mitochondrial targeting, as they are biocompatible. These ligands target mitochondria either by mimicking mitochondrial import sequences or by interacting with mitochondrial-specific proteins. Such ligands are specific in nature and showcase versatility in mitochondrial targeting (Chuah and Matsugami, 2016; Jean et al., 2016).

##### 8.3.1.2.1 Szeto-Schiller peptides

These are small, cell-permeable tetrapeptides; they selectively target the IMM and scavenge ROS (reactive oxygen species), thereby showcasing antioxidant properties. These peptides may have the potential to negate mitochondrial dysfunction that is characteristic of neurodegenerative diseases, ischemia-reperfusion injury, and other related disorders (Szeto, 2006, 2008).

##### 8.3.1.2.2 Mitochondria-penetrating peptides

These peptides can traverse the mitochondrial membranes more efficiently. Innovations have led to enhanced drug delivery directly into the mitochondrial matrix. They interact with mitochondrial import mechanism and guide therapeutic agents to reach intended intracellular site (Horton et al., 2008).

### 8.3.2 Nanocarrier-based approaches

Nanocarrier-based delivery systems enhance drug stability, bioavailability, and confer specificity, thereby making them ideal for mitochondrial targeting. Such systems, along with mitochondria-targeting ligands, help improve selective drug accumulation in the mitochondrial matrix (Battogtokh et al., 2018; Xu et al., 2025).

#### *8.3.2.1 Liposomes*

Liposomes are spherical vesicles composed of phospholipids or related constituents that form the outer layer in which both hydrophilic and hydrophobic drugs can be incorporated. In order to enhance mitochondrial targeting, liposomes are often conjugated with ligands such as TPP, mitochondrial localizing sequence (MLS) peptides, or other ligands. Mitochondria-targeted liposomes have been developed by incorporating specific ligands (Boddapati et al., 2005).

- **TPP-modified liposomes**: Liposomes with TPP moiety enhance their accumulation in mitochondria. Studies have demonstrated the effectiveness of TPP-modified liposomes in delivery of anticancer drugs into the mitochondria, which will further induce apoptosis in cancer cells (Boddapati et al., 2005).
- **Peptide-conjugated liposomes:** Incorporating mitochondrial-targeting peptides into liposomal structures has improved the delivery of therapeutic agents to mitochondria, offering the potential to treat mitochondrial diseases.

Advantages of using liposomes include biocompatibility and, hence, low toxicity, protection of encapsulated drugs from enzymatic degradation, and the capability to deliver a wide range of therapeutic agents.

Limitations include short circulation half-life and possible instability in physiological conditions (Sanchez-Aranguren et al., 2025).

#### 8.3.2.2 Nanoparticles

Consisting of polymer nanoparticles, metal nanoparticles, and the like shows potential in mitochondrial drug delivery. Nanoparticles: Bio-differential polymers such as poly lactic-co-glycolic acid (PLGA) are used to construct functionalized nanoparticles in the mitochondrial targeting league. These nanoparticles promise to deliver antioxidants and chemotherapeutic agents directly to mitochondria. High surface area and functionalization capabilities allow for efficient drug delivery. Latest innovations include functionalization of mesoporous silica nanoparticles (MSNs) using TPP or peptides to achieve mitochondrial targeting and improve the therapeutic efficacy of indifferent drugs. Functionalization with a mitochondrial targeting agent improves specificity. For example, TPP-functionalized polymers were used for targeted delivery of chemotherapeutic agents to mitochondria. Micelles formed by the self-assembly of amphiphilic molecules have been tested for mitochondrial drugs (Lin et al., 2020; Pathak and Kolishetti, 2015).

#### 8.3.2.3 Micelles

TPP-binding polymer micelles: These micelles demonstrate their ability to effectively deliver chemotherapy agents to mitochondria. Mitochondrial medications were studied for submission. Dendrimers are highly efficient in drug delivery and allow for improved targeting through their highly divergent structures and conjugation with TPP or peptide ligands. Due to their high surface area and functionality, carbon-based nanomaterials, including graphene oxide and carbon nanotubes, exhibit potential for the delivery of mitochondrial drugs (Lee et al., 2025; Sena Ozbay et al., 2022).

### 8.3.3 Prodrugs and their derivatives

Prodrugs are inactive drug precursors that experience enzymatic or chemical activation at the target. Mitochondria-targeted prodrugs are subjected to selective activation in mitochondria to ensure controlled drug release and reduced exotoxicity. These modifications promote selective accumulation within mitochondria prior to activation. For example, prodrug that targets the anticancer component mitochondria shows increased cytotoxicity in cancer cells, while healthy cells are spared. Recent developments focus on conjugated prodrugs containing mitochondrial target units (Singh, 2024; Zheng et al., 2018).

#### 8.3.3.1 Mitochondria-targeting prodrugs

TPP-binding prodrugs: These prodrugs accumulate by mitochondrial TPP units. After reaching the mitochondrial matrix, they are activated to release therapeutic measures and minimize systemic toxicity. Enzyme division within mitochondria releases active drugs and provides the potential for treating mitochondrial dysfunction (Kelso et al., 2012).

#### 8.3.3.2 Activation of prodrugs within mitochondria

Redox-sensitive prodrugs: These prodrugs use unique redox mechanisms of mitochondria, activated by mitochondrial ROS, ensuring targeted drug release, and mitochondrial esterases for activation. Certain mitochondrial enzymes can activate prodrugs and release therapeutic agents accurately within mitochondria. This approach improves drug efficacy and reduces the effectiveness of Odtate. By using mitochondrial-targeting ligands with nanocarriers and evidence-based nanoconjugates, researchers are developing more effective and selective therapies. However, there are challenges such as drug stability, off-target effects, and mitochondrial heterogeneity. Future research will focus on optimizing targeting strategies, minimizing toxicity, and improving clinical translation (Ma and Song, 2019; Obeid et al., 2017).

## 8.4 Advances in mitochondria-targeted drug delivery for colon cancer

Mitochondria-targeted drug delivery systems (MTDDS) constitute an innovative therapeutic method that targets cancer therapy resistance in colon cancer and other malignancies. Energy production, apoptosis regulation, and reactive oxygen species (ROS) generation occur within the mitochondria that act as the central element for cellular functions. Colon cancer cells use impaired mitochondria to advance tumors as they spread through the body and develop drug resistance, which makes these cellular components valuable targets for treatment (Hu et al., 2016; Lin et al., 2023). New findings in MTDDS are presented in this section by exploring preclinical reports as well as the current stages of clinical implementation.

### 8.4.1 Preclinical studies

Research conducted before clinical application established mitochondria-targeted strategies due to their better efficiency and selectivity in models of colon cancer. Triphenylphosphonium (TPP+) vectors function successfully as lipophilic cationic vectors for directing chemotherapeutic agents to mitochondria. The doxorubicin-TPP+ conjugate serves as an exemplary example that demonstrates improved mitochondrial accumulation that results in breakdown of membrane potential and pathway activation of apoptosis (Battogtokh et al., 2018; Zielonka et al., 2017).

The field of nanotechnology has achieved improved MTDDS by designing three distinct categories of nanocarriers, namely liposomes, dendrimers, and biodegradable polymeric nanoparticles. The carriers possess mitochondria-penetrating peptides (MPPs) and stimuli-responsive linkers that release their payload by sensing specific pH and redox signaling found inside tumors. The drug delivery systems achieve higher drug effectiveness while lowering the number of unintended toxicities and tackling drug-resistant behavior (Pathak and Kolishetti, 2015; Zeng et al., 2021).

The therapeutic efficacy of MTDDS has been confirmed through experimental tests that utilized both advanced spheroid cellular systems and patient-derived xenograft (PDX) models. The experimental models demonstrate that MTDDS have dual capability to elevate apoptotic indicators along with reduction of tumor size while minimizing systemic toxicities compared to conventional chemotherapy protocols (Lingappa and Akintunde, 2022).

### 8.4.2 Clinical translation

#### *8.4.2.1 Current challenges*

MTDDS require overcoming multiple problems when moving from preclinical success to clinical application in colon cancer management. The translation process for colon cancer applications faces multiple challenges that include manufacturing complications, safety aspects, biological limitations, and regulatory concerns.

A significant challenge stands in the way of transforming laboratory-made products into clinical application. Complex chemical reactions are needed for MTDDS conjugations because precise control is essential over reaction conditions. The therapeutic and targeting performance of mitochondria-directed drug delivery systems suffers major changes when important quality attribute characteristics, like particle size, zeta potential, and targeting ligand density, differ from batch to batch. According to Kleemann et al., a comprehensive analysis revealed that ±5% fluctuations in TPP conjugation efficiency yield significant alterations in mitochondrial uptake, thus underscoring the requirement for strict quality control measures during manufacturing (Hua et al., 2018).

Besides scalability, there exists another major manufacturing problem. Laboratory techniques used for producing MTDDS exist, but when implementing these processes at industrial levels, their operational parameters maintain inconsistent relationships. During the commercial manufacturing transition process, research teams encounter two difficulties: GMPuisine requires either therapeutic system reformulation or processing changes, which can affect both drug delivery characteristics and pharmacological properties. The manufacturing difficulties increase significantly when MTDDS systems contain dual elements of specific targeting ligands together with therapeutic payloads (Younis et al., 2022).

The safety concerns about new delivery systems function as a major obstacle in their advancement from laboratory to clinical translation. The therapeutic effectiveness of mitochondrial disruption as a cancer cell treatment brings risks of potential mitochondrial damage in healthy cells. Research conducted by Peterson and colleagues showed that TPP-based delivery systems exhibit higher accumulation in cancer mitochondria; however, they are absorbed by mitochondria that exist abundantly in cardiac and skeletal muscle tissues. Accurate dosage control is essential because it determines a therapeutic range where benefits against cancer cells exist while minimizing detrimental effects to healthy mitochondria. The development process encounters additional complications from immunogenicity issues. The repeated use of peptide-based targeting moieties and modified antimicrobial peptides for mitochondrial delivery triggers immune responses in recipients. The initial stage of testing multiple nanomedicine platforms has unveiled this phenomenon, which results in fast blood clearance and weaker therapeutic effectiveness for succeeding treatment administrations. Research now focuses on creating immune-untriggered alternatives and protocols to suppress immune system identification (Buchke and Sharma, 2022; Hua et al., 2018; Younis et al., 2022).

The obstacles caused by pharmacokinetic challenges prove difficult to overcome. An optimal MTDDS must achieve a sophisticated delivery process starting from administration and concluding at colon cancer cell mitochondria. Drug delivery pathways involve survival against reticuloendothelial system removal, as well as tumor site extravasation, followed by stromal penetration before endocytosis and reaching the mitochondria. Each sequential step introduces a risk that will significantly decrease the dosage delivered to its subcellular objective. The research by Venkataraman and colleagues analyzed intravenous MTDDS distribution in colon cancer mitochondria and determined that less than 0.1% reach this target (Cong and Thakur, 2023; Hammond et al., 2016).

Clinical translation of colon cancer treatments becomes more difficult due to its multiple forms of heterogeneity. Different molecular subtypes (CMS1–4) exhibit distinct metabolic profiles and mitochondrial characteristics. The metabolic dysregulation of CMS3 tumors produces anomalous mitochondrial membrane potentials when compared to CMS4 mesenchymal tumors, therefore influencing the effectiveness of membrane potential-based targeting approaches. This heterogeneity necessitates either subtype-specific targeting approaches or the development of delivery systems with broader applicability across tumor subtypes (Lin et al., 2015).

Regulatory pathways for novel MTDDS present additional challenges. MTDDS function as combination products that incorporate nanocarrier devices with drug components, so they encounter uncertain regulatory specifications during approval processes. The ability to identify suitable patients through companion diagnostics becomes more complex because it affects the potential use of mitochondrial-targeting strategies. To successfully address regulatory complexities, businesses must begin their engagement with regulatory organizations while planning their product development directions (Wen et al., 2015; Xie et al., 2020).

Various strategic methods continue to develop that help move medical discoveries toward clinical usage. The pursuit of clinical translation includes three main approaches: developing basic MTDDS with reduced components to ensure better manufacturing consistency, studying alternative target moieties with more favorable safety outcomes, and implementing theranostics machinery that combines diagnostic and therapeutic capabilities to monitor drug delivery patterns at early clinical trial stages (Shaham et al., 2025).

#### 8.4.2.2 Ongoing clinical trials

| Trial ID | Phase | Mitochondria-targeted system | Drug | Patient population | Primary endpoint | Status |
|---|---|---|---|---|---|---|
| NCT04852367 | I | TPP-conjugated liposomal formulation | Oxaliplatin | Metastatic colorectal cancer (mCRC), second line | MTD, Dose-limiting toxicitys | Recruiting |

| NCT05123781 | I/II | MitoQ-functionalized polymeric micelles | SN-38 | Advanced colorectal cancer (CRC), refractory | Safety, Objective response rate (ORR) | Active |
|---|---|---|---|---|---|---|
| NCT05362487 | II | Mito-Porter nanocarrier system | 5-Fluorouracil | Stage III CRC, adjuvant | Disease-free survival at two years | Recruiting |
| NCT05296382 | I | Szeto-Schiller-31 peptide-conjugated nanoparticles | Irinotecan | Advanced CRC with Kristen rat sarcoma viral oncogene mutations | Safety, PK/PD | Active |
| NCT04771689 | I/II | MITO-4G1 (TPP-modified dendrimer) | Doxorubicin | Refractory mCRC | Maximum tolerated dose, tumor response | Completed |
| NCT05498117 | I | Mitochondria-penetrating peptide | Docetaxel | Advanced solid tumors, including CRC | Safety, tissue distribution | Recruiting |
| NCT05673356 | II | Dequalinium-based liposomes | Cisplatin | MSI-H CRC | ORR, Progression-free survival (PFS) | Not yet recruiting |
| NCT05218759 | I | Rhodamine-123-labeled exosomes | miR-497 mimic | CRC with liver metastasis | Safety, biodistribution | Active |
| NCT05421169 | II | Mitochondrial tamoxifen | Tamoxifen derivative | Metastatic CRC (third line) | Disease control rate | Recruiting |

## 8.5 Mitochondria-targeted drug delivery

Medical advancements in mitochondrial research, coupled with optimum drug delivery techniques, enabled scientists to create innovative anticancer substances effective in mitochondrial cell space. These drugs have dual benefits of increasing treatment success while using cancer cells' metabolic weaknesses to boost their therapeutic value.

### 8.5.1 Mitochondria-targeted chemotherapeutics

#### *8.5.1.1 Doxorubicin*

The potential role of doxorubicin as an anthracycline antibiotic has been extensively evaluated for its intercalating properties in mitochondrial treatment of colon cancer. The delivery of doxorubicin to mitochondria is based on many different factors. Through nuclear DNA intercalation and topoisomerase II inhibition, doxorubicin performs its main action, but the agent produces detectable mitochondrial effects that enhance its cytotoxic properties. Doxorubicin exerts its effects on mitochondria by breaking down complex I electron transport chain complexes and creating reactive oxygen species as well as opening the mitochondrial permeability transition pore (MPTP) (Chen et al., 2023; Cho et al., 2013).

The creation of doxorubicin formulations specifically designed for mitochondria has pursued different methods that tackle delivery inefficiencies along with toxicity reduction needs. Research on TPP-conjugated compounds constitutes one of the most extensively studied methods among these strategies. Researchers created TPP-doxorubicin conjugates that accumulated 6.4 times higher amounts in the mitochondrial tissue of HCT116 colon cancer cells as opposed to pure doxorubicin administration. The enhanced delivery of the medicine to mitochondria through TPP conjugation resulted in diminished IC50 levels (reduction of 73%) along with improved drug effectiveness in P-glycoprotein-mediated resistance cells, which frequently pose treatment complications for colon cancer patients (Liu et al., 2021).

Nanocarrier technologies have advanced the delivery system of doxorubicin to mitochondria. Research showed polymeric nanoparticles that combined TPP with pH-triggered breakable linkages. These delivery systems maintain stability at normal body pH until they degrade precisely in acidic tumor regions to release doxorubicin-TPP conjugates that reach mitochondria. The delivery of doxorubicin using this method strengthened therapeutic efficiency while decreasing cardiotoxicity by 83% compared to free drug administration, while maintaining equivalent antitumor effects in orthotopic colon cancer models (Zhao et al., 2013).

The clinical development of mitochondria-targeted doxorubicin (MITO-DOX) has begun with the completion of the MITO-DOX phase I trial (NCT04771689). A TPP-modified dendrimer carrying doxorubicin served as the main experimental compound during clinical trials involving refractory metastatic CRC patients. The initial study data confirmed that the pharmacokinetic behavior showed better results than standard doxorubicin, thus indicating lower exposure risk for heart muscle tissue. The highest dose-level testing showed no significant changes in left ventricular ejection fraction through serial cardiac monitoring, thus providing evidence for mitochondrial targeting's protective effects on the heart (Xiong & Xiao, 2018).

Further development of MITO-DOX faces ongoing challenges to enhance its use as a colon cancer therapeutic agent. A distinctive challenge arises from tumor region heterogeneity in mitochondrial membrane potential, which causes necrotic and hypoxic areas to show reduced uptake of charge-based targeting systems. To overcome this problem, researchers are working on creating hypoxia-activated drugs with mitochondrial delivery components that activate exclusively under low-oxygen conditions typical of advanced tumors (Jiang et al., 2024; Liu et al., 2017).

#### 8.5.1.2 Paclitaxel

Medical professionals have found clinically relevant uses of paclitaxel as a microtubule-stabilizing agent obtained from Pacific yew tree (Taxus brevifolia), particularly in combination treatments for colon cancer. The main functioning process of paclitaxel includes stabilizing microtubules during cell division, leading to cell cycle arrest and apoptotic destruction. Recent research has demonstrated essential roles of paclitaxel in mitochondria, which can be made beneficial by targeted drug delivery methods (Komal et al., 2024).

The mitochondrial mechanisms of paclitaxel treatment involve blocking Bcl-2 family proteins as well as MPTP opening and disturbances of mitochondrial fission and fusion proteins. Standard drug formulations restrict the toxic potential of paclitaxel because the drug fails to accumulate sufficiently within mitochondria, although its mitochondrial mechanisms support the cytotoxic mechanism. Research on paclitaxel delivery systems for targeting mitochondria has shown significant advancements during the past 10 years. Researchers began their studies with TPP conjugation methods, which proved that HT29 colon cancer cells accumulated TPP-paclitaxel conjugates at enhanced rates (5.8-fold) compared to paclitaxel alone. The production of these conjugates forced developers to create complicated formulation plans because the compounds displayed decreased water solubility (Gu et al., 2016; Weaver, 2014).

The combination of mitochondria-directed paclitaxel delivery with metabolic regulator drugs produces synergistic effects, according to studies. When paclitaxel is combined with dichloroacetate (a pyruvate dehydrogenase kinase inhibitor) inside TPP-modified liposomes, the experiment demonstrated increased mitochondrial membrane potential. Targeted nanocarriers absorbed more efficiently into cancer cells because of metabolic changes, resulting in stronger accumulation of nanocarriers and higher cytotoxicity against KRAS-mutated colon cancer xenografts (Chen et al., 2020; Mhaidat et al., 2014).

The research advancement led to the initiation of clinical trials where MPP-linked paclitaxel nanoparticles are being tested on patients who have solid tumors, including CRC, at NCT05498117. Clinical research involving new pharmacodynamic evaluations explores tumor biopsy changes in mitochondrial network structure and mitochondrial apoptosis development to explain the therapeutic effects of targeted mitochondrial paclitaxel compounds.

#### 8.5.1.3 Novel chemotherapeutic agents

New agents that specifically target mitochondria emerged as the next generation of chemotherapeutics after established medicines were discussed previously. The compounds implement mitochondriotropic components directly into their main structure instead of using these ligands for targeting purposes, which leads to enhanced pharmacokinetic properties and better mitochondrial specificity (Wu et al., 2024).

The group of compounds called Mitocans has become a leading type among these purpose-built molecular agents. Research on colon cancer indicates that these compounds demonstrate strong promise by attacking three main mitochondrial components, including the electron transport chain, mitochondrial DNA, and the MPTP components. Neuzil and colleagues dedicated their study to vitamin E analogs that possess mitochondria-targeted properties via their use of mitochondria-targeted tocopheryl succinate (MitoVES). MitoVES shows remarkable selectivity for cancer mitochondria because it combines vitamin E redox regulation with mitochondrial-specific distribution, which exploits the higher membrane potential of cancer cells. The COLO205 xenograft studies revealed that MitoVES inhibited tumor growth substantially while causing minimal side effects to the rest of the body due to its selective mitochondrial accumulation properties (Bakalova et al., 2023; de et al., 2023).

Preclinical studies of CRC have indicated that F16 demonstrates exceptional performance as a rhodacyanine derivative that shows intrinsic mitochondriotropic properties. The compound focuses accumulation inside mitochondria using voltage-dependent membrane transport until it reaches the adenine nucleotide translocator to trigger apoptotic cell death through the permeability transition pore. Research from the Fantin laboratory (2023) proved F16 exhibited substantial anticancer properties against colon cancer organoids, which proved especially potent against KRAS/BRAF wild-type strains. The antitumor properties of F16 maintained effectiveness against tumors from patients whose chemotherapeutic-resistant metastatic lesions showed diminished responses to conventional drugs (Boyle et al., 2018; Zhao et al., 2024).

The field of photodynamic therapy experiences revitalized interest because of its recently developed mitochondrial-targeting capacity. The natural tendency of cationic porphyrin derivatives allows them to accumulate in mitochondria before light activation induces reactive oxygen species production. Researchers at Zheng and colleagues used mitochondria-targeted verteporfin to establish exceptional photodynamic effectiveness in orthotopic colon cancer models under near-infrared light activation through colonoscopy. Research data showed that this method eliminated complete tumors in 7 out of 10 tested animals while maintaining healthy mucosal tissue surrounding the target area, which might find clinical use in stage one cancers or surgical support protocols (Zhou et al., 2024).

Research investigators have identified mitochondrial mRNA translation inhibitors as one of the latest promising treatment methods. The molecules exclusively block mitochondrial ribosomes without affecting cytosolic protein generation. The research from Wang et al. (2024) produced tigecycline derivative compounds that selectively reached colon cancer mitochondria and then disrupted the expression of electron transport chain components that originate from mitochondrial DNA. Research demonstrated that inhibition from this therapy mainly affected cancer cells with high metabolic dependency on OXPHOS, potentially benefiting specific groups of patients.

Clinical research of novel agents is currently in early development, while MitoTams phase I/II trial for metastatic CRC (NCT05421169) stands as the most developed program. The European Society for Medical Oncology (ESMO) 2023 meeting featured initial data showing promising therapeutic effects and acceptable safety outcomes in patients who received prior extensive treatment. The research demonstrated disease control in five out of eight assessable participants, and it showed mainly two safety issues that included tiredness and short-term liver test increases. Scientists found reduced mitochondrial DNA amounts in peripheral blood mononuclear cells through correlative tests, which confirmed the mitochondria-targeted effects in actual human beings.

## 8.5.2 Combination therapies

### *8.5.2.1 Mitochondria-targeted drug plus immunotherapy*

CRC research points toward the most promising treatment advance by combining mitochondrial therapies with immunotherapeutic strategies. The combinatorial technique exploits mitochondrial toxicity as well as its ability to boost immune antitumor responses using various biological pathways (Sahin et al., 2022).

The immunological effects of mitochondrial damage now represent crucial therapeutic outcome factors that scientists increasingly recognize. The destruction of mitochondria leads to multiple damage-associated molecular pattern (DAMP) substances that activate immune detection of tumor cells. Both the cGAS-STING signaling pathway and inflammasome pathways become activated through different mitochondrial structures and compounds, including mitochondrial DNA (mtDNA) with its bacterial heritage as well as unmethylated cytosine-phosphate-guanine (CpG) motifs and cardiolipin. The combination of oxidative stress, along with both endogenous and exogenous danger signals, creates a condition called immunogenic cell death (ICD) that benefits antitumor immunity (Sahin et al., 2022; Wang et al., 2024; Xie & Jiang, 2024).

Research into mitochondrial dynamic intervention methods shows the most encouraging results when implemented with immunotherapy strategies. The cellular processes of fusion and fission in mitochondria function as key regulators of both cancer metabolic transformation and cellular reaction to stress. Cells from colon cancers became more sensitive to immunotherapy when researchers from Martins et al. applied mitochondria-targeted Drp1 inhibitors, which control cellular fission dynamics. The treatment led to better mitochondrial antigen display and elevated major histocompatibility complex class I expression while simultaneously decreasing the cancer cells' production of kynurenine and other immunosuppressive metabolites. The use of mitochondrial fission inhibitor therapy, together with anti-CTLA-4 immunotherapy, resulted in long-term complete response for about 40% of MC38 colon cancer-bearing animals.

## 8.6 Challenges and future perspectives

Major biological, technological, and translational issues still exist despite notable advancements in MTDDSs for treating colon cancer. Transforming mitochondria-targeted approaches into safe and effective therapeutic treatments requires overcoming these obstacles (Martínez-Reyes & Chandel, 2021; Pavlova & Thompson, 2016; Wallace, 2012).

### 8.6.1 Biological barriers and limitations

#### *8.6.1.1 Targeting specificity*

Delivery of drugs selectively to colon cancer cells in the mitochondria still remains a major issue. The majority of ligands that target mitochondria, particularly lipophilic cations, rely on cancer cells' increased capacity to form mitochondrial membranes in order to be retained. This is your sentence in the passive voice (Modica-Napolitano & Aprille, 2001; Murphy & Smith, 2007). Nevertheless, drug uptake by healthy organs, including the liver, kidneys, and heart, can also occur; this feature is not exclusive to tumor cells (Scaduto & Grotyohann, 1999; Vyas et al., 2016). Additionally, tumor heterogeneity and variations in mitochondrial metabolism among colon cancer subtypes result in varied therapy outcomes and further impair targeted precision (Liu et al., 2020; Martínez-Reyes & Chandel, 2021).

#### *8.6.1.2 Off-target effects*

Mitochondria are known to play a key role in energy production, redox regulation, and programmed cell death in normal cells. Systemic toxicity and altered cellular metabolism may be caused by off-target mitochondrial damage (Ott et al., 2007; Weinberg & Chandel, 2009). Elevated levels of chemotherapeutic agents accumulating in tissues may trigger oxidative stress and apoptosis in noncancerous tissues, limiting the effective therapeutic range (Fulda et al., 2010; Vyas et al., 2016). This problem is particularly significant for highly potent mitochondrial toxins and redox-active agents that lack specificity for tumor cells (Smith et al., 2011).

#### 8.6.1.3 Delivery system stability and pharmacokinetics

Keeping mitochondria-targeted nanocarriers stable in the bloodstream is still a major limitation for clinical use. Early drug release detection by the immune system and rapid clearance from body can result in less drug reaching the tumor site, reducing its effectiveness (Nombela et al., 2021; Torchilin, 2014). In addition, designing carriers that successfully reach the tumor site, enter cells, and escape endosomes targeting mitochondria makes formulation and pharmacokinetic optimization extremely complicated (Thomas & Weber, 2019; Varkouhi et al., 2011).

### 8.6.2 Emerging technologies

#### 8.6.2.1 CRISPR-Cas9 and mitochondrial genome editing

Emerging gene-editing methods offer the ability to directly modify mitochondrial pathways involved in the progression of colon cancer and resistance to therapies (Gammage et al., 2018). Traditional CRISPR-Cas9 systems are not very efficient in targeting mitochondrial DNA because guide RNAs cannot be effectively delivered into mitochondria (Taylor & Turnbull, 2005). However, recent developments in RNA-free mitochondrial base editors and protein-only genome-editing technologies offer potential for fixing harmful mitochondrial mutations and controlling mitochondrial activity (Mok et al., 2020).

#### 8.6.2.2 Mitochondrial transplantation

Mitochondrial transplantation is a novel approach that intends to restore mitochondrial function or change how cells use energy (McCully et al., 2017; Spees et al., 2006). While it is still on experimental basis, it may complement mitochondria-targeted therapies by altering tumor metabolism or making cancer cells more responsive to treatment (Islam et al., 2012). However, issues like concordance, suitability, optimized delivery, and large-scale production must be addressed before it can be used in patients (McCully et al., 2017).

### 8.6.3 Future directions

Upcoming research should aim to design delivery systems that are versatile and stimulus-sensitive, enhancing mitochondrial targeting while minimizing adverse effects (Galluzzi et al., 2013; Torchilin, 2014). Combining mitochondria-directed therapies with immune-based therapies, metabolism-modulating agents, and personalized medicine approach could enhance colon cancer treatment (Ferraris et al., 2021; Varkouhi et al., 2011). Developing advanced lab and animal models that reflect mitochondrial diversity and the complexity of the tumor environment will also be important for moving these strategies toward clinical use (Porporato et al., 2018; Wagner et al., 2006; Wallace, 2012).

## 8.7 Conclusion

Mitochondria-based drug delivery systems have demonstrated improved therapeutic outcomes and enhanced delivery efficiency of targeted molecules. Dysregulation of mitochondrial functions, such as redox balance and energy production, makes cells more susceptible to apoptosis and necrosis, ultimately resulting in cell death. Selective targeting of mitochondria offers several advantages over conventional or other organelle-based delivery strategies due to its critical role in tumor progression and resistance.

Modern microscopy-based techniques serve as qualitative and analytical tools to explore mitochondrial targeting in both in vitro and in vivo studies. Combining mitochondrial targeting with modulation of metabolic and immune functions can enhance sensitivity, leading to synergistic and additive therapeutic effects. This approach can overcome resistance to monotherapies and address tumor heterogeneity. Despite these advances, translational barriers must be considered when moving from bench to bedside. Key challenges include quantifying delivery efficiency, minimizing off-target effects, and overcoming biological barriers. Future generations of mitochondria-targeted therapies may incorporate stimuli-responsive, personalized, and precision-based approaches that target the tumor microenvironment more effectively.

In summary, mitochondria-targeted drug delivery represents an innovative strategy for colon cancer therapy. Integrating interdisciplinary approaches—including mitochondrial physiology, nanoscience, and clinical oncology—will be essential to advance these systems toward successful clinical translation.

## References

Ahn, J., Lee, B., Choi, Y., Jin, H., Lim, N. Y., Park, J., Kim, J. H., Bae, J., & Jung, J. H. (2018). Non-peptidic guanidinium-functionalized silica nanoparticles as selective mitochondria-targeting drug nanocarriers. *Journal of Materials Chemistry B, 6*, 5698–5707. https://doi.org/10.1039/C8TB01358F.

Aisu, Y., Oshima, N., Hyodo, F., Elhelaly, A. E., Masuo, A., Okada, T., Hisamori, S., Tsunoda, S., Hida, K., Morimoto, T., Miyoshi, H., Taketo, M. M., Matsuo, M., Neckers, L. M., Sakai, Y., & Obama, K. (2024). Dual inhibition of oxidative phosphorylation and glycolysis exerts a synergistic antitumor effect on colorectal and gastric cancer by creating energy depletion and preventing metabolic switch. *PLoS ONE, 19*, e0309700. https://doi.org/10.1371/journal.pone.0309700.

Al Kamzari, K. A. M., & Constantinou, C. (2025). Navigating the colorectal cancer maze: Unveiling pathways to diagnosis, management, pathophysiology and prevention. *Current Oncology Reports, 27*, 1115–1130. https://doi.org/10.1007/s11912-025-01707-w.

Alrushaid, N., Khan, F. A., Al-Suhaimi, E., & Elaissari, A. (2023). Progress and perspectives in colon cancer pathology, diagnosis, and treatments. *Diseases, 11*, 148. https://doi.org/10.3390/diseases11040148.

Bakalova, R., Lazarova, D., Sumiyoshi, A., Shibata, S., Zhelev, Z., Nikolova, B., Semkova, S., Vlaykova, T., Aoki, I., & Higashi, T. (2023). Redox-cycling "mitocans" as effective new developments in anticancer therapy. *International Journal of Molecular Sciences, 24*, 8435. https://doi.org/10.3390/ijms24098435.

Battogtokh, G., Cho, Y.-Y., Lee, J. Y., Lee, H. S., & Kang, H. C. (2018). Mitochondrial-targeting anticancer agent conjugates and nanocarrier systems for cancer treatment. *Frontiers in Pharmacology, 9*. https://doi.org/10.3389/fphar.2018.00922.

Biswas, S., Dodwadkar, N. S., Deshpande, P. P., & Torchilin, V. P. (2012). Liposomes loaded with paclitaxel and modified with novel *Triphenylphosphonium*-PEG-PE conjugate possess low toxicity, target mitochondria and demonstrate enhanced antitumor effects in vitro and in vivo. *Journal of Controlled Release, 159*, 393–402. https://doi.org/10.1016/j.jconrel.2012.01.009.

Boddapati, S. V., Tongcharoensirikul, P., Hanson, R. N., D'Souza, G. G. M., Torchilin, V. P., & Weissig, V. (2005). Mitochondriotropic liposomes. *Journal of Liposome Research, 15*, 49–58. https://doi.org/10.1081/lpr-64958.

Boyle, K. A., VanWickle, J., Zielonka, J., Cheng, G., Kalyanaraman, B., & Dwinell, M. B. (2018). Abstract 878: Mitochondria-targeted drugs stimulate mitophagy and abrogate colon cancer cell proliferation. *Cancer Research, 78*, 878. https://doi.org/10.1158/1538-7445.AM2018-878.

Boyle, P., & Leon, M. E. (2002). Epidemiology of colorectal cancer. *British Medical Bulletin, 64*, 1–25. https://doi.org/10.1093/bmb/64.1.1.

Buchke, S., Sharma, M., Bora, A., Relekar, M., Bhanu, P., & Kumar, J. (2022). Mitochondria-targeted, nanoparticle-based drug-delivery systems: Therapeutics for mitochondrial disorders. *Life, 12*, 657. https://doi.org/10.3390/life12050657.

Chen, D., Ge, S., Zuo, L., Wang, S., Liu, M., & Li, S. (2020). Adjudin-loaded redox-sensitive paclitaxel-prodrug micelles for overcoming multidrug resistance with efficient targeted Colon cancer therapy. *Drug Delivery, 27*, 1094–1105. https://doi.org/10.1080/10717544.2020.1797245.

Chen, F., Xue, Y., Zhang, W., Zhou, H., Zhou, Z., Chen, T., YinWang, E., Li, H., Ye, Z., Gao, J., & Wang, S. (2024). The role of mitochondria in tumor metastasis and advances in mitochondria-targeted cancer therapy. *Cancer and Metastasis Reviews, 43*, 1419–1443. https://doi.org/10.1007/s10555-024-10211-9.

Chen, R., Niu, M., Hu, X., & He, Y. (2023). Targeting mitochondrial dynamics proteins for the treatment of doxorubicin-induced cardiotoxicity. *Frontiers in Molecular Biosciences, 10*. https://doi.org/10.3389/fmolb.2023.1241225.

Cho, K., Shin, H.-W., Kim, Y.-I., Cho, C.-H., Chun, Y.-S., Kim, T.-Y., & Park, J.-W. (2013). Mad1 mediates hypoxia-induced doxorubicin resistance in colon cancer cells by inhibiting mitochondrial function. *Free Radical Biology & Medicine, 60*, 201–210. https://doi.org/10.1016/j.freeradbiomed.2013.02.022.

Chuah, J.-A., Matsugami, A., Hayashi, F., & Numata, K. (2016). Self-assembled peptide-based system for mitochondrial-targeted gene delivery: Functional and structural insights. *Biomacromolecules, 17*, 3547–3557. https://doi.org/10.1021/acs.biomac.6b01056.

Cong, B., Thakur, T., Uribe, A. H., Stamou, E., Gopinath, S., Maddocks, O., & Cagan, R. (2023). Colon cancer cells evade drug action by enhancing drug metabolism. *Oncogene*. https://doi.org/10.1101/2023.12.21.572817.

Ferraris, C., Rimicci, C., Garelli, S., Ugazio, E., & Battaglia, L. (2021). Nanosystems in cosmetic products: A brief overview of functional, market, regulatory and safety concerns. *Pharmaceutics, 13*, 1408. https://doi.org/10.3390/pharmaceutics13091408.

Fulda, S., Galluzzi, L., & Kroemer, G. (2010). Targeting mitochondria for cancer therapy. *Nature reviews. Drug discovery, 9*, 447–464. https://doi.org/10.1038/nrd3137.

Galluzzi, L., Kepp, O., Heiden, M. G. V., & Kroemer, G. (2013). Metabolic targets for cancer therapy. *Nature Reviews Drug discovery, 12*, 829–846. https://doi.org/10.1038/nrd4145.

Gammage, P. A., Moraes, C. T., & Minczuk, M. (2018). Mitochondrial genome engineering: The revolution may not be CRISPR-Ized. *Trends in Genetics, 34*, 101–110. https://doi.org/10.1016/j.tig.2017.11.001.

Ghosh, P., Vidal, C., Dey, S., & Zhang, L. (2020). Mitochondria targeting as an effective strategy for cancer therapy. *International Journal of Molecular Sciences, 21*, 3363. https://doi.org/10.3390/ijms21093363.

Gu, W., Zou, H., Li, L., Garcia Carcedo, I., Xu, Z. P., & Monteiro, M. (2016). Synergistic inhibition of colon cancer cell growth with nanoemulsion-loaded paclitaxel and PI3K/mTOR dual inhibitor BEZ235 through apoptosis. *International journal of nanomedicine*, 1947. https://doi.org/10.2147/IJN.S100744.

Hammond, W. A., Swaika, A., & Mody, K. (2016). Pharmacologic resistance in colorectal cancer: A review. *Therapeutic Advances in Medical Oncology, 8*, 57–84. https://doi.org/10.1177/1758834015614530.

Horton, K. L., Stewart, K. M., Fonseca, S. B., Guo, Q., & Kelley, S. O. (2008). Mitochondria-penetrating peptides. *Chemistry and Biology, 15*, 375–382. https://doi.org/10.1016/j.chembiol.2008.03.015.

Hu, T., Li, Z., Gao, C.-Y., & Cho, C. H. (2016). Mechanisms of drug resistance in colon cancer and its therapeutic strategies. *World Journal of Gastroenterology, 22*, 6876. https://doi.org/10.3748/wjg.v22.i30.6876.

Hua, S., de Matos, M. B. C., Metselaar, J. M., & Storm, G. (2018). Current trends and challenges in the clinical translation of nanoparticulate nanomedicines: Pathways for translational development and commercialization. *Frontiers in Pharmacology, 9*. https://doi.org/10.3389/fphar.2018.00790.

Islam, M. N., Das, S. R., Emin, M. T., Wei, M., Sun, L., Westphalen, K., Rowlands, D. J., Quadri, S. K., Bhattacharya, S., & Bhattacharya, J. (2012). Mitochondrial transfer from bone-marrow-derived stromal cells to pulmonary alveoli protects against acute lung injury. *Nature Medicine, 18*, 759–765. https://doi.org/10.1038/nm.2736.

Jean, S. R., Ahmed, M., Lei, E. K., Wisnovsky, S. P., & Kelley, S. O. (2016). Peptide-mediated delivery of chemical probes and therapeutics to mitochondria. *Accounts of Chemical Research, 49*, 1893–1902. https://doi.org/10.1021/acs.accounts.6b00277.

Jiang, H., Bao, Q., Yang, T., Yang, M., & Mao, C. (2024). Precision treatment of colon cancer using doxorubicin-loaded metal–organic-framework-coated magnetic nanoparticles. *ACS Applied Materials Interfaces, 16*, 49003–49012. https://doi.org/10.1021/acsami.4c08602.

Johnson, L. V., Walsh, M. L., & Chen, L. B. (1980). Localization of mitochondria in living cells with rhodamine 123. In: *Proceedings of the National Academy of Sciences of the United States of America, 77*, 990–994. https://doi.org/10.1073/pnas.77.2.990.

Kelso, G. F., Maroz, A., Cochemé, H. M., Logan, A., Prime, T. A., Peskin, A. V., Winterbourn, C. C., James, A. M., Ross, M. F., Brooker, S., Porteous, C. M., Anderson, R. F., Murphy, M. P., & Smith, R. A. J. (2012). A mitochondria-targeted macrocyclic Mn(II) superoxide dismutase mimetic. *Chemistry and Biology, 19*, 1237–1246. https://doi.org/10.1016/j.chembiol.2012.08.005.

Komal, B. P., Nanda, L., Singh, R., Bhatia, A., & Singh (2024). Paclitaxel in colon cancer management: From conventional chemotherapy to advanced nanocarrier delivery systems. *Naunyn-Schmiedeberg's Archives of Pharmacology, 397*, 9449–9474. https://doi.org/10.1007/s00210-024-03256-8.

Księżakowska-Łakoma, K., Żyła, M., & R. Wilczyński, J. (2014). Mitochondrial dysfunction in cancer. *Menopausal Review, 2*, 136–144. https://doi.org/10.5114/pm.2014.42717.

Lampidis, T. J., Bernal, S. D., Summerhayes, I. C., & Chen, L. B. (1983). Selective toxicity of rhodamine 123 in carcinoma cells in vitro. *Cancer Research, 43*, 716–720.

Lee, S. H., Lee, J.-S., Lee, W. C., Jeon, M., Lee, W., Jeon, J., Park, B. J., Kim, B., Park, J., Denora, N., Kim, S. E., Im, H.-J., & Lee, B. C. (2025). A three-step mitochondria-targeted photodynamic therapy platform integrating TSPO ligands and pH-responsive liposomes. *Health Nanotechnology, 1*, 8. https://doi.org/10.1186/s44301-025-00009-1.

Li, J., Pan, J., Wang, L., Ji, G., & Dang, Y. (2025). Colorectal cancer: Pathogenesis and targeted therapy. *MedComm (Beijing), 6*. https://doi.org/10.1002/mco2.70127.

Lin, C., Ng, H., Pan, W., Chen, H., Zhang, G., Bian, Z., Lu, A., & Yang, Z. (2015). Exploring different strategies for efficient delivery of colorectal cancer therapy. *International Journal of Molecular Sciences, 16*, 26936–26952. https://doi.org/10.3390/ijms161125995.

Lin, C.-S., Liu, L.-T., Ou, L.-H., Pan, S.-C., Lin, C.-I., & Wei, Y.-H. (2017). Role of mitochondrial function in the invasiveness of human colon cancer cells. *Oncology Reports*. https://doi.org/10.3892/or.2017.6087.

Lin, Y., Liu, J., Bai, R., Shi, J., Zhu, X., Liu, J., Guo, J., Zhang, W., Liu, H., & Liu, Z. (2020). Mitochondria-inspired nanoparticles with microenvironment-adapting capacities for on-demand drug delivery after ischemic injury. *ACS Nano, 14*, 11846–11859. https://doi.org/10.1021/acsnano.0c04727.

Lin, Y., Yang, B., Huang, Y., Zhang, Y., Jiang, Y., Ma, L., & Shen, Y.-Q. (2023). Mitochondrial DNA-targeted therapy: A novel approach to combat cancer. *Cell Insight, 2*, 100113. https://doi.org/10.1016/j.cellin.2023.100113.

Lingappa, A. F., Akintunde, O., Ewald, C., Froehlich, M., Ziari, N., Michon, M., Yu, S. F., Mallesh, S., Lin, J., Kitaygorodskyy, A., Solas, D., Reed, J. C., Lingappa, J. R., Mueller-Schiffmann, A., Korth, C., Prasad, D., Nalca, A., Aston, E., Fabbri, B., ... Lingappa, V. R. (2022). Small molecule protein assembly modulators with pan-cancer therapeutic efficacy. *Open Biology*. https://doi.org/10.1101/2022.09.28.509937.

Liu, G., Wang, M., He, H., & Li, J. (2021). Doxorubicin-loaded tumor-targeting peptide-decorated polypeptide nanoparticles for treating primary orthotopic colon cancer. *Frontiers in Pharmacology, 12*. https://doi.org/10.3389/fphar.2021.744811.

Liu, L., Wang, Y., Wu, J., Liu, J., Qin, Z., & Fan, H. (2020). N6-Methyladenosine: A potential breakthrough for human cancer. *Molecular Therapy Nucleic Acids, 19*, 804–813. https://doi.org/10.1016/j.omtn.2019.12.013.

Liu, Y., Zhang, X., Zhou, M., Nan, X., Chen, X., & Zhang, X. (2017). Mitochondrial-targeting lonidamine-doxorubicin nanoparticles for synergistic chemotherapy to conquer drug resistance. *ACS Appl. Mater. Interfaces, 9*, 43498–43507. https://doi.org/10.1021/acsami.7b14577.

Liu, Y., Wang, H., Zhang, S., Peng, N., Hai, S., Zhao, H., Liu, J., & Liu, W. (2025). The role of mitochondrial biogenesis, mitochondrial dynamics and mitophagy in gastrointestinal tumors. *Cancer Cell International, 25*, 46. https://doi.org/10.1186/s12935-025-03685-2.

Luo, Y., Ma, J., & Lu, W. (2020). The significance of mitochondrial dysfunction in cancer. *International Journal of Molecular Sciences, 21*, 5598. https://doi.org/10.3390/ijms21165598.

Ma, N., Song, A., Li, Z., & Luan, Y. (2019). Redox-sensitive prodrug molecules meet graphene oxide: An efficient graphene oxide-based nanovehicle toward cancer therapy. *ACS Biomaterials Science & Engineering, 5*, 1384–1391. https://doi.org/10.1021/acsbiomaterials.9b00114.

Martínez-Reyes, I., & Chandel, N. S. (2021). Cancer metabolism: Looking forward. *Nature Reviews. Cancer, 21*, 669–680. https://doi.org/10.1038/s41568-021-00378-6.

McCully, J. D., Cowan, D. B., Emani, S. M., & Del Nido, P. J. (2017). Mitochondrial transplantation: From animal models to clinical use in humans. *Mitochondrion, 34*, 127–134. https://doi.org/10.1016/j.mito.2017.03.004.

Mhaidat, N. M., Alzoubi, K. H., Al-azzam, S. I., & Alsaad, A. A. (2014). Caffeine inhibits paclitaxel-induced apoptosis in colorectal cancer cells through the upregulation of Mcl-1 levels. *Molecular Medicine Reports, 9*, 243–248. https://doi.org/10.3892/mmr.2013.1763.

Modica-Napolitano, J. S., & Aprille, J. R. (2001). Delocalized lipophilic cations selectively target the mitochondria of carcinoma cells. *Advanced Drug Delivery Reviews, 49*, 63–70. https://doi.org/10.1016/s0169-409x(01)00125-9.

Mok, B. Y., de Moraes, M. H., Zeng, J., Bosch, D. E., Kotrys, A. V., Raguram, A., Hsu, F., Radey, M. C., Peterson, S. B., Mootha, V. K., Mougous, J. D., & Liu, D. R. (2020). A bacterial cytidine deaminase toxin enables CRISPR-free mitochondrial base editing. *Nature, 583*, 631–637. https://doi.org/10.1038/s41586-020-2477-4.

Murphy, M. P., & Smith, R. A. J. (2007). Targeting antioxidants to mitochondria by conjugation to lipophilic cations. *Annual Review of Pharmacology and Toxicology, 47*, 629–656. https://doi.org/10.1146/annurev.pharmtox.47.120505.105110.

Neitzel, C., Demuth, P., Wittmann, S., & Fahrer, J. (2020). Targeting altered energy metabolism in colorectal cancer: Oncogenic reprogramming, the central role of the TCA Cycle and therapeutic opportunities. *Cancers (Basel), 12*, 1731. https://doi.org/10.3390/cancers12071731.

Nombela, P., Miguel-López, B., & Blanco, S. (2021). The role of m6A, m5C and Ψ RNA modifications in cancer: Novel therapeutic opportunities. *Molecular Cancer, 20*, 18. https://doi.org/10.1186/s12943-020-01263-w.

Obeid, M. A., Al Qaraghuli, M. M., Alsaadi, M., Alzahrani, A. R., Niwasabutra, K., & Ferro, V. A. (2017). Delivering natural products and biotherapeutics to improve drug efficacy. *Therapeutic Delivery, 8*, 947–956. https://doi.org/10.4155/tde-2017-0060.

Ott, M., Gogvadze, V., Orrenius, S., & Zhivotovsky, B. (2007). Mitochondria, oxidative stress and cell death. *Apoptosis, 12*, 913–922. https://doi.org/10.1007/s10495-007-0756-2.

Padder, R. A., Bhat, Z. I., Ahmad, Z., Singh, N., & Husain, M. (2021). DRP1 Promotes BRAFV600E-driven tumor progression and metabolic reprogramming in colorectal cancer. *Frontiers in Oncology, 10*. https://doi.org/10.3389/fonc.2020.592130.

Pang, B., & Wu, H. (2025). Metabolic reprogramming in colorectal cancer: A review of aerobic glycolysis and its therapeutic implications for targeted treatment strategies. *Cell Death Discovery, 11*, 321. https://doi.org/10.1038/s41420-025-02623-5.

Pathak, R. K., Kolishetti, N., & Dhar, S. (2015). Targeted nanoparticles in mitochondrial medicine. *Wiley Interdisciplinary Reviews: Nanomedicine and Nanobiotechnology, 7*, 315–329. https://doi.org/10.1002/wnan.1305.

Pavlova, N. N., & Thompson, C. B. (2016). The emerging hallmarks of cancer metabolism. *Cell Metabolism, 23*, 27–47. https://doi.org/10.1016/j.cmet.2015.12.006.

Porporato, P. E., Filigheddu, N., Pedro, J. M. B.-S., Kroemer, G., & Galluzzi, L. (2018). Mitochondrial metabolism and cancer. *Cell Research, 28*, 265–280. https://doi.org/10.1038/cr.2017.155.

Qiu, X., Wang, A., Wang, J., Zhang, Z., & Tao, L. (2025). Mitochondrial metabolic reprogramming in colorectal cancer: Mechanisms of resistance and future clinical interventions. *Cell Death Discovery, 11*, 375. https://doi.org/10.1038/s41420-025-02670-y.

Ross, M. F., Da Ros, T., Blaikie, F. H., Prime, T. A., Porteous, C. M., Severina, I. I., Skulachev, V. P., Kjaergaard, H. G., Smith, R. A. J., & Murphy, M. P. (2006). Accumulation of lipophilic dications by mitochondria and cells. *The Biochemical Journal, 400*, 199–208. https://doi.org/10.1042/BJ20060919.

Sahin, I. H., Ciombor, K. K., Diaz, L. A., Yu, J., & Kim, R. (2022). Immunotherapy for microsatellite stable colorectal cancers: challenges and novel therapeutic avenues. *American Society of Clinical Oncology Educational Book,* 242–253. https://doi.org/10.1200/EDBK_349811.

Sanchez-Aranguren, L., Al Tahan, M. A., Uppal, M., Juvale, P., & Marwah, M. K. (2025). Mitochondrial-targeted liposome-based drug delivery - therapeutic potential and challenges. *Journal of Drug Targeting, 33*, 575–586. https://doi.org/10.1080/1061186X.2024.2437440.

Scaduto, R. C., & Grotyohann, L. W. (1999). Measurement of mitochondrial membrane potential using fluorescent rhodamine derivatives. *Biophysical Journal, 76*, 469–477. https://doi.org/10.1016/S0006-3495(99)77214-0.

Sena Ozbay, H., Yabanoglu-Ciftci, S., Baysal, I., Gultekinoglu, M., Can Eylem, C., Ulubayram, K., Nemutlu, E., Topaloglu, R., & Ozaltin, F. (2022). Mitochondria-targeted CoQ10 loaded PLGA-b-PEG-TPP nanoparticles: Their effects on mitochondrial functions of COQ8B-/- HK-2 cells. *European Journal of Pharmaceutics and Biopharmaceutics, 173*, 22–33. https://doi.org/10.1016/j.ejpb.2022.02.018.

Shaham, S. H., Vij, P., & Tripathi, M. K. (2025). Advances in targeted and chemotherapeutic strategies for Colorectal cancer: Current insights and future directions. *Biomedicines, 13*, 642. https://doi.org/10.3390/biomedicines13030642.

Sibrian-Vazquez, M., Nesterova, I. V., Jensen, T. J., & Vicente, M. G. H. (2008). Mitochondria targeting by guanidine– and biguanidine–porphyrin photosensitizers. *Bioconjugate Chemistry, 19*, 705–713. https://doi.org/10.1021/bc700393u.

Singh, D. (2024). A sojourn on mitochondria targeted drug delivery systems for cancer: Strategies, clinical and future prospects. *Mitochondrion, 74*, 101826. https://doi.org/10.1016/j.mito.2023.101826.

Smith, R. A., Porteous, C. M., Coulter, C. V., & Murphy, M. P. (1999). Selective targeting of an antioxidant to mitochondria. *European Journal of Biochemistry, 263*, 709–716. https://doi.org/10.1046/j.1432-1327.1999.00543.x.

Smith, R. A. J., Hartley, R. C., & Murphy, M. P. (2011). Mitochondria-targeted small molecule therapeutics and probes. *Antioxidants & Redox Signaling, 15*, 3021–3038. https://doi.org/10.1089/ars.2011.3969.

Spees, J. L., Olson, S. D., Whitney, M. J., & Prockop, D. J. (2006). Mitochondrial transfer between cells can rescue aerobic respiration. In: *Proceedings of the National Academy of Sciences of the United States of America, 103*, 1283–1288. https://doi.org/10.1073/pnas.0510511103.

Sun, X., Zhan, L., Chen, Y., Wang, G., He, L., Wang, Q., Zhou, F., Yang, F., Wu, J., Wu, Y., Xing, J., He, X., & Huang, Q. (2018). Increased mtDNA copy number promotes cancer progression by enhancing mitochondrial oxidative phosphorylation in microsatellite-stable colorectal cancer. *Signal Transduction and Targeted Therapy, 3*, 8. https://doi.org/10.1038/s41392-018-0011-z.

Szeto, H. H. (2006). Cell-permeable, mitochondrial-targeted, peptide antioxidants. *The AAPS Journal, 8*, E277–E283. https://doi.org/10.1007/BF02854898.

Szeto, H. H. (2008). Mitochondria-targeted cytoprotective peptides for ischemia-reperfusion injury. *Antioxidants & Redox Signaling, 10*, 601–619. https://doi.org/10.1089/ars.2007.1892.

Taylor, R. W., & Turnbull, D. M. (2005). Mitochondrial DNA mutations in human disease. *Nature Reviews Genetics, 6*, 389–402. https://doi.org/10.1038/nrg1606.

Thomas, O. S., & Weber, W. (2019). Overcoming physiological barriers to nanoparticle delivery—Are we there yet? *Frontiers in Bioengineering and Biotechnology, 7*. https://doi.org/10.3389/fbioe.2019.00415.

Torchilin, V. P. (2014). Multifunctional, stimuli-sensitive nanoparticulate systems for drug delivery. *Nature reviews. Drug discovery, 13*, 813–827. https://doi.org/10.1038/nrd4333.

Varkouhi, A. K., Scholte, M., Storm, G., & Haisma, H. J. (2011). Endosomal escape pathways for delivery of biologicals. *Journal of Controlled Release, 151*, 220–228. https://doi.org/10.1016/j.jconrel.2010.11.004.

Vyas, S., Zaganjor, E., & Haigis, M. C. (2016). Mitochondria and cancer. *Cell, 166*, 555–566. https://doi.org/10.1016/j.cell.2016.07.002.

Wagner, V., Dullaart, A., Bock, A.-K., & Zweck, A. (2006). The emerging nanomedicine landscape. *Nature Biotechnology, 24*, 1211–1217. https://doi.org/10.1038/nbt1006-1211.

Wallace, D. C. (2012). Mitochondria and cancer. *Nature ReviewsCancer, 12*, 685–698. https://doi.org/10.1038/nrc3365.

Wang, M., Xue, L., Fei, Z., Luo, L., Zhang, K., Gao, Y., Liu, X., & Liu, C. (2024). Characterization of mitochondrial metabolism related molecular subtypes and immune infiltration in colorectal adenocarcinoma. *Scientific Reports, 14*, 24326. https://doi.org/10.1038/s41598-024-75482-2.

Weaver, B. A. (2014). How Taxol/paclitaxel kills cancer cells. *Molecular Biology of the Cell, 25*, 2677–2681. https://doi.org/10.1091/mbc.e14-04-0916.

Weinberg, F., & Chandel, N. S. (2009). Mitochondrial metabolism and cancer. *Annals of the New York Academy of Sciences, 1177*, 66–73. https://doi.org/10.1111/j.1749-6632.2009.05039.x.

Wen, H., Jung, H., & Li, X. (2015). Drug Delivery Approaches In Addressing Clinical Pharmacology-related Issues: Opportunities and challenges. *The AAPS Journal, 17*, 1327–1340. https://doi.org/10.1208/s12248-015-9814-9.

Wu, Z., Xiao, C., Long, J., Huang, W., You, F., & Li, X. (2024). Mitochondrial dynamics and colorectal cancer biology: Mechanisms and potential targets. *Cell Communication and Signaling, 22*, 91. https://doi.org/10.1186/s12964-024-01490-4.

Xie, Y., & Jiang, H. (2024). The exploration of mitochondrial-related features helps to reveal the prognosis and immunotherapy methods of colorectal cancer. *Cancer Reports, 7*. https://doi.org/10.1002/cnr2.1914.

Xie, Y.-H., Chen, Y.-X., & Fang, J.-Y. (2020). Comprehensive review of targeted therapy for colorectal cancer. *Signal Transduction and Targeted Therapy, 5*, 22. https://doi.org/10.1038/s41392-020-0116-z.

Xing, T., Xie, Y., Abel, P., & Tu, Y. (2023). Aberrantly upregulated Drp1 induces mitochondrial fission and promotes colorectal cancer cell migration. *The Journal of Pharmacology and Experimental Therapeutics, 385*, 36. https://doi.org/10.1124/jpet.122.522350.

Xiong, S., & Xiao, G. (2018). Reverting doxorubicin resistance in colon cancer by targeting a key signaling protein, steroid receptor coactivator. *Experimental and Therapeutic Medicine*. https://doi.org/10.3892/etm.2018.5912.

Xu, Z., Xie, Y., Chen, W., & Deng, W. (2025). Nanocarrier-based systems for targeted delivery: Current challenges and future directions. *MedComm (Beijing), 6*, e70337. https://doi.org/10.1002/mco2.70337.

Younis, M. A., Tawfeek, H. M., Abdellatif, A. A. H., Abdel-Aleem, J. A., & Harashima, H. (2022). Clinical translation of nanomedicines: Challenges, opportunities, and keys. *Advanced Drug Delivery Reviews, 181*, 114083. https://doi.org/10.1016/j.addr.2021.114083.

Yousif, L. F., Stewart, K. M., Horton, K. L., & Kelley, S. O. (2009). Mitochondria-penetrating peptides: Sequence effects and model cargo transport. *Chembiochem, 10*, 2081–2088. https://doi.org/10.1002/cbic.200900017.

Zeng, Z., Fang, C., Zhang, Y., Chen, C.-X., Zhang, Y.-F., & Zhang, K. (2021). Mitochondria-targeted nanocarriers promote highly efficient cancer therapy: A review. *Frontiers in Bioengineering and Biotechnology, 9*. https://doi.org/10.3389/fbioe.2021.784602.

Zhao, C., Feng, Q., Dou, Z., Yuan, W., Sui, C., Zhang, X., Xia, G., Sun, H., & Ma, J. (2013). Local targeted therapy of liver metastasis from colon cancer by galactosylated liposome encapsulated with doxorubicin. *PLoS ONE, 8*, e73860. https://doi.org/10.1371/journal.pone.0073860.

Zhao, Y., Guo, X., Zhang, L., Wang, D., & Li, Y. (2024). Mitochondria: A crucial factor in the progression and drug resistance of colorectal cancer. *Frontiers in Immunology, 15*. https://doi.org/10.3389/fimmu.2024.1512469.

Zhao, Z., Mei, Y., Wang, Z., & He, W. (2022). The effect of oxidative phosphorylation on cancer drug resistance. *Cancers (Basel), 15*, 62. https://doi.org/10.3390/cancers15010062.

Zheng, Y., Ji, X., Yu, B., Ji, K., Gallo, D., Csizmadia, E., Zhu, M., Choudhury, M. R., De La Cruz, L. K. C., Chittavong, V., Pan, Z., Yuan, Z., Otterbein, L. E., & Wang, B. (2018). Enrichment-triggered prodrug activation demonstrated through mitochondria-targeted delivery of doxorubicin and carbon monoxide. *Nature Chemistry, 10*, 787–794. https://doi.org/10.1038/s41557-018-0055-2.

Zhou, Y., Zhang, W., Wang, B., Wang, P., Li, D., Cao, T., Zhang, D., Han, H., Bai, M., Wang, X., Zhao, X., & Lu, Y. (2024). Mitochondria-targeted photodynamic therapy triggers GSDME-mediated pyroptosis and sensitizes anti-PD-1 therapy in colorectal cancer. *Journal for Immunotherapy of Cancer, 12*, e008054. https://doi.org/10.1136/jitc-2023-008054.

Zielonka, J., Joseph, J., Sikora, A., Hardy, M., Ouari, O., Vasquez-Vivar, J., Cheng, G., Lopez, M., & Kalyanaraman, B. (2017). Mitochondria-targeted triphenylphosphonium-based compounds: Syntheses, mechanisms of action, and therapeutic and diagnostic applications. *Chemical Reviews, 117*, 10043–10120. https://doi.org/10.1021/acs.chemrev.7b00042.

de, M., Rainho, A., Siqueira, P. B., de Amorim, Í. S. S., Mencalha, A. L., & Thole, A. A. (2023). Mitochondria in colorectal cancer stem cells - A target in drug resistance. *Cancer Drug Resistance, 6*, 273–283. https://doi.org/10.20517/cdr.2022.116.

CHAPTER

# 9 Revolutionizing colon cancer care: advances in carbon-based materials as theranostic agents

**Rahul Kumar[1], Molakpogu Ravindra Babu[2], Sakshi Patel[1], Arya Tripathi[2], Shaik Abdul Rahaman[2] and Manoj Gadewar[3]**

[1]*Faculty of Pharmacy, Vidya University, Meerut, Uttar Pradesh, India,*
[2]*Department of Pharmacy, School of Medical and Allied Sciences, Galgotias University, Greater Noida, Uttar Pradesh, India,*
[3]*Department of Pharmacology, JSS University, Greater Noida, Uttar Pradesh, India*

## 9.1 Introduction

Colon cancer is the most common and the deadliest cancer in the world, and one of the leading causes of cancer mortality worldwide. Despite improvements in screening programs and treatment protocols, the disease continues to impose an enormous health burden due to late detection, treatment heterogeneity, and resistance to standard treatments (Gallaher et al., 2018). It was projected that the United States will see around 1,918,030 new cancer cases and 609,360 fatalities resulting from this disease (Kuczynski & Gotay, 2024). The Global Cancer Observatory estimates over 30 million cancer-related fatalities annually by 2030, resulting in a significant financial strain on families and communities (Machida et al., 2019). It is essential to prioritize cancer prevention, screening, diagnosis, and treatment. Standard management of colon cancer, including surgical resection, chemotherapy, and radiation therapy, has limitations, including systemic toxicity, recurrence, and specificity to identified cancerous cells. An important step in any effective colon cancer care is early detection (Arruebo et al., 2011). Most symptoms emerge only in final stages, which makes treatment far less effective. Currently, there is a plethora of diagnostic tools utilized, which include colonoscopy, histopathology, and imaging modalities that have gained useful insights but are associated with drawbacks of invasiveness, high cost, and limited sensitivity in identifying microscopic lesions (Cheng et al., 2021). Moreover, the selection of drug-resistant cancer cells necessitates the generation of new therapeutic strategies that can be used to selectively kill malignant cells with less damage to healthy tissues.

To address these challenges, the approaches that combine precision medicine with latest technology need to be taken. In this context, carbon-based theranostic agents based on nanotechnology have emerged as a game changer in oncology with the power of combination diagnostic and therapeutic in one delivery system (Zhang et al., 2022).

### 9.1.1 Role of nanotechnology in cancer care

The goals of nanomedicine and biomedical research on the nanoscale have been realized through the nanotechnology revolution that has resulted in the capacity to control materials at the nanoscale,

Nano Drug Delivery Systems for Colon Cancer. DOI: https://doi.org/10.1016/B978-0-443-40443-6.00004-0

thereby improving their physicochemical properties for medical applications (Grodzinski et al., 2019). Nanoparticles functioning as multifunctional platforms to accommodate various strategies of imaging, targeted drug delivery, and therapy for cancer care represent exciting, potentially more precise, and less invasive methods for treatment of cancer (Sharma et al., 2021).

Systemic administration of drugs in chemotherapy results in toxicity to normal tissues and suboptimal drug concentrations at the tumor site (Hartshorn et al., 2018). This limitation is overcome by surface modifications using nanotechnology-based drug delivery systems that selectively lead to drug accumulation in cancer cells through the use of enhanced permeability and retention (EPR) effect (Kemp & Kwon, 2021). In addition, nanoparticles can be engineered to respond to particular tumor microenvironment events, pH, enzymes, or redox potential to deliver drugs in a controlled manner and increase therapy outcomes.

Nanotechnology has great potential in diagnostics to enhance the imaging techniques by improving contrast, resolution, and sensitivity in imaging the cancerous tissues (Goldberg, 2019). Conventionally, early detection, real-time tracking, and personalized treatment are facilitated by attaching biomarkers or contrast agents to nanoparticles (Jin et al., 2020). Thus, oncologists have the ability to monitor treatment response dynamically and adjust strategies in theranostic nanoplatforms, materials that combine diagnosis and therapy into a single system with the advent (Dessale et al., 2022). Several nanomaterials have been studied extensively, but among them, carbon-based nanoparticles possess the most important attributes, such as the unique structural and functional properties, biocompatibility, and the application capabilities, including diagnostic and therapeutic purposes (Taddie & Dessie, 2019).

### 9.1.2 Overview of carbon-based theranostic materials

Carbon-based nanomaterials such as graphene, fullerenes, carbon nanotubes (CNTs), and carbon quantum dots (CQDs) have now emerged as promising chancers for cancer theranostics (Hosseini et al., 2023). Exceptionally strong, with a high surface area, optically tunable, yet highly electrically and thermally conductive, such materials are ideal for imaging and therapy materials.

1. **Graphene and its derivatives:** Graphene-based materials have ultrathin structure, great drug-loading capacity, and superior photothermal properties, which could potentially be applied in targeted drug delivery and photothermal therapy (PTT) (Al-Ani et al., 2020). These results also confirm that functionalized graphene oxide (GO) is a promising platform for enhancing the bioavailability of chemotherapeutic agents through minimization of off-target toxicity.
2. **CNTs:** CNTs are cylindrical nanostructures that possess excellent conductivity and mechanical strength and are thus good candidates for drug delivery, biosensing, and even gene therapy purposes (Abreu et al., 2023). The efficiency of penetration of cell membranes allows for the release of drugs in the cytoplasm and enhances treatment efficacy.
3. **Fullerene (cage heads):** Cage-like carbon molecules with antioxidant properties; CQDs: Fluorescent carbon molecules that have practical applications for imaging. These materials have been analyzed for photodynamic therapy (PDT), in which they generate reactive oxygen species (ROS) to selectively kill cancer cells (Siringan et al., 2023).

There is potential for these carbon-based materials to transform colon cancer care by integrating them into theranostic technologies. This also makes them the next generation of tools in the hands of the oncologists to implement real-time and personalized treatment strategies with high-resolution imaging (Ma et al., 2023). It is becoming increasingly evident, as the research continues, regarding the potential of carbon-based theranostic agents for colon cancer. Further parts of this chapter delve into the mechanisms, applications, challenges, and perspectives for these state-of-the-art nanomaterials in oncological care (Sajjadi et al., 2021).

## 9.2 Materials derived from carbon for cancer diagnostics

The combination of nanotechnology in oncology has opened new frontiers in early cancer diagnosis and effective treatment. Carbon-based materials are among the various nanomaterials studied owing to their special structural, electrical, optical, and biochemical properties (Nayak et al., 2014). These materials exhibit high surface area for drug loading, tunable functionalities, and enhanced biocompatibility, and are excellent candidates for developing the cancer theranostics, a hybrid therapy and diagnosis platform in a single framework (Albert & Hsu, 2016). In this section (Table 9.1, the most promising carbon-based materials are graphene, CNT, fullerenes, and carbon nanonetworks, and their role as a revolutionizing material in colon cancer treatment is explored.

### 9.2.1 Graphene and its derivatives: the multitalented nanomaterial

Graphene, when stripped of its outer shell, has emerged as one of the most adaptable nanomaterials in cancer research (Mura et al., 2022). As an excellent candidate for various biomedical applications, including drug delivery, imaging, and PTT, its ultrathin structure, extremely high mechanical strength, and extremely high electrical conductivity are extremely attractive (Luo et al., 2016). Chemotherapy drugs such as doxorubicin and paclitaxel can be very effectively carried by functionalized graphene, which can deliver them to the tumor site in a controlled, targeted, and minimally toxic way. In addition, graphene's capability to absorb near-infrared (NIR) light can be utilized for PTT, whereby the absorbed energy becomes heat to selectively destroy cancer cells without affecting the healthy tissues (Sable et al., 2024).

In addition to drug delivery and therapy, graphene is essential to the field of gene therapy as a means to deliver small interfering RNA (siRNA) and microRNA (miRNA) and control cancer-related genes, inhibiting tumor growth. Both the medical and diagnostic applications of graphene are equally promising, especially in their application for medical imaging and biosensing (Wang et al., 2024). Such graphene derivatives can be used as highly effective contrast agents for magnetic resonance imaging (MRI) and fluorescence imaging to improve sensitivity and resolution for early cancer detection. In addition, graphene-based biosensors have the capability of detecting particular cancer biomarkers in the blood to aid in early, noninvasive colon cancer diagnosis (Li et al., 2022). However, despite all these advantages, toxicity and biodegradability still remain clinical translation barriers. Nevertheless, strategy of functionalization, such as PEGylation (PEGylation, i.e., attachment of polyethylene glycol), has been developed to improve the biocompatibility of graphene and would facilitate its safe clearance from the body for its use in cancer theranostics in future (Nicolle et al., 2021).

**Table 9.1 Chart outlining essential carbon-based materials used in theranostic procedures.**

| Material | Structure | Imaging modality | Therapeutic potential | Challenges |
|---|---|---|---|---|
| Graphene | 2D honeycomb | Magnetic resonance imaging (MRI), fluorescence | Drug delivery, phototherapy | Biocompatibility, stability |
| Carbon nanotube | Cylindrical tubes | Computed tomography, Raman spectroscopy | Gene therapy, photothermal therapy | Aggregation, toxicity |
| Fullerenes | Spherical | Positron emission tomography, optical imaging | Antioxidants, drug carriers | Functionalization challenges |
| Carbon dots | Spherical | Fluorescence, MRI | Bioimaging, drug delivery | Quantum yield variations |

### 9.2.2 Fullerenes and carbon quantum dots: small but mighty cancer fighters

Cage-like carbon molecules that are fullerenes (C60, C70, etc.) have unique electronic, photophysical, and antioxidant properties and are therefore very valuable in cancer theranostics. PDT with PDT agents is well known to be a promising approach for destroying cancer cells, as the PDT agents have the ability to generate ROS upon light activation (Gaur et al., 2021). Second, the hollow structure allows for drug encapsulation and protection from degradation of chemotherapy agents and delivery to target sites for further enhancing treatment efficacy (Siva et al., 2024). Like fullerenes, the fullerenes also work as radical scavengers to neutralize free radicals and prevent oxidative stress to healthy cells before being damaged by chemotherapy. Nevertheless, their low water solubility prevents bioavailability, thus functionalization is necessary in order to enhance the dispersibility and the biological interaction (Ramezani Farani et al., 2024).

As are the CQDs, nanoscale carbon particles (<10 nm), which have shown excellent biocompatibility and photostability. Real-time tumor imaging with bright fluorescence is provided by CQDs as substitutes for conventional quantum dots (QDs) and contains no toxicity (Dhas et al., 2022). This allows for targeted drug delivery, where drugs of interest and biomolecules are conjugated with them, enabling the selective cancer treatment with the minimal toxicity (Mahor et al., 2021). Additionally, CQDs play an important role in biosensing as they may be used to discriminate cancer biomarkers with high specificity and sensitivity, serving the purpose of early colon cancer detection. Because of their minimal toxicity and tunable optical properties, CQDs promise to be a next-generation tool for imaging and therapy of colon cancer, and add to the list of carbon-based nanomaterials with the potential to expand the use of such nanomaterials in precision oncology (Bayda, 2017).

### 9.2.3 Carbon nanotubes: the needle-like warriors against cancer

CNTs have excellent mechanical strength, good electrical conductivity, and high surface-to-volume ratio, and they are extremely effective for the delivery of drugs, imaging, and hyperthermia treatment processes (Komane et al., 2023). Two main forms of CNTs are: single-walled CNTs (SWCNTs), built from just one rolled carbon layer in the form of a molecule, and multi-walled CNTs (MWCNTs), composed of more than one concentric nanotube layer. This allows them to act as smart drug carriers, effectively penetrating into the cancer cells to perform as "nanoneedles" that will deliver anticancer drugs with high precision (Malarkey et al., 2009). CNTs show a hollow architecture with a great drug-loading capacity and controllable release, helping treatment efficacy while simultaneously decreasing systemic toxicity. Moreover, CNTs possess superior photothermal and photodynamic properties in absorbing NIR radiation and inducing local heat to kill cancer cells in PTT, as well as producing ROS in PDT (Xiang et al., 2020) for inducing cancer cell apoptosis. In addition, CNTs are effective radiotherapy sensitizers as they help increase oxidative stress in tumor cells, rendering them more susceptible to radiation-induced damage (Singhai et al., 2020).

Furthermore, CNTs have a large utility in addition to their therapeutic applications in the realm of cancer diagnostics. Functionalization of CNTs improves fluorescence and Raman imaging, and increases imaging sensitivity as well as real-time tumor tracking (Heller et al., 2020). In addition, electrochemical biosensors based on CNTs offer a minimally invasive diagnostic tool for detecting circulating tumor cells (CTCs) or tumor circulating markers (TCMs) released from the tumor in blood samples for early cancer detection (Comparetti et al., 2020). However, CNTs have enormous potential and face challenges for clinical application due to lack of information on biodegradability, their accumulation in organs in the body and their toxicity (Prajapati et al., 2019). Nevertheless, CNTs are being advanced for their biocompatibility, safety, and clearance through means that enhance their compatibility for use in next-generation cancer theranostics, such as PEGylation and peptide conjugation.

### 9.2.4 Hybrid carbon nanocomposites: the future of cancer theranostics

The hybrid carbon nanocomposites (graphene-CNT hybrids and CQDs functionalized fullerenes) provide a synergistic approach to improving theranostic performance for combating cancer (Kashyap et al., 2023). These composites utilize the unique properties of each of the constituents, including their special combination of large surface area, electrical conductivity, or fluorescence provided by graphene, CNTs, or CQDs, respectively, to enhance the performance of superior multifunctional applications (Rahman et al., 2015). Due to their high (regulated) drug delivery, such as higher drug-loading capacity, controlled release, and better cellular uptake, they are excellent carriers for chemotherapy. Moreover, hybrid nanocomposites merge the benefits of actual two-modal imaging and therapy into a solitary platform due to their ability to incorporate multiple features, that is, perfuse contrast enhancement, photothermal effects, and biosensing of a solitary program, advancing both research and treatment accuracy (Andoh et al., 2024).

But these advanced materials also possess great potential for emerging applications in colon cancer (Tang et al., 2022). The fluorescence governed by graphene-CQD composites has been used to guide surgery and diagnosis of tumors, allowing it to be resected with significantly more accuracy than simply relying on visual observations alone (Madni et al., 2018). Based on the enhanced ROS generation of CNT-fullerene hybrids, PDT efficiency can be improved. Likewise, graphene-CNT drug carriers also lead to extended circulating time, thus minimizing systemic toxicity while still allowing for targeted chemotherapy delivery. Consequently, hybrid carbon nanocomposites serve at the dawn of nanomedicine in developing personalized, precision-based colon cancer treatment (Sargazi et al., 2022).

## 9.3 Diagnostic applications of carbon-based materials

Carbon nanomaterials that are carbon-based revolutionized cancer diagnostics by improving imaging methods, earlier disease detection, and the accuracy of biomarker identification (Table 9.1) (Beniwal et al., 2023). Based on their unique physicochemical properties, that is, extremely high surface area, very high conductivity, and tunable optical properties, CNTs are excellent candidates for high-end diagnostic applications (Raval et al., 2023).

### 9.3.1 Enhanced imaging techniques (MRI, CT, PET, optical imaging)

Carbon-based nanomaterials, such as graphene, CNT, and CQDs, have greatly benefited the traditional imaging modalities, that is, MRI, computed tomography (CT), positron emission tomography (PET), and optical imaging (Darrigues et al., 2017). Due to their capability to enhance signal intensity and enable tumor visualization, functionalized GO and CNTs are very efficient contrast agents in MRI (Kekelidze et al., 2013). Carbon-based nanomaterials functionalized with metal nanoparticles (e.g., gold or gadolinium) enable excellent, high-resolution imaging while maintaining high biocompatibility in CT and PET imaging (Zhou et al., 2022). Another advantage of CQDs is the observation that they possess strong fluorescence and good photostability, which makes CQDs very suitable for optical imaging and thus for tracking tumors in real time along with guiding surgical procedures with strong precision.

### 9.3.2 Biosensors for early detection

Since time of response, sensitivity, and capacity to detect low concentrations of disease biomarkers (as low as a few molecules in the target cell), carbon-based biosensors are highly sensitive and powerful tools for early cancer detection (Zhang et al., 2023). For example, graphene-based electrochemical biosensors can report real time, noninvasive diagnostic capabilities in which minute electrical signal changes can be observed upon interaction with cancer-specific biomolecule (Quinchia et al., 2020). Similarly, CNT-based biosensors improve signal amplification to increase accuracy of detection of CTCs and genetic mutations linked to colon cancer. The great thing about these advanced biosensing platforms is that they allow extremely early intervention that dramatically increases the patient survival (Narayan et al., 2019).

### 9.3.3 Targeted biomarker detection

The identification of specific biomarkers that serve as indications of disease progression is one of the most important features of cancer diagnostics (Kamel et al., 2022). Selective binding with cancer proteins, nucleic acids, and metabolic byproducts is enhanced by carbon-based nanomaterials. Examples of such functionalized fullerenes and CQDs include those that can be conjugated to antibodies or aptamers to selectively recognize colon cancer biomarkers of carcinoembryonic antigen (CEA) or KRAS mutations (Gray et al., 2021). Graphene-based nanoprobes also increase the sensitivity of liquid biopsy techniques for the detection of circulating DNA and RNA fragments associated with cancer, as well as cancer cells in the bloodstream (Chuang et al., 2022). The first of these highly specific and sensitive detection strategies opens the door for personalized medicine by providing insight into what constitutes an individual's personalized treatment plan (Turano et al., 2019). By integrating carbon-based materials into diagnostic platforms, early detection and monitoring of colon cancer can be significantly advanced for improved treatment outcomes and increased patient care.

## 9.4 Therapeutic applications of carbon-based materials

Carbon-based nanomaterials have introduced novel therapeutic strategies for cancer targeting (including targeted drug delivery, PTT, PDT, and advanced gene therapy), and their development is revolutionizing cancer treatment (Fig. 9.1 (Mrówczyński, 2017). Because of their unique physicochemical properties, including a high surface area, tunable functionalization, and excellent biocompatibility, these materials are outstanding candidates for precision medicine.

### 9.4.1 Drug delivery and controlled release mechanisms

Carbon-based materials are known as an efficient drug delivery system for cancer therapy. Functionalization of graphene, CNTs, fullerenes, and CQDs with chemotherapeutic agents provides an improvement in solubility, stability, and targeted release at the tumor (Kang et al., 2018). One of our significant advantages is the use of these nanocarriers to achieve controlled and sustained release, so that the drug is released at a controlled rate and with minimal or no systemic toxicity, thereby increasing therapeutic efficacy. For example, functionalized CNTs serve as "nanoneedles" penetrating the cancer cells with great precision and allowing their release in the intracellular environment (Hani et al., 2021). Similarly, GO and fullerenes also exhibit a high drug-loading capacity as a result of having a wider surface area and $\pi$-$\pi$ interactions with the drug molecules (Wang et al., 2022). Additionally, the carbon-based nanocarriers can be designed to respond to the pH, temperature changes, or enzymatic activities in the tumor microenvironment to release drugs selectively into the cancer cells and prevent the damage to the normal tissues (Lee et al., 2020).

### 9.4.2 Photothermal and photodynamic therapy

Carbon-based nanomaterials are characterized by unique optical and electronic properties that make them excellent for PTT and PDT. In PTT, NIR light is absorbed by materials such as graphene, CNTs, and fullerenes, which are heated locally, causing their death through hyperthermia-mediated destruction of cancer cells (Mendes de Almeida Junior et al., 2024). It is a noninvasive and highly selective method since tumor tissues absorb more NIR light than normal cells. Fullerenes and CQDs are also employed as photosensitizers in PDT, where ROS are generated in response to light and result in oxidative stress-induced apoptosis of cancer cells (Hao et al., 2022). The PTT aids in the practical delivery of PDT by reducing the collateral damage to surrounding healthy tissues and synergistically increasing the therapeutic efficacy of PDT when administered in the same carbon-based nanoplatform. These light-activated treatments offer a possible replacement for chemotherapy and radiotherapy, especially for drug-insensitive tumors (Zhou et al., 2024).

### 9.4.3 Gene therapy and RNA interference approaches

Carbon-based nanomaterials are also important for gene therapy and RNA interference (RNAi)-based approaches, and they have constituted new ways for personalized medicine (Jebelli et al., 2021). siRNA, miRNA, or plasmid DNA can be delivered into cancer cells by functionalized CNTs, GO, and CQDs for therapeutic purposes by silencing oncogenes or reprogramming gene expression (Tangudu et al., 2015). Genetic material is protected during degradation, cellular uptake is enhanced, and genetic delivery is targeted, which aids in treatment success. As a case in point, we demonstrate that CNT-based gene delivery systems are effective at suppressing tumor growth when delivering tumor suppressor genes or RNAi molecules that downregulate the pathways to promote cancer (Kaymaz & Kosova, 2013).

Moreover, graphene-based nanocarriers can be used for CRISPR-Cas9 gene editing as the precision-targeted genetic modifications to eliminate oncogenic mutations. These advancements point the way for highly specific and minimally invasive treatment strategies for cancer using carbon-based nanomaterials, which offer immense potential in revolutionizing cancer gene therapy (Sil et al., 2023). Carbon-based nanomaterials are redefining how therapeutic approaches in oncology are used with their multifunctional capabilities, such as safer, more effective, and personalized colon cancer treatments and beyond (Lieberman & Sharp, 2015).

## 9.5 Challenges and biocompatibility considerations

Although the remarkable potential of carbon-based nanomaterials for cancer theranostics exists, their successful clinical translation is hindered by many challenges related to cytotoxicity, biodistribution, clearance, and immunogenicity (Alkatheeri et al., 2025). The biocompatibility of these materials is of high importance for their safe and effective use in colon cancer therapy (Table 9.2 Carbon-based nanomaterials promise much for colon cancer therapy; however, biocompatibility of these materials is a serious concern for safe and effective use (Adebayo et al., 2023). The efforts on minimizing toxicity, optimizing clearance mechanisms, and eliminating immune reactions through advanced functionalization and nanocarrier design continue as a very important ongoing work. Solving this problem will enable the incorporation of carbon-based theranostic agents into standard cancer treatment with improved safety, efficacy, and patient-tolerant therapy (Younis et al., 2024).

### 9.5.1 Cytotoxicity and safety concerns

Cytotoxicity of carbon-based nanomaterials is one of the main issues pertaining to them. These materials have high surface functionalization and penetration into cells; however, their interactions with biological systems result in oxidative stress, DNA damage, and apoptosis in healthy cells (Maghiari et al., 2020). Graphene and CNTs, in particular, seem to have been shown to produce ROS, resulting in inflammation and cytotoxicity at higher concentrations. Additionally, these nanomaterials have the potential to aggregate and accumulate in vital organs (liver, lungs, and kidneys), leading to increased long-term toxicity (Gor et al., 2023). In order to reduce the risks posed by carbon materials in medical applications, strategies of surface functionalization with biocompatible polymers (e.g., PEGylation, chitosan coating), as well as optimization of the dose, are being developed (Najmuddin et al., 2020) (Table 9.3).

### 9.5.2 Biodistribution and clearance mechanisms

Clearance of carbon-based nanomaterials from the body is required in an efficient way to prevent bioaccumulation and toxicity (Cabral et al., 2024). These materials range from small particles to large, from spherical to cuboidal shapes, and can have large or small surface charges and different degrees of functionalization, all of which have a strong influence on their biodistribution and excretion pathways (Polli & Balthasar, 2022). Materials of smaller sizes, such as CQDs, are typically excreted through kidneys via renal clearance, while larger nanoparticles, like MWCNTs and GO, tend to deposit in liver and spleen and are slowly cleared by reticuloendothelial system (RES) (Bourquin et al., 2018). A possible way to conquer these clearance challenges is by developing biodegradable carbon nanocomposites that can decompose into nontoxic byproducts. In addition, the delivery of these materials through ligand-modified nanocarriers is targeted to prevent off-target accumulation in healthy tissues (Nowak-Jary & Machnicka, 2023).

### 9.5.3 Immunogenicity and long-term effects

Carbon-based nanomaterials also elicit an important immune response. However, while some functionalized nanomaterials are immunogenic, most are not, and others will elicit an immune reaction by binding proteins and immune cells (Ruan et al., 2020). For example, it has been reported

**Table 9.2 Carbon-based substances for colon cancer diagnosis.**

| Carbon-based material | Imaging modality | Mechanism of action | Advantages | Limitations |
|---|---|---|---|---|
| Graphene oxide | Fluorescence, magnetic resonance imaging (MRI), Raman imaging | Improves picture contrast, optimized for biomarker identification | Elevated surface area, biocompatibility | Possible toxicity, aggregation |
| Carbon nanotubes | Computed tomography (CT), Raman spectroscopy | Intense Raman signals for chemical identification, increased X-ray absorption for contrast enhancement | High sensitivity, profound tissue penetration | Concerns with cytotoxicity and clearance difficulties |
| Fullerenes | Positron emission tomography (PET), optical imaging | Function as contrast agents and photosensitizers | Antioxidant characteristics, stability | Restricted aqueous solubility |
| Carbon dots | MRI, fluorescence imaging | Fluorescent labeling of neoplastic cells, paramagnetic characteristics for MRI contrast enhancement | Biocompatible, adjustable fluorescence | Brief fluorescence duration |
| Hybrid carbon nanocomposites | Multimodal imaging (MRI, CT, PET) | Integration of carbon nanomaterials with metal nanoparticles for improved imaging | Enhanced resolution, multifunctional capabilities | Intricacy in synthesis |

**Table 9.3 Obstacles and biocompatibility evaluations of carbon-based materials in colon cancer theranostics.**

| Biocompatibility concerns | Description | Potential solutions |
|---|---|---|
| Toxicity and cytotoxic effects | Generation of reactive oxygen species, oxidative stress, and possible DNA damage | Surface functionalization (e.g., PEGylation) and dosage optimization |
| Clearance and biodegradability | Gradual degradation may result in prolonged accumulation in tissues | Utilization of biodegradable coatings and hybrid nanomaterials |
| Immunogenicity | Induction of immunological response resulting in inflammation | Surface alterations with biocompatible polymers |
| Regulatory and safety concerns | Lack of defined methods for clinical translation | Extensive in vivo investigations, formulation of regulatory frameworks |

that CNTs and graphene can activate macrophages and inflammatory cytokine release, which could result in chronic inflammation (Bains et al., 2020). Moreover, carbon nanomaterials may be harmful to humans if exposed to nonbiodegradable carbon nanomaterials over a long period, causing risks of fibrosis, granuloma formation, and immune suppression (Hege et al., 2017). To overcome these, we need extensive in vivo studies and long-term toxicity assessments for evaluating the long-term biocompatibility of these materials. To facilitate future clinical translation, these structures may be engineered to be biodegradable, nonimmune generating, and coated with antiinflammatory coatings as well as targeted drug release profiles (He et al., 2020).

## 9.6 Clinical translation and future prospects

The area of carbon-based nanomaterials for cancer diagnosis and treatment has shown tremendous potential for revolutionary outcomes (Zhang et al., 2023). However, despite the success of their work in preclinical research, these treatments have not been translated to the clinic mainly because of regulatory, ethical, and safety issues (Parker et al., 2022). The first is a review of the current clinical and preclinical landscape, regulatory and ethical hurdles that need to be passed, and the future fashion that will truly define personalized cancer theranostics for colon cancer treatment (Mi et al., 2021).

### 9.6.1 Current clinical trials and preclinical studies

Scientific investigations at the preclinical level have confirmed the therapeutic potential of carbon-based nanomaterials, which include graphene, CNTs, fullerenes, and CQDs (Brown et al., 2016). Scientific examinations have proven these materials suitable for drug delivery while also performing imaging, PTT therapy, PDT therapy, and gene therapy applications (Lannagan et al., 2021). Preclinical research established that carbon-based nanomaterials adopt three key properties to enhance drug effectiveness, minimize adverse effects, and improve treatment accuracy.

GO nanocarriers with anticancer drug doxorubicin delivered to tumors demonstrated superior accumulation compared to traditional chemotherapy treatments along with reduced detrimental side effects (Narayanankutty, 2019). Stemming from research findings about CNTs and siRNA complexes that demonstrated their capability to combat colorectal cancer tumor development through oncogene silencing. The combination of fullerenes and CQDs in PDT applications displays strong phototoxicity levels against colon cancer cells while protecting normal tissues from harm (Rizzo et al., 2021).

Numerous carbon-based nanomaterials remain in clinical trials with different prospective cancer treatment applications (Hervieu et al., 2021). The pharmaceutical industry currently assesses human safety and efficacy for two categories of carbon nanomaterial pharmaceutical products through Phase I and Phase II trials (Crutcher et al., 2022). Current clinical trial research investigates how targeted drug delivery systems composed of GO and PEGylated CNTs perform in colorectal cancer treatment. Scientists test the pharmacokinetic behavior, together with the tumor-seeking properties and biological compatibility of the materials in clinical trials (Kawczyk-Krupka et al., 2016).

Medical use of these nanomaterials in clinics awaits the resolution of multiple obstacles that include extended toxicity concerns, varying drug distribution patterns, and the need for standardized quality production methods (Miura et al., 2014). The scientific community dedicates research efforts to create functionalization methods that both enhance biocompatibility and minimize immune responses so nanomaterials can be used safely in human applications.

## 9.6.2 Regulatory and ethical consideration

Strict regulatory frameworks, together with ethical considerations, control the clinical implementation process of carbon-based nanomaterials discovered through laboratory research (Purriños et al., 2022). Nanomedicine approval guidelines exist through regulatory bodies that include the US Food and Drug Administration (FDA), the European Medicines Agency (EMA), and the National Medical Products Administration (NMPA) in China. Currently existing rules need regular updates because these materials remain new and intricate in nature (Papin-Lefebvre et al., 2017).

Safety assessment, together with standardization, presents the major hurdle for regulatory approval. The properties of carbon-based nanomaterials depend on their size and structural shape, which results in distinct toxicity expressions compared to conventional pharmaceuticals (Kiran et al., 2024). Prior to receiving regulatory approval, researchers need to prove that their batches adhere to stability requirements and exhibit identical performance, along with demonstrating how the materials degrade safely over time (Sharma et al., 2024).

The field of nanomedicine requires immediate attention to both moral issues. The safety of patients, along with their consent, represents essential factors for introducing new nanotherapeutic products (Spisak et al., 2024). The worry about carbon-based nanoparticles penetrating biological barriers increases as scientists identify potential buildup of these particles within liver tissue along with spleen and lung tissue. Compliance with ethical standards requires thorough risk evaluations, together with complete clinical trial disclosure, along with thorough advisement of patients about danger factors versus therapeutic effects (del Moral-Martinez et al., 2025).

Environmental safety joins a list of regulatory issues that require examination. Rules of environmental protection determine how nanomaterials should be produced at scale because avoiding toxic effects on both aquatic and terrestrial ecosystems is necessary (Mascarenhas et al., 2024). The regulatory bodies of different countries establish environmentally sustainable manufacturing approaches for biomedical nanomaterials along with disposal protocols for these materials.

### 9.6.3 Future directions in personalized cancer and colon cancer theranostics

Nanotechnology progress will cause carbon-based theranostic agents to emerge as critical elements within personalized cancer care. Tremendous growth in the field will be determined by the upcoming research trends (Dey et al., 2023). The development of carbon-based nanomaterials shows positive prospects for cancer theranostics because they advance targeted therapy and imaging technology as well as patient-specific medicine. To achieve their full potential, clinical validation, along with strict adherence to regulatory compliance and ethical concern assessment, must be demonstrated rigorously (Bilgin et al., 2024). Technology advancements, alongside continuous medical research on these materials, create a promising future for carbon-based nanotherapeutics in colon cancer treatment, where effective patient-centered therapy is anticipated (Lim et al., 2021).

#### 9.6.3.1 Smart and responsive nanocarriers

Carbon-based nanocarriers of the future will utilize smart release systems that sense particular tumor environment characteristics, which include pH variations, temperature fluctuations, and enzyme activities (Mi, 2020). GO, along with CNTs, can release drugs only in response to tumor site stimuli, which minimizes both off-target delivery and systemic toxicity effects (Licciardi et al., 2019).

#### 9.6.3.2 Hybrid and multifunctional nanocomposites

The development of improved theranostic characteristics occurs when carbon nanomaterials receive integration with either metallic nanoparticles (gold, silver, or iron oxide) or organic molecules (polymers and peptides) (Zhang et al., 2022). These materials create the ability to perform imaging functions while administering therapy, while simultaneously monitoring treatment effectiveness in real time (Jiang et al., 2020).

#### 9.6.3.3 AI-driven nanomedicine optimization

Artificial intelligence (AI), together with machine learning systems, assists nanomedicine advancement through improved nanocarrier design while predicting adverse effects and developing individualized treatment sequences from specific patient information (Szymaszek et al., 2024). The use of AI-powered drug delivery models enables individuals to select nanocarrier options alongside therapeutic agents according to individual patient requirements (Xu et al., 2023).

#### 9.6.3.4 CRISPR-Cas9 and carbon-based gene therapy platforms

Scientists conduct research on carbon-based nanoparticles that may serve as carriers for CRISPR-Cas9 tools used to fix cancer-linked genetic abnormalities (Mostafavi & Zare, 2022). Such a technological breakthrough will open doors for treating colon cancer at its genetic origins through precise molecular medicine.

#### 9.6.3.5 Advancements in biosensing for early detection

The upcoming generation of carbon-based biosensors promises highly sensitive operation to find cancer biomarkers directly from blood or saliva samples (Thenrajan & Wilson, 2022). The bio-integrated nature of flexible electronics using graphene components allows continuous monitoring of cancer indicators to detect colon cancer early, which facilitates better prevention techniques (Kumarganesh et al., 2024).

#### 9.6.3.6 Regulatory evolution and clinical translation acceleration

The advancement of clinical trials will lead regulatory frameworks to change their approval processes for nanotherapeutics (Turner, 2024). Government regulatory bodies will launch special approval procedures for nanomedicine innovations, which will result in faster entry of life-saving therapies to patients (Porter et al., 2021).

#### 9.6.3.7 Sustainable and scalable manufacturing

The creation of carbon-based nanomaterials at large scale requires economic viability, consistency, and respect for environmental quality standards (Bigham et al., 2024). The field of green nanotechnology will develop sustainable creation tactics to produce nanotherapeutics with high clinical quality while minimizing environmental damage (Maher et al., 2016).

## 9.7 Conclusion

The combination of carbon-based materials with colon cancer therapeutic and diagnostic applications leads to substantial progress in cancer treatment. Our analysis has shown through this chapter how graphene, as well as its derivatives, CNTs, fullerenes, CQDs, and hybrid carbon nanocomposites, enhance diagnosis and treatment approaches for cancer. These exceptional materials provide extensive surface area, along with adjustable electronic properties and exceptional biocompatible features, which make them appropriate selection options for future theranostic systems.

Carbon-based nanomaterials provide enhanced resolution and sensitivity through MRI, as well as CT analysis, PET studies, and optical imaging diagnostics for detecting malignant tissues. Biosensor applications enabled by these materials enhance cancer detection methods, since they improve the ability to find specific biomarkers during early stages of cancer. The development of these technologies allows medical professionals to intervene earlier with patients who have colon cancer, thus leading to better treatment results.

Carbon-based nanomaterials show exceptional medical qualities for drug transportation because their targeted delivery methods reduce systemic damage while providing maximum therapeutic effects. The use of carbon-based nanomaterials in photothermal and photodynamic therapies enables doctors to develop new methods of cancer treatment through light-induced cell elimination in minimally invasive procedures. Carbon nanomaterials demonstrate effective treatment outcome enhancement potential when used in gene therapy and RNAi approaches to manage oncogenic pathways.

However, the advancement of these materials has not been able to avoid challenges in facilitating their safety and biocompatibility. The potential adverse effect issues must be systematically addressed, including cytotoxicity, biodistribution, clearance mechanisms, and immunogenic response, to mitigate them. Therefore, rigorous preclinical and clinical evaluation is essential to define a standardized approach for translating these materials to clinical practice.

Future carbon-based theranostic agents will therefore depend on the course of ongoing clinical trials and regulatory considerations. While personalized medicine is developing in various and personalized directions, such as AI, machine learning, and nanotechnology, the particular application in cancer diagnosis and treatment may become more efficient in integrating AI, machine learning, and nanotechnology, making interventions more patient specific.

Finally, it is worth mentioning that carbon-based theranostic materials can revolutionize colon cancer care. There is reason to believe that these could continue to be developed and refined to usher in a new era of precision oncology, where early detection and targeted therapy are the norm. The realization of the full potential of these innovative materials will be possible only by addressing current challenges through interdisciplinary research and collaboration with the aim to improve survival rates and quality of life for colon cancer patients worldwide.

## References


Abreu, S., Vale, N., & Soares, O. S. G. P. (2023). Combination of CNTs with classical drugs for treatment in human colorectal adenocarcinoma (HT-29) cell line. *Nanomaterials, 13*(13), 1933.

Adebayo, A. S., Agbaje, K., Adesina, S. K., & Olajubutu, O. (2023). Colorectal cancer: Disease process, current treatment options, and future perspectives. *Pharmaceutics, 15*(11), 2620.

Al-Ani, L. A., Kadir, F. A., Hashim, N. M., Julkapli, N. M., Seyfoddin, A., Lu, J., et al. (2020). The impact of curcumin-graphene based nanoformulation on cellular interaction and redox-activated apoptosis: An in vitro colon cancer study. *Heliyon, 6*(11).

Albert, K., & Hsu, H. Y. (2016). Carbon-based materials for photo-1. Carbon-based materials for photo-triggered theranostic applications. *Molecules. Triggered Theranostic Applications. Molecules, 21*(11), 1585.

Alkatheeri, A., Salih, S., Kamil, N., Alnuaimi, S., Abuzar, M., & Abdelrahman, S. S. (2025). Nano-radiopharmaceuticals in colon cancer: Current applications, challenges, and future directions. *Pharmaceuticals, 18*(2), 257.

Andoh, V., Ocansey, D. K. W., Naveed, H., Wang, N., Chen, L., Chen, K., et al. (2024). The Advancing role of nanocomposites in cancer diagnosis and treatment. *International Journal of Nanomedicine [Internet], 19*(null), 6099–6126. https://doi.org/10.2147/IJN.S471360, https://www.tandfonline.com/doi/abs/.

Arruebo, M., Vilaboa, N., Sáez-Gutierrez, B., Lambea, J., Tres, A., Valladares, M., et al. (2011). Assessment of the evolution of cancer treatment therapies. *Cancers (Basel), 3*(3), 3279–3330.

Bains, S. J., Abrahamsson, H., Flatmark, K., Dueland, S., Hole, K. H., Seierstad, T., et al. (2020). Immunogenic cell death by neoadjuvant oxaliplatin and radiation protects against metastatic failure in high-risk rectal cancer. *Cancer Immunology, Immunotherapy CII, 69*, 355–364.

Bayda S. Synthesis of fluorescent carbon nanoparticles (CNPs) and their applications in drug delivery. 2017.

Beniwal, S. S., Lamo, P., Kaushik, A., Lorenzo-Villegas, D. L., Liu, Y., & MohanaSundaram, A. (2023). Current status and emerging trends in colorectal cancer screening and diagnostics. *Biosensors*.

Bigham, A., Zarepour, A., Khosravi, A., Iravani, S., & Zarrabi, A. (2024). Sustainable nanomaterials for precision medicine in cancer therapy. *Materials Today Sustainability*100865.

Bilgin, G. B., Bilgin, C., Burkett, B. J., Orme, J. J., Childs, D. S., Thorpe, M. P., et al. (2024). Theranostics and artificial intelligence: New frontiers in personalized medicine. *Theranostics, 14*(6), 2367.

Bourquin, J., Milosevic, A., Hauser, D., Lehner, R., Blank, F., Petri-Fink, A., et al. (2018). Biodistribution, clearance, and long-term fate of clinically relevant nanomaterials. *Advanced Materials, 30*(19), 1704307.

Brown, K. M., Xue, A., Mittal, A., Samra, J. S., Smith, R., & Hugh, T. J. (2016). Patient-derived xenograft models of colorectal cancer in pre-clinical research: A systematic review. *Oncotarget, 7*(40), 66212.

Cabral, H., Li, J., Miyata, K., & Kataoka, K. (2024). Controlling the biodistribution and clearance of nanomedicines. *Nature Reviews Bioengineering, 2*(3), 214–232.

Cheng, Z., Li, M., Dey, R., & Chen, Y. (2021). Nanomaterials for cancer therapy: Current progress and perspectives. *Journal of Hematology & Oncology [Internet], 14*(1), 85. https://doi.org/10.1186/s13045-021-01096-0.

Chuang, J. P., Tsai, H. L., Chen, P. J., Chang, T. K., Su, W. C., Yeh, Y. S., et al. (2022). Comprehensive review of biomarkers for the treatment of locally advanced colon cancer. *Cells.*

Comparetti, E. J., Romagnoli, G. G., Gorgulho, C. M., de Albuquerque Pedrosa, V., & Kaneno, R. (2020). Anti-PSMA monoclonal antibody increases the toxicity of paclitaxel carried by carbon nanotubes. *Materials Science and Engineering: C, 116*, 111254.

Crutcher, M. M., Baybutt, T. R., Kopenhaver, J. S., Snook, A. E., & Waldman, S. A. (2022). Emerging drug targets for colon cancer: A preclinical assessment. *Expert Opinion on Therapeutic Targets, 26*(3), 207–216.

Darrigues, E., Dantuluri, V., Nima, Z. A., Vang-Dings, K. B., Griffin, R. J., Biris, A. R., et al. (2017). Raman spectroscopy using plasmonic and carbon-based nanoparticles for cancer detection, diagnosis, and treatment guidance. Part 2: Treatment. *Drug Metabolism Reviews [Internet], 49*(2), 253–283. https://doi.org/10.1080/03602532.2017.1307387.

del Moral-Martinez, M., Sánchez-Uceta, P., Clemente-Gonzalez, R., Moreno-SanJuan, S., Puentes-Pardo, J. D., Khaldy, H., et al. (2025). iNOS-produced nitric oxide from cancer cells as an intermediate of stemness regulation by PARP-1 in colorectal cancer. *Biomolecules, 15*(1), 125.

Dessale, M., Mengistu, G., & Mengist, H. M. (2022). Nanotechnology: A promising approach for cancer diagnosis, therapeutics and theragnosis. *International Journal of Nanomedicine, 17*, 3735.

Dey, A., Mitra, A., Pathak, S., Prasad, S., Zhang, A. S., Zhang, H., et al. (2023). Recent advancements, limitations, and future perspectives of the use of personalized medicine in treatment of colon cancer. *Technology in Cancer Research & Treatment, 22* 15330338231178404.

Dhas, N., Pastagia, M., Sharma, A., Khera, A., Kudarha, R., Kulkarni, S., et al. (2022). Organic quantum dots: An ultrasmall nanoplatform for cancer theranostics. *Journal of Controlled Release Official Journal of the Controlled Release Society, 348*, 798–824.

Gallaher, J. A., Enriquez-Navas, P. M., Luddy, K. A., Gatenby, R. A., & Anderson, A. R. A. (2018). Spatial heterogeneity and evolutionary dynamics modulate time to recurrence in continuous and adaptive cancer therapies. *Cancer Research, 78*(8), 2127–2139.

Gaur, M., Misra, C., Yadav, A. B., Swaroop, S., Maolmhuaidh, F.Ó., Bechelany, M., et al. (2021). Biomedical applications of carbon nanomaterials: Fullerenes, quantum dots, nanotubes, nanofibers, and graphene. *Materials (Basel), 14*(20), 5978.

Goldberg, M. S. (2019). Improving cancer immunotherapy through nanotechnology. *Nature Reviews. Cancer, 19*(10), 587–602.

Gor, R., Gharib, A., Dharshini Balaji, P., Madhavan, T., & Ramalingam, S. (2023). Inducing cytotoxicity in colon cancer cells and suppressing cancer stem cells by dolasetron and ketoprofen through inhibition of RNA binding protein PUM1. *Toxics, 11*(8), 669.

Gray, M., Marland, J. R. K., Murray, A. F., Argyle, D. J., & Potter, M. A. (2021). Predictive and diagnostic biomarkers of anastomotic leakage: A precision medicine approach for colorectal cancer patients. *Journal of Personalized Medicine.*

Grodzinski, P., Kircher, M., Goldberg, M., & Gabizon, A. (2019). Integrating nanotechnology into cancer care. *ACS Publications.*

Hani, U., Honnavalli, Y. K., Begum, M. Y., Yasmin, S., Osmani, R. A. M., & Ansari, M. Y. (2021). Colorectal cancer: A comprehensive review based on the novel drug delivery systems approach and its management. *Journal of Drug Delivery Science and Technology, 63*, 102532.

Hao, Y., Chung, C. K., Gu, Z., Schomann, T., Dong, X., Veld, R. V. H.in 't, et al. (2022). Combinatorial therapeutic approaches of photodynamic therapy and immune checkpoint blockade for colon cancer treatment. *Molecular Biomedicine, 3*(1), 26.

Hartshorn, C. M., Bradbury, M. S., Lanza, G. M., Nel, A. E., Rao, J., Wang, A. Z., et al. (2018). Nanotechnology strategies to advance outcomes in clinical cancer care. *ACS Nano, 12*(1), 24–43.

He, H., Liu, L., Liang, R., Zhou, H., Pan, H., Zhang, S., et al. (2020). Tumor-targeted nanoplatform for in situ oxygenation-boosted immunogenic phototherapy of colorectal cancer. *Acta Biomaterialia, 104*, 188–197.

Hege, K. M., Bergsland, E. K., Fisher, G. A., Nemunaitis, J. J., Warren, R. S., McArthur, J. G., et al. (2017). Safety, tumor trafficking and immunogenicity of chimeric antigen receptor (CAR)-T cells specific for TAG-72 in colorectal cancer. *Journal for ImmunoTherapy of Cancer, 5*, 1–14.

Heller, D. A., Jena, P. V., Pasquali, M., Kostarelos, K., Delogu, L. G., Meidl, R. E., et al. (2020). Banning carbon nanotubes would be scientifically unjustified and damaging to innovation. *Nature Nanotechnology, 15*(3), 164–166.

Hervieu, C., Christou, N., Battu, S., & Mathonnet, M. (2021). The role of cancer stem cells in colorectal cancer: From the basics to novel clinical trials. *Cancers (Basel), 13*(5), 1092.

Hosseini, S. M., Mohammadnejad, J., Najafi-Taher, R., Zadeh, Z. B., Tanhaei, M., & Ramakrishna, S. (2023). Multifunctional carbon-based nanoparticles: Theranostic applications in cancer therapy and diagnosis. *ACS Applied Bio Materials, 6*(4), 1323–1338.

Jebelli, A., Baradaran, B., Mosafer, J., Baghbanzadeh, A., Mokhtarzadeh, A., & Tayebi, L. (2021). Recent developments in targeting genes and pathways by RNAi-based approaches in colorectal cancer. *Medicinal Research Reviews, 41*(1), 395–434.

Jiang, Y., Guo, Z., Fang, J., Wang, B., Lin, Z., Chen, Z. S., et al. (2020). A multi-functionalized nanocomposite constructed by gold nanorod core with triple-layer coating to combat multidrug resistant colorectal cancer. *Materials Science and Engineering: C, 107*, 110224.

Jin, C., Wang, K., Oppong-Gyebi, A., & Hu, J. (2020). Application of nanotechnology in cancer diagnosis and therapy-a mini-review. *International Journal of Medical Sciences, 17*(18), 2964.

Kamel, F., Eltarhoni, K., Nisar, P., & Soloviev, M. (2022). Colorectal cancer diagnosis: The obstacles we face in determining a non-invasive test and current advances in biomarker detection. *Cancers*.

Kang, J. H., Hwang, J. Y., Seo, J. W., Kim, H. S., & Shin, U. S. (2018). Small intestine-and colon-specific smart oral drug delivery system with controlled release characteristic. *Materials Science and Engineering: C, 91*, 247–254.

Kashyap, B. K., Singh, V. V., Solanki, M. K., Kumar, A., Ruokolainen, J., & Kesari, K. K. (2023). Smart nanomaterials in cancer theranostics: Challenges and opportunities. *ACS Omega [Internet], 8*(16), 14290–14320. https://doi.org/10.1021/acsomega.2c07840.

Kawczyk-Krupka, A., Bugaj, A. M., Latos, W., Zaremba, K., Wawrzyniec, K., Kucharzewski, M., et al. (2016). Photodynamic therapy in colorectal cancer treatment—The state of the art in preclinical research. *Photodiagnosis and Photodynamic Therapy, 13*, 158–174.

Kaymaz, B. T., & Kosova, B. (2013). Advances in therapeutic approaches using RNA interference as a gene silencing tool. *Advanced Technologies in Biology & Medicine, 1*(108), 2.

Kekelidze, M., D'Errico, L., Pansini, M., Tyndall, A., & Hohmann, J. (2013). Colorectal cancer: Current imaging methods and future perspectives for the diagnosis, staging and therapeutic response evaluation. *World Journal of Gastroenterology WJG, 19*(46), 8502–8514.

Kemp, J. A., & Kwon, Y. J. (2021). Cancer nanotechnology: Current status and perspectives. *Nano Convergence, 8*(1), 34.

Kiran, N. S., Yashaswini, C., Maheshwari, R., Bhattacharya, S., & Prajapati, B. G. (2024). Advances in precision medicine approaches for colorectal cancer: From molecular profiling to targeted therapies. *ACS Pharmacology & Translational Science, 7*(4), 967–990.

Komane, P., Kumar, P., & Choonara, Y. (2023). Functionalised carbon nanotubes: Promising drug delivery vehicles for neurovascular disorder intervention. *AAPS PharmSciTech, 24*(7), 201.

Kuczynski, G., & Gotay, C. (2024). Cancer prevention: Principles and approaches. *Understanding cancer prevention through geospatial science: Putting cancer in its place,* 17–43.

Kumarganesh, S., Murugesan, M., Ganesh, C., Santhiyakumari, N., Thangavel, M., Jayaram, K., et al. (2024). Hybrid plasmonic biosensors with deep learning for colorectal cancer detection. *Plasmonics,* 1–15.

Lannagan, T. R. M., Jackstadt, R., Leedham, S. J., & Sansom, O. J. (2021). Advances in colon cancer research: In vitro and animal models. *Current Opinion in Genetics & Development, 66,* 50–56.

Lee, S. H., Bajracharya, R., Min, J. Y., Han, J. W., Park, B. J., & Han, H. K. (2020). Strategic approaches for colon targeted drug delivery: An overview of recent advancements. *Pharmaceutics, 12*(1), 68.

Li, J., Xiang, H., Zhang, Q., & Miao, X. (2022). Polysaccharide-based transdermal drug delivery. *Pharmaceuticals, 15*(5), 602.

Licciardi, M., Scialabba, C., Puleio, R., Cassata, G., Cicero, L., Cavallaro, G., et al. (2019). Smart copolymer coated SPIONs for colon cancer chemotherapy. *International Journal of Pharmaceutics, 556,* 57–67.

Lieberman, J., & Sharp, P. A. (2015). Harnessing RNA interference for therapy: The silent treatment. *JAMA the Journal of the American Medical Association, 313*(12), 1207–1208.

Lim, E. K., Kim, T., Paik, S., Haam, S., Huh, Y. M., & Leec, K. (2021). Nanomaterials for theranostics: Recent advances and future challenges. *Nanomater Neoplasms,* 587–775.

Luo, S., Yang, X., & Shi, C. (2016). Newly emerging theranostic agents for simultaneous cancertargeted imaging and therapy. *Current Medicinal Chemistry, 23*(5), 483–497.

Ma, J., Wang, G., Ding, X., Wang, F., Zhu, C., & Rong, Y. (2023). Carbon-based nanomaterials as drug delivery agents for colorectal cancer: Clinical preface to colorectal cancer citing their markers and existing theranostic approaches. *ACS Omega, 8*(12), 10656–10668.

Machida, T., Higuchi, Y., Nakano, S., Izumi, M., Ishige, S., Fujikawa, A., et al. (2019). *CLINICAL ARTICLE Journal of Neurosurgery, 130,* 1957–1964.

Madni, A., Noreen, S., Maqbool, I., Rehman, F., Batool, A., Kashif, P. M., et al. (2018). Graphene-based nanocomposites: Synthesis and their theranostic applications. *Journal of Drug Targeting [Internet], 26*(10), 858–883. https://doi.org/10.1080/1061186X.2018.1437920.

Maghiari, A. L., Coricovac, D., Pinzaru, I. A., Macașoi, I. G., Marcovici, I., Simu, S., et al. (2020). High concentrations of aspartame induce pro-angiogenic effects in ovo and cytotoxic effects in HT-29 human colorectal carcinoma cells. *Nutrients, 12*(12), 3600.

Maher, S., Kumeria, T., Wang, Y., Kaur, G., Fathalla, D., Fetih, G., et al. (2016). From the mine to cancer therapy: Natural and biodegradable theranostic silicon nanocarriers from diatoms for sustained delivery of chemotherapeutics. *Advanced Healthcare Materials, 5*(20), 2667–2678.

Mahor, A., Singh, P. P., Bharadwaj, P., Sharma, N., Yadav, S., Rosenholm, J. M., et al. (2021). Carbon-based nanomaterials for delivery of biologicals and therapeutics: A cutting-edge technology. *C, 7*(1), 19.

Malarkey, E. B., Fisher, K. A., Bekyarova, E., Liu, W., Haddon, R. C., & Parpura, V. (2009). Conductive single-walled carbon nanotube substrates modulate neuronal growth. *Nano Letters, 9*(1), 264–268.

Mascarenhas, M., Martins, M., Ribeiro, T., Afonso, J., Cardoso, P., Mendes, F., et al. (2024). Software as a medical device (SaMD) in digestive healthcare: Regulatory challenges and ethical implications. *Diagnostics, 14*(18), 2100.

Mendes de Almeida Junior, A., Ferreira, A. S., Camacho, S. A., Gontijo Moreira, L., de Toledo, K. A., Oliveira, J. O. N., et al. (2024). Enhancing phototoxicity in human colorectal tumor cells through nanoarchitectonics for synergistic photothermal and photodynamic therapies. *ACS Applied Materials & Interfaces, 16*(18), 23742–23751.

Mi, P., Miyata, K., Kataoka, K., & Cabral, H. (2021). Clinical translation of self-assembled cancer nanomedicines. *Advances in Therapy, 4*(1), 2000159.

Mi, P. (2020). Stimuli-responsive nanocarriers for drug delivery, tumor imaging, therapy and theranostics. *Theranostics, 10*(10), 4557.

Miura, K., Satoh, M., Kinouchi, M., Yamamoto, K., Hasegawa, Y., Philchenkov, A., et al. (2014). The preclinical development of regorafenib for the treatment of colorectal cancer. *Expert Opinion on Drug Discovery, 9*(9), 1087–1101.

Mostafavi, E., & Zare, H. (2022). Carbon-based nanomaterials in gene therapy. *OpenNano, 7*, 100062.

Mrówczyński, R. (2017). Polydopamine-based multifunctional (nano) materials for cancer therapy. *ACS Applied Materials & Interfaces, 10*(9), 7541–7561.

Mura, P., Maestrelli, F., Cirri, M., & Mennini, N. (2022). Multiple roles of chitosan in mucosal drug delivery: An updated review. *Marine Drugs, 20*(5), 335.

Najmuddin, S. U. F. S., Amin, Z. M., Tan, S. W., Yeap, S. K., Kalyanasundram, J., Ani, M. A. C., et al. (2020). Cytotoxicity study of the interleukin-12-expressing recombinant Newcastle disease virus strain, rAF-IL12, towards CT26 colon cancer cells in vitro and in vivo. *Cancer Cell International, 20*, 1–18.

Narayan, T., Kumar, S., Kumar, S., Augustine, S., Yadav, B. K., & Malhotra, B. D. (2019). Protein functionalised self assembled monolayer based biosensor for colon cancer detection. *Talanta [Internet], 201*, 465–473. https://www.sciencedirect.com/science/article/pii/S003991401930428X.

Narayanankutty, A. (2019). PI3K/Akt/mTOR pathway as a therapeutic target for colorectal cancer: A review of preclinical and clinical evidence. *Current Drug Targets, 20*(12), 1217–1226.

Nayak, T. R., Zhang, Y., & Cai, W. (2014). Cancer theranostics with carbon-based nanoplatforms. In *Cancer theranostics*. Elsevier, 347–361.

Nicolle, L., Journot, C. M. A., & Gerber-Lemaire, S. (2021). Chitosan functionalization: Covalent and non-covalent interactions and their characterization. *Polymers (Basel), 13*(23), 4118.

Nowak-Jary, J., & Machnicka, B. (2023). In vivo biodistribution and clearance of magnetic iron oxide nanoparticles for medical applications. *International Journal of Nanomedicine*, 4067–4100.

Papin-Lefebvre, F., Guillaume, E., Moutel, G., Launoy, G., & Berchi, C. (2017). General practitioners' preferences with regard to colorectal cancer screening organisation Colon cancer screening medico-legal aspects. *Health Policy, 121*(10), 1079–1084.

Parker, A. L., Benguigui, M., Fornetti, J., Goddard, E., Lucotti, S., Insua-Rodríguez, J., et al. (2022). Current challenges in metastasis research and future innovation for clinical translation. *Clinical & Experimental Metastasis, 39*(2), 263–277.

Polli, J. R., & Balthasar, J. P. (2022). Cell penetrating peptides conjugated to anti-carcinoembryonic antigen "catch-and-release" monoclonal antibodies alter plasma and tissue pharmacokinetics in colorectal cancer xenograft mice. *Bioconjugate Chemistry, 33*(8), 1456–1466.

Porter, R. J., Arends, M. J., Churchhouse, A. M. D., & Din, S. (2021). Inflammatory bowel disease-associated colorectal cancer: Translational risks from mechanisms to medicines. *Journal of Crohn's and Colitis, 15*(12), 2131–2141.

Prajapati, S. K., Jain, A., Shrivastava, C., & Jain, A. K. (2019). Hyaluronic acid conjugated multi-walled carbon nanotubes for colon cancer targeting. *International Journal of Biological Macromolecules [Internet], 123*, 691–703. https://www.sciencedirect.com/science/article/pii/S0141813018326710.

Purriños, N. C., Bendayán, I. M., & Sánchez, A. D. (2022). Ethical and legal aspects in CRC: Research and clinical assistance. In *Foundations of colorectal cancer*. Elsevier, 619–627.

Quinchia, J., Echeverri, D., Cruz-Pacheco, A. F., Maldonado, M. E., & Orozco, J. (2020). Electrochemical biosensors for determination of colorectal tumor biomarkers. *Micromachines*.

Rahman, M., Akhter, S., Ahmad, M. Z., Ahmad, J., Addo, R. T., Ahmad, F. J., et al. (2015). Emerging advances in cancer nanotheranostics with graphene nanocomposites: Opportunities and challenges. *Nanomedicine [Internet], 10*(15), 2405–2422. https://doi.org/10.2217/nnm.15.68.

Ramezani Farani, M., Lak, M., C. Cho, W., Kang, H., Azarian, M., Yazdian, F., et al. (2024). Carbon nanomaterials: A promising avenue in colorectal cancer treatment. *Carbon Letters, 34*(8), 2035–2053.

Raval, J. B., Mehta, V. N., & Kailasa, S. K. (2023). In S. K. Kailasa, & C. M. B. T. C. D.in A. C. Hussain (Eds.). *Chapter 21 - Carbon nanomaterials-based diagnostic tools* (pp. 279–291). Elsevier. https://www.sciencedirect.com/science/article/pii/B9780323983501000190.

Rizzo, G., Bertotti, A., Leto, S. M., & Vetrano, S. (2021). Patient-derived tumor models: A more suitable tool for pre-clinical studies in colorectal cancer. *Journal of Experimental & Clinical Cancer Research CR, 40*(1), 178.

Ruan, H., Leibowitz, B. J., Zhang, L., & Yu, J. (2020). Immunogenic cell death in colon cancer prevention and therapy. *Molecular Carcinogenesis, 59*(7), 783–793.

Sable, A. A., Kunwar, A., & Barik, A. (2024). Alginate and chitosan-based delivery systems for improving the bioavailability and therapeutic efficacy of curcumin. *Pharmaceutics, 16*(3), 423.

Sajjadi, M., Nasrollahzadeh, M., Jaleh, B., Soufi, G. J., & Iravani, S. (2021). Carbon-based nanomaterials for targeted cancer nanotherapy: Recent trends and future prospects. *Journal of Drug Targeting, 29*(7), 716–741.

Sargazi, S., ER, S., Mobashar, A., Gelen, S. S., Rahdar, A., Ebrahimi, N., et al. (2022). Aptamer-conjugated carbon-based nanomaterials for cancer and bacteria theranostics: A review. *Chemico-Biological Interactions [Internet], 361*, 109964. https://www.sciencedirect.com/science/article/pii/S0009279722001697.

Sharma, G., Singh, M., Sharma, K., Sharma, P., & Mehra, S. (2021). *Role of nanotechnology in cancer treatment. Importance & applications of nanotechnology*. MedDocs Publishers.

Sharma, S., Chakraborty, M., Yadav, D., Dhullap, A., Singh, R., Verma, R. K., et al. (2024). Strategic developments in polymer-functionalized liposomes for targeted colon cancer therapy: An updated review of clinical trial data and future horizons. *Biomacromolecules, 25*(9), 5650–5669.

Sil, S., Bertilla, J., & Rupachandra, S. (2023). A comprehensive review on RNA interference-mediated targeting of interleukins and its potential therapeutic implications in colon cancer. *3 Biotech, 13*(1), 18.

Singhai, N. J., Maheshwari, R., & Ramteke, S. (2020). CD44 receptor targeted 'smart'multi-walled carbon nanotubes for synergistic therapy of triple-negative breast cancer. *Colloids and Interface Science Communications, 35*, 100235.

Siringan, M. J., Dawar, A., & Zhang, J. (2023). Interactions between fullerene derivati1. Siringan MJ, Dawar A, Zhang J. Interactions between fullerene derivatives and biological systems. *Materials Chemistry Frontiers, 7*(11), 2153–2174.

Siva, V., Kannan, S., Murugan, A., Shameem, A. S., Thangarasu, S., & Raja, A. (2024). Evaluation of zero-dimensional buckyball-based fullerene molecules in targeted drug delivery applications in dreaded diseases. In *Carbon-based nanomaterials in biosystems*. Elsevier, 301–330.

Spisak, S., Chen, D., Likasitwatanakul, P., Doan, P., Li, Z., Bala, P., et al. (2024). Identifying regulators of aberrant stem cell and differentiation activity in colorectal cancer using a dual endogenous reporter system. *Nature Communications, 15*(1), 2230.

Szymaszek, P., Tyszka-Czochara, M., & Ortyl, J. (2024). Application of photoactive compounds in cancer theranostics: Review on recent trends from photoactive chemistry to artificial intelligence. *Molecules (Basel, Switzerland), 29*(13), 3164.

Taddie, W., & Dessie, D. (2019). The role of nanotechnology in medicine. *International Journal of Excellence, Innovation and Development, 2*(8).

Tang, L., Li, J., Pan, T., Yin, Y., Mei, Y., Xiao, Q., et al. (2022). Versatile carbon nanoplatforms for cancer treatment and diagnosis: Strategies, applications and future perspectives. *Theranostics, 12*(5), 2290–2321.

Tangudu, N. K., Verma, V. K., Clemons, T. D., Beevi, S. S., Hay, T., Mahidhara, G., et al. (2015). RNA interference using c-Myc–conjugated nanoparticles suppresses breast and colorectal cancer models. *Molecular Cancer Therapeutics, 14*(5), 1259–1269.

Thenrajan, T., & Wilson, J. (2022). Biosensors for cancer theranostics. *Biosens Bioelectron X, 12*, 100232.

Turano, M., Delrio, P., Rega, D., Cammarota, F., Polverino, A., Duraturo, F., et al. (2019). Promising colorectal cancer biomarkers for precision prevention and therapy. *Cancers*.

Turner, J. H. (2024). Theranostics: Timing is everything. In *Cancer Biotherapy & Radiopharmaceuticals*. Mary Ann Liebert, Inc., 611–618 Vol. 39.

Wang, K., Shen, R., Meng, T., Hu, F., & Yuan, H. (2022). Nano-drug delivery systems based on different targeting mechanisms in the targeted therapy of colorectal cancer. *Molecules (Basel, Switzerland), 27*(9), 2981.

Wang, X., Sun, H., & Mu, T. (2024). Materials and structure of polysaccharide-based delivery carriers for oral insulin: A review. *Carbohydrate Polymers, 323*, 121364.

Xiang, C., Zhang, Y., Guo, W., & Liang, X. J. (2020). Biomimetic carbon nanotubes for neurological disease therapeutics as inherent medication. *Acta Pharmaceutica Sinica B, 10*(2), 239–248.

Xu, M., Chen, Z., Zheng, J., Zhao, Q., & Yuan, Z. (2023). Artificial intelligence-aided optical imaging for cancer theranostics. In *Seminars in Cancer Biology*. Elsevier, 62–80.

Younis, M., Kumar Digra, S., Gupta, A., Anand, V., & Kumar, P. (2024). Potential of 2D materials: Novel insights and applications in colorectal cancer research. In *Two-dimensional hybrid composites: Synthesis, properties and applications*. Springer, 351–373.

Zhang, N., Wu, Y., Xu, W., Li, Z., & Wang, L. (2022). Synergic fabrication of multifunctional liposomes nanocomposites for improved radiofrequency ablation combination for liver metastasis cancer therapy. *Drug Delivery, 29*(1), 506–518.

Zhang, P., Xiao, Y., Sun, X., Lin, X., Koo, S., Yaremenko, A. V., et al. (2023). Cancer nanomedicine toward clinical translation: Obstacles, opportunities, and future prospects. *Med, 4*(3), 147–167.

Zhang, W., Lu, Y., Zang, Y., Han, J., Xiong, Q., & Xiong, J. (2022). Photodynamic therapy and multi-modality imaging of up-conversion nanomaterial doped with AuNPs. *International Journal of Molecular Sciences, 23*(3), 1227.

Zhang, X., Tan, X., Wang, P., & Qin, J. (2023). Application of polypyrrole-based electrochemical biosensor for the early diagnosis of colorectal cancer. *Nanomaterials*.

Zhou, J., Chen, L., Chen, L., Zhang, Y., & Yuan, Y. (2022). Emerging role of nanoparticles in the diagnostic imaging of gastrointestinal cancer. *Seminars in Cancer Biology [Internet], 86*, 580–594. https://www.sciencedirect.com/science/article/pii/S1044579X22001031.

Zhou, Y., Gao, Y., Yao, N., Lu, G., Dong, C., Wang, K., et al. (2024). Multi-modal triggered-release sonodynamic/chemo/phototherapy synergistic nanocarriers for the treatment of colon cancer. *Frontiers in Bioengineering and Biotechnology, 12*, 1439883.

CHAPTER

# Surface modified metal nanoparticles in colon cancer

# 10

**Paramvir Singh Tomar[1], Shikha Yadav[1] and Ananda Chettupalli[2]**

[1]*Department of Pharmacy, School of Medical and Allied Sciences, Galgotias University, Greater Noida, Uttar Pradesh, India,*

[2]*Department of Pharmaceutical Technology, School of Health and Medical Sciences, Adamas Knowledge City, Adamas University, North 24 Parganas, Kolkata, West Bengal, India*

## 10.1 Introduction to colon cancer

Colon cancer is classified as colorectal cancer (CRC). It typically begins as benign polyps in the rectum or colon, which may progress to malignant tumours if neglected. CRC is the second most frequent cancer in the world, with over 10% of all cancer diagnoses, and the third most common cancer overall (https://www.who.int/news-room/fact-sheets/detail/colorectal-cancer). CRC originates in the epithelial cells that line the colon and is categorized into five stages according to its advancement (Ibrahim et al., 2023). Incidence rates have declined in Western nations, primarily attributable to the extensive implementation of colonoscopy screening. Nevertheless, the prevalence of the illness among younger individuals is rising (Stoffel & Murphy, 2020). The ongoing advancements in emerging nations are expected to result in a global rise in CRC occurrences, projected to reach 2.5 million new instances by 2035 (Arnold et al., 2017; Bray et al., 2018; Ouakrim et al., 2015). Operative intervention and chemotherapeutic therapy are the primary treatment modalities for CRC, contingent upon the cancer's stage, tumor localization during diagnosis, and the specific features of the patients (Stein et al., 2011). Chemotherapy is typically administered at various phases of treatment and is frequently utilized as postoperative adjuvant therapy for individuals with advanced colorectal carcinoma. It is utilized as neoadjuvant chemotherapy prior to surgical procedures to shrink the tumor before excision (Cunningham et al., 2010). The accessibility of diverse chemotherapeutic protocols has enhanced the favorable prognosis for individuals diagnosed with advanced CRC in recent decades (Hu et al., 2016). While genetic, lifestyle, weight, and environmental factors may be correlated, the precise causes of this increase remain incompletely known (Dekker et al., n.d.). Despite the rate of response to contemporary systemic chemotherapy treatments reaching up to 50%, pharmacological resistance emerges in the majority of CRC patients, thereby constraining the therapeutic effectiveness of anticancer medicines and eventually resulting in chemotherapeutic ineffectiveness (Dallas et al., 2009). Although these methods have enhanced survival rates, they frequently entail considerable disadvantages, including restricted selectivity, systemic toxicity, drug resistance, and the incapacity to eliminate cancer cells entirely.

Nano Drug Delivery Systems for Colon Cancer. DOI: https://doi.org/10.1016/B978-0-443-40443-6.00024-6

### 10.1.1 Risk factors

Epidemiological studies regularly demonstrate robust correlations between the male sex, advancing age, and prevalence of disease. Genetic and genetic risk variables that cause the development of CRC (Fig. 10.1). A favorable familial history appears to contribute to around 10%–20% of CRC cases (Henrikson et al., 2015), involving risk levels that vary contingent upon the quantity and degree of affected kin and the era at which CRC is diagnosed (Schoen et al., 2015). Heritability estimates for CRC, derived from twin and family studies, range from 12% to 35% (Czene et al., 2002; Lichtenstein et al., 2000).

Nanomedicines are thoroughly investigated for oncological treatment. Nanomedicine formulations seek to enhance the equilibrium between medication efficacy and toxicity by facilitating the efficient delivery of drug molecules to diseased areas while minimizing their accumulation in healthy organs and tissues (Lammers, 2024). Metal NPs (MNPs) have drawn considerable interest in cancer therapy owing to their distinctive physicochemical characteristics among the numerous types of NPs. Nanotechnology encompasses a diverse array of nanomaterials possessing significant diagnostic and therapeutic capabilities. Various nanomaterials and nanoformulations are applicable in oncological therapy, particularly CRC (Dezfuli et al., 2024).

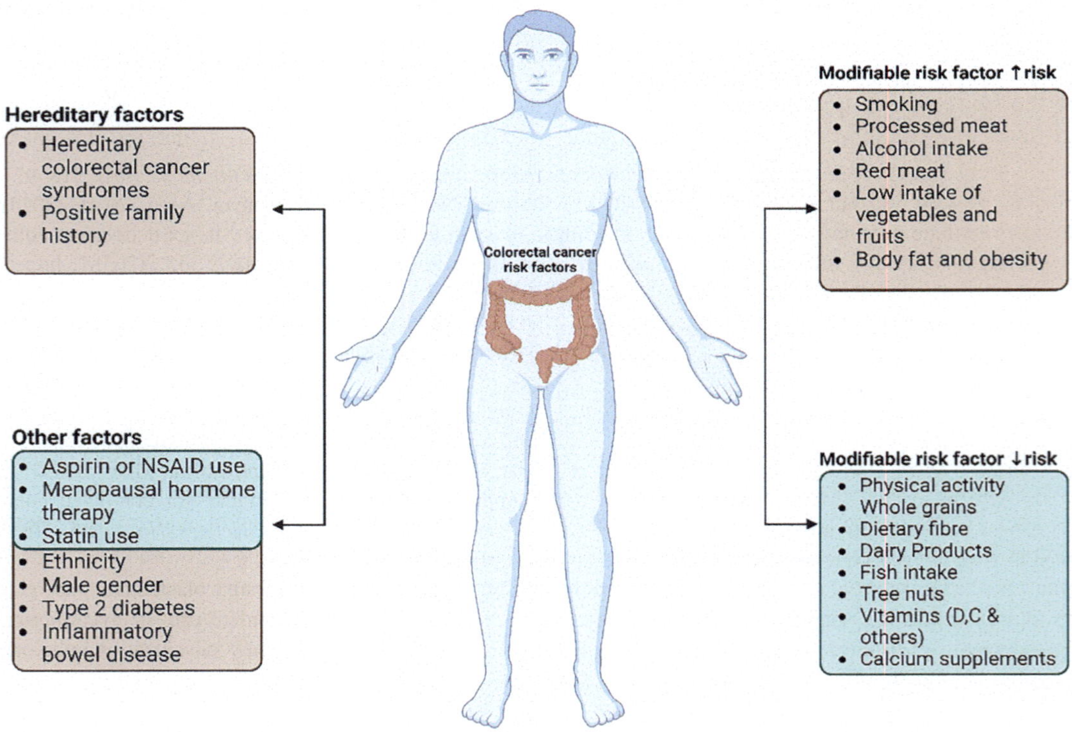


**FIGURE 10.1**

Enumeration of changeable and nonmodifiable risk factors for colorectal cancer.

Diverse multimodal strategies tailored to tumor-specific attributes and patient-related considerations are employed in the management of CRC, encompassing surgical intervention, local ablation of metastases, chemotherapy protocols (such as 5-fluorouracil, oxaliplatin, and irinotecan, either independently or in conjunction with specific agents), targeted therapies, immunotherapy, and radiation therapy (Bae et al., 2011; Zhao et al., 2018). Traditional medicinal medicines are associated with significant adverse effects, even while they yield short-term benefits (Rasouli, Mehdian, et al., 2021, Rasouli, Fallah, et al., 2021). Consequently, it is essential to ascertain efficient strategies to enhance the survival and prognosis of patients with CRC. The primary problem of effective chemotherapy for cancer is limited in its therapeutic use due to its toxicity to healthy tissues and the development of chemoresistance in CRC (Li et al., 2022; Lin et al., 2022).

In recent years, nanomaterials have experienced remarkable advancements, creating significant opportunities for biological applications, especially in the fight against cancer (Durmus et al., 2021). MNPs have drawn considerable interest in cancer therapy due to their distinctive physicochemical characteristics among the numerous types of NPs. Metal ions have been utilized in the field of medicine for approximately 5000 years (Orvig & Abrams, 1999). Metal-based chemicals interact with numerous biological molecules, including hormones, enzymes, proteins, and nucleic acids, rendering them essential for biological investigation (Kostova, 2006). When appropriately surface modified, MNPs can serve as an optimal platform for improving drug administration, imaging, and treatments, especially in colon cancer. Nanoparticle-based medication logistics frameworks have various benefits compared to traditional approaches. These systems are capable of augmenting the efficacy of pharmaceuticals and proteins by extending their longevity, improving drug dissolution, and facilitating targeted controlled release at designated sites. Compared with traditional methods, nanoparticle-based drug-delivery systems offer several advantages, leading to improved therapeutic outcomes (Liu et al., 2023; Shamloo et al., 2021, 2022; Shamloo et al., 2023).

## 10.2 Metal nanoparticles in cancer therapy

Nanotechnology possesses the potential to swiftly identify various molecular signals and biomarkers, resulting in progress in early detection, prognostics, therapeutic approaches, and diagnostics (Bakhtiari et al., 2023; Dessale et al., 2022; Laraib et al., 2022). The application of nanotechnology facilitates highly sensitive and specific quantifications, as well as the capacity for multiplexed evaluations (Timilsina et al., 2022). MNPs have garnered significant interest due to their multifaceted features, rendering them intriguing candidates for diverse uses, particularly in cancer therapy (Table 10.1) (Izhar et al., 2022).

These encompass gold, iron, iron oxide, copper, cerium oxide, silver, calcium, titanium, magnesium, zinc, nickel, barium, and bismuth-based NPs, as recorded in the scholarly publications (Khursheed et al., 2022; Xu et al., 2022). MNPs are crucial in modern oncological research platforms, garnering heightened fascination in this field. The following comparison of MNPs reveals that gold nanoparticles (AuNPs) demonstrate exceptional properties, placing them at the vanguard of research (Bansal et al., 2020). Other MNPs, such as silver NPs (AgNPs), have exhibited comparable efficacy to AuNPs (Ali et al., 2023). MNPs, consisting of metals like gold (Au), silver (Ag), iron oxide ($Fe_3O_4$), and platinum (Pt), exhibit unique physicochemical features that render them appropriate for biomedical applications. Their diminutive dimensions (often 1–100 nm) provide a substantial

**Table 10.1 Evaluating the distinct benefits of metal NPs (*MNPs*) versus non-MNPs in cancer therapy.**

| Standards | Metallic nanoparticles | Nonmetallic nanoparticles |
|---|---|---|
| *Concentration and distribution mechanisms* | Enhanced accuracy in tumor targeting attributable to magnetic characteristics | Enhanced permeability and retentioneffect |
| *Imaging* | Enhanced imaging modalities (e.g., magnetic resonance imaging and CT) | Restricted to particular imaging modalities (e.g., fluorescence) |
| *Pharmacological attributes* | Inherent therapeutic characteristics (e.g., photothermal therapy) | Frequently necessitate functionalization with pharmaceuticals or medicinal agents |
| *Stability* | Gold nanoparticles are extremely stable under physiological conditions | Materials play a role in stability |
| *Toxicity* | Possible cytotoxicity if inadequately designed | Reduced inherent toxicity is the norm |
| *Size and shape control* | Exact regulation of dimensions and form for enhanced performance | Effective regulation, however perhaps more intricate based on the material |
| *Modification of surfaces* | Facile surface modification with diverse ligands and targeting agents | Adaptable surface modification, although may necessitate intricate chemistry |
| *Biocompatibility* | Varies; can be optimized for enhanced biocompatibility | Overall satisfactory, however significantly reliant on the material and its functionalization |
| *Capacity for drug loading* | Significant capacity for medication encapsulation and regulated release | Varies, typically elevated with polymer-based nanoparticles |
| *Thermal production* | Effective in the treatment of hyperthermia (e.g., gold nanoparticles) | Restricted, predominantly utilized in pharmaceutical delivery rather than hyperthermia |
| *Expense* | Frequently costlier due to the materials and synthesis techniques employed | Typically, less expensive |
| *Ecological consequences* | Possible environmental issues if not adequately disposed of | Reduced environmental impact, with biodegradable alternatives accessible |

interfacial area-to-volume ratio, adjustable optical characteristics, and the capacity to traverse biological membranes, rendering them optimal candidates for drug administration and imaging applications (Batsalova et al., 2023). Current investigations, including initial investigations and preliminary tests, have indicated the potential efficacy of MNPs in malignancy therapy (Huang et al., 2020; Sharma et al., 2018). The application of metal-based oncological treatment offers possibilities for developing economically viable therapy alternatives, possibly exceeding the expenses linked to conventional therapies (Shi et al., 2020).

### 10.2.1 Distinct characteristics of metal nanoparticles

MNPs possess distinctive features that render them suitable for cancer treatment. These characteristics encompass size-dependent effects (Zhao et al., 2022), surface plasmon resonance (SPR) (Shipunova et al., 2022), magnetic properties (Mundekkad & Cho, 2022), and catalytic efficacy (Hariharan et al., 2022). These characteristics can be utilized for several purposes, including medication administration, imaging (Zhao et al., 2023), and tumor ablation (Lou et al., 2023).

#### *10.2.1.1 Surface plasmon resonance*

A notable characteristic of MNPs, particularly gold (Au) and silver (Ag), is their capacity to demonstrate SPR (Loiseau et al., 2021). Light interacting with thin metallic sheets can produce collective oscillations of free electrons, resulting in SPR (Soler & Lechuga, 2021). To induce SPR, particular phase-matching criteria must be satisfied, contingent upon the intricate relationship between the permittivity of the plasmonic material and that of the external medium (Enoch & Bonod, 2012). SPR transpires when conduction surface-bound electrons on MNPs vibrate in synchrony with incident light, resulting in augmented scattering and absorption of light at particular wavelengths (Shao et al., 2024). This phenomenon is contingent upon size, as NPs of varying dimensions and configurations display SPR at distinct frequencies (Shao et al., 2024).

- AuNPs: AuNPs have a pronounced SPR effect in the visible to near-infrared (NIR) spectrum, rendering them suitable for optical imaging and photothermal treatment (PTT) (Kumar & Lim, 2023). AuNPs could potentially be used to treat neoplasms and arthritis due to their antiinflammatory properties, attributed to their improved penetration as well as retention impact (Hu, Zhang, et al., 2020). To provide target-specific drug-delivery and tumor treatment, AuNP surfaces have the potential to modify the various antibody and antigen receptors, demonstrating remarkable therapeutic potential. Upon exposure to NIR light, AuNPs absorb and transmute the light energy into heat (Kim & Lee, 2017; Mendes et al., 2017), therefore selectively eliminating cancer cells via hyperthermia (Choi et al., 2022). In CRC, surface-modified AuNPs can be designed to target tumor tissues selectively (Kurokawa et al., 2022), facilitating localized heating while minimizing harm to adjacent healthy tissues (Jiang, Li, et al., 2023).
- AgNPs: AgNPs possess distinctive physical and chemical features, such as visual characteristics, exceptional high electrical conductivity, thermal stability, and antibacterial efficacy; this may provide a supplementary benefit in mitigating infections during cancer therapy (Al-zou'by et al., 2023; Gurunathan et al., 2015). AgNPs effectively scatter and absorb light, with the coloration of colloidal AgNPs contingent upon their morphology and size (Li et al., 2010).

Optimizing the size, shape, and surface chemistry of MNPs can enhance their SPR capabilities for diverse medicinal applications, such as augmenting imaging contrast, facilitating drug distribution, and improving the efficacy of photothermal tumor ablation (Monika and Katare, 2024).

#### *10.2.1.2 Magnetic properties*

Magnetic NPs, especially those made of iron oxide ($Fe_3O_4$), possess considerable potential in cancer treatment owing to their distinctive magnetic characteristics (Żuk et al., 2022). These characteristics facilitate their application in magnetic hyperthermia, magnetic resonance imaging (MRI), and targeted medication administration (Wulandari et al., 2023). The magnetic properties of these NPs enable external manipulation and exact control of their movement within the body through magnetic fields (Rukhsar et al., 2022).

- Magnetic hyperthermia: Hyperthermia (HT) has garnered significant attention in oncology research as a recently proposed noninvasive cancer therapy (Bucharskaya et al., 2022). Characterized as the targeted usage of thermal energy, HT is a reemerged technology that could be integrated with CT and RT. Hyperthermia of malignant neoplasms entails elevating the temperature of the tissue to 40°C–45°C to provoke protein denaturation and impair alternative biological macromolecules, hence inducing apoptosis (Szwed & Marczak, 2024). Additionally, HT can stimulate the initiation of the immunological response and improve circulation and oxygen delivery in neoplastic areas, which exhibit twice the blood flow of normal tissues (Farzin et al., 2022).
- MRI contrast agents: Iron oxide NPs (IONPs) have attracted considerable attention as contrast agents for MRI due to their remarkable magnetic characteristics and biocompatibility (Wang et al., 2024). The capacity of functionalized IONPs to preferentially concentrate in tumor tissues renders them very significant for the prompt identification of cancer, including CRC. These NPs improve contrast in MRI studies, facilitating the detection of tiny cancers or metastases that traditional imaging techniques may overlook (Sun, Fang, et al., 2008). IONPs can be coupled with specific targeting ligands to enhance their tumor-targeting efficacy. A study investigated the application of chlorotoxin-conjugated IONPs for targeted medication administration to tumors and enhancement of MRI contrast (Sun, Fang, et al., 2008).
- Magnetic targeting: As such, MNPs have been integrated for therapeutic purposes alongside diagnostic applications, primarily in oncology (Park et al., 2011). MNPs in drug-delivery systems enhance pharmacological accumulation at targeted areas via magnetic manipulation or generate hyperthermia through magnetic fields, resulting in apoptosis or the activation of pharmaceutical liberation from thermoresponsive magnetic carriers (Moros et al., 2019). Lately, MNPs have been extensively utilized in an experimental context study about cancer treatment, leading to the development of numerous magnetic formulations that demonstrate potential treatment outcomes for several forms of cancer (Baghban et al., 2021).

#### 10.2.1.3 Catalytic activity

Nanomaterials operate in alignment with enzyme kinetics and provide benefits over natural enzymes, such as cost-effectiveness, enhanced stability, simplicity of modification, and tunable catalytic activity (Wang et al., 2018; Wei & Wang, 2013; Wu et al., 2019). MNPs, such as platinum, palladium, and gold, exhibit remarkable catalytic properties advantageous for diverse applications, including cancer treatment (Klebowski et al., 2021). Their distinctive features allow them to engage in biological activities, functioning as catalysts that improve the efficacy of medicinal medicines (Karczmarska et al., 2022). NPs function as catalysts for particular metabolic events that can produce reactive oxygen species (ROS) or other harmful chemicals (Fu et al., 2021).

- Platinum NPs (PtNPs): Platinum-based drugs, such as carboplatin and cisplatin, are essential components of chemotherapy protocols for several malignancies, including testicular, ovarian, and lung cancers (Boeriu et al., 2022). PtNPs are capable of acting as catalysts in the production of ROS in the tumor environment. The production of ROS is significant since it can cause oxidative stress in malignant cells, leading to their destruction in the end. This is the main reason that gives platinum nanoparticles the significant antitumor activity they possess, establishing their role in various cancers (Burlec et al., 2023).

- AuNPs: Besides their SPR properties, AuNPs have catalytic features that promote several biological activities. AuNPs can induce different forms of biological activities because of their unique physicochemical characteristics (Grancharov et al., 2005). These nanostructures improve the rate of reactions and also present a bigger surface area for the activities of the catalysts, which is very important in biological mechanisms (Doty et al., 2005). The efficacy of AuNPs as catalysts in different biological reactions, which include oxidation and reduction reactions, is quite impressive (Moon et al., 2005).
- Palladium NPs (PdNPs): PdNPs display excellent potential and therefore are used in biological applications based on their many attributes, which include high catalytic activity, hydrogen storage, and sensing capabilities (Du, 2024). These properties enable them to absorb wavelengths of light and encourage chemical reactions, which in turn lead to the production of psychotropic chemicals. PdNPs can generate heat from light, which presents exciting opportunities for cancer treatment improvement in a novel way (Cheng et al., 2007). The effectiveness of catalytic processes is crucial, as they have the potential to improve the breakdown of harmful chemicals within the tumor microenvironment, thereby increasing the overall success of therapy (Hrkach et al., 2012).

#### *10.2.1.4 Elevated surface functionalization and surface area*

The distinctive physicochemical characteristics of MNPs, especially their diminutive size and substantial surface area-to-volume ratio, render them exceptionally effective for diverse medicinal applications, including drug administration (Burlec et al., 2023). These attributes augment the interaction of NPs with cells and tissues, hence boosting the effectiveness of drug-delivery mechanisms (Burlec et al., 2023). Initially, surface modification can enhance the stability of NPs against agglomeration (Grancharov et al., 2005). A secondary consideration for altering NPs is to enhance their compatibility with an alternative phase. Metallic particulates can become water soluble when appropriate groups are present (Doty et al., 2005). An additional instance is the incorporation of altered inorganic (nano-)fillers into organic polymers. The adjustment can prevent uniformity and congruence issues among the two phases, hence enhancing the mechanical characteristics of the composite (Moon et al., 2005). NPs, owing to their diminutive dimensions and extensive surface area, offer a beneficial framework for cancer treatment. The vast surface area enhances the adhesion of a larger quantity of therapeutic chemicals, which is essential for optimizing therapeutic efficacy. This attribute not only increases the loading capacity but also facilitates augmented pharmaceutical delivery directly to tumor locations (Du, 2024).

- Efficient drug loading: The pharmacological loading of NPs is highly varied and frequently contingent upon the production procedure. Polymeric formulations created using nanoprecipitation generally attain drug loading of merely 1%–2%, but NPs produced via emulsion/solvent evaporation have exhibited drug entrapment reaching up to 14% (Cheng et al., 2007; Hrkach et al., 2012). Alternative frequently utilized nanocarriers that contain chemotherapeutic agents, including micelles, microemulsions, and liposomes, attain approximately 10% drug loading (Dicko et al., 2010; Dong et al., 2009). MNPs, especially silver and gold, demonstrate a significant surface area-to-volume ratio, essential for their use in drug-delivery systems. This attribute enables a higher quantity of therapeutic molecules to be attached to their surfaces in comparison to larger particles. The augmented surface area enhances the interplay between NPs and biological system entities, hence boosting the efficacy of medication delivery (Kumar et al., 2024). The application of NPs can markedly improve the therapeutic index of medications like 5-fluorouracil in the treatment of malignant cells (Mirzaghavami et al., 2020). When incorporated into nanoparticle formulations, 5-FU can be administered at elevated levels directly to cancer cells, thereby addressing the obstacles associated with medication resistance (Mirzaghavami et al., 2020). The improvement of delivery systems is crucial for enhancing treatment effects (Liu et al., 2018).

- Surface functionalization: NPs can be utilized in oncology for two primary objectives: diagnostics and treatment (Borkowska et al., 2020; Tammaro et al., 2020). Surface modification of NPs constitutes an efficacious technique to address or mitigate problems associated with nanoparticle toxicity and absorption, as both phenomena are intricately connected to the surface makeup of the NPs (Sanità et al., 2020). Various types of nanomaterials possess unique chemical properties and functional groups on their surfaces, which may be utilized within the initial phases of functionalization. The initial phase of surface modification typically involves the application of homo- or hetero-bifunctional crosslinkers to incorporate an organic functional group (R-$NH_2$, R-COOH, etc.) beneficial for the binding of biological substances (Sanità et al., 2020). Noble metals, including gold, can be functionalized using crosslinkers with –SH or –$NH_2$ groups that interact with the metal to establish a covalent bond. These bifunctional linkers, such as thio-carboxylic acids, contain functional groups at the other end for ligand binding (Banihashem et al., 2020). Metal oxides can be readily altered by a ligand exchange technique that involves substituting the original surfaces with functional groups such as diols, amines, carboxylic acids, and thiols, which are advantageous for subsequent processes (Korpany et al., 2016). The predominant linkers for silica NPs are aminosilanes, which incorporate an amino group onto the nanoparticle surface for subsequent bioconjugation (Jung et al., 2012; Rea et al., 2014).

Table 10.2 delineates the categories of NPs, their chemical functionalities and/or compositions, along with the suitable substances or methodologies applicable for surface modification using crosslinkers.

#### 10.2.1.5 Photothermal and photodynamic therapy

MNPs, especially those exhibiting pronounced localized surface plasmon resonance (LSPR), are widely employed in PTT for oncological treatment (Jain et al., 2008). These NPs transform light into heat, effectively annihilating cancer cells when subjected to suitable light sources (Gunzer & Magdy, 2018). Their biocompatibility and little cytotoxicity augment their suitability for medical applications (Bagheri et al., 2023).

In PTT, NPs capture light and transmute it into heat, resulting in localized thermal injury and the eradication of cancer cells. In photodynamic therapy (PDT), NPs generate ROS upon activation by light, resulting in oxidative damage to cancer cells.

- PTT: PTT operates through the absorption of optical energy through photothermal agents such as gold, iron oxide, and graphene-based nanomaterials, subsequently transforming this energy into thermal energy upon exposure to light irradiation. This process necessitates absorbed light, which

**Table 10.2 Summary of the predominant methods used to modify the surface of NPs associated with nanomaterials.**

| Substance | Utilizable functional/ chemical groupings | Illustration of chemical substances and procedures appropriate for surface modification |
|---|---|---|
| Silica | –SiOH | X-Si($OC_2H_5$)$_3$ |
| Precious metals | –Au, –Ag (plasmonic metals) | X-SH, X-$NH_2$ |
| Metal oxide (MOx) | MOx | X-COOH, X-(OH)n, X-$NH_2$ (adsorption) |
| Carbon-based | $sp^2$ hybridized carbon | Oxidation, halogenation, cycloaddition |

is conveyed through electron-electron and electron-phonon relaxation, culminating in the creation of elevated nanoparticle (NP) lattices and the subsequent cooling of NP structures through phonon-phonon relaxation. This also amplifies heat scattering, resulting in a temperature increase of cancer cells adjacent to the NPs (Kadkhoda et al., 2022).

When the gold nanorods (AuNRs) internalized by the cell are stimulated by a light source with a wavelength around the AuNRs' localized surface plasmon (LSP) resonance, this enhanced absorption can provide a potent photothermal therapy (PTT) effect, resulting in the destruction of cancer cells (Hou et al., 2020).

- PDT: MNPs might augment the effectiveness of photodynamic treatment by functioning as photosensitizers that generate ROS upon light exposure (Amirjani et al., 2023). These NPs enhance the absorption of light energy and transform it into forms of oxidative stress that are harmful to cancer cells. The targeted application of these NPs enables accurate localization of tumor cells while reducing harm to adjacent healthy tissues (Amirjani et al., 2023). AgNPs are recognized for producing singlet oxygen when exposed to light irradiation. Singlet oxygen is a highly reactive variant of oxygen capable of inducing cellular apoptosis by destroying cellular constituents, such as lipids, proteins, and DNA (Amirjani et al., 2023; Mota et al., 2021). The generation of singlet oxygen enhances the efficacy of PDT, rendering it a potent instrument in cancer treatment.

### 10.2.2 Obstacles in the utilization of metal nanoparticles

Despite their favorable properties, metallic nanoparticles (NPs) face significant hurdles regarding their clinical utilization. The key challenges include safety concerns for therapeutic exploitation, due to possible toxicity, and rapid elimination by the immune system, which makes it difficult for these NPs to reach effective levels within the tissues (Gao et al., 2015). Additionally, this may cause nonspecific biodistribution, leading to undesirable interactions and side effects (Liu et al., 2016).

## 10.3 Modification of metal nanoparticle surfaces for colon cancer

In addition to their application in targeted drug delivery, the modification of nanoparticle surfaces enhances their properties so that they can selectively interact with certain molecular receptors found in different cells (Darmostuk et al., 2015; El-Hammadi et al., 2017). The alteration of the surface of the nanoparticles is possible in diverse ways, for instance, by combining various biomolecule and ligand attachments (Osborne et al., 1997). This increases the stability and biocompatibility of the nanoparticles and hence promotes better biological activity interactions (Hashkavayi et al., 2017). Polymers are essential in the surface modification of the nanoparticles, especially in stabilizing them in water solution (Jalalian et al., 2013). Polymer coatings prevent agglomeration and facilitate the controlled release of pharmaceutical agents (Yuan et al., 2017). Moreover, these coatings have the potential to enhance cellular absorption and reduce nonspecific binding, thereby leading to improved therapeutic efficacy (Gong et al., 2012). Modified metal nanoparticles exhibit extensive utility in biomedical fields, including drug delivery, imaging techniques, and cancer treatment (Cheema, 2013).

### 10.3.1 Targeting ligands

The surface modification of NPs with tumor-targeting ligands, including peptides, folic acid, and antibodies, or the integration of these ligands into NPs by disulfide bonding, may facilitate targeted intra-tumor medication release (Wang et al., 2012).

- Folic acids: 5-fluorouracil (5FU) is a widely recommended chemotherapeutic drug for the therapy of CRC. It is an analog of uracil with a fluorine atom substituting the hydrogen at the C-5 position (He et al., 2008; Masloub et al., 2016). Owing to its brief plasma half-life (15–20 minutes), elevated dosages of 5FU (400–600 mg $m^{-2}$) are required to attain therapeutic drug levels (Sun, Zhang, et al., 2008). To improve the acknowledgment and uptake of NPs by target tissues, their surfaces can be changed with targeting ligands, including folic acid, integrins, antibodies, transferrin, and polysaccharides (Gao et al., 2015; Sriraman et al., 2016). Folate receptors (FRs) serve as remarkable targets for targeted medication delivery for tumors, as they are overexpressed on the surfaces of numerous malignant cells while being present at lesser frequencies in normal tissues (Zhang & Yao, 2012). Additionally, folic acid, a water-soluble vitamin, serves as the natural ligand for the folate receptor and can be included in NPs to enhance targeted efficacy against tumor cells, owing to its intrinsic elevated affinity, nontoxicity, and diminutive size (Liu et al., 2016). Certain research has indicated that the utilization of folate-targeted NPs enhances the antitumor efficacy of 5FU against cancer cells. It was determined that 5FU-loaded folate-polylactic-co-glycolic acid (PLGA) demonstrated significantly greater toxicity compared to 5FU-loaded PLGA (El-Hammadi et al., 2017).
- Aptamers: Nucleic acid aptamers represent brief RNA or DNA sequences extracted from extensive combinatorial databases by the Systematic Evolution of Ligands by Exponential enrichment (SELEX) methodology. They attach to a diverse range of targets, from tiny molecules to whole cells, with remarkable affinity and selectivity (Darmostuk et al., 2015; Osborne et al., 1997). The primary issue in treating CRC is the low response rate to existing medicines in advanced stages of the illness (Hashkavayi et al., 2017). Consequently, early identification of CRC can be advantageous in disease treatment. At present, colonoscopy is the definitive standard for

CRC detection to detect precancerous lesions (Jalalian et al., 2013). Moreover, some important biomarkers exist whose precise and sensitive detection is crucial for illness prediction and the monitoring of cancer spread and recurrence throughout treatment (Yuan et al., 2017). Certain biomarkers possess potential as radiation response indicators, and their discovery may aid in predicting the success of cancer treatment (Gong et al., 2012). Aptamers, due to their binding properties, might possess numerous applications in pertaining to the observation and treatment of patients' diagnostics, including immunohistochemistry, in vivo tumor imaging, and biosensors for the identification of neoplastic cells and cancer-related biomarkers (Alibolandi et al., 2017; Calzada et al., 2017; Mosafer et al., 2017; Shi et al., 2014).

- Peptides: Integrins are a class of adhesion receptors for the extracellular matrix, significantly influencing various physiological processes, including inflammation, cellular proliferation regulation, cellular migration, and angiogenesis (Hwang & Varner, 2004; Jin & Varner, 2004). Moreover, the overexpression of integrins correlates with enhanced cellular infiltration and metastasis (Pilch et al., 2002). A significant amount of expressiveness was seen in pancreatic, colorectal, and breast cancers of individuals with metastatic diseases (Funahashi et al., 2003; Vonlaufen et al., 2001). The $\alpha v\beta 3$ integrin can be recognized by many compounds featuring the tripeptide sequence Arg-Gly-Asp (RGD) (Hynes, 2002; Jin & Varner, 2004) and, notably, by conformationally constrained cyclic peptides demonstrating greater selectivity, potency, and stability compared to their linear counterparts (Haubner et al., 1997). Nano systems modified with these ligands were created for precise imaging (Arosio et al., 2011; Di Pietro et al., 2016; Kim et al., 2018; Sánchez et al., 2018; Zhu et al., 2018), medication delivery (Mo et al., 2016; Murugan et al., 2016), and phototherapy (Kim et al., 2017; Shi et al., 2018). RGD peptides selectively attach to integrin receptors, which are essential in the adherence and proliferation of neoplastic cells. These receptors are overexpressed in multiple malignancies, including CRC, rendering them appealing targets for therapeutic approaches (Fu et al., 2019; Liolios et al., 2021). The specificity of RGD for integrins promotes improved targeting of malignant tissues (Chakravarty and Chakraborty, 2015). These altered NPs can markedly enhance the efficacy of medication uptake in cancer cells, facilitating a more potent therapeutic response (Sheikh et al., 2022). The design of these NPs improves specificity, guaranteeing that medications are preferentially administered to tumor locations (Zamora et al., 2018).

## 10.3.2 Polymer coatings

Polymers are frequently employed to encapsulate MNPs, enhancing their stability, biocompatibility, and circulation duration. A standard nanoparticle formulation in circulation is opsonized and swiftly eliminated through the mononuclear phagocyte system (MPS), which is responsible for various parameters, including size, surface characteristics, and morphology, mostly influenced by polymer properties (You et al., 2016). The application of biodegradable polymeric NPs for the delivery of medicinal drugs and biologically active molecules has demonstrated considerable curative efficacy, particularly in light of recent developments and rigorous examination of NP characteristics (Prabhu et al., 2015). The findings demonstrated that the optimal dimensions (often between 10 and 200 nm) and stealth characteristics of NPs made from polymeric material facilitate their extravasation through permeable blood vessels and selective aggregation in tumors via the enhanced permeability and retention (EPR) effect (You et al., 2016). The macromolecules utilized for the fabrication of these NPs in question may originate from either organic or synthetic sources. These polymeric NPs possess the capability to enhance medication bioavailability, regulate pharmaceutical release properties, prolong circulation duration, and diminish nonspecific toxicity within the medical field (Singh et al., 2014). Active and passive targeting capabilities of polymeric NPs enable precise targeting of specific tissue regions and improve intracellular medication penetration into tumors while preserving healthy tissues (Kamaly et al., 2012). The remarkable characteristics of biodegradable polymer NPs render them an optimal platform for regulated and targeted medication administration in colon cancer therapy.

Numerous polymers, including PLGA, polylactic acid (PLA), PCL, poly(D,L-lactide), chitosan, and polyethylene glycol (PEG), have been engineered for both passive and active, precise administration of medicinal agents (Prabhu et al., 2015).

### 10.3.3 Antibody conjugation

To enhance the effectiveness of the treatment, targeted ligands may be infused into the multifaceted pharmaceutical delivery system. Monoclonal antibodies (Abs) are the most extensively researched ligands, as they can specifically target diverse tumor types by recognizing biological macromolecules in both healthy and malignant cells, provided they are well selected (Patel et al., 2018; Singh et al., 2018). The deployment of an anti-CD133 antibody is essential for cancer therapy aimed at specific genes for therapeutic intervention, as the cell surface marker CD133 (prominin-1) has been identified in various malignancies, including brain, prostate, and colorectal cancer (CRC). CD133, a five-domain transmembrane glycoprotein, serves as a cell surface marker found in cancer stem cells (CSCs). CSCs are a cohort of cancerous cells linked to tumor growth and metastasis, found in CRC (Belov et al., 2010). This section aims to generate antibody-focused multisubstance and multifunction medication delivery nanoparticle conjugates utilizing superparamagnetic IONPs and oxaliplatin, both encapsulated in PLGA and PEGylated PLGA NPs via the double emulsion approach (Moreno et al., 2008). Surface modification of PLGA NPs was accomplished through conjugation with an anti-CD133 monoclonal antibody (conjugated to Alexa Fluor 488), facilitating the targeting of the PLGA NPs along with their cargo (e.g., OXA), thereby augmenting cargo efficacy and enhancing distribution and concentration at the site of action within CRC cells (Carter et al., 2016; Patel et al., 2018). This methodology aims to mitigate the IPN induced by cumulative doses of OXA, a constraining element in treatment, while simultaneously enhancing therapeutic efficacy through the provision of two distinct agents (i.e., OXA and IONPs) within a single multicomponent particle (Zumaya et al., 2022).

## 10.4 Therapeutic uses of surface-modified metal nanoparticles in colorectal cancer

Among the different categories of NPs, metallic NPs, specifically those made of iron oxide, silver, and gold—exhibit significant potential use as medicine dispensers (Akbari et al., 2020). Their success can be attributed to their distinctive qualities suited for this purpose. A notable characteristic is their elevated surface area-to-volume ratio. This facilitates the inclusion of several drug molecules, thereby enhancing the payload of the therapeutic agent administered and influencing the treatment's efficacy (Banu et al., 2021).

Metallic NPs are extensively investigated for their potential applications in cancer treatment. A promising strategy entails the active targeting of NPs to reach cancerous cells through functionalization with ligands that identify numerous receptors on the surfaces of carcinoma cells (Paswan, 2023).

## 10.4.1 Drug delivery

Modifying the surface chemistry can substantially boost the interaction between NPs and the drug, improving the therapeutic formulation's overall stability (Talha, 2024). This alteration facilitates enhanced regulation of drug release profiles, which is crucial for optimizing therapeutic efficacy while reducing adverse effects (Yan, 2015). The principal issue in colon cancer therapy is to ensure that chemotherapeutic drugs attain sufficient concentrations at the tumor site while minimizing systemic damage. Conventional chemotherapy agents, including 5-fluorouracil (5-FU), irinotecan, and oxaliplatin, frequently induce considerable side effects due to their impact on normal, healthy cells (Alnaffakh & Talei, 2023). MNPs, when surface modified with targeting moieties and functional groups, can transform medication delivery by enhancing drug stability, facilitating targeted administration, and providing controlled release.

### *10.4.1.1 Gold nanoparticle*

Advancements in nanotechnology have facilitated the examination of its physicochemical features, rendering it an ideal material for nanoparticle production. AuNPs, or colloids of gold, typically range in size from 1 to 100 nm and exhibit several morphological forms, comprising nanospheres, nanorods, nanoshells, and nanocages (Kyzioł et al., 2021). They represent the most stable metallic NPs exhibiting distinctive features, contingent on their dimensions, configuration, hollowness/porosity, and additional factors, which are absent in mass substances. They are favored over other metallic NPs due to various advantages, including facile synthesis. Experimental variables may regulate the dimensions and morphology, have a substantial surface-to-volume ratio, demonstrate distinctive optical characteristics, and display exceptional compatibility with a wide range of chemical and biological substances (Chen et al., 2018; Kalimuthu et al., 2020). AuNPs are promising candidates for cancer diagnosis and therapy owing to their physicochemical features and inventive potential, as evidenced by various reports on their application in cancer treatment (Table 10.3) (Soto et al., 2021).

Functionalized AuNPs encapsulating 5-FU or doxorubicin have demonstrated increased cytotoxic efficacy against colon cancer cells. The encapsulated medications attain elevated concentrations within the tumor and facilitate prolonged release, enhancing the efficacy of cancer cell eradication relative to traditional techniques (Giráldez-Pérez et al., 2021; Mulens-Arias et al., 2021). This focused strategy may enhance patient outcomes by diminishing drug resistance frequently seen with conventional chemotherapy treatments (Park et al., 2024).

### *10.4.1.2 Silver nanoparticles*

Among the various varieties of NPs, AgNPs have distinctive properties that render them effective for the treatment of certain ailments (Satapathy et al., 2013). AgNPs exhibit therapeutic potential uses due to their distinctive features, serving as potent comprehensive-range antibacterial agents within fabrics,

**Table 10.3 Utilization of gold NPs (*AuNPs*) for cancer therapy.**

| Therapeutic modality | The mechanism | References |
|---|---|---|
| Photothermal | Utilization as photothermal agents to facilitate an improved conversion of light into heat. AuNPs of diverse geometries exhibit a broad range of maximum absorption (spanning from ultraviolet to near-infrared) and possess elevated absorption segments, necessitating reduced dose durations and lower laser power compared to traditionally utilized dyes. | Baruah et al. (2018), Rahimi-Moghaddam et al. (2019) |
| Radiofrequency | The process of nanoparticle-enhanced heating remains inadequately comprehended; however, various ways of radio sensitization have been suggested. The disparity in energy absorption characteristics of gold (Au) relative to connective tissues facilitates physical dosage amplification and promotes catalytic efficacy through AuNP-mediated electron transfer from surface-attached donor groups to $O_2$, resulting in the generation of superoxide radicals. | Lu et al. (2020) |
| Transporters for medications | AuNPs function as medication transport workers to enhance the drug distribution system of the medication, hence minimizing nonspecific adverse effects and facilitating greater concentrated medication delivery dosages. The effective use of AuNPs is their utilization as carriers for the transport of chemicals into cells. | Lydia et al. (2020) |
| Therapeutics | Reducing the process of angiogenesis, which is involved in the development and proliferation of new veins and arteries for vascular supply. Surface modification using targeted ligands may improve action specificity while concurrently diminishing indiscriminate connections to healthy tissues and cells. | Peng and Liang (2019) |

culinary storage receptacles, antiseptic aerosols, catheters, dressings, and oncological treatments (De Matteis et al., 2018; Qasim et al., 2014). AgNPs are promising medicines for cancer treatment. A minimal dosage of AgNPs causes DNA damage and chromosomal anomalies without significant harmful consequences (AbdelRahim et al., 2017; Zhang et al., 2016). AgNPs are acknowledged for their intrinsic anticancer characteristics, primarily via the development of oxidative damage in neoplastic cells (Hamida et al., 2023). Oxidative damage can cause cellular damage, induce apoptosis, and significantly impede cell proliferation (Motafeghi et al., 2023). As a result, AgNPs have emerged as effective agents against multiple cancer types, positioning them as a focal point in therapeutic research (Hamida et al., 2023). Pharmacological resistance constitutes a widespread impediment in the management of colon malignancy, frequently failing chemotherapy (Zeng et al., 2022). This resistance stems from multiple processes, including the activation of efflux pumps and modifications in drug metabolism (Montazami et al., 2015). AgNPs can be designed to simultaneously transport small interfering RNA (siRNA) and chemotherapeutic agents, tackling the challenge of multidrug resistance (Sun et al., 2023). The application of siRNA to inhibit genes associated with resistance can augment the susceptibility of cancer cells to chemoresistance (Lin et al., 2021).

#### 10.4.1.3 Iron oxide nanoparticles

IONPs own distinctive magnetic attributes that allow them to rapidly react to applied magnetic fields. This responsiveness renders them exemplary choices for applications necessitating precision targeting in drug-delivery systems (Kumar et al., 2018). IONPs in drug delivery are largely distinguished by their capacity to be guided to tumor locations using an external magnetic field. Such a monitoring system serves to increase the amount of drug in a particular area of neoplastic tissues and, therefore, enhance treatment (Shubhra, 2023). The magnetic response is important for precise drug distribution, which is essential given the need for targeting cancers that are difficult to reach safely (Thalmayer et al., 2023). Cancer drugs can be easily loaded onto IONPs and brought to the site of the tumor, delivering these medications directly. This functionalization ensures that the agents remain protected until they reach the desired site, enhancing their action on cancer cells while limiting the effect on normal tissues. Therefore, this approach maximizes the therapeutic benefit in the management of cancer (Shubhra, 2023). Incorporating polymers such as chitosan and dextran on the surface of IONPs has been shown to enhance drug loading capacity and retention significantly (Ullah et al., 2022). These coatings serve to prevent the premature degradation of the incorporated drug, which would be detrimental to the drug's activity once in the systemic circulation, where the drug does not need to act (Ullah et al., 2022). These changes allow nanoparticles to maintain their form until they reach the therapeutic site. Further functionalization of the polymeric-coated IONPs with specific targeting ligands is, however, very necessary for selective internalization to cancer cells (Udaipuria & Bhattacharya, 2024). This anchors the NPs to colon cancer cells and directs the therapeutic application of the NPs (Udaipuria & Bhattacharya, 2024). Such specific distribution of drugs also helps to restrict the damaging effects on normal tissues, therefore reducing unwanted side effects that are commonly associated with the conventional chemotherapy regimen.

#### 10.4.1.4 Platinum nanoparticles

Platinum nanoparticles (NPs) are considered versatile delivery agents in drug-delivery modalities, permitting individualized treatment (Gevariya et al., 2023). These NPs have considerable drug-loading ability and can be tailored to enhance their interaction with cancer cells (Gevariya et al., 2023). Their unique properties allow for better delivery of chemotherapeutics to tumor sites, thus minimizing the general toxicity (Gevariya et al., 2023). Cisplatin and other similar platinum drugs can be linked to the PtNPs to increase their effectiveness against cancer cells (Ms et al., 2015). This improvement solves the problems presented by cisplatin, such as the production of significant side effects and the issue of resistance (Predarska et al., 2022). Most importantly, they enhance the effectiveness of these drugs, which makes PtNPs a breakthrough in cancer therapy. Platinum-based NPs increase the delivery of cisplatin into cancerous cells, thus addressing one of the major challenges of treatment failure. PtNPs could improve the in-cell retention of cisplatin, which in turn improves the effectiveness of the drug against resistant colon carcinoma cells. This enhanced formulation ensures that a higher volume of the active substance reaches the site, which is very important for the therapeutic action to be achieved. Functionalized PtNPs can additionally be accustomed to carry siRNA or micro-RNA (miRNA) against genes associated with resistance to drugs. This facilitates the process of silencing or reducing the expression of the resistance genes, which in turn sensitizes the cancer cells to platinum treatment (Jiang, Yang, et al., 2023). The combination of the therapeutic effect of PtNPs and gene-silencing techniques can greatly improve cancer treatment regimens.

#### 10.4.1.5 Multifunctional metal nanoparticles for combination therapy

Features of the surface of metallic nanoparticles provide a possibility for the simultaneous administration of several chemotherapeutic agents. This tactic not only concerns therapeutic approaches used in the defeat of cancer cells but also enhances the overall anticancer treatment. The delivery of several drugs at once, embedded in a single nanoparticle system, increases the chances of effective delivery to the cancer cells, especially in cases of treatment failure (Li et al., 2018). Combination therapy is a contemporary approach in cancer therapy that combines classical chemotherapeutic agents with gene-silencing agents like siRNA or miRNA to combat cancer (Liang et al., 2024). This two-pronged approach aims to fill the gaps that are often left out in the typical medications, thereby enhancing their effectiveness (Girigoswami & Girigoswami, 2023). The subsequent technique is the codelivery of nanoparticles, where an NP fills in the anticancer syndicate and an immune-modulating drug. This enables tumor-specific drug delivery, thereby increasing the local levels of both drugs and their action against the tumor, respectively (Yang et al., 2022). This method aims to improve the effectiveness of cancer treatment by adding to a chemotherapeutic drug, which may be 5-fluorouracil (5-FU), additional forces such as immune checkpoint blockers or cytokines, rather than traditional methods based on the monotherapy of cytostatics only (Hu, Shang, et al., 2020).

## 10.5 Future directions and challenges

Surface-modified MNPs applications in CRC are looking ahead toward combination treatments with other methods. That is, NPs would also be administered together with chemotherapy, RT, or with immunotherapeutic approaches to better direct cancer cells while minimizing damage to the surrounding healthy tissue. There is a great opportunity for personalized medicine delivery. For instance, NPs can be designed according to the tumor profile of the individual patient, which enhances the treatment outcome as hoped. Moreover, the introduction of the use of biodegradable NPs is also anticipated to counter the long-term toxicity and environmental concerns. Targeting strategies to overcome drug resistance toward chemotherapeutic drugs, such as CRC chemoresistance, are, for example, an active area of research. The invention of a smart drug-delivery system that responds to changes in the tumor environment, for example, the temperature, pH, and controlled release, results in better enhancement of the therapy.

Yet, several obstacles remain. The potential toxicity of MNPs is a major concern that calls for more extensive surface modifications to enhance biocompatibility without compromising effectiveness. Additionally, immune clearance often causes NPs to be eliminated from the body before they reach the targeted tumor, thereby limiting their therapeutic potential. This problem must be addressed through surface modifications that allow for immuno-evasion. Moreover, scalability and the high cost of fabricating nanoparticles present insurmountable issues in the clinical use of these therapies. In addition, the regulatory requirements dictate that all new therapies must undergo rigorous safety and effectiveness evaluations, so new treatments are unlikely to be available soon. It is necessary to resolve these challenges if the surface-modified MNPs are to be fully effective in the treatment of CRC.

## 10.6 Conclusions

Surface-modified MNPs indicate great promise in treating CRC and indeed offer significant advantages over standard therapy. Their unique features, such as enhanced drug-delivery efficiency, the ability to restrict the side effects of therapy, and the possibility to combine them with existing therapies, make them powerful tools for improving treatment results. NPs are used for cancer cell-specific targeting, which helps to prevent systemic toxicity and other complications, like drug resistance. However, although these developments have been made, some issues, like iatrogenic toxicity, too fast immune clearance, and high costs of nanoparticle production, remain unaddressed. Therefore, ongoing studies are critical to improving the biocompatibility, safety, and efficacy of these systems. Moreover, through further improvement and development of large-scale manufacturing processes, surface-modified MNPs hold a promise for superiority when it comes to the therapy of CRC, thus giving patients more progressive and personalized treatment options.

## References


AbdelRahim, K., Mahmoud, S. Y., Ali, A. M., Almaary, K. S., Mustafa, A. E. Z. M. A., & Husseiny, S. M. (2017). Extracellular biosynthesis of silver nanoparticles using *Rhizopus stolonifer*. *Saudi Journal of Biological Sciences, 24*(1), 208–216.

Akbari, M., Morad, R., & Maaza, M. (2020). First principal study of silver nanoparticle interactions with antimalarial drugs extracted from *Artemisia annua* plant. *Journal of Nanoparticle Research, 22*(11), 331.

Ali, S., Chen, X., Shi, W., Huang, G., Yuan, L.ming, Meng, L., et al. (2023). Recent advances in silver and gold nanoparticles-based colorimetric sensors for heavy metal ions detection: A review. *Critical Reviews in Analytical Chemistry, 53*(3), 718–750.

Alibolandi, M., Taghdisi, S. M., Ramezani, P., Hosseini Shamili, F., Farzad, S. A., Abnous, K., et al. (2017). Smart AS1411-aptamer conjugated pegylated PAMAM dendrimer for the superior delivery of camptothecin to colon adenocarcinoma in vitro and in vivo. *International Journal of Pharmaceutics [Internet], 519*(1), 352–364. https://www.sciencedirect.com/science/article/pii/S0378517317300534.

Alnaffakh, N., & Talei, M. R. (2023). Metal nanoparticles as novel drug delivery systems: A review of current challenges and opportunities. *Iraqi Journal of Natural Sciences and Nanotechnology [Internet]*. https://api.semanticscholar.org/CorpusID:268675156.

Al-zou'by, J. Y., Alzoubi, F. Y., Migdadi, A. B., & Al-Zboon, K. (2023). Evaluating the impacts of manufactured silver nanoparticles dispersed in various wastewaters on biochemical oxygen demand kinetics of the resulting wastewaters. *Nanotechnology for Environmental Engineering, 8*(1), 119–129.

Amirjani, A., Shokrani, P., Sharif, S. A., Moheb, H., Ahmadi, H., Ahmadiani, Z. S., et al. (2023). Plasmon-enhanced nano-photosensitizers: Game-changers in photodynamic therapy of cancers. *Journal of Materials Chemistry B [Internet]*. https://api.semanticscholar.org/CorpusID:257424368.

Arnold, M., Sierra, M. S., Laversanne, M., Soerjomataram, I., Jemal, A., & Bray, F. (2017). Global patterns and trends in colorectal cancer incidence and mortality. *Gut, 66*(4), 683–691.

Arosio, D., Manzoni, L., Araldi, E. M. V., & Scolastico, C. (2011). Cyclic RGD functionalized gold nanoparticles for tumor targeting. *Bioconjugate Chemistry, 22*(4), 664–672.

Bae, K. H., Chung, H. J., & Park, T. G. (2011). Nanomaterials for cancer therapy and imaging. *Molecules and Cells, 31*(4), 295–302.

Baghban, R., Afarid, M., Soleymani, J., & Rahimi, M. (2021). Were magnetic materials useful in cancer therapy? *Biomedicine & Pharmacotherapy, 144*, 112321.

Bagheri, S., Farokhnezhad, M., & Esmaeilzadeh, M. (2023). Transition metal dichalcogenide coated gold nanoshells for highly effective photothermal therapy. *Physical Chemistry Chemical Physics [Internet]*. https://api.semanticscholar.org/CorpusID:265162404.

Bakhtiari, M. A., Tavoosi, S. N., Shaygani, H., Tohidloo, M., Seifi, S., & Shamloo, A. (2023). Effects of different wheels on the mobility of thermally driven fullerenes-based nanotrucks. *Sensors and Actuators. A, Physical, 363*, 114769.

Banihashem, S., Nezhati, M. N., & Panahia, H. A. (2020). Synthesis of chitosan-grafted-poly (N-vinylcaprolactam) coated on the thiolated gold nanoparticles surface for controlled release of cisplatin. *Carbohydrate Polymers, 227*, 115333.

Bansal, S. A., Kumar, V., Karimi, J., Singh, A. P., & Kumar, S. (2020). Role of gold nanoparticles in advanced biomedical applications. *Nanoscale Advances, 2*(9), 3764–3787.

Banu, A., Gousuddin, M., & Yahya, E. B. (2021). Green synthesized monodispersed silver nanoparticles' characterization and their efficacy against cancer cells. *Biomedical Research and Therapy, 8*(8), 4476–4482.

Baruah, D., Goswami, M., Yadav, R. N. S., Yadav, A., & Das, A. M. (2018). Biogenic synthesis of gold nanoparticles and their application in photocatalytic degradation of toxic dyes. *Journal of Photochemistry and Photobiology. B, Biology, 186*, 51–58.

Batsalova, T., Vasil'kov, A., Moten, D., Voronova, A., Teneva, I., Naumkin, A., et al. (2023). Bimetallic gold–iron oxide nanoparticles as carriers of methotrexate: Perspective tools for biomedical applications. *Applied Sciences [Internet], 13*(23). https://www.mdpi.com/2076-3417/13/23/12894.

Belov, L., Zhou, J., & Christopherson, R. I. (2010). Cell surface markers in colorectal cancer prognosis. *International Journal of Molecular Sciences, 12*(1), 78–113.

Boeriu, E., Boc, A. G., Borda, A., Negrean, R. A., Feciche, B., Boeriu, A. I., et al. (2022). Insights on lipomatosis after platinum-based chemotherapy use in pediatric oncology: A case report. *Medicina (B Aires) [Internet], 58*(12). https://www.mdpi.com/1648-9144/58/12/1715.

Borkowska, M., Siek, M., Kolygina, D. V., Sobolev, Y. I., Lach, S., Kumar, S., et al. (2020). Targeted crystallization of mixed-charge nanoparticles in lysosomes induces selective death of cancer cells. *Nature Nanotechnology, 15*(4), 331–341.

Bray, F., Ferlay, J., Soerjomataram, I., Siegel, R. L., Torre, L. A., & Jemal, A. (2018). Global cancer statistics 2018: GLOBOCAN estimates of incidence and mortality worldwide for 36 cancers in 185 countries. *CA: A Cancer Journal for Clinicians, 68*(6), 394–424.

Bucharskaya, A. B., Khlebtsov, N. G., Khlebtsov, B. N., Maslyakova, G. N., Navolokin, N. A., Genin, V. D., et al. (2022). Photothermal and photodynamic therapy of tumors with plasmonic nanoparticles: Challenges and prospects. *Materials, 15*(4), 1606.

Burlec, A. F., Corciova, A., Boev, M., Batir-Marin, D., Mircea, C., Cioanca, O., et al. (2023). Current overview of metal nanoparticles' synthesis, characterization, and biomedical applications, with a focus on silver and gold nanoparticles. *Pharmaceuticals [Internet], 16*(10). https://www.mdpi.com/1424-8247/16/10/1410.

Calzada, V., Moreno, M., Newton, J., González, J., Fernández, M., Gambini, J. P., et al. (2017). Development of new PTK7-targeting aptamer-fluorescent and -radiolabelled probes for evaluation as molecular imaging agents: Lymphoma and melanoma in vivo proof of concept. *Bioorganic & Medicinal Chemistry [Internet], 25*(3), 1163–1171. https://www.sciencedirect.com/science/article/pii/S0968089616308203.

Carter, T., Mulholland, P., & Chester, K. (2016). Antibody-targeted nanoparticles for cancer treatment. *Immunotherapy, 8*(8), 941–958.

Chakravarty, R., Chakraborty, S., & Dash, A. (2015). Molecular imaging of breast cancer: Role of RGD peptides. *Mini-Reviews in Medicinal Chemistry [Internet], 15*(13), 1073–1094. https://api.semanticscholar.org/CorpusID:40360904.

Cheema, T. A. (2013). *Synthesis and surface modification of metal oxide nanoparticles for functional nanocomposite and thin film applications*. https://www.api.semanticscholar.org/CorpusID:103022026.

Chen, H., Zhou, K., & Zhao, G. (2018). Gold nanoparticles: From synthesis, properties to their potential application as colorimetric sensors in food safety screening. *Trends Food Sciences Technology, 78*, 83–94.

Cheng, J., Teply, B. A., Sherifi, I., Sung, J., Luther, G., Gu, F. X., et al. (2007). Formulation of functionalized PLGA–PEG nanoparticles for in vivo targeted drug delivery. *Biomaterials, 28*(5), 869–876.

Choi, H. W., Lim, J. H., Kim, C. W., Lee, E., Kim, J. M., Chang, K., et al. (2022). Near-infrared light-triggered generation of reactive oxygen species and induction of local hyperthermia from indocyanine green encapsulated mesoporous silica-coated graphene oxide for colorectal cancer therapy. *Antioxidants [Internet], 11*(1). https://www.mdpi.com/2076-3921/11/1/174.

Cunningham, D., Atkin, W., Lenz, H. J., Lynch, H. T., Minsky, B., Nordlinger, B., et al. (2010). Colorectal cancer. *The Lancet [Internet], 375*(9719), 1030–1047. https://doi.org/10.1016/S0140-6736(10)60353-4.

Czene, K., Lichtenstein, P., & Hemminki, K. (2002). Environmental and heritable causes of cancer among 9.6 million individuals in the Swedish family-cancer database. *International Journal of Cancer. Journal International du Cancer, 99*(2), 260–266.

Dallas, N. A., Xia, L., Fan, F., Gray, M. J., Gaur, P., Van Buren, G., et al. (2009). Chemoresistant colorectal cancer cells, the cancer stem cell phenotype, and increased sensitivity to insulin-like growth factor-I receptor inhibition. *Cancer Research, 69*(5), 1951–1957.

Darmostuk, M., Rimpelova, S., Gbelcova, H., & Ruml, T. (2015). Current approaches in SELEX: An update to aptamer selection technology. *Biotechnol Advances [Internet], 33*(6, Part 2), 1141–1161. https://www.sciencedirect.com/science/article/pii/S0734975015000336.

De Matteis, V., Cascione, M., Toma, C. C., & Leporatti, S. (2018). Silver nanoparticles: Synthetic routes, in vitro toxicity and theranostic applications for cancer disease. *Nanomaterials, 8*(5), 319.

Dekker, E., Tanis, P. J., Vleugels, J. L. A., Kasi, P. M., & Wallace, M. B. (n.d.). *Colorectal cancer*. https://www.thelancet.com/.

Dessale, M., Mengistu, G., & Mengist, H. M. (2022). Nanotechnology: A promising approach for cancer diagnosis, therapeutics and theragnosis. *International Journal of Nanomedicine, 17*, 3735.

Dezfuli, A. A. Z., Abu-Elghait, M., & Salem, S. S. (2024). Recent insights into nanotechnology in colorectal cancer. *Applied Biochemistry and Biotechnology, 196*(7), 4457–4471.

Di Pietro, P., Zaccaro, L., Comegna, D., Del Gatto, A., Saviano, M., Snyders, R., et al. (2016). Silver nanoparticles functionalized with a fluorescent cyclic RGD peptide: A versatile integrin targeting platform for cells and bacteria. *RSC Advances, 6*(113), 112381–112392.

Dicko, A., Kwak, S., Frazier, A. A., Mayer, L. D., & Liboiron, B. D. (2010). Biophysical characterization of a liposomal formulation of cytarabine and daunorubicin. *International Journal of Pharmaceutics, 391*(1–2), 248–259.

Dong, X., Mattingly, C. A., Tseng, M. T., Cho, M. J., Liu, Y., Adams, V. R., et al. (2009). Doxorubicin and paclitaxel-loaded lipid-based nanoparticles overcome multidrug resistance by inhibiting P-glycoprotein and depleting ATP. *Cancer Research, 69*(9), 3918–3926.

Doty, R. C., Tshikhudo, T. R., Brust, M., & Fernig, D. G. (2005). Extremely stable water-soluble Ag nanoparticles. *Chemistry of Materials, 17*(18), 4630–4635.

Du, Y. (2024). Nanoparticles in breast cancer therapy. *MedScien [Internet]*. https://www.api.semanticscholar.org/CorpusID:270330822.

Durmus, I. M., Deveci, I., & Karakurt, S. (2021). Synthesis of silica based nanoparticles against the proliferation of human prostate cancer. *Anti-Cancer Agents in Medicinal Chemistry (Formerly Current Medicinal Chemistry-Anti-Cancer Agents), 21*(18), 2553–2562.

El-Hammadi, M. M., Delgado, Á. V., Melguizo, C., Prados, J. C., & Arias, J. L. (2017). Folic acid-decorated and PEGylated PLGA nanoparticles for improving the antitumour activity of 5-fluorouracil. *International Journal of Pharmaceutics, 516*(1–2), 61–70.

Enoch, S., & Bonod, N. (2012). *Plasmonics: from basics to advanced topics*. Springer, 167.

Farzin, L., Saber, R., Sadjadi, S., Mohagheghpour, E., & Sheini, A. (2022). Nanomaterials-based hyperthermia: A literature review from concept to applications in chemistry and biomedicine. *Journal of Thermal Biology, 104*, 103201.

Fu, L. H., Wan, Y., Qi, C., He, J., Li, C., Yang, C., et al. (2021). Nanocatalytic theranostics with glutathione depletion and enhanced reactive oxygen species generation for efficient cancer therapy. *Advanced Materials [Internet], 33*. https://api.semanticscholar.org/CorpusID:230485555.

Fu, S., Xu, X., Ma, Y., Zhang, S., & Zhang, S. (2019). RGD peptide-based non-viral gene delivery vectors targeting integrin αvβ3 for cancer therapy. *Journal of Drug Targeting [Internet], 27*, 1–11. https://api.semanticscholar.org/CorpusID:4578249.

Funahashi, H., Takeyama, H., Sawai, H., Furuta, A., Sato, M., Okada, Y., et al. (2003). Alteration of integrin expression by glial cell line-derived neurotrophic factor (GDNF) in human pancreatic cancer cells. *Pancreas, 27*(2), 190–196.

Gao, Y., Li, Z., Xie, X., Wang, C., You, J., Mo, F., et al. (2015). Dendrimeric anticancer prodrugs for targeted delivery of ursolic acid to folate receptor-expressing cancer cells: Synthesis and biological evaluation. *European Journal of Pharmaceutical Sciences, 70*, 55–63.

Gevariya, D., Priya, L., Mehta, S., Patel, V., Bhuva, D., Panjwani, D., et al. (2023). Bio-functional mesoporous silica nanoparticles as nano-structured carriers in cancer theranostic review on recent advancements. *Current Drug Targets [Internet]*. https://api.semanticscholar.org/CorpusID:260971014.

Giráldez-Pérez, R. M., Grueso, E., Domínguez, I., Pastor, N., Kuliszewska, E., Prado-Gotor, R., et al. (2021). Biocompatible DNA/5-fluorouracil-gemini surfactant-functionalized gold nanoparticles as promising vectors in lung cancer therapy. *Pharmaceutics [Internet], 13*(3). https://www.mdpi.com/1999-4923/13/3/423.

Girigoswami, A., & Girigoswami, K. (2023). Potential applications of nanoparticles in improving the outcome of lung cancer treatment. *Genes (Basel) [Internet], 14*(7). https://www.mdpi.com/2073-4425/14/7/1370.

Gong, P., Shi, B., Zheng, M., Wang, B., Zhang, P., Hu, D., et al. (2012). PEI protected aptamer molecular probes for contrast-enhanced in vivo cancer imaging. *Biomaterials [Internet], 33*(31), 7810–7817. https://www.sciencedirect.com/science/article/pii/S0142961212007806.

Grancharov, S. G., Zeng, H., Sun, S., Wang, S. X., O'Brien, S., Murray, C. B., et al. (2005). Bio-functionalization of monodisperse magnetic nanoparticles and their use as biomolecular labels in a magnetic tunnel junction-based sensor. *The Journal of Physical Chemistry. B, 109*(26), 13030–13035.

Gunzer, F., & Magdy, M. (2018). Simulating the optical properties of isolated and dimer nanoshells for photothermal therapy. *2018 Advances in Science and Engineering Technology International Conferences (ASET) [Internet]*, 1–6. https://api.semanticscholar.org/CorpusID:49188157.

Gurunathan, S., Park, J. H., Han, J. W., & Kim, J. H. (2015). Comparative assessment of the apoptotic potential of silver nanoparticles synthesized by *Bacillus tequilensis* and *Calocybe indica* in MDA-MB-231 human breast cancer cells: Targeting p53 for anticancer therapy. *International Journal of Nanomedicine*, 4203–4223.

Hamida, R. S., Albasher, G., & Bin-Meferij, M. M. (2023). Apoptotic responses mediated by nostoc-synthesized silver nanoparticles against ehrlich ascites carcinoma tumor-bearing mice. *Journal of Nanomaterials [Internet]*. https://api.semanticscholar.org/CorpusID:256050400.

Hariharan, K., Patel, P., & Mehta, T. A. (2022). Surface modifications of gold nanoparticles: Stabilization and recent applications in cancer therapy. *Pharmaceutical Development and Technology [Internet], 27*, 665–683. https://api.semanticscholar.org/CorpusID:250642172.

Hashkavayi, A. B., Raoof, J. B., Ojani, R., & Kavoosian, S. (2017). Ultrasensitive electrochemical aptasensor based on sandwich architecture for selective label-free detection of colorectal cancer (CT26) cells. *Biosensors and Bioelectronics [Internet], 92*, 630–637. https://www.sciencedirect.com/science/article/pii/S095656631631048X.

Haubner, R., Finsinger, D., & Kessler, H. (1997). Stereoisomeric peptide libraries and peptidomimetics for designing selective inhibitors of the αvβ3 integrin for a new cancer therapy. *Angewandte Chemie International Edition in English, 36*(13-14), 1374–1389.

He, W., Du, Q., Cao, D. Y., Xiang, B., & Fan, L. F. (2008). Study on colon-specific pectin/ethylcellulose film-coated 5-fluorouracil pellets in rats. *International Journal of Pharmaceutics, 348*(1–2), 35–45.

Henrikson, N. B., Webber, E. M., Goddard, K. A., Scrol, A., Piper, M., Williams, M. S., et al. (2015). Family history and the natural history of colorectal cancer: Systematic review. *Genetics in Medicine, 17*(9), 702–712.

Hou, Y.jing, Yang, X.xin, Liu, R.qi, Zhao, D., Guo, C.xu, Zhu, A.chao, et al. (2020). Pathological mechanism of photodynamic therapy and photothermal therapy based on nanoparticles. *International Journal of Nanomedicine,* 6827–6838.

Hrkach, J., Von Hoff, D., Ali, M. M., Andrianova, E., Auer, J., Campbell, T., et al. (2012). Preclinical development and clinical translation of a PSMA-targeted docetaxel nanoparticle with a differentiated pharmacological profile. *Science Translational Medicine, 4*(128) 128ra39-128ra39.

Hu, Q., Shang, L., Wang, M., Tu, K., Hu, M., Yu, Y., et al. (2020). Co-delivery of paclitaxel and interleukin-12 regulating tumor microenvironment for cancer immunochemotherapy. *Advanced Healthcare Materials [Internet], 9*. https://api.semanticscholar.org/CorpusID:216645998.

Hu, T., Li, Z., Gao, C. Y., & Cho, C. H. (2016). Mechanisms of drug resistance in colon cancer and its therapeutic strategies. In *World Journal of Gastroenterology, 22*, Baishideng Publishing Group Co, 6876–6889.

Hu, X., Zhang, Y., Ding, T., Liu, J., & Zhao, H. (2020). Multifunctional gold nanoparticles: A novel nanomaterial for various medical applications and biological activities. *Frontiers in Bioengineering and Biotechnology, 8*, 990.

Huang, H., Feng, W., Chen, Y., & Shi, J. (2020). Inorganic nanoparticles in clinical trials and translations. *Nano Today, 35*, 100972.

Hwang, R., & Varner, J. (2004). The role of integrins in tumor angiogenesis. *Hematology/Oncology Clinics, 18*(5), 991–1006.

Hynes, R. O. (2002). A reevaluation of integrins as regulators of angiogenesis. *Nature Medicine, 8*(9), 918–921.

Ibrahim, I. A. A., Alzahrani, A. R., Alanazi, I. M., Shahzad, N., Shahid, I., Falemban, A. H., et al. (2023). Carbohydrate polymers-based surface modified nano delivery systems for enhanced target delivery to colon cancer - A review. *International Journal of Biological Macromolecules [Internet], 253*, 126581. https://www.sciencedirect.com/science/article/pii/S0141813023034773.

Izhar, F., Imran, M., Izhar, H., Latif, S., Hussain, N., Iqbal, H. M. N., et al. (2022). Recent advances in metal-based nanoporous materials for sensing environmentally-related biomolecules. *Chemosphere, 307*, 135999.

Jain, P. K., Huang, X., El-Sayed, I. H., & El-Sayed, M. A. (2008). Noble metals on the nanoscale: Optical and photothermal properties and some applications in imaging, sensing, biology, and medicine. *Accounts of Chemical Research [Internet], 41*(12), 1578–1586. https://api.semanticscholar.org/CorpusID:206827895.

Jalalian, S. H., Taghdisi, S. M., Shahidi Hamedani, N., Kalat, S. A. M., Lavaee, P., ZandKarimi, M., et al. (2013). Epirubicin loaded super paramagnetic iron oxide nanoparticle-aptamer bioconjugate for combined colon cancer therapy and imaging in vivo. *European Journal of Pharmaceutical Sciences [Internet], 50*(2), 191–197. https://www.sciencedirect.com/science/article/pii/S0928098713002406.

Jiang, R., Li, J., Li, L., Lu, Y., Zhang, H., & Li, M. (2023). Engineering near-infrared plasmonic spiky gold nanostructures for highly efficient surface-enhanced raman spectroscopy-guided cancer hyperthermia therapy. *Advanced Functional Materials [Internet], 34*. https://api.semanticscholar.org/CorpusID:262025423.

Jiang, X., Yang, Q., Qi, R., & Yan, L. (2023). Nanoparticle-mediated delivery of satraplatin to overcome cisplatin drug resistance. *Journal of Functional Biomaterials [Internet], 14*(7). https://www.mdpi.com/2079-4983/14/7/387.

Jin, H., & Varner, J. (2004). Integrins: roles in cancer development and as treatment targets. *British Journal of Cancer, 90*(3), 561–565.

Jung, H. S., Moon, D. S., & Lee, J. K. (2012). Quantitative analysis and efficient surface modification of silica nanoparticles. *Journals of Nanomater, 2012*(1), 593471.

Kadkhoda, J., Tarighatnia, A., Barar, J., Aghanejad, A., & Davaran, S. (2022). Recent advances and trends in nanoparticles based photothermal and photodynamic therapy. *Photodiagnosis Photodynamic Therapy, 37*, 102697.

Kalimuthu, K., Cha, B. S., Kim, S., & Park, K. S. (2020). Eco-friendly synthesis and biomedical applications of gold nanoparticles: A review. *Microchemical Journal, 152*, 104296.

Kamaly, N., Xiao, Z., Valencia, P. M., Radovic-Moreno, A. F., & Farokhzad, O. C. (2012). Targeted polymeric therapeutic nanoparticles: Design, development and clinical translation. *Chemical Society Reviews, 41*(7), 2971–3010.

Karczmarska, A., Adamek, M., El Houbbadi, S., Kowalczyk, P., & Laskowska, M. (2022). Carbon-supported noble-metal nanoparticles for catalytic applications—A review. *Crystals (Basel) [Internet], 12*(5). https://www.mdpi.com/2073-4352/12/5/584.

Khursheed, R., Dua, K., Vishwas, S., Gulati, M., Jha, N. K., Aldhafeeri, G. M., et al. (2022). Biomedical applications of metallic nanoparticles in cancer: Current status and future perspectives. *Biomedicine & Pharmacotherapy, 150*, 112951.

Kim, H. S., & Lee, D. Y. (2017). Photothermal therapy with gold nanoparticles as an anticancer medication. *Journal of Pharmaceutical Investigation [Internet], 47*(1), 19–26. https://doi.org/10.1007/s40005-016-0292-6.

Kim, M., Sahu, A., Kim, G. B., Nam, G. H., Um, W., Shin, S. J., et al. (2018). Comparison of in vivo targeting ability between cRGD and collagen-targeting peptide conjugated nano-carriers for atherosclerosis. *Journal of Controlled Release, 269*, 337–346.

Kim, S. K., Lee, J. M., Oh, K. T., & Lee, E. S. (2017). Extremely small-sized globular poly (ethylene glycol)-cyclic RGD conjugates targeting integrin αvβ3 in tumor cells. *International Journal of Pharmaceutics, 528*(1–2), 1–7.

Klebowski, B., Stec, M., Depciuch, J., Gałuszka, A., Pajor-Swierzy, A., Baran, J., et al. (2021). Gold-decorated platinum and palladium nanoparticles as modern nanocomplexes to improve the effectiveness of simulated anticancer proton therapy. *Pharmaceutics [Internet], 13*(10). https://www.mdpi.com/1999-4923/13/10/1726.

Korpany, K. V., Mottillo, C., Bachelder, J., Cross, S. N., Dong, P., Trudel, S., et al. (2016). One-step ligand exchange and switching from hydrophobic to water-stable hydrophilic superparamagnetic iron oxide nanoparticles by mechanochemical milling. *Chemical Communications, 52*(14), 3054–3057.

Kostova, I. (2006). Platinum complexes as anticancer agents. *Recent Pat Anticancer Drug Discover, 1*(1), 1–22.

Kumar, P., Agnihotri, S., & Roy, I. (2018). Preparation and characterization of superparamagnetic iron oxide nanoparticles for magnetically guided drug delivery. *International Journal of Nanomedicine [Internet], 13*, 43–46. https://api.semanticscholar.org/CorpusID:54776958.

Kumar, P., Chandra, P., Verma, N., Diksha, Sharma, A., Mani, M., et al. (2024). Gold nanoparticles: An emerging novel technology for targeted delivery system for site-specific diseases. *Current Drug Therapy [Internet]*. https://api.semanticscholar.org/CorpusID:271361715.

Kumar, P. P. P., & Lim, D. K. (2023). Photothermal effect of gold nanoparticles as a nanomedicine for diagnosis and therapeutics. In *Pharmaceutics, 15*. Multidisciplinary Digital Publishing Institute (MDPI).

Kurokawa, H., Taninaka, A., Yoshitomi, T., Shigekawa, H., & Matsui, H. (2022). Near-infrared light irradiation of porphyrin-modified gold nanoparticles promotes cancer-cell-specific cytotoxicity. *Molecules [Internet], 27*(4). https://www.mdpi.com/1420-3049/27/4/1238.

Kyzioł, A., Łukasiewicz, S., Sebastian, V., Kuśtrowski, P., Kozieł, M., Majda, D., et al. (2021). Towards plant-mediated chemistry–Au nanoparticles obtained using aqueous extract of Rosa damascena and their biological activity in vitro. *Journal of Inorganic Biochemistry, 214*, 111300.

Lammers, T. (2024). Nanomedicine tumor targeting. In *Advanced Materials, 36*. John Wiley and Sons Inc.

Laraib, U., Sargazi, S., Rahdar, A., Khatami, M., & Pandey, S. (2022). Nanotechnology-based approaches for effective detection of tumor markers: A comprehensive state-of-the-art review. *International Journal of Biological Macromolecules, 195*, 356–383.

Li, K., Dai, Y., Chen, W., Yu, K., Xiao, G., Richardson, J. J., et al. (2018). Self-assembled metal-phenolic nanoparticles for enhanced synergistic combination therapy against colon cancer. *Advanced Biosystem [Internet], 3*. https://api.semanticscholar.org/CorpusID:139698242.

Li, S., Liu, J., Zheng, X., Ren, L., Yang, Y., Li, W., et al. (2022). Tumorigenic bacteria in colorectal cancer: Mechanisms and treatments. *Cancer Biology & Medicine, 19*(2), 147.

Li, W. R., Xie, X. B., Shi, Q. S., Zeng, H. Y., Ou-Yang, Y. S., & Chen, Y. B. (2010). Antibacterial activity and mechanism of silver nanoparticles on *Escherichia coli*. *Applied Microbiology and Biotechnology, 85*, 1115–1122.

Liang, Y., Sun, X. Y., Xu, D. Z., Gao, Y., Tang, Q., Lu, Z., et al. (2024). Codelivery of CPT and siPHB1 with GSH/ROS dual-responsive hybrid nanoparticles based on a [12]aneN3-derived lipid for synergistic lung cancer therapy. *ACS Applied Bio Materials [Internet]*. https://api.semanticscholar.org/CorpusID:269327173.

Lichtenstein, P., Holm, N. V., Verkasalo, P. K., Iliadou, A., Kaprio, J., Koskenvuo, M., et al. (2000). Environmental and heritable factors in the causation of cancer—Analyses of cohorts of twins from Sweden, Denmark, and Finland. *New England Journal of medicine, 343*(2), 78–85.

Lin, D., Zhang, H., Liu, R., Deng, T., Ning, T., Bai, M., et al. (2021). iRGD-modified exosomes effectively deliver CPT1A siRNA to colon cancer cells, reversing oxaliplatin resistance by regulating fatty acid oxidation. *Molecular Oncology, 15*(12), 3430–3446.

Lin, J. F., Hu, P. S., Wang, Y. Y., Tan, Y. T., Yu, K., Liao, K., et al. (2022). Phosphorylated NFS1 weakens oxaliplatin-based chemosensitivity of colorectal cancer by preventing PANoptosis. *Signal Transduction and Targeted Therapy, 7*(1), 54.

Liolios, C., Sachpekidis, C., Kolocouris, A., Dimitrakopoulou-Strauss, A., & Bouziotis, P. (2021). PET diagnostic molecules utilizing multimeric cyclic RGD peptide analogs for imaging integrin αvβ3 receptors. *Molecules [Internet], 26*(6). https://www.mdpi.com/1420-3049/26/6/1792.

Liu, M. C., Ma, X. Q., Xu, Y., Peng, L. H., Han, M., & Gao, J. Q. (2016). Liquid chromatography–tandem mass spectrometry evaluation of the pharmacokinetics of a diacid metabolite of norcantharidin loaded in folic acid-targeted liposomes in mice. *Journal of Pharmaceutical and Biomedical Analysis, 119*, 76–83.

Liu, R., Luo, C., Pang, Z., Zhang, J., Ruan, S., Wu, M., et al. (2023). Advances of nanoparticles as drug delivery systems for disease diagnosis and treatment. *Chinese Chemical Letters, 34*(2), 107518.

Liu, W., Zhu, Y., Wang, F., Li, X., Liu, X., Pang, J., et al. (2018). Galactosylated chitosan-functionalized mesoporous silica nanoparticles for efficient colon cancer cell-targeted drug delivery. *Royal Society Open Science [Internet], 5*. https://api.semanticscholar.org/CorpusID:2133298.

Loiseau, A., Zhang, L., Hu, D., Salmain, M., Mazouzi, Y., Flack, R., et al. (2021). *Core–shell gold/silver nanoparticles for localized surface plasmon resonance-based naked-eye toxin biosensing*. https://api.semanticscholar.org/CorpusID:249343426

Lou, W., Xie, L.ting, Xu, L., Xu, M., Xu, F., Zhao, Q., et al. (2023). Present and future of metal nanoparticles in tumor ablation therapy. *Nanoscale [Internet]*. https://www.api.semanticscholar.org/CorpusID:264092726.

Lu, S. L., Liu, W., Cheng, J. C., Lin, L. C., Wang, C. R., & Li, P. (2020). *Advanced Drug Delivery Reviews, 21*, 1–16.

Lydia, D. E., Khusro, A., Immanuel, P., Esmail, G. A., Al-Dhabi, N. A., & Arasu, M. V. (2020). Photo-activated synthesis and characterization of gold nanoparticles from *Punica granatum* L. seed oil: An assessment on antioxidant and anticancer properties for functional yoghurt nutraceuticals. *Journal of Photochemistry and Photobiology. B, Biology, 206*, 111868.

Masloub, S. M., Elmalahy, M. H., Sabry, D., Mohamed, W. S., & Ahmed, S. H. (2016). Comparative evaluation of PLGA nanoparticle delivery system for 5-fluorouracil and curcumin on squamous cell carcinoma. *Archives of Oral Biology, 64*, 1–10.

Mendes, R., Pedrosa, P., Lima, J. C., Fernandes, A. R., & Baptista, P. V. (2017). Photothermal enhancement of chemotherapy in breast cancer by visible irradiation of Gold Nanoparticles. *Scientific Reports [Internet], 7*(1), 10872. https://doi.org/10.1038/s41598-017-11491-8.

Mirzaghavami, P. S., Khoei, S., Khoee, S., & Shirvalilou, S. (2020). *Folic acid conjugated PEG–PCL–PEG triblock copolymer nanoparticles for enhanced delivery of 5-fluorouracil to HT29 colon cancer cells*. https://www.api.semanticscholar.org/CorpusID:238950086

Mo, J., He, L., Ma, B., & Chen, T. (2016). Tailoring particle size of mesoporous silica nanosystem to antagonize glioblastoma and overcome blood–brain barrier. *ACS Applied Material Interfaces, 8*(11), 6811–6825.

Monika, kr, Katare, R., & Tripathi, S. (2024). Study on synthesis and characterization of metal nanoparticles for biomedical applications. *RESEARCH REVIEW International Journal of Multidisciplinary [Internet]*. https://www.api.semanticscholar.org/CorpusID:271053462.

Montazami, N., Kheir Andish, M., Majidi, J., Yousefi, M., Yousefi, B., Mohamadnejad, L., et al. (2015). siRNA-mediated silencing of MDR1 reverses the resistance to oxaliplatin in SW480/OxR colon cancer cells. *Cellular and Molecular Biology [Internet], 61*(2), 98–103. https://www.api.semanticscholar.org/CorpusID:841409.

Moon, J. H., Shul, Y. G., Han, H. S., Hong, S. Y., Choi, Y. S., & Kim, H. T. (2005). A study on UV-curable adhesives for optical pick-up: I. Photo-initiator effects. *International Journal of Adhesion and Adhesives, 25*(4), 301–312.

Moreno, D., De Ilarduya, C. T., Bandrés, E., Buñuales, M., Azcona, M., García-Foncillas, J., et al. (2008). Characterization of cisplatin cytotoxicity delivered from PLGA-systems. *European Journal of Pharmaceutics and Biopharmaceutics, 68*(3), 503–512.

Moros, M., Idiago-López, J., Asín, L., Moreno-Antolín, E., Beola, L., Grazú, V., et al. (2019). Triggering antitumoural drug release and gene expression by magnetic hyperthermia. *Advanced Drug Delivery Reviews, 138*, 326–343.

Mosafer, J., Abnous, K., Tafaghodi, M., Mokhtarzadeh, A., & Ramezani, M. (2017). In vitro and in vivo evaluation of anti-nucleolin-targeted magnetic PLGA nanoparticles loaded with doxorubicin as a theranostic agent for enhanced targeted cancer imaging and therapy. *European Journal of Pharmaceutics and Biopharmaceutics [Internet], 113*, 60–74. https://www.sciencedirect.com/science/article/pii/S0939641116309651.

Mota, D. R., Lima, G. A. S., Oliveira, H. P. M., & Pellosi, D. S. (2021). Pluronic-loaded silver nanoparticles/photosensitizers nanohybrids: Influence of the polymer chain length on metal-enhanced photophysical properties. *Photochemistry Photobiology [Internet], 98*. https://api.semanticscholar.org/CorpusID:236430723.

Motafeghi, F. S., Gerami, M., Mortazavi, P., Khayambashi, B., Ghassemi-Barghi, N., & Shokrzadeh, M. (2023). Green synthesis of silver nanoparticles, graphene, and silver-graphene nanocomposite using *Melissa officinalis* ethanolic extract: Anticancer effect on MCF-7 cell line. *Iranian Journal of Basic Medical Sciences [Internet], 26*, 57–68. https://api.semanticscholar.org/CorpusID:255340229.

Ms, S., Eb, A., Gayathri, Varma, H., & Mohanan, P. V. (2015). *Attenuation of cisplatin induced toxicity by melatonin, loaded on dextran modified iron oxide nanoparticles: An in vitro study*. https://api.semanticscholar.org/CorpusID:86208651.

Mulens-Arias, V., Nicolás-Boluda, A., Pinto, A., Balfourier, A., Carn, F., Silva, A. K. A., et al. (2021). Tumor-selective immune-active mild hyperthermia associated with chemotherapy in colon peritoneal metastasis by photoactivation of fluorouracil-gold nanoparticle complexes. *ACS Nano [Internet].* https://www.api.semanticscholar.org/CorpusID:231790770.

Mundekkad, D., & Cho, W. C. (2022). Mitophagy induced by metal nanoparticles for cancer treatment. *Pharmaceutics [Internet], 14*(11). https://www.mdpi.com/1999-4923/14/11/2275.

Murugan, C., Rayappan, K., Thangam, R., Bhanumathi, R., Shanthi, K., Vivek, R., et al. (2016). Combinatorial nanocarrier based drug delivery approach for amalgamation of anti-tumor agents in breast cancer cells: An improved nanomedicine strategy. *Scientific Reports, 6*(1), 1–22.

Orvig, C., & Abrams, M. J. (1999). Medicinal inorganic chemistry: Introduction. *Chemical Reviews, 99*(9), 2201–2204.

Osborne, S. E., Matsumura, I., & Ellington, A. D. (1997). Aptamers as therapeutic and diagnostic reagents: Problems and prospects. *Current Opinion in Chemical Biology[Internet], 1*(1), 5–9. https://www.sciencedirect.com/science/article/pii/S1367593197801020.

Ouakrim, D. A., Pizot, C., Boniol, M., Malvezzi, M., Boniol, M., Negri, E., et al. (2015). Trends in colorectal cancer mortality in Europe: Retrospective analysis of the WHO mortality database. *BMJ (Clinical Research ed.), 351.*

Park, J. O., Stephen, Z., Sun, C., Veiseh, O., Kievit, F. M., Fang, C., et al. (2011). Glypican-3 targeting of liver cancer cells using multifunctional nanoparticles. *Molecular Imaging Official Journal of the Society for Molecular Imaging, 10*(1), 2010–7290.

Park, S. M., Kim, D. Y., Lee, K. H., Shin, Y. I., Han, S. C., & Kwon, S. M. (2024). Anti-Tumor Efficacy Of Oleuropein-loaded ZnO/Au mesoporous silica nanoparticle in 5-FU-resistant colorectal cancer Cells. *International Journal of Nanomedicine [Internet], 19*, 2675–2690. https://api.semanticscholar.org/CorpusID:268422814.

Paswan, S. K. (2023). Optimization, and evaluation of hyaluronic acid conjugated PLGA nanoparticles of etoposide for cancer treatment. *Asian Journal of Pharmaceutics (AJP), 17*(1).

Patel, J., Amrutiya, J., Bhatt, P., Javia, A., Jain, M., & Misra, A. (2018). Targeted delivery of monoclonal antibody conjugated docetaxel loaded PLGA nanoparticles into EGFR overexpressed lung tumour cells. *Journal of Microencapsulation, 35*(2), 204–217.

Peng, J., & Liang, X. (2019). *Medicine*, 98, 15313.

Pilch, J., Habermann, R., & Felding-Habermann, B. (2002). Unique ability of integrin αvβ3 to support tumor cell arrest under dynamic flow conditions* 210. *Journal of Biological Chemistry, 277*(24), 21930–21938.

Prabhu, R. H., Patravale, V. B., & Joshi, M. D. (2015). Polymeric nanoparticles for targeted treatment in oncology: Current insights. *International Journal of Nanomedicine,* 1001–1018.

Predarska, I., Saoud, M., Drača, D., Morgan, I., Komazec, T., Eichhorn, T., et al. (2022). Mesoporous silica nanoparticles enhance the anticancer efficacy of platinum (iv)-phenolate conjugates in breast cancer cell lines. *Nanomaterials [Internet], 12*(21). https://www.mdpi.com/2079-4991/12/21/3767.

Qasim, M., Lim, D. J., Park, H., & Na, D. (2014). Nanotechnology for diagnosis and treatment of infectious diseases. *Journal of Nanoscience and Nanotechnology, 14*(10), 7374–7387.

Rahimi-Moghaddam, F., Sattarahmady, N., & Azarpira, N. (2019). Gold-curcumin nanostructure in photothermal therapy on breast cancer cell line: 650 and 808 nm diode lasers as light sources. *Journal of Biomedical Physics & Engineering, 9*(4), 473.

Rasouli, M., Fallah, N., & Bekeschus, S. (2021). Combining nanotechnology and gas plasma as an emerging platform for cancer therapy: Mechanism and therapeutic implication. *Oxidative Medicine and Cellular Longevity, 2021*(1), 2990326.

Rasouli, M., Mehdian, H., Hajisharifi, K., Amini, E., Ostrikov, K., & Robert, E. (2021). Plasma-activated medium induces apoptosis in chemotherapy-resistant ovarian cancer cells: High selectivity and synergy with carboplatin. *Plasma Processes and Polymers, 18*(9), 2100074.

Rea, I., Martucci, N. M., De Stefano, L., Ruggiero, I., Terracciano, M., Dardano, P., et al. (2014). Diatomite biosilica nanocarriers for siRNA transport inside cancer cells. *Biochimica et Biophysica Acta (BBA)-General Subjects, 1840*(12), 3393–3403.

Rukhsar, M., Ahmad, Z., Rauf, A., Zeb, H., Ur-Rehman, M., & Hemeg, H. A. (2022). An overview of iron oxide ($Fe_3O_4$) nanoparticles: From synthetic strategies, characterization to antibacterial and anticancer applications. *Crystals (Basel) [Internet], 12*(12). https://www.mdpi.com/2073-4352/12/12/1809.

Sánchez, A., Ovejero Paredes, K., Ruiz-Cabello, J., Martínez-Ruíz, P., Pingarrón, J. M., Villalonga, R., et al. (2018). Hybrid decorated core@ shell janus nanoparticles as a flexible platform for targeted multimodal molecular bioimaging of cancer. *ACS Applied Material Interfaces, 10*(37), 31032–31043.

Sanità, G., Carrese, B., & Lamberti, A. (2020). Nanoparticle surface functionalization: How to improve biocompatibility and cellular internalization. In *Frontiers in Molecular Biosciences, 7*. Frontiers Media S.A.

Satapathy, S. R., Mohapatra, P., Preet, R., Das, D., Sarkar, B., Choudhuri, T., et al. (2013). Silver-based nanoparticles induce apoptosis in human colon cancer cells mediated through p53. *Nanomedicine Nanotechnology, Biology, and Medicine, 8*(8), 1307–1322.

Schoen, R. E., Razzak, A., Kelly, J. Y., Berndt, S. I., Firl, K., Riley, T. L., et al. (2015). Incidence and mortality of colorectal cancer in individuals with a family history of colorectal cancer. *Gastroenterology, 149*(6), 1438–1445.

Shamloo, A., Boroumand, A., Ebrahimi, S., Kalantarnia, F., Maleki, S., & Moradi, H. (2021). Modeling of an ultrasound system in targeted drug delivery to abdominal aortic aneurysm: A patient-specific in silico study based on ligand–receptor binding. *IEEE Transactions on Ultrasonics, Ferroelectrics, and Frequency Control, 69*(3), 967–974.

Shamloo, A., Ebrahimi, S., Ghorbani, G., & Alishiri, M. (2022). Targeted drug delivery of magnetic microbubble for abdominal aortic aneurysm: An in-silico study. *Biomechanics and Modeling in Mechanobiology, 21*(2), 735–753.

Shamloo, A., Naseri, T., Rahbary, A., Bakhtiari, M. A., Ebrahimi, S., & Mirafzal, I. (2023). In-silico study of drug delivery to atherosclerosis in the human carotid artery using metal–organic frameworks based on adhesion of nanocarriers. *Scientific Reports, 13*(1), 21481.

Shao, E., Tuersun, P., Wumaier, D., Li, S., & Abudula, A. (2024). Analysis and optimization of light absorption and scattering properties of metal nanocages. *Nanomaterials [Internet]*. https://api.semanticscholar.org/CorpusID:273126646.

Sharma, A., Goyal, A. K., & Rath, G. (2018). Recent advances in metal nanoparticles in cancer therapy. *Journal of Drug Targeting, 26*(8), 617–632.

Sheikh, A., Alhakamy, N. A., Md, S., & Kesharwani, P. (2022). Recent progress of RGD modified liposomes as multistage rocket against cancer. *Frontiers in Pharmacology [Internet], 12*. https://www.frontiersin.org/journals/pharmacology/articles/10.3389/fphar.2021.803304.

Shi, H., He, X., Cui, W., Wang, K., Deng, K., Li, D., et al. (2014). Locked nucleic acid/DNA chimeric aptamer probe for tumor diagnosis with improved serum stability and extended imaging window in vivo. *Analytica Chimica Acta [Internet], 812*, 138–144. https://www.sciencedirect.com/science/article/pii/S0003267013015729.

Shi, S., Zhang, L., Zhu, M., Wan, G., Li, C., Zhang, J., et al. (2018). Reactive oxygen species-responsive nanoparticles based on PEGlated prodrug for targeted treatment of oral tongue squamous cell carcinoma by combining photodynamic therapy and chemotherapy. *ACS Applied Material Interfaces, 10*(35), 29260–29272.

Shi, Z., Yang, W., Gu, Y., Liao, T., & Sun, Z. (2020). Metal-nitrogen-doped carbon materials as highly efficient catalysts: Progress and rational design. *Advanced Science, 7*(15), 2001069.

Shipunova, V. O., Belova, M. M., Kotelnikova, P. A., Shilova, O. N., Mirkasymov, A. B., Danilova, N. V., et al. (2022). Photothermal therapy with HER2-targeted silver nanoparticles leading to cancer remission. *Pharmaceutics [Internet], 14*(5). https://www.mdpi.com/1999-4923/14/5/1013.

Shubhra, Q. T. H. (2023). Iron oxide nanoparticles in magnetic drug targeting and ferroptosis-based cancer therapy. *Medical Review [Internet], 3*, 444–447. https://api.semanticscholar.org/CorpusID:262054793.

Singh, A., Talekar, M., Tran, T. H., Samanta, A., Sundaram, R., & Amiji, M. (2014). Combinatorial approach in the design of multifunctional polymeric nano-delivery systems for cancer therapy. *Journal of Material Chemistry B, 2*(46), 8069–8084.

Singh, S., Tank, N. K., Dwiwedi, P., Charan, J., Kaur, R., Sidhu, P., et al. (2018). Monoclonal antibodies: A review. *Current Clinical Pharmacology, 13*(2), 85–99.

Soler, M., & Lechuga, L. M. (2021). Principles, technologies, and applications of plasmonic biosensors. *Journal of Applied Physics, 129*(11).

Soto, K. M., Mendoza, S., López-Romero, J. M., Gasca-Tirado, J. R., & Manzano-Ramírez, A. (2021). Gold nanoparticles: Synthesis, application in colon cancer therapy and new approaches - review. In *Green Chemistry Letters and Reviews, 14*, Taylor and Francis Ltd., 663–676.

Sriraman, S. K., Salzano, G., Sarisozen, C., & Torchilin, V. (2016). Anti-cancer activity of doxorubicin-loaded liposomes co-modified with transferrin and folic acid. *European Journal of Pharmaceutics and Biopharmaceutics, 105*, 40–49.

Stein, A., Atanackovic, D., & Bokemeyer, C. (2011). Current standards and new trends in the primary treatment of colorectal cancer. *European Journal of Cancer, 47*, S312–S314.

Stoffel, E. M., & Murphy, C. C. (2020). Epidemiology and mechanisms of the increasing incidence of colon and rectal cancers in young adults. *Gastroenterology, 158*(2), 341–353.

Sun, C., Fang, C., Stephen, Z., Veiseh, O., Hansen, S. J., Lee, D., et al. (2008). Tumor-targeted drug delivery and MRI contrast enhancement by chlorotoxin-conjugated iron oxide nanoparticles. *Nanomedicine [Internet], 3*(4), 495–505. https://api.semanticscholar.org/CorpusID:20461365.

Sun, W., Zhang, N., Li, A., Zou, W., & Xu, W. (2008). Preparation and evaluation of N3-O-toluyl-fluorouracil-loaded liposomes. *International Journal of Pharmaceutics, 353*(1–2), 243–250.

Sun, X., Zhao, P., Lin, J., Chen, K. M., & Shen, J. (2023). Recent advances in access to overcome cancer drug resistance by nanocarrier drug delivery system. *Cancer Drug Resistance [Internet], 6*, 390–415. https://api.semanticscholar.org/CorpusID:259439258.

Szwed, M., & Marczak, A. (2024). Application of nanoparticles for magnetic hyperthermia for cancer treatment—The current state of knowledge. In *Cancers, 16*. Multidisciplinary Digital Publishing Institute (MDPI).

Talha, M. (2024). Dopamine functionalized coatings for corrosion protection of metallic implants and advanced drug delivery: A review. *Corrosion Reviews [Internet], 42*, 163–183. https://api.semanticscholar.org/CorpusID:267691302.

Tammaro, O., Costagliola di Polidoro, A., Romano, E., Netti, P. A., & Torino, E. (2020). A microfluidic platform to design multimodal PEG-crosslinked hyaluronic acid nanoparticles (PEG-cHANPs) for diagnostic applications. *Scientific Reports, 10*(1), 6028.

Thalmayer, A. S., Zeising, S., Lübke, M., & Fischer, G. (2023). Towards steering magnetic nanoparticles in drug targeting using a linear halbach array. *Advances in Radio Science [Internet]*. https://api.semanticscholar.org/CorpusID:257678556.

Timilsina, S. S., Durr, N., Yafia, M., Sallum, H., Jolly, P., & Ingber, D. E. (2022). Ultrarapid method for coating electrochemical sensors with antifouling conductive nanomaterials enables highly sensitive multiplexed detection in whole blood. *Advanced Healthcare Materials, 11*(8), 2102244.

Udaipuria, N., & Bhattacharya, S. (2024). Novel carbohydrate polymer-based systems for precise drug delivery in colon cancer: Improving treatment effectiveness with intelligent biodegradable materials. *Biopolymers [Internet]*e23632. https://api.semanticscholar.org/CorpusID:272967618.

Ullah, S., Azad, A. K., Nawaz, A., Shah, K. U., Iqbal, M., Albadrani, G. M., et al. (2022). 5-Fluorouracil-loaded folic-acid-fabricated chitosan nanoparticles for site-targeted drug delivery cargo. *Polymers (Basel) [Internet], 14*(10). https://www.mdpi.com/2073-4360/14/10/2010.

Vonlaufen, A., Wiedle, G., Borisch, B., Birrer, S., Luder, P., & Imhof, B. A. (2001). Integrin αvβ3 expression in colon carcinoma correlates with survival. *Modern Pathology, 14*(11), 1126–1132.

Wang, A. Z., Langer, R., & Farokhzad, O. C. (2012). Nanoparticle delivery of cancer drugs. *Annual Review of Medicine, 63*(1), 185–198.

Wang, J., Liu, W., Luo, G., Li, Z., Zhao, C., Zhang, H., et al. (2018). Synergistic effect of well-defined dual sites boosting the oxygen reduction reaction. *Energy Environmental Sciences, 11*(12), 3375–3379.

Wang, S., Li, J., Chen, L., Zeng, J., & Gao, M. (2024). Fe2+-Dominated relaxometric properties of iron oxide nanoparticles as MRI contrast agents. *The Journal of Physical Chemistry Letters [Internet], 15*(34), 8861–8866. https://doi.org/10.1021/acs.jpclett.4c01876.

Wei, H., & Wang, E. (2013). Nanomaterials with enzyme-like characteristics (nanozymes): Next-generation artificial enzymes. *Chemical Society Reviews, 42*(14), 6060–6093.

Wu, J., Wang, X., Wang, Q., Lou, Z., Li, S., Zhu, Y., et al. (2019). Nanomaterials with enzyme-like characteristics (nanozymes): Next-generation artificial enzymes (II). *Chemical Society Reviews, 48*(4), 1004–1076.

Wulandari, R., Paradita, D., Mufti, N., Abadi, M. T. H., Taufiq, A., & Mujamilah (2023). Fe3O4 nanoparticles prepared by coprecipitations method for hyperthermia therapy. *Key Engineering Materials [Internet], 940*, 73–80. https://api.semanticscholar.org/CorpusID:256438078.

Xu, J. J., Zhang, W. C., Guo, Y. W., Chen, X. Y., & Zhang, Y. N. (2022). Metal nanoparticles as a promising technology in targeted cancer treatment. *Drug Delivery, 29*(1), 664–678.

Yan, L. (2015). *Layered double hydroxide and metal–organic framework nanoparticles for drug delivery and bioimaging*. https://api.semanticscholar.org/CorpusID:136269091.

Yang, T., Zhang, Z., Zhang, J., Li, Y., Li, W., Liang, H., et al. (2022). Developing a gallium (III) agent based on the properties of the tumor microenvironment and lactoferrin: Achieving two-agent co-delivery and multi-targeted combination therapy of cancer. *Journal of Medicinal Chemistry [Internet]*. https://www.api.semanticscholar.org/CorpusID:255044182.

You, X., Kang, Y., Hollett, G., Chen, X., Zhao, W., Gu, Z., et al. (2016). Polymeric nanoparticles for colon cancer therapy: Overview and perspectives. In *Journal of Materials Chemistry B, 4*, Royal Society of Chemistry, 7779–7792.

Yuan, B., Jiang, X., Chen, Y., Guo, Q., Wang, K., Meng, X., et al. (2017). Metastatic cancer cell and tissue-specific fluorescence imaging using a new DNA aptamer developed by Cell-SELEX. *Talanta [Internet], 170*, 56–62. https://www.sciencedirect.com/science/article/pii/S0039914017304071.

Zamora, A., Gandioso, A., Massaguer, A., Buenestado, S., Calvis, C., Hernández, J. L., et al. (2018). Toward angiogenesis inhibitors based on the conjugation of organometallic platinum (II) complexes to RGD peptides. *ChemMedChem [Internet], 13*. https://api.semanticscholar.org/CorpusID:49344318.

Zeng, C., Qi, G., Shen, Y., Li, W., Zhu, Q., Yang, C., et al. (2022). DPEP1 promotes drug resistance in colon cancer cells by forming a positive feedback loop with ASCL2. *Cancer Medicine [Internet], 12*, 412–424. https://api.semanticscholar.org/CorpusID:249433950.

Zhang, X. F., Liu, Z. G., Shen, W., & Gurunathan, S. (2016). Silver nanoparticles: Synthesis, characterization, properties, applications, and therapeutic approaches. *International Journal of Molecular Sciences, 17*(9), 1534.

Zhang, Z., & Yao, J. (2012). Preparation of irinotecan-loaded folate-targeted liposome for tumor targeting delivery and its antitumor activity. *AAPS PharmSciTech, 13*, 802–810.

Zhao, C. Y., Cheng, R., Yang, Z., & Tian, Z. M. (2018). Nanotechnology for cancer therapy based on chemotherapy. *Molecules (Basel, Switzerland), 23*(4), 826.

Zhao, R., Xiang, J., Wang, B., Chen, L., & Tan, S. (2022). Recent advances in the development of noble metal NPs for cancer therapy. *Bioinorganic Chemistry and Applications [Internet], 2022*. https://api.semanticscholar.org/CorpusID:246403619.

Zhao, X., He, S., Li, B., Liu, B., Shi, Y., Cong, W., et al. (2023). DUCNP@Mn-MOF/FOE as a highly selective and bioavailable drug delivery system for synergistic combination cancer therapy. *Nano Letters [Internet]*. https://api.semanticscholar.org/CorpusID:255970369.

Zhu, T., Xiong, J., Xue, Z., Su, Y., Sun, F., Chai, R., et al. (2018). A novel amphiphilic fluorescent probe BODIPY–O-CMC–cRGD as a biomarker and nanoparticle vector. *RSC Advances, 8*(36), 20087–20094.

Żuk, M., Podgórski, R., Ruszczyńska, A., Ciach, T., Majkowska-Pilip, A., Bilewicz, A., et al. (2022). Multifunctional nanoparticles based on iron oxide and gold-198 designed for magnetic hyperthermia and radionuclide therapy as a potential tool for combined HER2-positive cancer treatment. *Pharmaceutics [Internet], 14*(8). https://www.mdpi.com/1999-4923/14/8/1680.

Zumaya, A. L. V., Rimpelová, S., Štějdířová, M., Ulbrich, P., Vilčáková, J., & Hassouna, F. (2022). Antibody conjugated PLGA nanocarriers and superparmagnetic nanoparticles for targeted delivery of oxaliplatin to cells from colorectal carcinoma. *International Journal of Molecular Sciences, 23*(3).

CHAPTER

# 11 Nano-based phyto drug delivery systems for the management of colon cancer

**Charmi Dudhat[1], Mohit Shah[1], Mehul Luhar[2], Dikshitkumar Modi[3] and Pranav Shah[4]**

[1]*Department of Pharmaceutics, Institute of Pharmacy, Nirma University, Ahmedabad, Gujarat, India,*
[2]*Department of Pharmaceutics, Vidhyadeep Institute of Pharmacy, Vidhyadeep University, Surat, Gujarat, India,*
[3]*Pharmacy, S. S. Agrawal Institute of Pharmacy, Navsari, Gujarat, India,*
[4]*Dr. Chunibhai Vallabhbhai Patel College of Pharmacy, UKA Tarsadia University (NAAC Accredited), Bardoli, Surat, Gujarat, India*

## 11.1 Introduction

Colorectal cancer (CRC) is one of the most prevalent malignancies globally and represents a growing public health challenge due to its increasing incidence and mortality rates. It is the second most common cause of cancer-related fatalities and the third most common cancer in terms of diagnosis, characterized by uncontrolled proliferation of cells in the colon region of the body (Anonymous. Colorectal Cancer, n.d.; Klimeck et al., 2023). Originating from the epithelial lining of the colon or rectum, CRC develops through a complex, multistep process involving genetic mutations, epigenetic modifications, chromosomal instability, microsatellite instability, and the interplay of environmental and lifestyle components (Bardhan & Liu, 2013). Key regulatory genes such as adenomatous polyposis coli (APC), Kirsten rat sarcoma viral oncogene homolog (KRAS), tumor protein 53 (TP53), suppressor of mothers against decapentaplegic (SMAD4), and phosphatidylinositol-4,5-bisphosphate 3-kinase catalytic subunit alpha (PIK3CA) play pivotal roles in tumor onset and progression by altering cell cycle regulation, apoptosis, and DNA repair mechanisms (Al-Joufi et al., 2022; Pancione et al., 2012).

Critical risk factors associated with colon cancer include age, lifestyle choices, family history, environment, heredity, alcohol consumption, clinical characteristics, inflammatory bowel disease, smoking, and genetic predisposition. CRC risk increases significantly after the age of 50 due to cumulative genetic damage and a decline in cellular repair capacity (Bener et al., 2024; Roshandel et al., 2024). Dietary patterns high in processed meat and low in fiber increase carcinogenic exposure in the colon through the formation of nitrosamines and heterocyclic amines (Tudosie et al., 2022). Symptoms of colon cancer typically emerge in the middle and advanced stages, including blood in stool, anemia, abdominal pain, changes in bowel habits, and unintended weight loss. Some factors responsible for the development of tumors are diet, smoking, a sedentary lifestyle, and lack of physical activity (Chu et al., 2025; Hani et al., 2021). The diagnosis of colon cancer is confirmed using colonoscopy and tissue biopsy techniques, along with advanced imaging tools such as computed tomography (CT), magnetic resonance imaging (MRI), positron emission tomography (PET) scans,

Nano Drug Delivery Systems for Colon Cancer. DOI: https://doi.org/10.1016/B978-0-443-40443-6.00016-7

fecal occult blood test (FOBT), and biomarker profiling, such as carcinoembryonic antigen (CEA) and circulating tumor DNA (ctDNA) (Kwan et al., 2025; Zhang et al., 2023).

Major clinical issues still include recurrence, metastasis, and treatment resistance, even with advancements in surgical procedures, chemotherapy, radiation, and targeted biological therapies (Ashique et al., 2024; Hervieu et al., 2021). Metastasis to the liver and lungs is common due to the colon's venous drainage through the portal system (Dromain et al., 2014). Tumor recurrence and multidrug resistance (MDR) are frequently associated with cancer stem cells (CSC), epithelial-mesenchymal transition (EMT), and overexpression of drug efflux transporters such as P-glycoprotein (Ashique et al., 2024).

In recent years, the incidence and mortality rates have steadily increased in developing nations, driven by changes in food habits and lifestyle (Benarba & Pandiella, 2018). It accounts for 10% of all cancers and holds the second position globally in cancer-related deaths. In 2020, over 1.9 million cases were registered, with 9.3 lakh deaths attributed to colon cancer. By 2040, it is projected that new cases of colon cancer will rise by 63%, with a 73% increase in annual deaths (Morgan et al., 2023; Xi & Xu, 2021). The current treatments for colon cancer comprise surgery, cryotherapy, chemotherapy, radiotherapy, monoclonal antibodies, and personalized or targeted therapies (Huo et al., 2025). However, these medical treatments often face challenges, such as nausea, vomiting, immunosuppression, limited therapeutic efficacy, and hematopoietic suppression. Chemotherapeutic agents, such as 5-fluorouracil, capecitabine, irinotecan, trifluridine, and oxaliplatin, are associated with adverse effects such as constipation, diarrhea, thrombocytopenia, mucositis, vomiting, neutropenia, and gastrointestinal complications. These effects vary depending on the therapeutic regimens, including drug dosage, duration of treatment, route of administration, and disease severity. Although potent chemotherapeutic drugs are available to treat colon cancer, their limited specificity and off-target biological activity frequently result in different dose-limiting adverse effects, which highlights the need for more targeted and effective therapies (Adebayo et al., 2023; Han et al., 2024). To resolve the problems of current treatment challenges, phytotherapy plays an innovative role based on plant-derived bioactive components for chemoprotective mechanism activity. Polyphenols, alkaloids, flavonoids, terpenoids, and organosulfur compounds are examples of naturally occurring secondary metabolites found in plants, known as phytochemicals. They are plant nonnutrient materials that demonstrate chemoprotective activity and reduce the adverse effects of radiotherapy (Esmeeta et al., 2022; Wall-Medrano et al., 2024). These pharmacologically active constituents are obtained from flowers, stems, bark, roots, and leaves. They work by reducing cancer cell metabolism and controlling enzyme activity, either by inhibiting or enhancing their overexpression. Various phytochemicals showcase antitumor activity (including cell proliferation, apoptosis, invasion, and metastasis) through the mechanistic activity of cancer cells. Additionally, they promote the activities of tumor cell apoptosis, as well as antiinflammatory, immunomodulatory, and antiangiogenic effects. Fig. 11.1 describes various molecular pathways that are helpful in the regulation of cancer cells (Das et al., 2023). Furthermore, phytochemicals reduce chemotherapy-induced toxicity, enhance immune response, and improve overall treatment tolerance. The role of dietary phytochemicals in the prevention and treatment of colon cancer has drawn more interest in recent years (Choudhari et al., 2020).

Compounds such as curcumin, resveratrol, quercetin, sulforaphane, epigallocatechin gallate (EGCG), berberine, and genistein have been extensively studied for their ability to induce cell cycle arrest, promote apoptosis, inhibit angiogenesis, suppress EMT, and regulate oxidative stress and

**FIGURE 11.1 Different molecular checkpoints in colon cancer.**

The figure describes various molecular pathways that are helpful for regulating cancer cells.

inflammatory signaling (Atuahene et al., 2025; Islam et al., 2022; McCubrey et al., 2017). They also enhance chemotherapy responsiveness by downregulating nuclear factor kappa B (**NF-κb**), cyclooxygenase-2 (**COX-2**), Signal Transducer and Activator of Transcription 3 (**STAT3**), and pro-inflammatory cytokines, while restoring redox balance and immune surveillance (Fu et al., 2025; Islam et al., 2022). Despite their therapeutic potential, the clinical application of phytochemicals is hindered by poor water solubility, low systemic bioavailability, rapid metabolism, and instability under physiological conditions, which limits their pharmacokinetic and pharmacodynamic performance (Ghosh et al., 2025). These challenges highlight the need for advanced nanotechnology-based delivery systems to maximize their therapeutic efficacy and enable clinical translation. However, phytochemicals offer a multitargeted strategy to address the multifactorial character of colon cancer by concurrently interacting with several signaling pathways, in contrast to single-target pharmacological medicines. Additionally, they frequently exhibit little toxicity to healthy cells, which qualifies them for long-term therapeutic or preventive use (Kim et al., 2024; Yao et al., 2020). The molecular mechanisms, clinical difficulties, and potential future paths in the development of phytochemicals for the treatment of colon cancer are covered in this chapter.

## 11.2 Phytochemicals in colon cancer treatment

Phytoconstituents, nowadays, are gaining significant attention for their potential role in the prevention and treatment of colon cancer. The medicinal plants contain phytochemicals derived from stems, roots, barks, seeds, and leaves. This nutritional composition package, including magnesium, iron, phosphate, vitamin B12, sulfur, and cysteine, helps in improving cancer chemotherapy due to its direct effect on metabolism and its ability to protect the normal mitochondrial cells from death (Esmeeta et al., 2022; Macharia et al., 2022). These herbs contain distinct bioactive components, as shown in Fig. 11.2, and have unique cancer treatment characteristics. They help regulate various

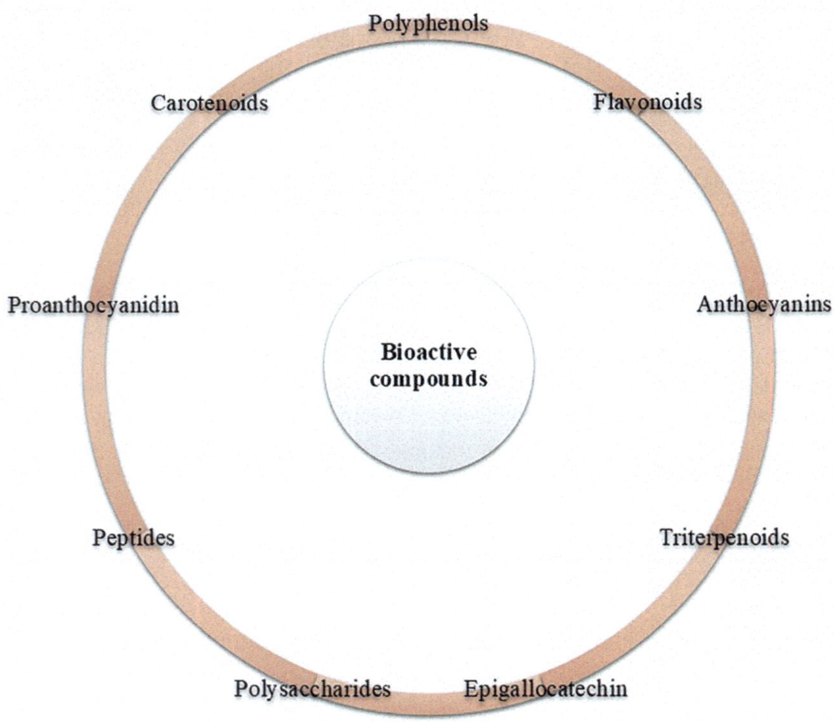


**FIGURE 11.2 Major characteristics of bioactive compounds.**

Different herbs contain distinct bioactive components.

checkpoints of cancer development, such as regulation of the cell cycle, DNA damage repair, controlling cell state transition period (epithelial to mesenchymal), and angiogenesis (Imtiaz et al., 2025).

Apoptosis is initiated via the mitochondrial pathway, where cytochrome C is released into the cytosol. This release triggers the activation of caspases, which in turn activate apoptotic factors, leading to the systemic dismantling of cellular components and programmed cell death. A few of these compounds may help induce pro-apoptotic proteins, that is, caspase-3 and BCL2-associated X protein (BAX), further inhibiting apoptosis factors. Moreover, they also attempt to block the NF-κb pro-inflammatory molecular nuclear factor by interfering with an Ik-B degradation pathway (Ahmed et al., 2019; Chauhan et al., 2021; Esmeeta et al., 2022). The widely used phytochemicals include curcumin (turmeric), resveratrol (grapes), quercetin (apple), genistein (soy), and sulforaphane (cruciferous vegetables). These traditional herbal substances are composed of antiangiogenic, antiproliferative, antiinvasive, antioxidant, antiinflammatory, and antimutagenic activities. The advantages of plant-derived substances are that they help to regulate cellular metabolism and stem cell self-renewal process, and they are safer to use without any side effects (Ganesan et al., 2020; Opara et al., 2025; Rashmy et al.,. n.d.; Talib et al., 2022).

Phytochemicals participate in multiple signaling pathways associated with colon tumor progression and suppression, such as Wingless/integrated-beta catenin signaling pathway (Wnt/β-catenin), phosphoinositide 3-kinase/protein kinase B pathway (PI3K/Akt), mitogen-activated protein kinase/extracellular signal-regulated kinase pathway (MAPK/ERK), Janus kinase/signal transducer and activator of transcription pathway (JAK/STAT), and transforming growth factor-beta pathway (TGF-β) (Paul et al., 2024; Wang et al., 2022). Through the modulation of these cellular mechanisms, they inhibit tumor cell survival, prevent abnormal gene expression, and promote tumor regression. Many phytochemicals function as epigenetic regulators by influencing histone acetylation, methylation, and microRNA expression, thereby modifying cancer-associated transcriptional activity (Ganapathy & Ezekiel, 2019). They also play a pivotal role in suppressing CSCs, which are primarily responsible for recurrence, metastasis, and chemoresistance. By downregulating stemness-associated proteins such as cluster of differentiation 44 (CD44), aldehyde dehydrogenase 1 (ALDH1), and Nanog homeobox protein, phytochemicals assist in reducing tumor cell renewal capacity and increasing sensitivity to standard chemotherapeutics (Taylor & Jabbarzadeh, 2017).

Furthermore, phytochemicals inhibit angiogenesis by reducing the expression of **vascular** endothelial growth factor (VEGF) and matrix metalloproteinases (MMP-2 and MMP-9), which are crucial for tumor invasion and metastasis. Antioxidant phytochemicals neutralize reactive oxygen species (ROS), prevent oxidative DNA damage, and restore cellular redox balance (Fernando et al., 2019). This activity also enhances immune cell activation and modulates the inflammatory response, creating an antitumor microenvironment (TME). Many phytochemicals also exhibit autophagy-inducing capabilities, helping to eliminate damaged or malignant cells without harming normal tissues (Paudel et al., 2023).

Several studies have shown that phytochemicals may be used synergistically with chemotherapeutic drugs to enhance therapeutic efficiency and reduce side effects. For instance, curcumin increases the cytotoxic activity of 5-fluorouracil by inhibiting NF-κB signaling and suppressing drug resistance gene expression (Zhang et al., 2017). Similarly, resveratrol enhances radiosensitivity and reduces oxidative stress-induced tissue damage (Agbele et al., 2020). Quercetin has demonstrated the potential to inhibit P-glycoprotein efflux pumps, thereby increasing intracellular retention of chemotherapeutic agents and effectively overcoming MDR (Zhou et al., 2020). Sulforaphane enhances detoxification enzymes and upregulates phase II metabolic enzymes that promote carcinogen elimination (Hao et al., 2020). Despite their high therapeutic promise, many phytochemicals face challenges such as poor solubility, weak membrane permeability, rapid metabolism, and limited oral bioavailability. This necessitates the development of modern delivery systems, including nanoemulsions, polymeric nanoparticles, solid lipid nanoparticles (SLNs), liposomes, phytosomes, and microencapsulation strategies to improve pharmacokinetic performance. These nanocarriers provide sustained drug release, increase tumor targeting via enhanced permeability and retention (EPR) effect, and reduce systemic toxicity (Dutt et al., 2023).

Overall, phytochemicals represent an evolving and powerful approach in colon cancer therapy due to their multitarget action, natural availability, safety margin, and ability to complement existing treatment strategies. By modulating key signaling cascades and enhancing apoptosis while reducing inflammation and oxidative stress, they provide an innovative and integrative therapeutic platform with the potential to transform future oncology practices.

## 11.3 Mechanistic basis of phytochemicals in colon cancer

Different processes that affect cellular signaling, apoptosis, inflammation, and oxidative equilibrium are how phytochemicals carry out their anticancer effects. The pathways listed below are especially pertinent to the development of colon cancer:

1. **Mechanisms of antioxidants and antiinflammation:** The development and spread of colon cancer are fueled by ongoing inflammation and oxidative damage. Curcumin, quercetin, and resveratrol are examples of phytochemicals that neutralize ROS and inhibit pro-inflammatory mediators such as NF-κB, COX-2, and inducible nitric oxide synthase (iNOS). This lessens inflammatory signals, genetic instability, and DNA damage. Chronic inflammation activates cytokines like TNF-α, IL-1β, and IL-6 within the TME, which stimulate oncogenic pathways and promote tumor initiation. Phytochemicals restore redox balance by enhancing antioxidant enzymes such as glutathione peroxidase, catalase, and superoxide dismutase (SOD), thereby controlling oxidative stress and preventing DNA mutations (Bardelčíková et al., 2023; Majumdar et al., 2009).
2. **Cell cycle arrest and apoptosis induction:** Apoptosis resistance is one of the characteristics of cancer. By downregulating antiapoptotic proteins (Bcl-2, survivin) and upregulating pro-apoptotic proteins (Bax, Bak, caspases), phytochemicals induce apoptosis again. Certain compounds, such as apigenin, genistein, and berberine, suppress unchecked proliferation by causing cell cycle arrest at the G1 or G2/M phases. Many phytochemicals also reactivate p53 signaling, enhance cytochrome c release from mitochondria, and activate caspase cascades, thereby promoting programmed cell death. Additionally, some activate extrinsic apoptosis pathways through Fas and tumor necrosis factor-related apoptosis-inducing ligand (TRAIL) receptors, making tumor cells more sensitive to immune destruction (Abraha & Ketema, 2016; Xu et al., 2016).
3. **Modulation of signaling pathways:** The process of signaling pathway modulation oncogenic signaling cascades such as the MAPK/ERK, JAK/STAT, PI3K/Akt/mTOR, and Wnt/β-catenin pathways are targeted by phytochemicals. Wnt/β-catenin signaling, crucial for CRC proliferation

and metastasis, is inhibited by curcumin and luteolin, preventing nuclear translocation of β-catenin and suppressing downstream oncogenes such as **cellular myelocytomatosis (c-Myc)** and **Cyclin-D1**. Resveratrol and EGCG inhibit PI3K/Akt signaling, enhancing chemosensitivity, while flavonoids such as apigenin modulate MAPK signaling, reducing inflammatory gene transcription and metastatic potential (Imtiaz et al., 2025; Paul et al., 2024).

4. **Epigenetic regulation:** By modifying DNA methylation, histone acetylation, and microRNA expression, phytochemicals affect the expression of genes. Tumor suppressor genes that have been silenced by hypermethylation can be reactivated by sulforaphane and resveratrol. For example, genistein regulates chromatin remodeling enzymes such as DNA methyltransferase (DNMTs) and histone deacetylases (HDACs), restoring tumor suppressor activity. Curcumin modulates miR-21 and miR-34a expression, inhibiting invasion and promoting apoptosis, showcasing phytochemicals' role as natural epigenetic regulators (dos Santos et al., 2020).
5. **Inhibition of angiogenesis and metastasis:** Angiogenesis and metastasis are essential for the progression of tumors. Flavonoids like apigenin and kaempferol reduce neovascularization and invasion by blocking MMP-2, MMP-9, and VEGF. Phytochemicals regulate EMT by reducing N-cadherin and vimentin levels and increasing E-cadherin expression, suppressing metastatic spread. They also target CSCs, reducing expression of stemness markers such as **CD44** and**ALDH1,** which lowers recurrence and therapy resistance (Luo et al., 2009).

Overall, phytochemicals act synergistically to disrupt tumor growth, metastasis, and recurrence by targeting the TME and multiple signaling networks. Their ability to modulate oxidative stress, restore apoptosis, suppress angiogenesis, and regulate oncogenic signaling highlights their promise as complementary and integrative therapeutic agents in colon cancer treatment.

## 11.4 Prominent phytochemicals studied in colon cancer

In preclinical models of colon cancer, a variety of phytochemicals have shown strong anticancer properties. Among the most extensively researched are:

1. Curcumin—It is a yellow polyphenol obtained from *Curcuma longa* (turmeric), regulates several signaling pathways, such as COX-2, STAT3, and NF-κB. It decreases angiogenesis, suppresses proliferation, and encourages apoptosis. Poor solubility and bioavailability, however, limit its therapeutic use; as a result, formulations involving nanoparticles, liposomes, and phospholipids have been developed to improve systemic absorption. In addition to its role in modulating inflammatory pathways, curcumin inhibits EMT by downregulating transcription factors including Snail, Slug, and Twist, thereby reducing metastatic spread. Curcumin also interferes with CSC maintenance by regulating surface markers such as CD44 and epithelial cell adhesion molecule (EpCAM), further reducing recurrence rates. It enhances chemotherapeutic sensitivity when used in combination with drugs such as 5-fluorouracil and oxaliplatin by lowering drug resistance-mediating proteins (P-glycoprotein, accelerated blood clearance [ABC] transporters). Recent investigations highlight curcumin-loaded polymeric nanoparticles and curcumin phytosomes as promising approaches to improve targeting, circulation time, and tissue bioavailability in CRC therapy (Ojo et al., 2022; Shnaikat et al., 2024).

2. Berberine—In colon cancer cells, berberine, an isoquinoline alkaloid present in *Berberis* species, has pro-apoptotic and antiproliferative properties. It causes autophagy, alters the composition of the gut microbiota, and targets the AMP-activated protein kinase (AMPK) and Wnt/β-catenin pathways. 5-fluorouracil therapy improves chemotherapeutic efficacy while lowering toxicity. Berberine has also been reported to induce mitochondrial apoptosis by activating caspase-8 and caspase-9, leading to suppression of tumor growth. It regulates glucose and lipid metabolism, counteracting metabolic advantages cancer cells rely on for rapid proliferation. Another important role of berberine is its capacity to modulate gut microbiota composition by increasing beneficial strains such as *Bifidobacterium* and *Lactobacillus*, which contribute to reduced inflammation and enhanced intestinal barrier integrity. Its antimetastatic role is further demonstrated by suppression of MMP-2, MMP-9, and VEGF expression, which restricts tumor angiogenesis and invasion. Novel delivery platforms including berberine-loaded liposomes, nanoemulsions, and SLNs are being explored to overcome limitations associated with poor solubility and gastrointestinal degradation (Sun et al., 2022).
3. Resveratrol—Resveratrol, a stilbene polyphenol found in grapes and red wine, suppresses angiogenesis by downregulating VEGF and activating the tumor suppressor p53 and AMPK. Analogs and nanoformulations hold promise for enhancing therapeutic potential despite their limited bioavailability. Resveratrol exerts antioxidant and antiinflammatory effects by reducing ROS levels and inhibiting NF-κB signaling. It disrupts cancer cell proliferation and promotes apoptosis by activating intrinsic mitochondrial pathways and enhancing caspase-3 and caspase-7 activity. Resveratrol inhibits COX-2 and IL-6 production, reducing tumor-promoting inflammation and improving chemotherapy responsiveness. It also blocks the PI3K/Akt/mTOR pathway, which is crucial for tumor growth and survival. Due to its ability to inhibit EMT and CSC characteristics, resveratrol plays an essential role in preventing metastasis and recurrence. Novel nanoencapsulation approaches such as resveratrol-loaded polymeric micelles and resveratrol-phospholipid complexes have demonstrated improved absorption and increased tissue targeting (Roshani et al., 2022).
4. Sulforaphane—which is included in cruciferous vegetables like broccoli, suppresses the characteristics of CSC and their ability to spread by inhibiting HDACs and inducing phase II detoxification enzymes. Additionally, it works in concert with chemotherapy drugs. Sulforaphane activates Nrf2-dependent cytoprotective pathways, stimulating antioxidant defense mechanisms that reduce oxidative DNA damage. It also plays a key role in epigenetic regulation through suppression of DNMTs and modulation of oncogenic microRNAs. By inhibiting HDACs, sulforaphane reactivates tumor suppressor genes and enhances apoptosis signaling. It prevents angiogenesis and metastasis by inhibiting VEGF and focal adhesion kinase/akt kinase (FAK/Akt) pathways. Sulforaphane has shown significant potential in enhancing chemotherapeutic efficacy when coadministered with agents like cisplatin, paclitaxel, and 5-fluorouracil. Its safety profile and dietary accessibility make it an attractive candidate for long-term chemoprevention.
5. Flavonoids (quercetin, apigenin, luteolin)—These substances control a number of signaling cascades, including the MAPK and PI3K/Akt pathways. Specifically, quercetin reduces angiogenesis, oxidative stress, and inflammation while improving the chemotherapeutic response. Apigenin and luteolin target the expression of cyclin D1 and β-catenin to have antiproliferative actions. Flavonoids enhance immune system regulation by modulating T-cell and macrophage activity. They also suppress EMT progression and cancer migration by inhibiting TGF-β signaling

and reducing MMP expression. Quercetin, in particular, sensitizes resistant cancer cells to chemotherapy by inhibiting drug efflux transporters such as P-gp, enabling higher intracellular drug concentrations. Luteolin promotes autophagic cell death and regulates apoptosis-related proteins. Apigenin interferes with tumor metabolism by reducing glycolytic enzyme expression, thereby inhibiting cancer cell nutrient utilization (Tuli et al., 2023).

6. EGCG—EGCG, the main catechin in green tea, increases apoptosis, inhibits DNA methyltransferase activity, and obstructs metastatic routes. Its promise as a chemopreventive drug in populations at high risk for CRC is demonstrated by clinical trials. EGCG suppresses inflammation and oxidative stress by modulating cytokines and antioxidant pathways. It inhibits Akt and ERK phosphorylation, reducing tumor cell proliferation. EGCG also prevents adhesion and invasion of cancer cells by suppressing extracellular matrix (ECM) degradation enzymes and integrin signaling. Furthermore, it induces mitochondrial dysfunction and triggers autophagy-mediated death in colon cancer cells. Nanoformulations such as EGCG-loaded poly (lactic-co-glycolic acid) (PLGA) nanoparticles and lipid-based carriers have improved its metabolic stability and therapeutic potency (Das et al., 2024).

## 11.5 Limitations of phytochemicals in traditional formulations

Bioactive constituents are abundant, yet addressing formulation challenges and advancing these compounds to clinical trials remains a significant hurdle. One key issue with herbal components is achieving commercialization without compromising their effectiveness or bioavailability. Achieving consistent results is often challenging due to the natural variability of plant-derived compounds. Research suggests that phytochemicals exhibit greater effectiveness when extracted through semisynthetic pathways. However, this approach reduces yield and contributes to the depletion of plant resources, raising sustainability concerns. Improving extraction methods, advanced formulation technologies, and strategies to enhance stability and scalability are essential for bringing herbal-based products into clinical and commercial applications (Ojo et al., 2022; Unnikrishnan Meenakshi et al., 2024).

The other challenges faced by phytochemicals as chemopreventive agents are that some have toxic properties or may even act as tumor promoters; clinical records have not yet documented significant toxicity profiles for most phytochemicals. This underscores the importance of careful selection of appropriate phytochemicals for cancer prevention (Choudhari et al., 2020). Regulatory challenges include a lack of specific Food and Drug Administration (FDA) protocols for testing and approving traditional plants and their compounds (Anonymous. New Plant Variety Regulatory Information| FDA, n.d.). Additionally, certain phytochemicals act as antinutrients, impairing the body's ability to absorb vital nutrients. Challenges in formulation development include issues with solubility, extract purity, and bioavailability, which hinder the efficacy of many plant-derived compounds (Pathaw et al., 2022). Low solubility and limited absorption often restrict the therapeutic action of many bioactive compounds. The concentration of extract plays a vital role in determining the appropriate dosage and method of drug delivery. Additionally, plant materials with low oral bioavailability or poor permeation rates in transdermal delivery result in reduced absorption at a target site, ultimately lowering the bioavailability. A major limitation of phytochemicals in conventional dosage forms is their **instability** under physiological environments such as pH variations, enzymatic degradation, heat, and light sensitivity. Many compounds undergo rapid first-pass metabolism in the liver and intestine, significantly reducing their systemic exposure (Aqil et al., 2013; Cao et al., 2020). For instance, curcumin and resveratrol degrade quickly and are rapidly eliminated, leading to insufficient therapeutic levels. Similarly, hydrophobic compounds exhibit poor solubility in aqueous biological fluids, preventing efficient membrane permeation.

## 11.6 How nanotechnology can overcome limitations

Management of colon cancer requires a multidisciplinary approach to cure it. Worldwide, nanotechnology has gained significant effectiveness in enhancing the bioavailability of drugs for colon cancer from acute to chronic conditions. The nanoformulations may be beneficial in treating this cancer due to their unique characteristics, such as submicron size and site-specific drug delivery. They can target multiple sites simultaneously through dual-drug loading, increasing specificity, triggering drug release, and actively influencing the molecular mechanism of cancer cells (Kemp & Kwon, 2021). There are various types of nanocarriers, such as polymer and lipid bases, dendrimers, metallic, micelles, peptides, and hydrogels, and they have different structures and chemical properties. Due to their nano-size (100 nm) feature, they provide a large surface area to penetrate, enable targeted drug delivery, and reduce the adverse effects. Nanocarriers such as liposomes, niosomes, and polymeric nanoparticles help to encapsulate these drugs to enhance solubility. These improve the absorption and bioavailability (Kumari Singh et al., 2024).

The submicron dimension of these carriers further supports efficient epithelial barrier penetration and improves drug uptake at the tumor site, translating to superior therapeutic efficacy. Nanotechnology also enables controlled and sustained release, ensuring a prolonged therapeutic window. Moreover, ligand-functionalized nanocarriers allow combinational therapies through coencapsulation of agents within systems such as cubosomes, transfersomes, metallic nanoparticles (gold, silver, platinum), SLNs, and nanostructured lipid carriers (NLCs). This multifunctional approach improves precision and reduces off-target toxicity.

One of the most significant advantages of nanotechnology is its ability to protect phytoactive molecules from premature degradation in the gastrointestinal tract. Encapsulation within a nanocarrier matrix prevents enzymatic hydrolysis, oxidation, and photodegradation, thereby maintaining the pharmacological integrity of unstable phytochemicals like curcumin, quercetin, and resveratrol. Additionally, stimuli-responsive nanocarriers can be engineered to release their payload selectively in response to acidic pH, elevated ROS levels, enzyme overexpression, or inflammatory mediators within the TME. Nanocarriers also improve intracellular delivery via receptor-mediated endocytosis, overcoming the diffusion limitations associated with free phytochemicals. Surface modifications such as PEGylation enhance systemic circulation time by preventing immune recognition and clearance by the reticuloendothelial system (RES). Furthermore, active targeting strategies using ligands like folic acid, hyaluronic acid, transferrin, aptamers, and antibodies enable selective binding to overexpressed receptors on colon cancer cells, including CD44, epidermal growth factor receptor (EGFR), folate receptor-α, and CXCR4, ensuring precise tumor localization and minimal impact on healthy colon tissue.

Nanotechnology also paves the way for codelivery platforms, where phytochemicals and conventional chemotherapeutic drugs (such as 5-FU, irinotecan, or oxaliplatin) are encapsulated together to achieve synergistic effects, overcome MDR, and suppress survival pathways such as PI3K/Akt/mTOR and NF-κB, leading to improved apoptosis and reduced tumor relapse. In addition to cytotoxic effects, nanoplatforms can modulate the TME by inhibiting angiogenesis, altering ECM architecture, and reprogramming tumor-associated macrophages (TAMs) to an antitumor phenotype.

Nanocarriers also exploit the EPR effect, enabling passive accumulation at tumor sites due to leaky vasculature and impaired lymphatic drainage. This can be further enhanced through colon-specific delivery strategies, including pH-responsive coatings (Eudragit, chitosan derivatives), microbiota-activated polysaccharide carriers, mucoadhesive nanoparticles, and time-programmed release systems, ensuring that phytochemicals bypass early absorption and release selectively in the colon. Thus, integrating nanotechnology with phytochemical-based therapy represents a paradigm shift in colon cancer management by enhancing solubility, protecting active molecules from degradation, improving membrane permeability, enabling targeted localization, modulating immune response, bypassing MDR mechanisms, and improving patient compliance. This innovative approach lays the groundwork for next-generation phyto-nanomedicine platforms with high translational potential for colon cancer therapy.

## 11.7 Need for phytochemical-based nanoformulations in colorectal cancer therapy

Phytoconstituents, including vitamins, flavonoids, minerals, terpenes, alkaloids, gums, oils, glycosides, lignins, saponins, podophyllotoxin, taxol, vinca alkaloids, and polyphenols, act through various mechanisms to regulate specific enzymes, proteins, or metabolites in CRC cells without causing damage to normal cells (More et al., 2021). Phytochemicals exhibit antiinflammatory, antioxidant, and anticancer properties. However, their therapeutic application is limited due to low aqueous solubility, instability in physiological conditions, poor intestinal absorption, rapid metabolism, and low bioavailability (Anonymous. Recent Advance of in Vitro Models in Natural Phytochemicals Absorption and Metabolism - Huang, 2021 EFood - Wiley Online Library, n.d.). To address these challenges, researchers are increasingly focusing on phytochemical-based nanoformulations for CRC treatment. These nanoformulations enhance the delivery and efficacy of natural compounds, making them a promising area of study in CRC therapy (Mohapatra et al., 2022).

The use of nanotechnology in healthcare, also known as nanomedicine, has become more popular domain in a number of studies compared to conventional treatments because of the special attributes of nanoparticles (NPs), including their high surface-to-volume ratios, submicron sizes, and controlled, target-specific drug delivery (Boisseau & Loubaton, 2011). To improve therapeutic efficacy, reduce adverse effects, and overcome drug resistance, nanoparticles (NPs) can be optimized by modifying their composition, size, shape, and surface properties (Kundu et al., 2019). Nano-based drug delivery systems exhibit significantly altered bioactivity and toxicity profiles, making them a promising approach for cancer treatment. They enhance the solubility of phytochemicals with poor water solubility, extend their half-life by protecting them from degradation under physiological conditions through encapsulation, facilitate intestinal absorption by penetrating biological membranes, enable the simultaneous delivery of one or more drugs, and support controlled, target-specific drug delivery with improved bioavailability. Since these advancements change the pharmacokinetics and biodistribution of phytochemicals, they increase their therapeutic index by decreasing toxicity and increasing efficacy, making phytochemical-based nanoformulations a highly useful platform for cancer therapy and enabling safer and more effective treatment options. Phytochemicals and antibodies can be combined to create antibody–drug conjugates (ADCs) to deliver potent cytotoxic phytochemicals to cancerous cells in a targeted way. These conjugates' nanometer-scale diameters allow high-potency cytotoxic phytochemicals to reach a wider therapeutic window since ADCs are more selective for tumors than for other organs. Nanovehicles target tumors with EPR effects, as most solid tumors exhibit poor lymphatic drainage and structural abnormalities in their blood vessels, allowing nanocarriers and macromolecules to preferentially leak from blood vessels near the tumor while the impaired lymphatic drainage permits accumulation around cancer cells. The EPR effect is primarily influenced by factors such as size and biocompatibility; however, for the EPR effect to be effective, phytochemicals must remain in circulation for an extended period to ensure adequate drug distribution. Due to their low molecular weights, most phytochemicals are rapidly cleared from the body and tend to be widely distributed in healthy tissues and organs, but nano drug delivery systems overcome these limitations through controlled release and prolonged circulation time.

Conventional therapeutic options for CRC, including 5-fluorouracil, oxaliplatin, and irinotecan, commonly cause severe adverse effects such as immunosuppression, neurotoxicity, and gastrointestinal toxicity, which limit treatment response and patient compliance. In contrast, phytochemicals demonstrate multitargeted anticancer properties with minimal toxicity to normal tissues. However, their poor pharmacokinetic behavior restricts their clinical applicability, which highlights the need for nanoformulation-based enhancement strategies. Various nanocarrier systems, such as polymeric nanoparticles, SLNs, NLCs, liposomes, nanoemulsions, dendrimers, metallic nanoparticles, and carbon nanotubes, have been explored for the targeted and efficient delivery of phytoconstituents in CRC therapy. These nanocarriers can be functionalized with ligands such as folic acid, transferrin, peptides, and antibodies for active targeting, enabling selective delivery to cancer cells that overexpress specific receptors.

Phytochemicals, including curcumin, quercetin, resveratrol, berberine, EGCG, and ellagic acid, have demonstrated remarkable CRC inhibitory potential when formulated using advanced nanodelivery systems. For instance, nanoencapsulation of curcumin in polymeric nanoparticles, micelles, and liposomes significantly improves its solubility, cellular uptake, systemic retention, and tumor apoptosis effects. Similarly, quercetin-loaded nanoparticles show superior suppression of tumor proliferation and angiogenesis compared to free quercetin. Phytochemicals encapsulated in nanoparticles can also overcome MDR, which is mediated by efflux transport proteins such as P-glycoprotein (P-gp), by downregulating these mechanisms and enhancing intracellular drug accumulation. Nanoformulations also enable colon-specific delivery via pH-sensitive, mucoadhesive, and enzyme-responsive systems designed to release drugs specifically in the colonic environment. In addition, nanotheranostics integrating phytochemicals with metallic nanoparticles enable real-time imaging and combinational therapies such as photothermal ablation and photodynamic therapy, improving precision and treatment monitoring (Hallajzadeh et al., 2020). Therefore, phytochemical-based nanoformulations represent a transformative approach for CRC management by enhancing therapeutic efficacy, reducing toxicity, overcoming resistance, enabling active and passive targeting, and offering multifunctional diagnostic and therapeutic potential. Continuous advancements in nanotechnology, clinical trials, biosafety testing, and large-scale production are critical for translating these novel strategies into clinically approved treatment platforms capable of significantly improving survival outcomes and patient quality of life.

## 11.8 Types of nano-based phyto drug delivery systems

Nanomaterials have gained global interest for their potential to improve conventional CRC treatment strategies. Some of the most promising nanomaterial-based therapies include photodynamic therapy, magnetic hyperthermia, targeted drug delivery, and gene therapy systems. The use of phytochemicals such as curcumin, genistein, quercetin, thymoquinone, ursolic acid, berberine, withaferin A, piperlongumine, nimbolide, lycopene, sulforaphane, resveratrol, and B-carotene in colorectal CRC treatment is largely focused on the development of customized targeted drug delivery systems and advanced therapeutic strategies. Certain nanostructures are more effective for specific drug delivery and tumor-screening applications than others. Examples include carbon nanotubes, liposomes, polymeric nanoparticles, metal–organic frameworks, silica nanoparticles, lipid nanoparticles, lipid bilayers, dendrimers, quantum dots, and nanocapsules, some of which are discussed below. Moreover, these targeted drug delivery systems have paved the way for groundbreaking nanoformulations, which have recently become a key focus of scientific research.

The selection of an appropriate nanocarrier system depends on multiple factors, such as the physicochemical properties of the phytochemical, the desired route of administration, drug-loading capacity, structural stability, biodistribution, and the requirement for controlled or stimuli-responsive release. Among these, polymeric nanoparticles are widely explored due to their biocompatibility, biodegradability, and high encapsulation efficiency. They can provide sustained release and protect phytochemicals from enzymatic or oxidative degradation in the gastrointestinal tract. Lipid-based carriers, including SLNs and NLCs, are also attractive options for colon-targeted cancer therapy because of their ability to incorporate both hydrophobic and hydrophilic molecules, controlled release profiles, and improved intestinal permeability. Liposomes, composed of phospholipid bilayers, are particularly effective for encapsulating hydrophobic phytochemicals, such as curcumin and quercetin, allowing prolonged circulation and enhanced accumulation in tumor tissues through EPR effects.

Carbon-based nanomaterials such as carbon nanotubes, graphene oxide, and fullerenes are also used as carriers due to their exceptional thermal, mechanical, and electrical properties. Their hollow structure allows efficient drug loading, and their ability to penetrate cell membranes promotes enhanced intracellular transport. Quantum dots, on the other hand, are nano-sized fluorescent semiconductor particles that enable simultaneous imaging and therapy, making them suitable for theranostic applications. Their integration with phytochemicals can provide real-time tumor detection and targeted ablation through photothermal or photodynamic mechanisms.

Stimuli-responsive nanoformulations have emerged as an advanced approach, capable of releasing phytochemicals specifically at the tumor site in response to biological or external triggers such as pH, temperature, ROS, enzymes, or light irradiation. This intelligent release minimizes systemic toxicity, increases drug accumulation at tumor tissues, and improves therapeutic efficacy. pH-responsive nanoparticles dissolve selectively in the slightly acidic TME, whereas redox-sensitive nanocarriers degrade rapidly in the presence of elevated glutathione levels typical of cancer cells.

Additionally, colon-specific delivery strategies are gaining significant attention because they enable phytochemicals to bypass early-stage metabolism and degradation in the stomach and small intestine. pH-sensitive polymers such as Eudragit S100 and Eudragit L100 dissolve only in the higher pH environment of the colon, allowing targeted release. Mucoadhesive nanoparticles enhance retention at the colon site, increasing localized drug effects and reducing systemic exposure. Another important development in nano-based phytochemical delivery is combinational nanotherapy, in which phytochemicals and conventional chemotherapeutics are coencapsulated to achieve synergistic effects. Such systems allow dose reduction of toxic chemotherapies while maintaining or improving anticancer efficacy. For example, codelivery of curcumin with 5-fluorouracil or oxaliplatin through polymeric nanoparticles significantly reduces MDR mechanisms and enhances apoptosis. These multifunctional nanoplatforms can integrate targeting ligands, diagnostic imaging agents, and controlled release features, representing a powerful future direction in personalized cancer therapy.

Overall, nano-based phyto drug delivery systems represent a transformative trend in CRC therapy due to their adaptability, high payload capacity, biological safety, tumor-targeting ability, and potential to overcome major therapeutic challenges such as solubility, toxicity, metabolism, and MDR. Continued innovations in nanomaterial engineering, surface functionalization, and clinical translational research will further enhance their utility, paving the way for safer, more efficient, and highly individualized cancer treatments. Table 11.1 provides an overview of the various types of nano-based phyto drug delivery systems.

Although phytochemicals, which are naturally occurring substances found in plants and include polyphenols, flavonoids, flavonolignans, and others, have intriguing therapeutic potential, their clinical application is frequently impeded by issues with solubility, stability, bioavailability, and targeting. Phytosomes (also known as phyto-phospholipid complexes), lipid-based vesicles (such as liposomes, ethosomes, and transferosomes), SLNs, NLCs, polymeric nanoparticles, nanoemulsions, and hybrid/inorganic carriers are popular nano-based delivery systems that have been developed to get around these restrictions (Islam et al., 2022).

Nanoparticles and microparticles have been extensively used to encapsulate phytochemical compounds like catechins, Cuscuta chinensis extract, camptothecin, curcumin, hypericin, triptolide, and tetrandrine, improving their stability and bioavailability. Liposomal and proliposomal systems have been created for catechins, camptothecin, curcumin, silymarin, and vincristine, enhancing solubility and targeted distribution. SLNs are commonly used to transport camptothecin, curcumin, tetrandrine, triptolide, and podophyllotoxin, providing controlled release and enhanced pharmacokinetic characteristics. Polymeric conjugates are mainly utilized for camptothecin and podophyllotoxin, improving drug stability and biocompatibility. In formulations based on microemulsions, bioactives like triptolide and babchi oil (psoralen) are included to enhance skin permeability and absorption. Transdermal gels and patches have been developed with guarana extract and khellin to promote prolonged dermal drug release. Solid dispersions and matrix tablets enhance the solubility and oral bioavailability of silymarin, whereas polymeric cardiac stents have been investigated for localized curcumin delivery in cardiovascular applications. Moreover, scaffolds infused with tetrandrine have been created for tissue engineering and targeted drug delivery (Hani et al., 2021).

Phytosomes represent a specialized class of nano-assemblies where phytochemicals or herbal extracts are chemically complexed with phospholipids—most commonly phosphatidylcholine—in a specific stoichiometric ratio. Unlike conventional encapsulation systems, phytosomes form true molecular complexes through hydrogen bonding or similar interactions between the polar head of the

**Table 11.1 Types of nano-based phyto drug delivery systems.**

| S. No. | Phytoconstituent alone or combine with | Carrier | Outcome | References |
|---|---|---|---|---|
| 1. | Curcumin | Liposomes | Liposomes increased the cytotoxic activity of curcumin in colorectal cancer (CRC) cells, showing that they are more effective than free curcumin | Pandelidou et al. (2011) |
| 2. | Curcumin, piperine | Emulsomes | Enhance their limited absorption and provide a synergistic anticancer effect in vitro colon cancer model. Piperine emulsomes augment the anticancer effects of curcumin on HCT116 cells when combined with CurcuEmulsomes | Bolat et al. (2020) |
| 3. | Curcumin, piperine | Poly(allylamine hydrochloride) nanocapsules | Evaluated on the Caco-2 cell line, reducing vitality with increasing piperine (PIP) concentration. In addition, PIP enhanced cytotoxicity while enhancing accumulation in colon cancer cells | Tefas et al. (2017) |
| 4. | Curcumin, doxorubicin | Liposome | More significant cytotoxic effect than doxorubicin-loaded liposomes at higher curcumin concentrations | Tefas et al. (2017) |
| 5. | Curcumin, catechin | Liposome | Dual-loaded with curcumin and catechin is described in research that shows a synergistic effect on colon cancer cells. In addition, HT29 and Caco-2 were used to test the anticancer effect regarding cytotoxicity, in which it was found that, using a concentration-dependent method, the cytotoxic result of dual-loaded liposomes on HT29 and Caco-2 cells was greater compared to the two single compounds | Hong et al. (2020) |
| 6. | Resveratrol | Poly (lactic-co-glycolic acid) (PLGA)-PEG NPs coated with chitosan | Examined with human COLO205-Luc colon cancer in an orthotopic implantation and xenograft model in athymic mice to investigate the antiproliferative and antiangiogenic properties of resveratrol and its nanoformulation. The outcomes showed that NP-RSV has significant anticancer and antiangiogenic properties | Sudha et al. (2020) |

(*Continued*)

**Table 11.1 Types of nano-based phyto drug delivery systems.** ***Continued***

| S. No. | Phytoconstituent alone or combine with | Carrier | Outcome | References |
|---|---|---|---|---|
| 7. | Resveratrol | Liposomes | Improve the chemotherapeutic effects of resveratrol on colon cancer cells | Soo et al. (2016) |
| 8. | Resveratrol | Mesoporous silica nanoparticles | MCM-48-reticuloendothelial system (RES) was found to enhance dose-dependent cytotoxicity in cell lines. The increased apoptosis rate caused by MCM-48-RES was validated by immunoblotting proteins involved in the apoptosis pathway | Summerlin et al. (2016) |
| 9. | Resveratrol | Gold nanoparticles | Ability to target HT29 cells in a mouse cancer model. Cellular absorption of the resveratrol gold nanoparticle was shown to be much greater than that of pure gold nanoparticles or resveratrol alone. Moreover, resveratrol gold nanoparticles delivered to rats with colon tumors produced more enhanced in vivo targeting than resveratrol 99 m technetium-tagged resveratrol | Kamal et al. (2018) |
| 10. | Resveratrol | Mesoporous silica nanoparticles | Improved the permeability of a monolayer of human colon cancer cells. Compared to the suspension and solution, resveratrol encapsulation exhibits a more substantial antiinflammatory impact | Juère et al. (2017) |
| 11. | Resveratrol | PEG-polylactic acid-based polymeric nanoparticles | Restrain glucose metabolism and the growth of a tumor, leading to an increase in apoptotic cell death | Peñalva et al. (2018); Qiao et al. (2020) |
| 12. | Tea polyphenols | Chitosan nanoparticles | Epigallocatechin-3-gallate (Chit-nanoEGCG) significantly increased poly (ADP-ribose) polymerase cleavage in tumor cells of mice studied with Chit-nanoEGCG in comparison to groups studied with epigallocatechin-3-gallate and controls, improved BCL2-associated X protein expression with a simultaneous reduction in the expression of Bcl-2 protein, and activated caspases | Khan et al. (2014); Siddiqui et al. (2014) |

| | | | | |
|---|---|---|---|---|
| 13. | Tea polyphenols | Graphene oxide sheets | Cure the HT29 and SW48 colon cancer cells. Efficient ligands for cargo binding to the surface of cancerous cells, according to cell granularity studies utilizing scanning electron microscopy and flow cytometry. More significant photothermal apoptosis is observed in giant metastatic cancer cells (SW48) compared to smaller metastatic cancer cells (SW48) by about 20%. Compared to similar nanoparticles made of carbon, the GT-photo-destruction rGO's efficiency factor is at least two orders of magnitude higher | Kadkhoda et al. (2022); Rahman et al. (2012) |
| 14. | Berberine | Nanoparticles based on chitosan, alginate, dextran, and PLGA | Oral administration and sustained/controlled Brb release were increased | Li et al. (2015) |
| 15 | Berberine | Solid lipid nanoparticles, micelles, nanostructured lipid carriers, and liposomes | Shows a significant anticancer effect | Hallajzadeh et al. (2020) |
| 16. | Berberine | Dendrimer, graphene, gold, and silver nanoparticle | Shows potential against cancer cells | Hallajzadeh et al. (2020); Li et al. (2015) |
| 17. | Quercetin | Chitosan-functionalized copper oxide nanoparticle | Quercetin-encapsulated chitosan-functionalized copper oxide nanoparticle (CuO-ChNPs-Q) demonstrated potent anticancer activity in vitro and in vivo. CuO-ChNPs-Q demonstrated cytotoxic effect against liver, breast, and CRC cells | Elsayed et al. (2021) |

phospholipid and the hydrophilic region of the phytoconstituent. This interaction produces a more lipophilic complex, which facilitates better membrane penetration, enhances gastrointestinal absorption, and significantly improves systemic bioavailability compared to standard herbal extracts. According to Chauhan et al., phytosomes also provide protection to bioactive plant molecules from degradation in the gastrointestinal environment, improve solubility, and enable effective therapeutic outcomes even at reduced doses (Benarba & Pandiella, 2018; Islam et al., 2022; Yang et al., 2024).

A well-documented example is the development of silymarin phytosomes, where silymarin—a flavonolignan derived from Silybum marianum—is incorporated into a phospholipid complex. As reported in Pharmaceuticals (2022), these silymarin-loaded phytosomes, with an average particle size of approximately 218 nm, demonstrated significantly greater aqueous solubility compared to unformulated silymarin. They also exhibited an enhanced rate and extent of in vitro drug release. Furthermore, in an in vivo study using a rat model of $CCl_4$-induced hepatic injury, the silymarin phytosomal formulation produced a markedly stronger hepatoprotective response and achieved nearly six times higher systemic bioavailability than the pure compound.

Lipid-based vesicular carriers such as liposomes, ethosomes, and transfersomes are extensively used for delivering phytochemicals. Liposomes are spherical vesicles composed of phospholipid bilayers that can encapsulate hydrophilic compounds in their aqueous core and lipophilic substances within the lipid layer. Ethosomes, enriched with ethanol, and transfersomes, known for their exceptional deformability, are specifically designed to enhance skin and transdermal drug delivery by facilitating deeper penetration across the epidermal barrier. These vesicular systems offer several advantages, including protection of unstable phytoconstituents from degradation, extended circulation time, controlled drug release, and improved bioavailability. Nevertheless, they may encounter limitations such as physical instability (leakage, fusion, or oxidation), difficulties in batch reproducibility, and relatively high production costs. As highlighted by Chauhan et al., liposomes and ethosomes remain among the most promising nanocarriers for improving the delivery efficiency of phytochemicals.

SLNs and NLCs constitute an important category of lipid-based nanocarriers. SLNs are composed of lipids that remain solid at both room and body temperatures and are stabilized by surfactants, while NLCs combine solid and liquid lipids, thereby enhancing drug-loading capacity and enabling more controlled release profiles. These lipid matrices are particularly effective for encapsulating lipophilic phytochemicals, offering protection against degradation, improved physicochemical stability, enhanced bioavailability, and the potential for sustained or controlled drug release. In the case of anticancer plant-derived compounds such as curcumin, quercetin, resveratrol, silymarin, and naringenin, both SLNs and NLCs have demonstrated superior therapeutic performance through increased cellular uptake, greater in vitro stability, enhanced efficacy, and reduced systemic toxicity compared to conventional formulations.

Polymeric nanoparticles, formulated using natural or synthetic biodegradable polymers such as PLGA, chitosan, and alginate, offer another versatile platform for phyto-drug delivery. These carriers enable precise control over drug release patterns—ranging from initial burst release to sustained release—while also allowing surface modification with targeting ligands or stimuli-responsive coatings (e.g., pH- or redox-sensitive materials) to achieve site-specific delivery. They provide strong protection for phytochemicals against enzymatic and chemical degradation and permit controlled biodegradation within the body. Despite their advantages, several challenges persist, including the need to ensure polymer biocompatibility and safety, prevent potential toxicity, maintain consistent particle size and drug-loading efficiency, and meet regulatory standards. Although lipid-based systems often receive greater attention, polymeric nanoparticles are increasingly recognized in Elsevier literature as a significant and adaptable approach for enhancing the delivery and therapeutic efficacy of phytochemicals.

Nanoemulsions and microemulsions are colloidal systems consisting of one immiscible liquid dispersed within another (commonly oil dispersed in water), stabilized by surfactants and cosurfactants, with droplet sizes typically ranging from a few tens to a few hundred nanometers. These systems are particularly advantageous for hydrophobic phytochemicals, whose low water solubility limits their absorption, as nanoemulsions can significantly enhance solubility, improve dissolution rates, and consequently increase bioavailability. They are relatively easy to formulate and can offer a quicker onset of therapeutic action. However, drawbacks include thermodynamic instability—manifesting as droplet coalescence or phase separation—possible toxicity from surfactant components, and limited control over drug release compared to solid or vesicular nanocarriers. Despite these limitations, nanoemulsion-based systems are widely discussed in the literature as effective strategies for enhancing the delivery and absorption of phytochemicals.

Hybrid and inorganic nanocarriers—such as metal or metal oxide nanoparticles, mesoporous silica particles, and lipid–polymer hybrid systems—provide unique functional advantages in phyto-drug delivery. These carriers can respond to external or physiological stimuli like pH, temperature, or redox conditions, and often possess multifunctional properties, including imaging and theranostic potential or magnetic field-guided targeting. Despite these promising features, their practical application is limited by concerns regarding biocompatibility, potential toxicity, slow clearance from the body, long-term accumulation risks, and stringent regulatory challenges that must be addressed before clinical translation.

Across all these nanocarrier platforms, certain recurring principles can be observed. The primary objective is to enhance the bioavailability of phytochemicals by improving their solubility, protecting them from enzymatic and chemical degradation, and increasing their permeability across biological barriers such as the intestinal epithelium and skin. Numerous review articles consistently highlight that phytosomes and lipid-based nanocarriers stand out as the most effective systems in achieving these improvements, due to their superior ability to stabilize and facilitate the absorption of plant-derived bioactives.

## 11.9 Clinical challenges and future perspectives

Nano-based phyto drug delivery systems are emerging as a transformative approach in the management of colon cancer, leveraging the therapeutic potential of herbal drugs through advanced nanotechnology. Phytochemicals, derived from medicinal plants, are known for their anticancer, antiinflammatory, and antioxidant properties. However, their clinical application has been limited by inherent drawbacks such as poor water solubility, low stability in physiological conditions, limited bioavailability, rapid metabolism, and nonspecific distribution, which often result in suboptimal therapeutic outcomes. Nanotechnology offers innovative solutions to overcome these challenges by engineering nanocarriers that can encapsulate phytochemicals, protecting them from premature degradation and enabling targeted and controlled delivery to the colon.

Nanocarrier systems, including liposomes, SLNs, phytosomes, polymeric nanoparticles, nanoemulsions, and nanocapsules, have demonstrated significant potential in improving the therapeutic efficacy of phytochemicals. These nanosystems enhance drug solubility, prolong systemic circulation, and ensure site-specific drug release at the tumor site, minimizing off-target effects and reducing systemic toxicity. For colon cancer specifically, nanocarriers can be designed to exploit the unique microenvironment of the colon, such as pH variations or enzymatic activity, to achieve precise drug release. Additionally, surface modifications of nanocarriers with ligands such as antibodies, peptides, or aptamers enable active targeting of overexpressed receptors on cancer cells, further enhancing the therapeutic effect.

Although preclinical studies provide strong evidence of efficacy, clinical translation of phytochemicals into oncology practice faces significant challenges:

1. **Poor bioavailability and pharmacokinetics**—Numerous phytochemicals have low intestinal absorption, fast metabolism, and poor solubility. For instance, even at high oral dosages, curcumin and resveratrol have modest plasma concentrations. To enhance pharmacokinetics, sophisticated delivery systems such as polymeric micelles, liposomes, and nanoparticles are being created.
2. **Variability and lack of standardization**—The absence of standardization and variability in consistency and reproducibility are impacted by the concentration of bioactive chemicals, which varies according to the plant source, harvest season, and extraction techniques. For clinical use, standardization and quality control are consequently essential.
3. **Insufficient clinical evidence**—There is not enough clinical evidence. Few extensive randomized clinical trials have confirmed efficacy in humans, despite hundreds of in vitro and animal studies showing anticancer potential. Without long-term objectives, the majority of data are restricted to observational or small pilot studies.
4. **Drug–herb interactions**—Through cytochrome P450 regulation, phytochemicals may change medication metabolism and impact the effectiveness of chemotherapy. When using herbal remedies in conjunction with prescription medications, careful pharmacovigilance is required.
5. **Regulatory and economic barriers**—The lack of patentability, ambiguous regulatory pathways, and limited financial incentives hinder investment in clinical trials and commercial development.

Despite these advancements, the clinical translation of nano-based phyto drug delivery systems faces significant challenges. Regulatory hurdles are a primary concern, as the safety and efficacy of these systems must be rigorously established through extensive preclinical and clinical trials. Manufacturing complexities, including scalability, reproducibility, and quality control, encounter additional barriers. Stability issues, particularly during storage and transportation, can also compromise the effectiveness of nanoformulations. Furthermore, interpatient variability, influenced by factors such as genetic differences, gut microbiota composition, and disease progression, adds another layer of complexity to the clinical application of these systems. The high production cost of nano-based systems also limits their accessibility, especially in resource-constrained settings.

Looking ahead, future perspectives in this field focus on addressing these challenges to fully harness the potential of nano-based phyto drug delivery systems for colon cancer management. Research should prioritize the development of biocompatible and biodegradable nanocarriers to ensure patient safety. Stimuli-responsive systems that respond to specific triggers, such as pH, temperature, or enzymes, can further refine site-specific drug delivery. Integrating nanotechnology with personalized medicine approaches can help treat individual patients based on their unique genetic and physiological profiles. Moreover, efforts to optimize cost-effective and scalable manufacturing processes will be critical for the widespread adoption of these systems.

## 11.10 Conclusion

In conclusion, nano-based phyto drug delivery systems present a promising and innovative solution for the treatment of colon cancer by overcoming the limitations of traditional phytochemical therapies, such as poor solubility, low bioavailability, systemic toxicity, and site-specific targeting. The use of advanced nanocarriers like liposomes, nanoparticles, and nanoemulsions enhances the stability, controlled release, and tumor-specific delivery of phytochemicals, thereby improving their therapeutic efficacy. Despite their significant potential, challenges such as high production costs, scalability issues, regulatory hurdles, and the need for extensive toxicity studies remain. Among these, phytosomes show particular promise for hydrophilic or polar phytoconstituents through lipid complexation, while SLNs and NLCs are better suited for lipophilic molecules. Vesicular systems, on the other hand, are especially advantageous for topical and transdermal drug delivery. Although preclinical and in vitro findings are highly encouraging—with examples such as silymarin phytosomes demonstrating significantly improved pharmacological outcomes—clinical translation remains relatively limited. To advance these technologies toward real-world therapeutic use, future efforts must prioritize improving formulation stability, developing scalable production processes, conducting comprehensive toxicity assessments, and implementing robust, well-structured clinical evaluations.

Future research should focus on addressing these challenges through interdisciplinary collaboration, technological advancements, and the development of streamlined regulatory frameworks. Additionally, integrating nanotechnology with personalized medicine, artificial intelligence, and biodegradable materials will further refine these systems, ensuring they are safe, efficient, and accessible. Ultimately, nano-based phyto drug delivery systems offer a novel approach to colon cancer treatment, combining the natural anticancer properties of phytochemicals with the precision of nanotechnology, with the potential to revolutionize cancer therapy and improve patient outcomes worldwide.

Among these, phytosomes show particular promise for hydrophilic or polar phytoconstituents through lipid complexation, while SLNs and NLCs are better suited for lipophilic molecules.

A promising and comprehensive strategy for the prevention and treatment of colon cancer is the use of phytochemicals. Compounds like curcumin, berberine, resveratrol, sulforaphane, and quercetin have shown strong anticancer benefits with low toxicity through antioxidant, antiinflammatory, and signaling regulation methods. However, inadequate clinical validation, lack of uniformity, and poor bioavailability continue to hinder clinical translation.

Large-scale randomized clinical trials, biomarker-guided therapy, and sophisticated delivery technologies are the future. A new paradigm in colon cancer treatment that balances precision oncology and natural product chemistry may be established by combining phytochemicals with contemporary pharmaceutical techniques. With more research being conducted worldwide, phytochemicals have the potential to be a key component of colon cancer prevention and treatment plans, providing safer and more long-lasting treatment options.

## References

Abraha, A. M., & Ketema, E. B. (2016). Apoptotic pathways as a therapeutic target for colorectal cancer treatment. *World Journal of Gastrointestinal Oncology, 8*(8), 583. https://doi.org/10.4251/WJGO.V8.I8.583.

Adebayo, A. S., Agbaje, K., Adesina, S. K., et al. (2023). Colorectal cancer: Disease process, current treatment options, and future perspectives. *Pharmaceutics, 15*(11). https://doi.org/10.3390/PHARMACEUTICS15112620.

Agbele, A. T., Fasoro, O. J., Fabamise, O. M., et al. (2020). Protection against ionizing radiation-induced normal tissue damage by resveratrol: A systematic review. *The Eurasian Journal of Medicine, 52*(3), 298. https://doi.org/10.5152/EURASIANJMED.2020.20143.

Ahmed, K., Zaidi, S. F., Cui, Z. G., et al. (2019). Potential proapoptotic phytochemical agents for the treatment and prevention of colorectal cancer. *Oncology Letters, 18*(1), 487. https://doi.org/10.3892/OL.2019.10349.

Al-Joufi, F. A., Setia, A., Salem-Bekhit, M. M., et al. (2022). Molecular pathogenesis of colorectal cancer with an emphasis on recent advances in biomarkers, as well as nanotechnology-based diagnostic and therapeutic approaches. *Nanomater, 12*(1), 169. https://doi.org/10.3390/NANO12010169.

Anonymous. Colorectal Cancer. (n.d.). https://www.who.int/news-room/fact-sheets/detail/colorectal-cancer .

Anonymous. (n.d.). New plant variety regulatory information | FDA. https://www.fda.gov/food/food-new-plant-varieties/new-plant-variety-regulatory-information.

Anonymous. (n.d.). Recent advance of in vitro models in natural phytochemicals absorption and metabolism - huang 2021 EFood - Wiley online library. https://iadns.onlinelibrary.wiley.com/doi/10.53365/efood.k/146945.

Aqil, F., Munagala, R., Jeyabalan, J., et al. (2013). Bioavailability of phytochemicals and its enhancement by drug delivery systems. *Cancer Letters, 334*(1), 133. https://doi.org/10.1016/J.CANLET.2013.02.032.

Ashique, S., Bhowmick, M., Pal, R., et al. (2024). Multi drug resistance in Colorectal Cancer- approaches to overcome, advancements and future success. *Advances in Cancer Biology - Metastasis, 10*, 100114. https://doi.org/10.1016/J.ADCANC.2024.100114.

Atuahene, D., Mahama, K., Sam, B. A., et al. (2025). Dietary targeting of cancer pathways: Role of bioactive compounds and nutraceuticals. *Food Humanit, 5*, 100748. https://doi.org/10.1016/J.FOOHUM.2025.100748.

Bardelčíková, A., Šoltys, J., & Mojžiš, J. (2023). Oxidative stress, inflammation and colorectal cancer: An overview. *Antioxidants, 12*(4), 901. https://doi.org/10.3390/ANTIOX12040901.

Bardhan, K., & Liu, K. (2013). Epigenetics and colorectal cancer pathogenesis. *Cancers (Basel), 5*(2), 676. https://doi.org/10.3390/CANCERS5020676.

Benarba, B., & Pandiella, A. (2018). Colorectal cancer and medicinal plants: Principle findings from recent studies. *Biomedicine and Pharmacotherapy, 107*, 408–423. https://doi.org/10.1016/J.BIOPHA.2018.08.006.

Bener, A., Öztürk, A. E., Dasdelen, M. F., et al. (2024). Colorectal cancer and associated genetic, lifestyle, cigarette, nargileh-hookah use and alcohol consumption risk factors: A comprehensive case-control study. *Oncology Reviews, 18*, 1449709. https://doi.org/10.3389/OR.2024.1449709.

Boisseau, P., & Loubaton, B. (2011). Nanomedicine, nanotechnology in medicine. *Comptes Rendus Physique, 12*(7), 620–636. https://doi.org/10.1016/J.CRHY.2011.06.001.

Bolat, Z. B., Islek, Z., Demir, B. N., et al. (2020). Curcumin- and piperine-loaded emulsomes enhance anticancer activity. *Frontiers in Bioengineering and Biotechnology, 8*, 50. https://doi.org/10.3389/fbioe.2020.00050.

Cao, J., Cao, J., Wang, H., et al. (2020). Solubility improvement of phytochemicals using (natural) deep eutectic solvents and their bioactivity evaluation. *Journal of Molecular Liquids, 318*, 113997. https://doi.org/10.1016/J.MOLLIQ.2020.113997.

Chauhan, A., Islam, A. U., Prakash, H., et al. (2021). Phytochemicals targeting NF-κB signaling: Potential anti-cancer interventions. *Journal of Pharmaceutical Analysis, 12*(3), 394. https://doi.org/10.1016/J.JPHA.2021.07.002.

Choudhari, A. S., Mandave, P. C., Deshpande, M., et al. (2020). Phytochemicals in cancer treatment: From preclinical studies to clinical practice. *Frontiers in Pharmacology, 10*, 497776. https://doi.org/10.3389/FPHAR.2019.01614/FULL.

Chu, A. H., Lin, K., Croker, H., et al. (2025). Dietary-lifestyle patterns and colorectal cancer risk: Global cancer update programme (CUP global) systematic literature review. *The American Journal of Clinical Nutrition, 121*(5), 986–998. https://doi.org/10.1016/J.AJCNUT.2025.01.014.

Das, A., Adhikari, S., Deka, D., et al. (2023). An updated review on the role of nanoformulated phytochemicals in colorectal cancer. *Medicina, 59*(4), 685. https://doi.org/10.3390/MEDICINA59040685.

Das, T., Mondal, S., Das, S., et al. (2024). Enhanced anticancer activity of (-)-epigallocatechin-3-gallate (EGCG) encapsulated NPs toward colon cancer cell lines. *Free Radical Research, 58*(10), 565–582. https://doi.org/10.1080/10715762.2024.2360013.

Dromain, C., Caramell, C., Dartigues, P., et al. (2014). Liver, lung and peritoneal metastases in colorectal cancers: Is the patient still curable? What should the radiologist know. *Diagnostic and Interventional Imaging, 95*(5), 513–523. https://doi.org/10.1016/J.DIII.2014.03.011.

Dutt, Y., Pandey, R. P., Dutt, M., et al. (2023). Liposomes and phytosomes: Nanocarrier systems and their applications for the delivery of phytoconstituents. *Coordination Chemistry Reviews, 491*, 215251. https://doi.org/10.1016/J.CCR.2023.215251.

Elsayed, A. M., Sherif, N. M., Hassan, N. S., et al. (2021). Novel quercetin encapsulated chitosan functionalized copper oxide nanoparticles as anti-breast cancer agent via regulating p53 in rat model. *International Journal of Biological Macromolecules, 185*, 134–152. https://doi.org/10.1016/J.IJBIOMAC.2021.06.085.

Esmeeta, A., Adhikary, S., Dharshnaa, V., et al. (2022). Plant-derived bioactive compounds in colon cancer treatment: An updated review. *Biomedicine and Pharmacotherapy, 153*, 113384. https://doi.org/10.1016/J.BIOPHA.2022.113384.

Fernando, W., Rupasinghe, H. P. V., & Hoskin, D. W. (2019). Dietary phytochemicals with anti-oxidant and pro-oxidant activities: A double-edged sword in relation to adjuvant chemotherapy and radiotherapy? *Cancer Letters, 452*, 168–177. https://doi.org/10.1016/J.CANLET.2019.03.022.

Fu, D., Fu, H., Wang, Y., et al. (2025). Therapeutic potentials of plant-based antioxidants on colorectal cancer: Challenges and perspectives (Review). *International Journal of Oncology, 67*(6), 107. https://doi.org/10.3892/IJO.2025.5813.

Ganapathy, A., & Ezekiel, U. (2019). Phytochemical modulation of MiRNAs in colorectal cancer. *Medicines, 6*(2), 48. https://doi.org/10.3390/MEDICINES6020048.

Ganesan, K., Jayachandran, M., & Xu, B. (2020). Diet-derived phytochemicals targeting colon cancer stem cells and microbiota in colorectal cancer. *International Journal of Molecular Sciences, 21*(11), 3976. https://doi.org/10.3390/IJMS21113976.

Ghosh, S., Solanki, R., Bhatia, D., et al. (2025). Nanomaterials for delivery of medicinal plant extracts and phytochemicals: Potential applications and future perspectives. *Plant Nano Biology, 12*, 100161. https://doi.org/10.1016/J.PLANA.2025.100161.

Hallajzadeh, J., Maleki Dana, P., Mobini, M., et al. (2020). Targeting of oncogenic signaling pathways by berberine for treatment of colorectal cancer. *Medical Oncology, 37*(6). https://doi.org/10.1007/S12032-020-01367-9.

Han, C. J., Ning, X., Burd, C. E., et al. (2024). Chemotoxicity and associated risk factors in colorectal cancer: A systematic review and meta-analysis. *Cancers (Basel), 16*(14), 2597. https://doi.org/10.3390/CANCERS16142597/S1.

Hani, U., Honnavalli, Y. K., Begum, M. Y., et al. (2021). Colorectal cancer: A comprehensive review based on the novel drug delivery systems approach and its management. *Journal of Drug Delivery Science and Technology, 63*. https://doi.org/10.1016/J.JDDST.2021.102532.

Hao, Q., Wang, M., Sun, N. X., et al. (2020). Sulforaphane suppresses carcinogenesis of colorectal cancer through the ERK/Nrf2-UDP glucuronosyltransferase 1A metabolic axis activation. *Oncology Reports, 43*(4), 1067. https://doi.org/10.3892/OR.2020.7495.

Hervieu, C., Christou, N., Battu, S., et al. (2021). The role of cancer stem cells in colorectal cancer: From the basics to novel clinical trials. *Cancers (Basel), 13*(5), 1092. https://doi.org/10.3390/CANCERS13051092.

Hong, S. C., Park, K. M., Hong, C. R., et al. (2020). Microfluidic assembly of liposomes dual-loaded with catechin and curcumin for enhancing bioavailability. *Colloids and Surfaces A: Physicochemical and Engineering Aspects, 594*. https://doi.org/10.1016/J.COLSURFA.2020.124670 124670.

Huo, Z., Liu, G., & Li, J. (2025). Recent research progress and clinical status of immunotherapy for colorectal cancer. *Journal of Advanced Research*. https://doi.org/10.1016/J.JARE.2025.06.089.

Imtiaz, I., Schloss, J., & Bugarcic, A. (2025). Interplay between traditional and scientific knowledge: Phytoconstituents and their roles in lung and colorectal cancer signaling pathways. *Biomolecules, 15*(3), 380. https://doi.org/10.3390/BIOM15030380.

Islam, M. R., Akash, S., Rahman, M. M., et al. (2022). Colon cancer and colorectal cancer: Prevention and treatment by potential natural products. *Chemico-biological Interactions, 368*, 110170. https://doi.org/10.1016/J.CBI.2022.110170.

Juère, E., Florek, J., Bouchoucha, M., et al. (2017). In vitro dissolution, cellular membrane permeability, and anti-inflammatory response of resveratrol-encapsulated mesoporous silica nanoparticles. *Molecular Pharmaceutics, 14*(12), 4431–4441. https://doi.org/10.1021/ACS.MOLPHARMACEUT.7B00529.

Kadkhoda, J., Tarighatnia, A., Barar, J., et al. (2022). Recent advances and trends in nanoparticles based photothermal and photodynamic therapy. Photodiagnosis and Photodynamic Therapy, *37*. https://doi.org/10.1016/J.PDPDT.2021.102697.

Kamal, R., Chadha, V. D., & Dhawan, D. K. (2018). Physiological uptake and retention of radiolabeled resveratrol loaded gold nanoparticles (99mTc-Res-AuNP) in colon cancer tissue. *Nanomedicine: Nanotechnology, Biology and Medicine, 14*(3), 1059–1071. https://doi.org/10.1016/J.NANO.2018.01.008.

Kemp, J. A., & Kwon, Y. J. (2021). Cancer nanotechnology: Current status and perspectives. *Nano Convergence, 8*(1). https://doi.org/10.1186/S40580-021-00282-7.

Khan, N., Bharali, D. J., Adhami, V. M., et al. (2014). Oral administration of naturally occurring chitosan-based nanoformulated green tea polyphenol EGCG effectively inhibits prostate cancer cell growth in a xenograft model. *Carcinogenesis, 35*(2), 415–423. https://doi.org/10.1093/CARCIN/BGT321.

Kim, J. H., Dareowolabi, B. O., Thiruvengadam, R., et al. (2024). Application of nanotechnology and phytochemicals in anticancer therapy. *Pharmaceutics, 16*(9), 1169. https://doi.org/10.3390/PHARMACEUTICS16091169.

Klimeck, L., Heisser, T., Hoffmeister, M., et al. (2023). Colorectal cancer: A health and economic problem. *Best Practice and Research. Clinical Gastroenterology, 66*, 101839. https://doi.org/10.1016/J.BPG.2023.101839.

Kumari Singh, S., Solanki, R., & Patel, S. (2024). Nano pharmaceutical delivery in combating colorectal cancer. *Medicine in Drug Discovery, 21*. https://doi.org/10.1016/j.medidd.2023.100173 100173.

Kundu, P., Singh, D., Singh, A., et al. (2019). Cancer nanotheranostics: A nanomedicinal approach for cancer therapy and diagnosis. *Anticancer Agents Med Chem, 20*(11), 1288–1299. https://doi.org/10.2174/1871520619666190820145930.

Kwan, A. S. H., Uwishema, O., Mshaymesh, S., et al. (2025). Advances in the diagnosis of colorectal cancer: the application of molecular biomarkers and imaging techniques: a literature review. *Annals of Medicine and Surgery, 87*(1), 192. https://doi.org/10.1097/MS9.0000000000002830.

Li, W., Hua, B., Saud, S. M., et al. (2015). Berberine regulates AMP-activated protein kinase signaling pathways and inhibits colon tumorigenesis in mice. *Molecular Carcinogenesis, 54*(10), 1096–1109. https://doi.org/10.1002/MC.22179.

Luo, H., Rankin, G. O., Liu, L., et al. (2009). Kaempferol inhibits angiogenesis and VEGF expression through both HIF dependent and independent pathways in human ovarian cancer cells. *Nutrition and Cancer, 61*(4), 554. https://doi.org/10.1080/01635580802666281.

Macharia, J. M., Mwangi, R. W., Rozmann, N., et al. (2022). Medicinal plants with anti-colorectal cancer bioactive compounds: Potential game-changers in colorectal cancer management. *Biomedicine and Pharmacotherapy, 153*, 113383. https://doi.org/10.1016/J.BIOPHA.2022.113383.

Majumdar, A. P. N., Banerjee, S., Nautiyal, J., et al. (2009). Curcumin synergizes with resveratrol to inhibit colon cancer. *Nutrition and Cancer, 61*(4), 544. https://doi.org/10.1080/01635580902752262.

McCubrey, J. A., Lertpiriyapong, K., Steelman, L. S., et al. (2017). Effects of resveratrol, curcumin, berberine and other nutraceuticals on aging, cancer development, cancer stem cells and microRNAs. *Aging (Albany NY), 9*(6), 1477. https://doi.org/10.18632/AGING.101250.

Mohapatra, P., Singh, P., Singh, D., et al. (2022). Phytochemical based nanomedicine: A panacea for cancer treatment. *OpenNano, 7*, 100055. https://doi.org/10.1016/j.onano.2022.100055.

More, M. P., Pardeshi, S. R., Pardeshi, C. V., et al. (2021). Recent advances in phytochemical-based nano-formulation for drug-resistant cancer. *Medicine in Drug Discovery, 10*. https://doi.org/10.1016/j.medidd.2021.100082 100082.

Morgan, E., Arnold, M., Gini, A., et al. (2023). Global burden of colorectal cancer in 2020 and 2040: Incidence and mortality estimates from GLOBOCAN. *Gut, 72*(2), 338–344. https://doi.org/10.1136/GUTJNL-2022-327736.

Ojo, O. A., Adeyemo, T. R., Rotimi, D., et al. (2022). Anticancer properties of curcumin against colorectal cancer: A review. *Frontiers in Oncology, 12*, 881641. https://doi.org/10.3389/FONC.2022.881641.

Opara, J. K., Sanchez, S., Thomas, S. M., et al. (2025). Natural products targeting cancer stem cells: A promising therapeutic approach. *Open Explor, 3*, 1008106. https://doi.org/10.37349/EDS.2025.1008106.

Pancione, M., Remo, A., & Colantuoni, V. (2012). Genetic and epigenetic events generate multiple pathways in colorectal cancer progression. *Pathology Research International, 2012*, 509348. https://doi.org/10.1155/2012/509348.

Pandelidou, M., Dimas, K., Georgopoulos, A., et al. (2011). Preparation and characterization of lyophilised egg PC liposomes incorporating curcumin. *Journal of Nanoscience and Nanotechnology, 11*(2), 1259–1266. https://doi.org/10.1166/jnn.2011.3093.

Pathaw, N., Devi, K. S., Sapam, R., et al. (2022). A comparative review on the anti-nutritional factors of herbal tea concoctions and their reduction strategies. *Frontiers in Nutrition, 9*, 988964. https://doi.org/10.3389/FNUT.2022.988964/FULL.

Paudel, S., Mishra, N., & Agarwal, R. (2023). Phytochemicals as immunomodulatory molecules in cancer therapeutics. *Pharmaceuticals, 16*(12), 1652. https://doi.org/10.3390/PH16121652.

Paul, J. K., Azmal, M., Haque, A. S. N. B., et al. (2024). Phytochemical-mediated modulation of signaling pathways: A promising avenue for drug discovery. *Advances in Redox Research, 13*. https://doi.org/10.1016/J.ARRES.2024.100113 100113.

Peñalva, R., Morales, J., González-Navarro, C. J., et al. (2018). Increased oral bioavailability of resveratrol by its encapsulation in casein nanoparticles. *International Journal of Molecular Sciences, 19*(9). https://doi.org/10.3390/IJMS19092816.

Qiao, L., Han, M., Gao, S., et al. (2020). Research progress on nanotechnology for delivery of active ingredients from traditional Chinese medicines. Journal of Materials Chemistry B, *8*(30), 6333–6351. https://doi.org/10.1039/D0TB01260B.

Rahman, M., Ahmad, M. Z., Kazmi, I., et al. (2012). Advancement in multifunctional nanoparticles for the effective treatment of cancer. *Expert Opinion on Drug Delivery, 9*(4), 367–381. https://doi.org/10.1517/17425247.2012.668522.

Rashmy Madathil, S., Kannappan, P., Muthusami, S., et al. (n.d.). Phytotherapeutics in colorectal cancer with emphasis on target molecular signaling pathways: A review. *Indian Journal of Pharmaceutical Sciences*.

Roshandel, G., Ghasemi-Kebria, F., & Malekzadeh, R. (2024). Colorectal cancer: Epidemiology, risk factors, and prevention. *Cancers, 16*(8), 1530. 10.3390/CANCERS16081530.

Roshani, M., Jafari, A., Loghman, A., et al. (2022). Applications of resveratrol in the treatment of gastrointestinal cancer. *Biomedicine and Pharmacotherapy, 153*, 113274. https://doi.org/10.1016/J.BIOPHA.2022.113274.

dos Santos, P. W.da S., Machado, A. R. T., De Grandis, R. A., et al. (2020). Transcriptome and DNA methylation changes modulated by sulforaphane induce cell cycle arrest, apoptosis, DNA damage, and suppression of proliferation in human liver cancer cells. *Food and Chemical Toxicology: An International Journal Published for the British Industrial Biological Research Association, 136*, 111047. https://doi.org/10.1016/J.FCT.2019.111047.

Shnaikat, S. G., Shakya, A. K., & Bardaweel, S. K. (2024). Formulation, development and evaluation of hyaluronic acid-conjugated liposomal nanoparticles loaded with regorafenib and curcumin and their in vitro evaluation on colorectal cancer cell lines. *Saudi Pharmaceutical Journal: SPJ, 32*(7), 102099. https://doi.org/10.1016/J.JSPS.2024.102099.

Siddiqui, I. A., Bharali, D. J., Nihal, M., et al. (2014). Excellent anti-proliferative and pro-apoptotic effects of (-)-epigallocatechin-3-gallate encapsulated in chitosan nanoparticles on human melanoma cell growth both in vitro and in vivo. Nanomedicine: Nanotechnology, Biology and Medicine, *10*(8), 1619–1626. https://doi.org/10.1016/j.nano.2014.05.007.

Soo, E., Thakur, S., Qu, Z., et al. (2016). Enhancing delivery and cytotoxicity of resveratrol through a dual nanoencapsulation approach. *Journal of Colloid and Interface Science, 462*, 368–374. https://doi.org/10.1016/J.JCIS.2015.10.022.

Sudha, T., El-Far, A. H., Mousa, D. S., et al. (2020). Resveratrol and its nanoformulation attenuate growth and the angiogenesis of xenograft and orthotopic colon cancer models. *Molecules, 25*(6). https://doi.org/10.3390/MOLECULES25061412.

Summerlin, N., Qu, Z., Pujara, N., et al. (2016). Colloidal mesoporous silica nanoparticles enhance the biological activity of resveratrol. *Colloids Surfaces B Biointerfaces, 144*, 1–7. https://doi.org/10.1016/J.COLSURFB.2016.03.076.

Sun, Q., Yang, H., Liu, M., et al. (2022). Berberine suppresses colorectal cancer by regulation of Hedgehog signaling pathway activity and gut microbiota. *Phytomedicine, 103*, 154227. https://doi.org/10.1016/J.PHYMED.2022.154227.

Talib, W. H., Awajan, D., Hamed, R. A., et al. (2022). Combination anticancer therapies using selected phytochemicals. *Molecules, 27*(17), 5452. https://doi.org/10.3390/MOLECULES27175452.

Taylor, W. F., & Jabbarzadeh, E. (2017). The use of natural products to target cancer stem cells. *American Journal of Cancer Research, 7*(7), 1588.

Tefas, L. R., Sylvester, B., Tomuta, I., et al. (2017). Development of antiproliferative long-circulating liposomes. *Drug Design, Development and Therapy, 11*, 1605–1621. https://doi.org/10.2147/DDDT.S129008.

Tudosie, M. S., Pauna, A., Stefani, C., et al. (2022). Diet and food chemicals increasing the risk of colorectal cancer—Literature review. *Journal of Mind and Medical Sciences, 9*(1), 118–124. https://doi.org/10.22543/7674.91.P118124.

Tuli, H. S., Garg, V. K., Bhushan, S., et al. (2023). Natural flavonoids exhibit potent anticancer activity by targeting microRNAs in cancer: A signature step hinting towards clinical perfection. *Translational Oncology, 27*, 101596. https://doi.org/10.1016/J.TRANON.2022.101596.

Unnikrishnan Meenakshi, D., Narde, G. K., Ahuja, A., et al. (2024). Therapeutic applications of nanoformulated resveratrol and quercetin phytochemicals in colorectal cancer—An updated review. *Pharmaceutics, 16*(6), 761. https://doi.org/10.3390/PHARMACEUTICS16060761.

Wall-Medrano, A., Ruiz-Rodríguez, G., Olivas-Aguirre, M., et al. (2024). Nanovectorization of metformin with phytochemicals for cancer therapy: A scoping review. *Journal of Drug Delivery Science and Technology, 101*. https://doi.org/10.1016/J.JDDST.2024.106277 106277.

Wang, M., Liu, X., Chen, T., et al. (2022). Inhibition and potential treatment of colorectal cancer by natural compounds via various signaling pathways. *Frontiers in Oncology, 12*, 956793. https://doi.org/10.3389/FONC.2022.956793/FULL.

Xi, Y., & Xu, P. (2021). Global colorectal cancer burden in 2020 and projections to 2040. *Translational Oncology, 14*(10), 101174. https://doi.org/10.1016/J.TRANON.2021.101174.

Xu, M., Wang, S., Song, Y. U., et al. (2016). Apigenin suppresses colorectal cancer cell proliferation, migration and invasion via inhibition of the Wnt/β-catenin signaling pathway. *Oncology Letters, 11*(5), 3075–3080. https://doi.org/10.3892/ol.2016.4331.

Yang, Y., Zhao, M., Kuang, Q., et al. (2024). A comprehensive review of phytochemicals targeting macrophages for the regulation of colorectal cancer progression. *Phytomedicine, 128*, 155451. https://doi.org/10.1016/j.phymed.2024.155451.

Yao, Y., Zhou, Y., Liu, L., et al. (2020). Nanoparticle-based drug delivery in cancer therapy and its role in overcoming drug resistance. *Frontiers in Molecular Biosciences, 7*. https://doi.org/10.3389/FMOLB.2020.00193.

Zhang, P., Lai, Z. L., Chen, H. F., et al. (2017). Curcumin synergizes with 5-fluorouracil by impairing AMPK/ULK1-dependent autophagy, AKT activity and enhancing apoptosis in colon cancer cells with tumor growth inhibition in xenograft mice. *Journal of Experimental and Clinical Cancer Research, 36*(1), 190. https://doi.org/10.1186/S13046-017-0661-7.

Zhang, Y., Wang, Y., Zhang, B., et al. (2023). Methods and biomarkers for early detection, prediction, and diagnosis of colorectal cancer. *Biomedicine and Pharmacotherapy, 163*, 114786. https://doi.org/10.1016/J.BIOPHA.2023.114786.

Zhou, Y., Zhang, J., Wang, K., et al. (2020). Quercetin overcomes colon cancer cells resistance to chemotherapy by inhibiting solute carrier family 1, member 5 transporter. *European Journal of Pharmacology, 881*. https://doi.org/10.1016/j.ejphar.2020.173185.

CHAPTER

# 12 Role of polymeric nanoparticles for the treatment of colon cancer

**Devyani Rajput[1], Shivangi Sharma[2] and Neeraj Mishra[3]**

[1]*Department of Pharmacognosy, Nims Institute of Pharmacy, Nims University, Jaipur, Rajasthan, India,*
[2]*Department of Applied Chemistry, Amity School of Applied Science, Amity University, Gwalior, Madhya Pradesh, India,*
[3]*Department of Pharmaceutics, Amity Institute of Pharmacy, Amity University, Gwalior, Madhya Pradesh, India*

## 12.1 Introduction

Colorectal cancer, also known as colon cancer, is a frequently occurring cancer that originates in the colon or rectum. The risk of developing colon cancer is influenced by various factors, including age, sedentary lifestyle, smoking, excessive alcohol consumption, obesity, family history of colon cancer or polyps, inherited genetic conditions, and a history of inflammatory bowel disease (IBD) (Crohn's disease). Initially, colon cancer often does not display any symptoms. However, as the disease progresses, individuals may notice persistent changes in their bowel habits, such as constipation or diarrhea (Esmaelbeygi et al., 2015; Garg et al., 2023; You et al., 2016). They may also experience fatigue, weight loss, rectal bleeding, abdominal pain, and a feeling of incomplete bowel movement. In India, the mortality rate for men with colon cancer is approximately 18.8%, and for women, it is 14.6%. The prevalence of the disease is 45.2% in men and 38.0% in women (Bhattacharya et al., 2024; Meng et al., 2021; Mishra & Chaurasia, 2017).

Certain dietary factors, including a diet high in fat and red meat and low in fiber, play a role in the global prevalence of colon cancer. The disease can develop from dysplastic precursor lesions, persistent inflammation, high-risk genetic conditions, and occasionally other causes. Colon cancer can present with various nonspecific symptoms, such as fatigue, weight loss, abdominal pain, and alterations in bowel habits. In rare cases, more specific symptoms like obstruction, intestinal bleeding, or perforation may arise, necessitating immediate surgical intervention (Halder et al., 2021; Ravani & Sharma, 2022).

The current treatment paradigm is plagued by issues such as the negative effects of treatments, ineffective adjuvant therapy, and the lack of precise patient selection methods. The development of effective colon-targeted delivery systems faces several challenges, including the diverse physical and chemical properties of drugs, formulation and process variables, and variations in the gastrointestinal tract's physiological conditions. Surgery is a common treatment for colorectal cancer, especially in stages 0 to II, while stages beyond that may require additional adjuvant therapy, chemotherapy, and targeted therapy after surgery. Determining the purpose of laparoscopic resection in patients with superficial rectal cancer is a crucial surgical concern (Dotan & Cohen, 2011). As colorectal cancer is common in both cases, it is important to develop broadly applicable approaches that will impact

Nano Drug Delivery Systems for Colon Cancer. DOI: https://doi.org/10.1016/B978-0-443-40443-6.00017-9

training methods and strategies. Pathological examination of tumor resection material is a vital tool for the postoperative care and prognosis of patients with colorectal cancer (Dotan & Cohen, 2011; Meropol, 2011).

Moreover, challenges arise when it comes to analyzing and documenting these samples, whether it is due to difficulties in applying existing standards or introducing innovative concepts. Accurate pathology reports are essential for colorectal cancer (CC) patients to receive the right prognosis and treatment recommendations. The increasing number of CC patients also faces barriers to survival due to developing resistance to treatment regimens. In recent decades, new chemotherapy treatments have improved the overall survival rate for individuals with advanced colon cancer. While current systemic chemotherapies effectively benefit up to 50% of patients, nearly all CC patients eventually develop resistance to drugs, diminishing the effectiveness of anticancer medications and leading to chemotherapy failure (Banerjee et al., 2017; Dey et al., 2023; Meropol, 2011).

By using dual-drug loading to target numerous components, triggered release to increase specificity, and physical modalities to eliminate malignant cells, nanoformulations can disrupt resistance mechanisms. Because tumors have leaky blood vessels and inadequate lymphatic drainage, nanoscale carriers can pass through the tumor endothelium and passively accumulate within the tumor (Jiang et al., 2024; Tiwari et al., 2024). Additionally, due to the unique physico-chemical characteristics of nanomaterials, they can be used in sensitive diagnostic procedures that improve patient prognosis and enable early cancer detection. The field of cancer diagnostics is gradually shifting from complex and invasive techniques to highly sensitive point-of-care liquid biopsies. In these biomarker detection processes, nanomaterials have proven to be highly useful (Narayana et al., 2024).

Some characteristics also enable imaging methods employed for tumor surveillance and surgical guidance to advance significantly, allowing for more precise surgical resection and improved treatment monitoring. Using radiosensitizers, nanomaterials can provide tumors with a highly targeted, consistent radiation dose while protecting surrounding healthy tissue (Dezfuli et al., 2024; Narayana et al., 2024). The adaptability and functionality of nanomaterials offer various uses in radiation, imaging, diagnostics, and cancer medication therapies. Patients' prognosis and quality of life can be enhanced by early identification, reduced radiation dosage, and greater therapy specificity, all of which can help minimize the systemic toxicities linked to conventional approaches (Bhattacharya et al., 2024; Dezfuli et al., 2024).

## 12.2 Polymeric nanoparticles: an overview

### 12.2.1 Definition and types of polymeric nanoparticles

Polymeric nanoparticles (PNPs) have become a highly researched approach in modern medicine for encapsulating drugs. Several factors have contributed to the rise of PNP research, such as personalized drug delivery, improved bioavailability, controlled drug release from a single dose, and drug stability until reaching the intended administration site (Bhattacharya et al., 2024; Madej et al., 2022). There are currently various methods for producing PNPs. The choice of synthesis process will impact key parameters in the application, such as particle size, system uniformity, and drug integration into the nanocarriers (Kesharwani et al., 2019; Madej et al., 2022). In addition to the synthesis process, polymers, whether natural or synthetic, are another crucial element. Natural polymers have shown outstanding properties, including ease of surface modification, biocompatibility, and biodegradability. Chitosan, collagen, albumin, and gelatin systems have been extensively studied and have demonstrated significant potential for use as biopolymers. On the other hand, synthetic polymers undergo complex modifications to achieve specific attributes, such as improved specificity, greater biological availability, reduced toxicity, and favorable pharmacokinetics (Madej et al., 2022). The interplay between the composition, surface properties, and biological systems of these polymers continues to drive the advancement of new and improved technologies. In terms of polymer selection, "smart" polymers are notable for their ability to respond to biological phenomena, such as pH and temperature changes, as well as external stimuli, enabling active targeting (Khan et al., 2024).

Polymers with hydrophilic properties, such as double esters, hydrazones, and vinyl esters, are commonly utilized in acidic pH environments like lysosomes, inflammatory sites, and malignant tissues. These polymers degrade and release the active substance. Nanoparticles are an innovative advancement in pharmaceuticals, offering solutions to traditional delivery limitations, introducing new approaches to diagnosis and treatment, and enhancing our understanding of biological processes (Khan et al., 2024; You et al., 2016; Younis et al., 2022). The exploration of nanoparticles for biomedical purposes, including controlled drug release and antibacterial properties, is receiving significant attention. The use of nanoparticles for delivering antibiotics directly to the infection site is a promising approach, especially for controlled release, which reduces the necessary dosage for achieving the desired clinical effect. Biodegradable nanoparticles are increasingly favored for this application. The term "nanoparticle" encompasses nanospheres and nanocapsules, which differ in structure and composition. Medication is either absorbed onto the polymeric wall or dissolved in the oily core of the nanocapsule, which is surrounded by a polymeric shell. Conversely, the medication can be evenly and physically adsorbed onto nanospheres made of a polymeric matrix without oil in their composition (Younis et al., 2022). PNPs are important as drug delivery systems with a diameter of at least 100 nm, most of which range from 100 to 500 nm. Controlled and targeted drug delivery systems utilize biodegradable synthetic polymers, particularly those derived from polyglycolic acid (PGA), polylactic acid (PLA), and polylactic-co-glycolic acid (PLGA), which exhibit biodegradable and biocompatible properties (Vashishat et al., 2023). Despite their low immunogenicity, biocompatibility, biodegradability, and protein, antibacterial activity and polysaccharide biopolymers such as cellulose, chitosan, alginate, starch, collagen, and hyaluronic acid (HA) are increasingly gaining popularity. Synthetic polymers offer several advantages over natural polymers and other materials due to their versatility in creating polymeric materials and their ability to be customized based on specific medical conditions and patient needs (Vashishat et al., 2023).

The use of new polymerization techniques enables the development of innovative polymers with distinct properties related to the polymer's molecular architecture or morphology for biomedical applications (Kim et al., 2019). Natural polymers, such as proteins, polysaccharides, and DNA macromolecules, are already present in living organisms, making them more biocompatible. The fate of nanoparticles in vivo is determined by their interactions with the surrounding biological components. Optimal drug delivery devices should be easily absorbed by phagocytic cells in an acidic medium to prevent drug degradation or loss of function. Nanostructured systems have demonstrated enhanced efficacy in treatments, as they can enter biological compartments that are inaccessible to standard pharmacological formulations due to their smaller size compared to cells

(Kim et al., 2019; Vashishat et al., 2023). These systems are effective both extracellularly and intracellularly. Traditional therapies face the challenge of drugs only acting extracellularly since the cell membrane acts as a barrier to prevent drug entry into the cell. Additionally, traditional medication versions require daily high-dosage administrations and lack sustained effects. Noteworthy aspects of nanoparticulate devices include the regulated, targeted release of medication (antibiotic) throughout the body, targeting bacteria in all biological compartments.

The medication's effectiveness and appropriate application prevent germs from developing resistance, extending the drug's half-life. PNPs provide enclosed active ingredients that improve bioavailability and effectiveness while also being biodegradable and biocompatible (Prabhu et al., 2015; Sharma and Singh, 2022, 2023a, 2023b; Sharma et al., 2022, 2023a, 2023b; Sharma, 2024; Sharma, Choubey, et al., 2024; Sharma, Das, et al., 2023; Sharma, Singh, et al., 2023). Their high dispersion in water makes them suitable for use in solutions. These polymeric structures can be designed to have responsive properties, enabling stable targeted drug delivery in environments with high pH and temperature, such as gastric juices and blood. Therefore, a broad range of parameters related to the selection of polymer and the synthesis method must be investigated to ensure that PNPs exhibit all the necessary drug-release properties. These parameters encompass surface properties, stability, morphology, size, uniformity, and interactions with the surrounding environment (Kim et al., 2019; Prabhu et al., 2015).

### 12.2.2 Advantages of polymeric nanoparticles in drug delivery

Nanotechnology is increasingly being recognized as an innovative approach for treating cancer and delivering medications. Compared to traditional drug delivery methods, nanotechnology-based drug delivery systems offer numerous advantages, such as targeted delivery to specific organs, prolonged circulation periods, and controlled release throughout the body. Nanotechnology can potentially overcome drug resistance, particularly in cases of multidrug resistance (MDR) (Meng et al., 2021; You et al., 2016). This allows for a significant increase in the effectiveness of chemotherapy, as it enables a higher concentration of medication to accumulate in the cytoplasm. PNPs are particularly well suited for medication delivery due to the wide range of raw materials used in their production. Additionally, these nanoparticles can be manufactured on a large scale and remain stable during storage. Moreover, they offer the potential for minor surface modifications to optimize factors like circulation half-life and tissue distribution. By altering the polymer structure, the loading and release kinetics can be controlled. It is important to note other advantages of PNPs, such as the ease of creating customized surfaces to target specific proteins and cells and the potential to develop nanodevices that respond to stimuli. Several PNPs are currently undergoing clinical trials (Begines et al., 2020; Pridgen et al., 2015).

## 12.3 Polymeric nanoparticles in colon cancer therapy

### 12.3.1 Role in chemotherapy

Historically, chemotherapy has been employed to treat metastatic disease. However, recent combinations have been shown to extend survival time. Patients with isolated, resectable liver metastases or isolated recurrence at the anastomosis site should still undergo assessment for surgery (Ahlawat et al., 2018). Older patients who are in good health and younger patients tolerate palliative chemotherapy well, with similar benefits in terms of survival and quality of life, and no significant adverse differences. Chemotherapy has led to increased survival in both treatable and incurable cases of colon cancer (Ahlawat et al., 2018; Rychahou et al., 2018). A considerable advancement in colon cancer treatment has been the effort to reduce relapse rates and improve overall survival through adjuvant therapy, as three-quarters of US-diagnosed patients are suitable for surgical resection. Medications initially proven effective in the metastatic setting have also been evaluated for adjuvant use, and a few new medications are undergoing assessment for both metastatic disease and adjuvant therapy (Rizi et al., 2022).

### 12.3.2 Polymeric nanoparticles for targeted therapy

In the gastrointestinal tract's colon or rectum, colorectal cancer is caused by epithelial cells. The primary method of treating early-stage colon cancer is surgery, which involves removing tumors and lymph nodes (Zheng et al., 2018). Chemotherapy is also an effective treatment, using anticancer drugs to halt, slow, or reverse cancer growth. It can be combined with surgery to lower risk and shrink the tumor. Chemotherapeutic drugs like irinotecan, oxaliplatin, capecitabine, and 5-fluorouracil are commonly used for colon cancer treatment (Baião et al., 2020). Additional treatments for colon cancer include immunotherapy, stem cell transplantation, and radiation therapy. Tumor tissues typically have an acidic microenvironment and leaky vasculature with poor lymphatic drainage. Factors such as cancer metastasis, the local microenvironment, and tumor size significantly impact colon cancer treatment decisions. Given these physiological characteristics, encapsulating anticolon cancer drugs in polymer-based nanoparticles may be preferable to overcome biological obstacles. Nanoparticles containing anticancer drugs must navigate biological barriers and precisely target tumor locations at specific concentrations for treating a particular malignancy (Baião et al., 2020; Cai et al., 2021; Masood, 2016). Additionally, the active ingredients should specifically destroy cancer cells to improve treatment efficacy and minimize adverse effects. Advancements in nanoparticle engineering allow for manipulating surface ligands, size, and shape to maximize therapeutic efficacy (Cai et al., 2021). Careful consideration of synthesis techniques, polymer types for nanocarrier creation, coupling strategies, and targeting ligand selections is necessary to achieve the intended nanostructure. Previous studies have shown that positively charged surfaces enhance cellular absorption, while hydrophilic surfaces increase nanoparticle circulation time. Polymer systems offer significant flexibility in customizing and optimizing nanoparticles for delivering effective medications and expediting their use in clinical practice. However, careful system design is crucial for optimal outcomes due to the numerous variables that can be optimized. Two forms of nanoparticle-targeted delivery for colon cancer treatment are passive and active. Nanoparticle systems can passively target tumor vasculature due to their distinct pathophysiologic features, such as the fast proliferation of endothelial cells leading to atypical and "leaky" microvasculature in tumor tissues (Yakati et al., 2022; Zhong et al., 2014).

Tumor tissues often have inefficient lymphatic drainage. The winding nature of the capillaries and the compromised lymphatic system, due to the well-documented "enhanced permeation and retention" (EPR) effect, can facilitate the escape of macromolecules and nanoparticles into the tumor interstitium. It is crucial to analyze the factors governing the EPR effect in order to create a targeted drug delivery system for tumors. However, the diverse pathophysiology of large tumors and the absence of an EPR effect in the central region of metastatic tumors limit the accumulation of nanoparticles throughout tumors. Nanoparticle delivery systems with targeting ligands are currently garnering significant attention for their capacity to achieve "active" targeting in different types of cancer (Russo et al., 2016; Van Vlerken & Amiji, 2006; Zhong et al., 2014).

The addition of targeting ligands, such as peptides, nucleic acid strands, antibodies, and other small molecules, to nanoparticles is a well-known method for enabling active targeting. Strengthening the binding to receptors overexpressed on the surface of tumor cells, these targeted ligands can enhance nanoparticle retention and cellular uptake (Yu et al., 2019). This approach reduces side effects and improves therapeutic efficacy compared to an untargeted NP system by concentrating medication in healthy tissue. When developing actively targeted nanoparticles for ligand biocompatibility, factors such as drug administration, binding affinity, cell specificity, and ligand surface density and arrangement need to be taken into account to achieve optimal results. With advancements in ligand and nanoparticle optimization, actively targeted NP delivery can be a valuable complement to passive targeting, further enhancing the effectiveness of cancer treatment. Both passive and active targeting play crucial roles in delivering therapeutic nanoparticles to specific tumor locations in colon cancer treatment. To achieve success in treating various cancers, the use of small interfering RNA (siRNA) to silence the human vascular endothelial growth factor (VEGF) and reduce angiogenesis is widespread. These siRNAs, short double-stranded RNA fragments, have a sequence-specific ability to silence a target mRNA (Prasad & Dangi, 2023; Yoon et al., 2011; Yu et al., 2019).

To effectively treat colon cancer, it is important for CCDs (colon cell line/normal colon epithelium - e.g., CCD-18Co) to be capable of penetrating biological barriers within the body and reaching the specific tumor tissues while minimizing volume loss and systemic circulation. Additionally, CCDs should be designed to destroy specific cancer cells within the tumor tissue using a prolonged or controlled-release mechanism without causing harm to surrounding normal cells. These fundamental strategies are also associated with improving patient survival and quality of life by increasing intracellular CCD concentration and reducing dose-limiting toxicities. Targeting through a nanoparticulate carrier can significantly decrease drug toxicity by facilitating precise delivery to the intended action area. Both passive and active methods are utilized in targeting PNPs (Phatak et al., 2023; Prasad & Dangi, 2023).

### 12.3.3 Use in combination therapies

The use of combination therapy involves administering multiple chemotherapy drugs simultaneously or sequentially. In clinical settings, this approach is often referred to as medication cocktail therapy (You et al., 2016). Combination chemotherapy is designed to address the challenges of multidrug resistance and the toxic effects of cancer drugs. While anticancer agents can be harmful to both normal and cancer cells, their detrimental effects are especially noticeable in rapidly dividing cells, such as those in the bone marrow, hair follicles, and the lining of the gastrointestinal tract (Madej et al., 2022). The side effects of anticancer drugs can be severe and even life-threatening. Research has shown that using multiple anticancer drugs can reduce the toxicity associated with high doses and improve treatment outcomes. Combining anticancer drugs with different mechanisms of action and toxicity profiles can create synergistic or additive therapeutic effects. Conversely, drugs with similar mechanisms of action and toxicity profiles can be combined at lower doses to achieve the same therapeutic effects while also altering toxicity and the risk of multidrug resistance (Bhaskaran & Kumar, 2021; Naeem et al., 2015).

### 12.3.4 Mechanism of polymeric nanoparticles in the treatment of cancer

PNPs play a vital role in improving the solubility of medications, protecting drugs from degradation, and precisely targeting the tumor site, making them a promising option for delivering drugs to treat colon cancer. The mechanism of action of PNPs in colon cancer treatment involves several crucial stages (Fig. 12.1) (Gulbake et al., 2016; Meng et al., 2021).

PNPs are specifically designed to contain therapeutic agents such as siRNA and monoclonal antibodies, as well as chemotherapy drugs like irinotecan and 5-fluorouracil. By encapsulating drugs, their solubility, bioavailability, and stability are significantly enhanced, and they are less susceptible to early breakdown in the gastrointestinal system. Polymeric NPs can undergo surface modification with ligands such as folic acid, antibodies, or peptides to selectively target receptors (epidermal growth factor receptor [EGFR], folate receptors, etc.) that are overexpressed on colon cancer cells. This targeted approach enhances the selective accumulation of NPs in tumor tissues, reduces off-target effects, and lowers systemic toxicity (Cabeza et al., 2020; Udompornmongkol & Chiang, 2015).

The EPR effect is related to tumor tissues with leaky blood vessels and inadequate lymphatic drainage. PNPs exploit the EPR effect to accumulate in the tumor tissue because of their size and surface characteristics, thereby enhancing localized drug delivery. When they reach the tumor site, cancer cells absorb the PNPs through endocytosis, which could involve passive absorption or receptor-mediated endocytosis. The PNPs can then release their payload, such as drugs or therapeutic agents, into the cytoplasm by breaking down the polymer matrix or reacting to stimuli, such as pH or enzymes. To maintain therapeutic drug concentrations in the tumor microenvironment for extended periods, PNPs are engineered to offer controlled and sustained drug release. In the acidic tumor environment, pH-responsive nanoparticles release the medication more effectively, improving targeting and reducing damage to healthy tissues. Once released, the medication, like chemotherapy, eliminates cancer cells by inhibiting DNA synthesis, initiating apoptosis, or halting cell division. Additionally, PNPs can transport immuno-modulating agents to enhance the immune system, siRNA to silence oncogenes, and nanoparticles with photothermal or photodynamic properties for combined therapies (Cabeza et al., 2020; Dristant et al., 2023).

PNPs can simultaneously administer multiple drugs, allowing for combination therapies that can target different pathways at the same time, which is advantageous for treating colon cancer that is resistant to treatment. Additionally, nanoparticles can carry P-glycoprotein inhibitors and other drugs blocking drug efflux pumps to prevent cancer cells from expelling chemotherapy drugs. The frequent utilization of biodegradable polymers, such as chitosan, polyethylene glycol (PEG), and PLGA, in these nanoparticles ensures the safe breakdown and elimination of the carrier from the body, thus reducing long-term toxicity (Dristant et al., 2023; Hosseinifar et al., 2018; Klippstein et al., 2015; Priyadarshi et al., 2021; Ramezani et al., 2020; Reimondez-Troitiño et al., 2019; Shad et al., 2020;

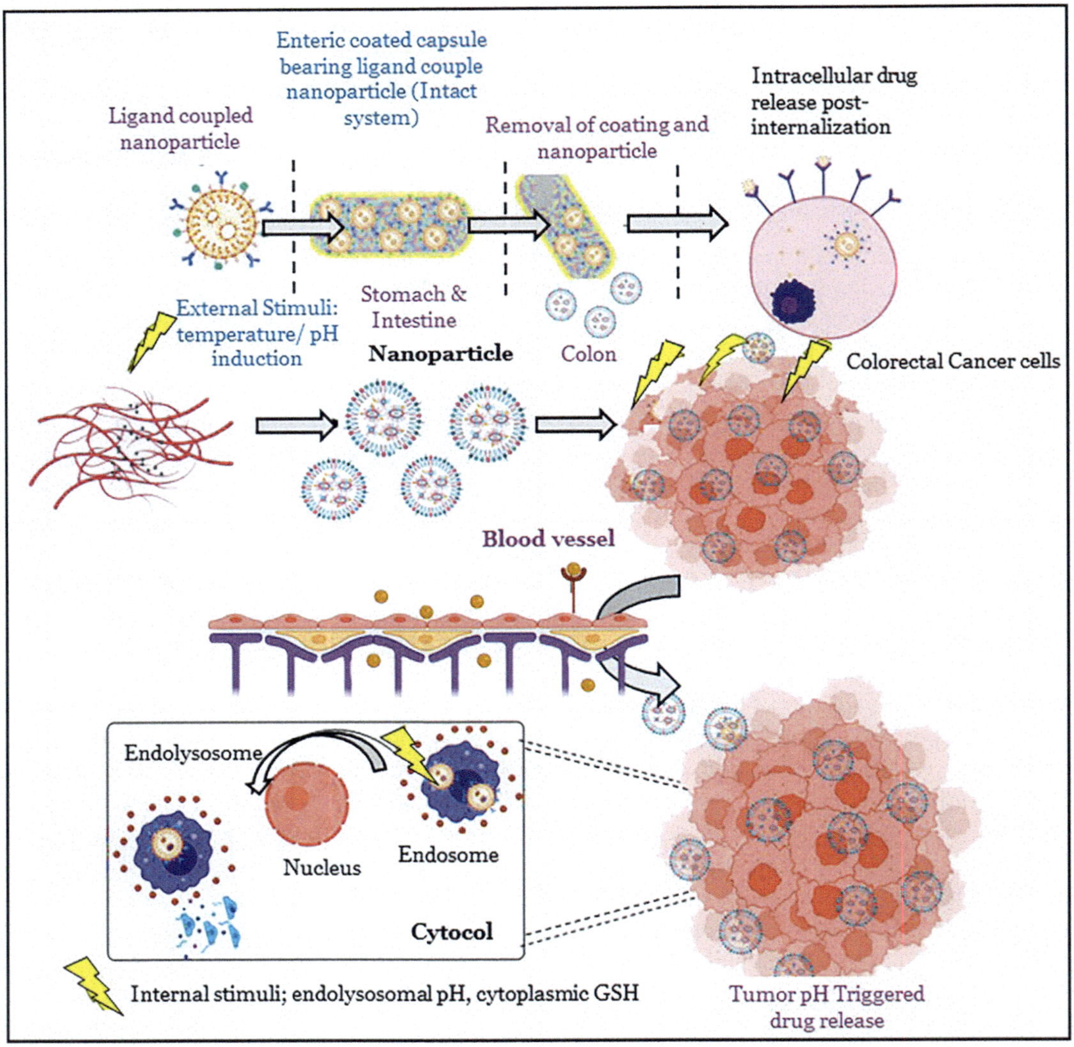


**FIGURE 12.1**

Schematic representation of various targeting approaches for colon cancer.

Simón-Gracia et al., 2016; Venkatesan et al., 2011; Wei et al., 2020; Xu et al., 2015; Yang et al., 2015) (Table 12.1).

### 12.3.5 Advantages of polymeric nanoparticles in colon cancer treatment

Improved drug solubility and stability in the harsh gastrointestinal environment.

**Table 12.1 Some examples of PNCs used for the treatment of colorectal cancer (CC).**

| Types of PNCs | Anticancer drug | Polymer | Model (in vitro and/or in vivo) | Effect of using PNCs | References |
|---|---|---|---|---|---|
| Polymeric micelles | Quercetin | Monomethoxy poly(ethylene glycol)–poly(ε-caprolactone) | Mouse colon cancer cell line CT26 BALB/c mice and Sprague Dawleyrats in female form | The medication's release was sustained in a laboratory setting.<br>There was a significant 37% decrease in IC50, and cells showed increased absorption.<br>Upon intravenous infusion, the medication's t1/2 and Cmax were increased.<br>After three weeks of treatment with Qu-loaded personalized medicined (PMs), tumor volumes in living organisms decreased, and the apoptotic index increased to 62.7% from 28.23% with the free medication. | Xu et al. (2015) |
| | Curcumin | Monomethyl poly (ethylene glycol)-poly (ε-caprolactone)-poly (trimethylene carbonate) | CT26 cell line; female BALB/c mice | In an in vitro study, cells treated with CUR-loaded polymeric micelles (PMs) exhibited an 81% increase in apoptosis induction, a 12% decrease in $IC_{50}$, and enhanced cellular absorption.<br>Furthermore, there was a noticeable reduction in in vivo tumor growth, as well as decreased angiogenesis and tumor proliferation. The | Yang et al. (2015) |

(*Continued*)

**Table 12.1 Some examples of PNCs used for the treatment of colorectal cancer (CC). *Continued***

| Types of PNCs | Anticancer drug | Polymer | Model (in vitro and/or in vivo) | Effect of using PNCs | References |
|---|---|---|---|---|---|
| | | | | apoptosis index rose by 16.05% after CUR-loaded PMs treatment, compared to 9.07% with the free medication. | |
| Polymersomes | Doxorubicin (DOX) | Mixture of transferrin-binding peptide (TBP)-functionalized (Tf@TBP-Ps) and nonfunctionalized (Ps) poly (ethylene glycol)-b-poly (trimethylene carbonate-co-dithiolane trimethylene carbonate) | Human colon tumor/Human colorectal carcinoma (HCT) 116 cells; female Balb/c nude mice | The uptake of polymersomes by cells increased due to higher TBP densities and overexpression of transferrin receptor. Tf@TBP-Ps-DOX demonstrated 2.5 times greater potency against HCT 116 cells compared to Ps-DOX, with an IC50 value.<br>After 24 h, the accumulation of Tf@TBP-Ps-DOX in tumors in vivo was twice as much as that of Ps-DOX. Tf@TBP-Ps-DOX exhibited a 6.7% increase in t1/2 and a 15% increase in area under the curve (AUC). At a dosage of 8 mg DOX $kg^{-1}$, Tf@TBP-Ps-DOX significantly surpassed Ps-DOX in inhibiting tumor growth. | Wei et al. (2020) |
| | 7-ethyl-10-hydroxy camptothecin (SN38) | AS1411 aptamer-polyethylene glycol-peptide-polylactic acid (PLA) triblock copolymer (peptide=PVGLIG) | C26 mouse colon adenocarcinoma and Chinese hamster ovary (CHO); female BALB/c mice | The release of SN38 was overseen by the MMP-2 enzyme.<br>Apt-SN38-polymersomes showed significantly higher cytotoxicity than SN38-polymersomes for C26 cells at | Ramezani et al. (2020) |

| | | | | | |
|---|---|---|---|---|---|
| | | | | low SN38 concentrations (0.15–1.25 μg $mL^{-1}$) but did not impact CHO cells. Apt-SN38 polymersomes resulted in a noticeable regression of tumor growth and a slight decrease in body weight. | |
| | Paclitaxel (PTX) | Internalizing RGD peptide (iRGD)-functionalized poly (oligoethylene glycol methacrylate)-poly(2-(diisopropylamino)ethyl methacrylate) (P[(OEG)10MA] 20-PDPA90; POEGMA-PDPA) | CT26 cell line; balb/c mice | Cells that express neuropilin-1 showed improved drug targeting, internalization, and cytotoxicity when exposed to PTX-loaded iRGD-polymersomes. Following intraperitoneal administration to mice with peritoneal tumors induced by CT26, iRGD-polymersomes exhibited increased tumor-specific accumulation and penetration, effectively preventing tumor growth compared to untargeted polymersomes. | Simón-Gracia et al. (2016) |
| Nanogels | 5-FU | Alginate-β-Cyclodextrin (AL-CD) | HT 29 cells | Diffusion and polymer breakdown regulated the release of 5-FU. Unlike free 5-FU, drug-loaded AL-CD nanogels demonstrated increased cytotoxicity, cellular uptake, and twofold improved apoptosis in HT 29 (Human colon adenocarcinoma - e.g., HT-29) cells. | Hosseinifar et al. (2018) |

*(Continued)*

**Table 12.1 Some examples of PNCs used for the treatment of colorectal cancer (CC).** *Continued*

| Types of PNCs | Anticancer drug | Polymer | Model (in vitro and/or in vivo) | Effect of using PNCs | References |
|---|---|---|---|---|---|
| | Oxaliplatin (OXA) | Alginate nanogels coated with hyaluronic acid (HA) and functionalized with folic acid (F/HA/AL) | HT 29 and HEK293 cells | OXA release from F/HA/AL nanogels was regulated for seven days. Drug-loaded F/HA/AL exhibited a higher rate of apoptosis and resulted in a 42% decrease in IC50 compared to free OXA. | Shad et al. (2020) |
| Nanocapsules | CUR | PEGylated polylactic-co-glycolic acid | CT26 cells; female balb/c mice | The release of CUR was controlled by nanocapsules, unlike free medication. Nanocapsules significantly reduced the viability of CUR-treated cells at low concentrations (10 nM) while enhancing cellular absorption in a dose-dependent manner, in contrast to cells treated with free medication.<br>Animals treated with CUR-loaded nanocapsules exhibited a twofold increase in tumor accumulation and a halving of tumor volume compared to those treated with free medication. | Klippstein et al. (2015) |
| | CUR and miR-145 | Protamine | SW480 human CC cells | The rate of cell division and migratory ability decreased significantly. | Reimondez-Troitiño et al. (2019) |

| | | | | | |
|---|---|---|---|---|---|
| | | | | There was an increased uptake of miR-145 inside the cells. The cytotoxicity of CUR increased. | |
| | Celecoxib | Hydroxyapatite-chitosan | HCT 15 and HT 29 cells; nude mice | Both HCT 15 and HT 29 cells showed time-dependent absorption of cytoplasm; this led to a notable decrease in cell proliferation and increased apoptosis. Additionally, there was a significant decrease in tumor growth following treatment with celecoxib-loaded nanoparticles. | Venkatesan et al. (2011) |
| Dendrimers | Gallic acid (GA) | GA-modified PAMAM | HCT 116 and HT 29, CC, MCF7 human breast carcinoma, and NIH 3 T3 (National Institutes of Health cell lines - e.g., NIH/3T3) mouse embryonic fibroblast cells | PAMAM-GA conjugate is taken up by HCT 116 cells in a manner that depends on the dose, and this occurs in less than 6 h.<br>The IC50 in HCT 116 cells is significantly decreased to 3.68 mM by the PAMAM-GA compound, while GA causes 69 mM of toxicity at the same dose in normal NIH 3 T3 cells.<br>The growth of HCT 116, HT 29, and MCF7 cells after treatment with PAMAM-GA is substantially decreased.<br>Following treatment with PAMAM-GA, there is a noticeable decrease in cell migration and an increase in cell death in HCT 116 by activating apoptosis signaling. | Priyadarshi et al. (2021) |

Enhanced tumor targeting due to ligand modification and the EPR effect.
Reduced systemic side effects through localized drug delivery.
Ability to deliver multiple therapeutic agents for combination therapy.
Controlled-release profiles, allowing for sustained drug release.

PNPs improve the treatment of colon cancer by enabling controlled drug release, decreasing toxicity, and targeting drugs, resulting in more effective and safer therapies (Aghebati-Maleki et al., 2020; Yu et al., 2012).

### 12.3.6 Strategies involved in polymeric nanoparticles in the treatment of cancer

PNPs that are intelligent and responsive, known as smart polymeric nanoparticles (SPNs), are an advanced form of drug delivery systems. They can react to specific triggers in their surroundings to release therapeutic substances, which can be internal (such as pH, enzyme, temperature) or external (like light, magnetic fields). SPNs can adjust their behavior based on the conditions they encounter, leading to more precise drug delivery, reduced side effects, and improved therapeutic effectiveness, especially for conditions like cancer (Aghebati-Maleki et al., 2020; Avramović et al., 2020; Calzoni et al., 2019).

The use of responsive nanoparticles to stimuli is a new approach to drug delivery. These systems offer benefits such as increased intracellular drug delivery, targeted delivery, longer circulation durations, and the controlled release of drugs based on dose, location, and time (Bigaj-Jozefowska & Grześkowiak, 2022; Indoria et al., 2020; Juan et al., 2020). The operation of these systems relies on the ability of specific cellular or extracellular stimuli of physical, chemical, or biochemical origin to change the composition or structural conformation of the nanocarriers, thereby releasing active substances into a particular biological environment. The effectiveness of these systems is significantly enhanced when stimulus-sensitive nanocarriers play a crucial role in disease pathology by selectively responding to pathogenic triggers and minimizing side effects (Banik et al., 2016; Conte et al., 2016). The stimuli can be internal, including pH, temperature, enzyme activity, host recognition, and antigen-antibody interactions, or external, such as light, magnetic or electric fields, and ultrasound. The latest studies have focused on stimuli related to various subjects. It is important to note that these nanosystems can also adjust the drug's release profile and therapeutic effectiveness in response to multiple inputs occurring simultaneously, at the exact location, or in separate modalities (Guo et al., 2022; Mohanty et al., 2020).

Nanocarriers with multifunctional capabilities are sophisticated drug delivery systems designed to carry out multiple tasks simultaneously, improving the efficiency and accuracy of medical treatments, particularly in addressing conditions like colon cancer. These nanocarriers integrate various functions, including drug encapsulation, targeting, imaging, and responsiveness to stimuli, to enhance drug delivery and treatment outcomes. The main purpose of multifunctional nanocarriers is to deliver drugs effectively to the intended location, safeguarding the therapeutic substance from degradation in the blood and ensuring its controlled release. These carriers can encase a range of drugs, such as chemotherapy medications, gene therapy agents, proteins, or nucleic acids like mRNA and siRNA. Multifunctional nanocarriers can be modified to target specific cells or tissues using either active or passive targeting methods. They can also be engineered to release the therapeutic payload in response to specific environmental cues, such as pH, temperature, enzymes, or external triggers like magnetic fields or light. This functionality guarantees that the drug is only released at the target site, such as tumor tissue, while remaining stable in other parts of the body, thereby reducing side effects (Kamaly et al., 2012; Swain et al., 2016; Yıldırım et al., 2023).

## 12.4 Clinical studies

### 12.4.1 Overview of preclinical studies

Studies in the preclinical stage have focused on the use of PNPs to improve drug delivery, increase the effectiveness of anticancer drugs, and decrease the adverse effects of traditional treatments in colon cancer. These investigations utilize various polymeric materials to create nanoparticles and evaluate their performance in in vitro (cell culture) and in vivo (animal) models. The objective is to refine nanoparticle formulations for clinical use and assess their potential in addressing the challenges of colon cancer therapy, including drug resistance, limited bioavailability, and systemic toxicity (Chaurasia et al., 2016; Egusquiaguirre et al., 2012; Farinha et al., 2022).

Preclinical research on PNPs for the treatment of colon cancer has demonstrated significant potential in enhancing drug delivery, combating drug resistance, and amplifying the therapeutic effectiveness of both conventional and innovative treatments. These studies underscore the capability of PNPs to offer targeted, controlled-release, and stimuli-responsive delivery systems that can address the constraints of current therapies. Nevertheless, additional refinement and exploration are necessary to ensure their safety, effectiveness, and suitability for clinical application (Sharma, Chakraborty, et al., 2024; Wong et al., 2019).

### 12.4.2 Clinical trials and outcomes

Despite extensive preclinical research in this field, the clinical application of polymeric nanocarriers (PNCs) for treating colorectal cancer is still in its early stages (Table 12.2) (Operti et al., 2021). PNC formulation is currently limited by variability between batches, lack of reproducibility, and potential toxicity associated with using organic solvents. The optimization and scaling up of PNC production for large-scale manufacturing are complex processes that require standardized protocols. Additionally, many PNCs face challenges related to long-term stability at room temperature, which complicates their practical use. Other hurdles include the absorption of PNCs by the reticuloendothelial system, their biodistribution, trafficking through the complex tumor microenvironment, and cellular uptake (Bhaskaran & Kumar, 2021; Haider et al., 2022; Truong-Dinh Tran et al., 2016).

### 12.4.3 Case studies in colon cancer treatment

The utilization of these advanced drug delivery systems in preclinical and early clinical research offers valuable insights. The studies aim to improve therapeutic outcomes, enhance drug delivery, and reduce toxicity. They demonstrate various PNP systems for colon cancer therapy.

**Table 12.2 Examples of clinical trials using PNCs to treat and diagnose CC.**

| Title | Identifier | Treatment | Stage |
|---|---|---|---|
| Targeted polymeric nanoparticles loaded with cetuximab and decorated with somatostatin analog for colon cancer | NCT03774680 | Cetuximab NPs/oral approved anticancer drug | Phase I |
| Pharmacokinetic, safety, and efficacy study of nanoparticle paclitaxel in patients with peritoneal cancer/a Phase I study of intraperitoneal nanoparticle paclitaxel in patients with peritoneal malignancies | NCT00666991 | Nanoparticulate paclitaxel | Phase I |
| Neoadjuvant chemoradiotherapy with CRLX-101 and capecitabine for rectal cancer | NCT02010567 | CRLX101 (a nanopharmaceutical formulation of camptothecin)/ capecitabine/radiotherapy/surgery | Phase I/ Phase II |
| Study of CRLX101 (NLG207) in the treatment of advanced solid tumors | NCT00333502 | Camptothecin (CPT) conjugated to a linear, cyclodextrin-based polymer | Phase I/ Phase II |
| A study to evaluate ONM-100, an intraoperative fluorescence imaging agent for the detection of cancer | NCT03735680 | ONM-100 (a polymer micelle covalently conjugated to indocyanine green) | Phase 2 |

In a study by Hashmi et al. (2022), it was reported that a 34-year-old male from Ecuador, with no previous medical or surgical history, was referred to the hospital by his primary care physician after presenting with low hemoglobin levels of 3.1 g $dL^{-1}$. The patient denied any history of smoking, alcohol consumption, or gastrointestinal bleeding. He had been experiencing symptoms such as dizziness, fatigue, headaches, and early satiety for eight months prior to immigrating to the United States from Ecuador a year earlier. Upon admission to the hospital, the patient underwent an examination for symptomatic anemia and received three blood transfusions, resulting in the restoration of his hemoglobin level to normal. There was no reported family history of gastrointestinal cancers. The investigation for anemia revealed a diagnosis of iron-deficient anemia. A computed tomography (CT) scan of the abdomen and pelvis identified a large lymph node in the abdomen and a 9 cm circumferential wall thickening of the proximal ascending colon. Metastasis was ruled out through CT scans of the chest and triple-phase liver imaging (Hashmi et al., 2022).

According to Zhang et al. (2024), in 2018, our hospital admitted an 80-year-old man with hemochromatosis who was diagnosed with advanced colon cancer. The patient's family did not agree with the standard surgical treatment. Over the course of 2018–21, we successfully provided the patient with various minimally invasive endoscopic therapies, such as argon plasma coagulation (Zhang et al., 2024).

In 2004, Benson and colleagues described the case of Mrs. B, an African American woman who was in good health but began experiencing vague gastrointestinal symptoms. These symptoms included off-and-on cramping and abdominal pain, which she thought were due to heartburn, as well as constipation and bloating. Her symptoms worsened over the next few months, and in February 2006, she visited her primary care doctor. She reported a 15-pound weight loss in the past year, attributing it to gastrointestinal problems and reduced appetite. A laboratory test revealed a carcinoembryonic antigen (CEA) level of 200 ng $mL^{-1}$ and grade 2 anemia (hemoglobin 9.4 g $dL^{-1}$). A colonoscopy confirmed a 6 cm mass in the sigmoid colon. Subsequent CT scans of the abdomen and chest revealed multiple liver lesions, but lung metastases were ruled out. During the left colectomy, a palpable tumor was found in the left lobe of the liver. 6 out of the 18 removed lymph nodes tested positive for the original tumor, which had spread into the subserosa and infiltrated the muscularis propria. It was determined that Mrs. B had stage IV (T3, N2, and M1) colon cancer (Benson III et al., 2004).

Furthermore, a 46-year-old woman named Mrs. K visited the doctor because of severe cramps and bloody diarrhea. She had been experiencing bright red blood per rectum for the past six months. A colonoscopy revealed a 19-cm-long obstructive lesion in her sigmoid colon. Her liver and renal function tests came back normal, and her CEA level was 1.1 ng $mL^{-1}$. However, an abdominal CT scan showed an 8-cm-long, nonresectable liver lesion. Mrs. K underwent laparoscopic sigmoid colectomy and was admitted to the hospital. Pathology results indicated a poorly differentiated adenocarcinoma invading the pericolonic adipose tissue and penetrating the muscularis propria. After testing 15 out of 15 lymph nodes positive for tumors, she was diagnosed with stage IV cancer (T4, N2, M1) (Wilkes, 2005; Yamamoto et al., 2005).

Ms. W, a 48-year-old woman who had recently divorced, complained of bright red blood in her rectum and acute, cramping abdominal discomfort while on vacation in Hawaii. She sought medical attention, and a CT scan revealed two liver tumors. An urgent colonoscopy uncovered a partially obstructive, 5-cm lesion in her descending colon. Her metastatic workup yielded negative results overall. Her preoperative baseline CEA level was 35.6 ng $mL^{-1}$. After discussing treatment options, it was decided that she would undergo colectomy, bevacizumab, and neoadjuvant chemotherapy with hopes of making her liver lesions treatable (Berry et al., 2006; Borner et al., 2002; Cassidy et al., 2004).

In November 2004, a 56-year-old man named Mr. J had a colonoscopy that confirmed he had colon cancer. He underwent surgery to remove most of the tumor. Following six rounds of treatment with FOLFOX and bevacizumab, he was declared "cancer-free" in July 2005. However, a positron emission tomography (PET)/CT scan conducted during his three-month follow-up revealed tumor recurrence in the stomach. Subsequent CT scans showed liver progression, and he received FOLFIRI treatment for six months. A PET scan revealed peritoneal carcinomatosis, ascites, multiple suspicious lung nodules, and severe liver disease with many lesions, a tumor in the porta hepatis, left portal vein blockage, and biliary ductal dilatation. Mr. J reported experiencing anorexia for five days, left-sided abdominal pain for six weeks, and constipation (Cunningham et al., 2004; Kopetz & Hoff, 2005; Saltz et al., 2005).

## 12.5 Challenges and limitations

The diagnosis of colorectal cancer using biomarkers is challenging due to numerous difficulties and limitations associated with this approach. Identifying reliable biomarkers for colorectal cancer and validating their effectiveness is a laborious and complex task. Many potential biomarkers identified in preclinical studies have not successfully transitioned into practical clinical tools due to various obstacles in the process. The development of biomarkers for colorectal cancer has faced challenges such as inadequate reproducibility, flaws in study design, and small sample sizes. Colorectal cancer is a complex disease with significant heterogeneity, encompassing numerous subtypes and a wide range of molecular characteristics. It is essential to consider this heterogeneity when creating biomarkers for colorectal cancer to ensure they accurately capture the diverse characteristics of the disease. Failure to account for this heterogeneity may result in the development of biomarkers that are limited to specific patient subsets or provide limited clinical insights (Khan & Lengyel, 2023).

Biomarkers need to achieve high sensitivity and specificity, which can be challenging because these two factors often need to be balanced. Biomarkers that are highly sensitive but not specific enough can lead to false positives, while those that are highly specific but not sensitive enough can result in false negatives. It is crucial to find the right equilibrium between sensitivity and specificity for accurate disease diagnosis (Khan & Lengyel, 2023).

Ensuring the standardization and reproducibility of biomarker tests is essential to produce consistent and reliable results across different platforms and laboratories. Without adequate standardization, the clinical usefulness of biomarkers may be limited due to significant outcome variability. To address this issue, robust protocols, quality control measures, and validation studies need to be implemented. These measures are vital to ensure the reproducibility and comparability of results obtained in various scenarios. Biomarker testing should be accessible, cost-effective, and suitable for use in routine clinical settings. However, some biomarker assays may be costly, require sophisticated techniques, or necessitate specialized equipment, posing challenges for widespread use (Garg et al., 2023; Khan & Lengyel, 2023).

The identification, application, and validation of biomarkers in clinical settings are crucial for guiding treatment decisions, monitoring regimens, and diagnosing colorectal cancer. Extensive clinical trials and other validation studies are necessary to determine the clinical importance of biomarkers and evaluate their impact on patient outcomes. Ethical and legal considerations arise when using biomarkers in clinical practice, particularly regarding patient privacy, consent, data sharing, and the potential for stigmatization or discrimination. To ensure the responsible and ethical use of biomarkers, comprehensive policies and regulations must be established. Despite the presence of barriers and limitations, ongoing research and technological advancements offer hope for overcoming these challenges and enhancing the utilization of biomarkers in the detection and treatment of colorectal cancer (Garg et al., 2023).

Here, we seek to draw attention to the difficulties this distinct population encounters along the cancer care continuum, beginning with prevention, detection, diagnosis, treatment, survivability, and end-of-life care (Fig. 12.2) (Alzaabi et al., 2022).

## 12.6 Future perspectives of polymeric nanoparticles for treatment of colon cancer

Research in nanomedicine has significantly improved therapeutic efficacy and created new treatment options for colorectal cancer (CC). Nanomedicine has led to advancements in novel nanobiomaterials

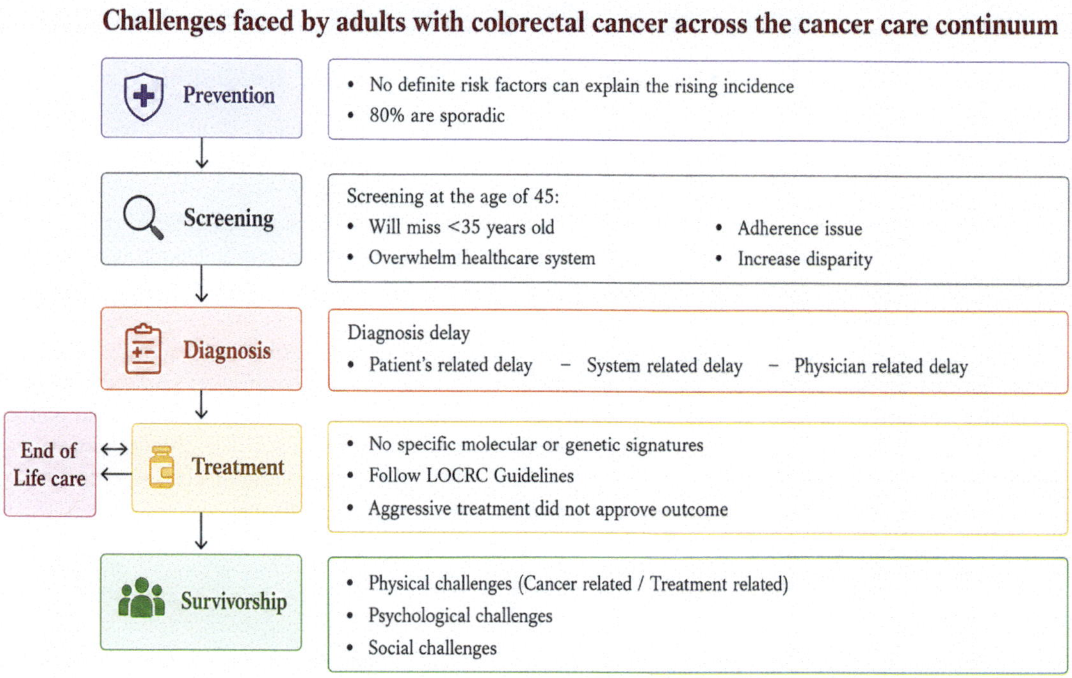


**FIGURE 12.2**

Challenges faced by young adults with colorectal cancer across the cancer care continuum.

and nanotherapeutics, which have enhanced the potential for colon-targeted drug delivery systems (CDDS). Chemotherapy for CC often faces limitations due to multidrug resistance, but nanotechnology can enhance the effectiveness of traditional treatments by improving drug bioavailability and reducing systemic side effects. Conventional drug delivery methods pose challenges due to nonspecific distribution and uncontrollable release, leading to higher toxicity and requiring larger doses. CDDS, however, allows for a lower drug dose, protection of the drug until it reaches the target site, reduced enzymatic activity, and improved absorption with minimal systemic side effects. This system supports local and systemic drug delivery, enhancing therapeutic effectiveness, bioavailability, and reducing toxicity.

Despite their potential, nanotherapeutics face significant obstacles, including challenges with large-scale production, lack of regulation, biological barriers, and inconsistencies in nanoparticle size and characteristics. These issues can lead to undesired biodistribution due to nanoparticle properties like size and surface chemistry variations. Addressing these problems requires intelligent, targeted drug delivery systems that can control drug release and accumulation in specific diseased areas for a longer duration. The clinical success of nanomedicine is still limited, emphasizing the need for more human data and standards for in vivo models to improve clinical outcomes. Future goals for nanotherapeutics include developing nanoparticles that release drugs in response to specific stimuli, reducing toxicity, and ensuring that the drugs accumulate in target tissues rather than in healthy areas.

## 12.7 Conclusion

### 12.7.1 Impact of polymeric nanoparticles on colon cancer treatment

Nanocarriers, particularly polymer-based systems, have shown significant promise in oncology, offering a novel approach to cancer treatment. These systems, already in clinical use, demonstrate their therapeutic potential, especially in delivering anticancer agents with higher efficacy and fewer side effects than traditional chemotherapy. Polymeric micelles, in particular, protect and concentrate chemotherapeutic drugs at tumor sites, reducing systemic toxicity and improving patient outcomes. However, the safety and targeting efficiency of these polymeric platforms need thorough validation before further clinical application.

PNPs have emerged as a targeted drug delivery system for treating colorectal cancer. PNPs can encapsulate 5-fluorouracil (5-FU) and deliver it directly to cancer cells by utilizing surface receptors or ligands. This method reduces harm to healthy cells while enhancing drug accumulation at cancer sites. PNPs can also carry therapeutic genes, providing a versatile platform for chemotherapy and gene therapy. While these strategies show potential to improve cancer treatment efficacy and reduce side effects, further research is necessary to optimize PNP designs for clinical use.

## References

Aghebati-Maleki, A., Dolati, S., Ahmadi, M., Baghbanzhadeh, A., Asadi, M., Fotouhi, A., Yousefi, M., & Aghebati-Maleki, L. (2020). Nanoparticles and cancer therapy: Perspectives for application of nanoparticles in the treatment of cancers. *Journal of Cellular Physiology, 235*, 1962–1972.

Ahlawat, J., Henriquez, G., & Narayan, M. (2018). Enhancing the delivery of chemotherapeutics: Role of biodegradable polymeric nanoparticles. *Molecules (Basel, Switzerland), 23*, 2157.

Alzaabi, A., Alharrasi, A., Almusalami, A., Almahyijari, N., Al Hinai, K., Aladawi, H., & Al-Shamsi, H. O. (2022). Early onset colorectal cancer: Challenges across the cancer care continuum. *Annals of Medicine and Surgery, 82*.

Avramović, N., Mandić, B., Savić-Radojević, A., & Simić, T. (2020). Polymeric nanocarriers of drug delivery systems in cancer therapy. *Pharmaceutics, 12*, 298.

Baião, A., Sousa, F., Oliveira, A. V., Oliveira, C., & Sarmento, B. (2020). Effective intracellular delivery of bevacizumab via PEGylated polymeric nanoparticles targeting the CD44v6 receptor in colon cancer cells. *Biomaterials Science, 8*, 3720–3729.

Banerjee, A., Pathak, S., Subramanium, V. D., Dharanivasan, G., Murugesan, R., & Verma, R. S. (2017). Strategies for targeted drug delivery in treatment of colon cancer: Current trends and future perspectives. *Drug Discovery Today, 22*, 1224–1232.

Banik, B. L., Fattahi, P., & Brown, J. L. (2016). Polymeric nanoparticles: The future of nanomedicine. *Wiley Interdisciplinary Reviews: Nanomedicine and Nanobiotechnology, 8*, 271–299.

Begines, B., Ortiz, T., Pérez-Aranda, M., Martínez, G., Merinero, M., Argüelles-Arias, F., & Alcudia, A. (2020). Polymeric nanoparticles for drug delivery: Recent developments and future prospects. *Nanomaterials, 10*, 1403.

Benson III, A. B., Ajani, J. A., Catalano, R. B., Engelking, C., Kornblau, S. M., Martenson Jr, J. A., Mccallum, R., Mitchell, E. P., O'dorisio, T. M., & Vokes, E. E. (2004). Recommended guidelines for the treatment of cancer treatment-induced diarrhea. *Journal of Clinical Oncology, 22*, 2918–2926.

Berry, S., Cunningham, D., Michael, M., Dibartolomeo, M., Rivera, F., Kretzschmar, A., Mazier, M., Lutiger, B., & Vancutsem, E. (2006). Preliminary safety of bevacizumab with first-line Folfox, Capox, Folfiri and capecitabine for mCRC—First BEA Trial. *Journal of Clinical Oncology, 24*, 3534.

Bhaskaran, N. A., & Kumar, L. (2021). Treating colon cancers with a non-conventional yet strategic approach: An overview of various nanoparticulate systems. *Journal of Controlled Release, 336*, 16–39.

Bhattacharya, S., Prajapati, B. G., & Singh, S. (2024). Polymeric nanoparticles in colorectal cancer. In *Colorectal Cancer*. Elsevier.

Bigaj-Jozefowska, M. J., & Grześkowiak, B. F. (2022). Polymeric nanoparticles wrapped in biological membranes for targeted anticancer treatment. *European Polymer Journal, 176*, 111427.

Borner, M. M., Dietrich, D., Stupp, R., Morant, R., Honegger, H., Wernli, M., Herrmann, R., Pestalozzi, B. C., Saletti, P., & Hanselmann, S. (2002). Phase II study of capecitabine and oxaliplatin in first-and second-line treatment of advanced or metastatic colorectal cancer. *Journal of Clinical Oncology, 20*, 1759–1766.

Cabeza, L., Perazzoli, G., Mesas, C., Jiménez-Luna, C., Prados, J., Rama, A. R., & Melguizo, C. (2020). Nanoparticles in colorectal cancer therapy: Latest in vivo assays, clinical trials, and patents. *AAPS PharmSciTech, 21*, 1–15.

Cai, X., Wang, X., He, M., Wang, Y., Lan, M., Zhao, Y., & Gao, F. (2021). Colon-targeted delivery of tacrolimus using pH-responsive polymeric nanoparticles for murine colitis therapy. *International Journal of Pharmaceutics, 606*, 120836.

Calzoni, E., Cesaretti, A., Polchi, A., Di Michele, A., Tancini, B., & Emiliani, C. (2019). Biocompatible polymer nanoparticles for drug delivery applications in cancer and neurodegenerative disorder therapies. *Journal of Functional Biomaterials, 10*, 4.

Cassidy, J., Tabernero, J., Twelves, C., Brunet, R., Butts, C., Conroy, T., Debraud, F., Figer, A., Grossmann, J., & Sawada, N. (2004). XELOX (capecitabine plus oxaliplatin): Active first-line therapy for patients with metastatic colorectal cancer. *Journal of Clinical Oncology, 22*, 2084–2091.

Chaurasia, S., Chaubey, P., Patel, R. R., Kumar, N., & Mishra, B. (2016). Curcumin-polymeric nanoparticles against colon-26 tumor-bearing mice: Cytotoxicity, pharmacokinetic and anticancer efficacy studies. *Drug Development and Industrial Pharmacy, 42*, 694–700.

Conte, C., Maiolino, S., Pellosi, D. S., Miro, A., Ungaro, F., & Quaglia, F. (2016). Polymeric nanoparticles for cancer photodynamic therapy. In *Light-responsive nanostructured systems for applications in nanomedicine* (pp. 61–112).

Cunningham, D., Humblet, Y., Siena, S., Khayat, D., Bleiberg, H., Santoro, A., Bets, D., Mueser, M., Harstrick, A., & Verslype, C. (2004). Cetuximab monotherapy and cetuximab plus irinotecan in irinotecan-refractory metastatic colorectal cancer. *New England Journal of Medicine, 351*, 337–345.

Dey, A., Mitra, A., Pathak, S., Prasad, S., Zhang, A. S., Zhang, H., Sun, X.-F., & Banerjee, A. (2023). Recent advancements, limitations, and future perspectives of the use of personalized medicine in treatment of colon cancer. *Technology in Cancer Research & Treatment, 22*, 15330338231178403.

Dezfuli, A. A. Z., Abu-Elghait, M., & Salem, S. S. (2024). Recent insights into nanotechnology in colorectal cancer. *Applied Biochemistry and Biotechnology, 196*, 4457–4471.

Dotan, E., & Cohen, S. J. (2011). Challenges in the management of stage II colon cancer. In *Seminars in oncology*. Elsevier, 511–520.

Dristant, U., Mukherjee, K., Saha, S., & Maity, D. (2023). An overview of polymeric nanoparticles-based drug delivery system in cancer treatment. *Technology in Cancer Research & Treatment, 22*.

Egusquiaguirre, S. P., Igartua, M., Hernández, R. M., & Pedraz, J. L. (2012). Nanoparticle delivery systems for cancer therapy: Advances in clinical and preclinical research. *Clinical and Translational Oncology, 14*, 83–93.

Esmaelbeygi, E., Khoei, S., Khoee, S., & Eynali, S. (2015). Role of iron oxide core of polymeric nanoparticles in the thermosensitivity of colon cancer cell line HT-29. *International Journal of Hyperthermia, 31*, 489–497.

Farinha, P., Pinho, J. O., Matias, M., & Gaspar, M. M. (2022). Nanomedicines in the treatment of colon cancer: A focus on metallodrugs. *Drug Delivery and Translational Research,* 1–18.

Garg, R., Gaur, M., Prajapati, B. G., Agrawal, G., Bhattacharya, S., & Elossaily, G. M. (2023). Polymeric nanoparticles approach and identification and characterization of novel biomarkers for colon cancer. *Results in Chemistry, 6*, 101167.

Gulbake, A., Jain, A., Jain, A., Jain, A., & Jain, S. K. (2016). Insight to drug delivery aspects for colorectal cancer. *World Journal of Gastroenterology, 22*, 582.

Guo, K., Xiao, N., Liu, Y., Wang, Z., Tóth, J., Gyenis, J., Thakur, V. K., Oyane, A., & Shubhra, Q. T. (2022). Engineering polymer nanoparticles using cell membrane coating technology and their application in cancer treatments: Opportunities and challenges. *Nano Materials Science, 4*, 295–321.

Haider, M., Zaki, K. Z., El Hamshary, M. R., Hussain, Z., Orive, G., & Ibrahim, H. O. (2022). Polymeric nanocarriers: A promising tool for early diagnosis and efficient treatment of colorectal cancer. *Journal of Advanced Research, 39*, 237–255.

Halder, T., Mehta, P., & Acharya, N. (2021). Trends in the functional food market and nutraceutical product development. In *Nutraceuticals for aging and anti-aging*. CRC Press.

Hashmi, S. S. H., Shady, A., Atallah-Vinograd, J., Cummings, D., Maranino, A., & Harley, J. (2022). Young-onset colon cancer: a case report. *Cureus, 14*.

Hosseinifar, T., Sheybani, S., Abdouss, M., Hassani Najafabadi, S. A., & Shafiee Ardestani, M. (2018). Pressure responsive nanogel base on alginate-cyclodextrin with enhanced apoptosis mechanism for colon cancer delivery. *Journal of Biomedical Materials Research. Part A, 106*, 349–359.

Indoria, S., Singh, V., & Hsieh, M.-F. (2020). Recent advances in theranostic polymeric nanoparticles for cancer treatment: A review. *International Journal of Pharmaceutics, 582*, 119314.

Jiang, J., Cui, X., Huang, Y., Yan, D., Wang, B., Yang, Z., Chen, M., Wang, J., Zhang, Y., & Liu, G. (2024). Advances and prospects in integrated nano-oncology. *Nano Biomedicine & Engineering, 16*.

Juan, A., Cimas, F. J., Bravo, I., Pandiella, A., Ocaña, A., & Alonso-Moreno, C. (2020). An overview of antibody conjugated polymeric nanoparticles for breast cancer therapy. *Pharmaceutics, 12*, 802.

Kamaly, N., Xiao, Z., Valencia, P. M., Radovic-Moreno, A. F., & Farokhzad, O. C. (2012). Targeted polymeric therapeutic nanoparticles: Design, development and clinical translation. *Chemical Society Reviews, 41*, 2971–3010.

Kesharwani, P., Paknikar, K. M., & Gajbhiye, V. (2019). *Polymeric nanoparticles as a promising tool for anti-cancer therapeutics*. Academic Press.

Khan, S. H., Anees, M., Zofair, S. F. F., Rasool, F., Khan, M. A., Moin, S., & Younus, H. (2024). Fucoidan based polymeric nanoparticles encapsulating epirubicin: A novel and effective chemotherapeutic formulation against colorectal cancer. *International Journal of Pharmaceutics, 664*, 124622.

Khan, S. Z., & Lengyel, C. G. (2023). Challenges in the management of colorectal cancer in low-and middle-income countries. *Cancer Treatment and Research Communications, 35*, 100705.

Kim, K.-T., Lee, J.-Y., Kim, D.-D., Yoon, I.-S., & Cho, H.-J. (2019). Recent progress in the development of poly (lactic-co-glycolic acid)-based nanostructures for cancer imaging and therapy. *Pharmaceutics, 11*, 280.

Klippstein, R., Wang, J. T. W., El-Gogary, R. I., Bai, J., Mustafa, F., Rubio, N., Bansal, S., Al-Jamal, W. T., & Al-Jamal, K. T. (2015). Passively targeted curcumin-loaded pegylated PLGA nanocapsules for colon cancer therapy in vivo. *Small (Weinheim an der Bergstrasse, Germany), 11*, 4704–4722.

Kopetz, S., & Hoff, P. M. (2005). Cytotoxic chemotherapy for advanced colorectal cancer. Recent advances in management. *Oncology, 19*, 11–17 *Williston Park, NY.*

Madej, M., Kurowska, N., & Strzalka-Mrozik, B. (2022). Polymeric nanoparticles—Tools in a drug delivery system in selected cancer therapies. *Applied Sciences, 12*, 9479.

Masood, F. (2016). Polymeric nanoparticles for targeted drug delivery system for cancer therapy. *Materials Science and Engineering: C, 60*, 569–578.

Meng, F., Yun, Z., Yan, G., Wang, G., & Lin, C. (2021). Engineering of anticancer drugs entrapped polymeric nanoparticles for the treatment of colorectal cancer therapy. *Process Biochemistry, 111*, 36–42.

Meropol, N. J. (2011). Ongoing challenge of stage II colon cancer. *American Society of Clinical Oncology.*

Mishra, B., & Chaurasia, S. (2017). Design of novel chemotherapeutic delivery systems for colon cancer therapy based on oral polymeric nanoparticles. *Therapeutic Delivery, 8*, 29–47.

Mohanty, A., Uthaman, S., & Park, I.-K. (2020). Utilization of polymer-lipid hybrid nanoparticles for targeted anti-cancer therapy. *Molecules (Basel, Switzerland), 25*, 4377.

Naeem, M., Choi, M., Cao, J., Lee, Y., Ikram, M., Yoon, S., Lee, J., Moon, H. R., Kim, M.-S., & Jung, Y. (2015). Colon-targeted delivery of budesonide using dual pH-and time-dependent polymeric nanoparticles for colitis therapy. *Drug Design, Development and Therapy,* 3789–3799.

Narayana, S., Gowda, B. J., Hani, U., Shimu, S. S., Paul, K., Das, A., Ashique, S., Ahmed, M. G., Tarighat, M. A., & Abdi, G. (2024). Inorganic nanoparticle-based treatment approaches for colorectal cancer: Recent advancements and challenges. *Journal of Nanobiotechnology, 22*, 427.

Operti, M. C., Bernhardt, A., Grimm, S., Engel, A., Figdor, C. G., & Tagit, O. (2021). PLGA-based nanomedicines manufacturing: Technologies overview and challenges in industrial scale-up. *International Journal of Pharmaceutics, 605*, 120807.

Phatak, N., Bhattacharya, S., Shah, D., Manthalkar, L., Sreelaya, P., & Jain, A. (2023). CD44 targeted delivery of hyaluronic acid-coated polymeric nanoparticles against colorectal cancer. *Nanomedicine: Nanotechnology, Biology, and Medicine, 18*, 1613–1634.

Prabhu, R. H., Patravale, V. B., & Joshi, M. D. (2015). Polymeric nanoparticles for targeted treatment in oncology: Current insights. *International Journal of Nanomedicine,* 1001–1018.

Prasad, S., & Dangi, J. (2023). Targeting efficacy and anticancer activity of polymeric nanoparticles of SN-38 on colon cancer cell lines. *Future Journal of Pharmaceutical Sciences, 9*, 12.

Pridgen, E. M., Alexis, F., & Farokhzad, O. C. (2015). Polymeric nanoparticle drug delivery technologies for oral delivery applications. *Expert Opinion on Drug Delivery, 12*, 1459–1473.

Priyadarshi, K., Shirsath, K., Waghela, N. B., Sharma, A., Kumar, A., & Pathak, C. (2021). Surface modified PAMAM dendrimers with gallic acid inhibit, cell proliferation, cell migration and inflammatory response to augment apoptotic cell death in human colon carcinoma cells. *Journal of Biomolecular Structure and Dynamics, 39*, 6853–6869.

Ramezani, P., Abnous, K., Taghdisi, S. M., Zahiri, M., Ramezani, M., & Alibolandi, M. (2020). Targeted MMP-2 responsive chimeric polymersomes for therapy against colorectal cancer. *Colloids and surfaces B: Biointerfaces, 193*, 111135.

Ravani, A., & Sharma, H. P. (2022). Meat based functional foods. *Functional Foods,* 235–287.

Reimondez-Troitiño, S., González-Aramundiz, J. V., Ruiz-Bañobre, J., López-López, R., Alonso, M. J., Csaba, N., & De La Fuente, M. (2019). Versatile protamine nanocapsules to restore miR-145 levels and interfere tumor growth in colorectal cancer cells. *European Journal of Pharmaceutics and Biopharmaceutics, 142*, 449–459.

Rizi, H. A. Y., Shin, D. H., & Rizi, S. Y. (2022). Polymeric nanoparticles in cancer chemotherapy: A narrative review. *Iranian Journal of Public Health, 51*, 226.

Russo, A., Maiolino, S., Pagliara, V., Ungaro, F., Tatangelo, F., Leone, A., Scalia, G., Budillon, A., Quaglia, F., & Russo, G. (2016). Enhancement of 5-FU sensitivity by the proapoptotic rpL3 gene in p53 null colon cancer cells through combined polymer nanoparticles. *Oncotarget, 7*, 79670.

Rychahou, P., Bae, Y., Reichel, D., Zaytseva, Y. Y., Lee, E. Y., Napier, D., Weiss, H. L., Roller, N., Frohman, H., & Le, A.-T. (2018). Colorectal cancer lung metastasis treatment with polymer–drug nanoparticles. *Journal of Controlled Release, 275*, 85–91.

Saltz, L., Lenz, H.-J., Hochster, H., Wadler, S., Hoff, P., Kemeny, N., Hollywood, E., Gonen, M., Wetherbee, S., & Chen, H. (2005). Randomized phase II trial of cetuximab/bevacizumab/irinotecan (CBI) versus cetuximab/bevacizumab (CB) in irinotecan-refractory colorectal cancer. *Journal of Clinical Oncology, 23*, 3508.

Shad, P. M., Karizi, S. Z., Javan, R. S., Mirzaie, A., Noorbazargan, H., Akbarzadeh, I., & Rezaie, H. (2020). Folate conjugated hyaluronic acid coated alginate nanogels encapsulated oxaliplatin enhance antitumor and apoptosis efficacy on colorectal cancer cells (HT29 cell line). *Toxicology in Vitro, 65*, 104756.

Sharma, S., & Singh, S. (2022). Synthetic routes to quinoline-based derivatives having potential anti-bacterial and anti-fungal properties. *Current Organic Chemistry, 26*, 1453–1469.

Sharma, S., & Singh, S. (2023a). The biological and pharmacological potentials of Indole-based heterocycles. *Letters in Organic Chemistry, 20*, 711–729.

Sharma, S., & Singh, S. (2023b). Research article molecular docking study for binding affinity of 2H-thiopyrano [2, 3-b] quinoline derivatives against CB1a. *Extraction, 2*, 3.

Sharma, S. (2024). Molecular docking study for binding affinity of Indole derivatives against solution structure of the antimicrobial peptide Btd-2 [3, 4]. *Journal of Molecular Chemistry, 4*, 686.

Sharma, S., Chakraborty, M., Yadav, D., Dhullap, A., Singh, R., Verma, R. K., Bhattacharya, S., & Singh, S. (2024). Strategic developments in polymer-functionalized liposomes for targeted colon cancer therapy: An updated review of clinical trial data and future horizons. *Biomacromolecules*.

Sharma, S., Choubey, R., Gupta, M., & Singh, S. (2024). Heterocyclic-based analogues against sarcine-ricin loop RNA from *Escherichia coli*: In silico molecular docking study and machine learning classifiers. *Medicinal Chemistry, 20*, 452–465.

Sharma, S., Das, A. K., & Singh, S. (2023). Bioactive indole heterocycles and their synthetic routes: A comprehensive review. *Current Trends in Biotechnology and Pharmacy, 17*, 1141–1158.

Sharma, S., Mishra, A. K., & Singh, S. (2022). Molecular docking study of 2, 3, 4-trisubstituted-2, 3, 4, 9-tetrahydrothiopyrano[2, 3-b] indole derivatives with TRPV channels: Possible new analgesics. *ETJRI*.

Sharma, S., Monga, Y., Gupta, A., & Singh, S. (2023a). 2-Oxindole and related heterocycles: synthetic methodologies for their natural products and related derivatives. *RSC Advances, 13*, 14249–14267.

Sharma, S., Monga, Y., Gupta, A., & Singh, S. (2023b). In silico screening and molecular docking study of quinoline based compounds with Human kallikrein 7 in complex with 1, 4-diazepane-7-one 1-acetamide derivative receptor target for potential antibacterials. *Journal of Molecular Chemistry, 3*, 585.

Sharma, S., Singh, S., & Yadav, D. (2023). Quinoline-based anti-oncogenic molecules: Synthesis and biological evaluation. *Medicinal Chemistry, 19*, 848–858.

Simón-Gracia, L., Hunt, H., Scodeller, P., Gaitzsch, J., Kotamraju, V. R., Sugahara, K. N., Tammik, O., Ruoslahti, E., Battaglia, G., & Teesalu, T. (2016). iRGD peptide conjugation potentiates intraperitoneal tumor delivery of paclitaxel with polymersomes. *Biomaterials, 104*, 247–257.

Swain, S., Kumar Sahu, P., Beg, S., & Manohar Babu, S. (2016). Nanoparticles for cancer targeting: Current and future directions. *Current Drug Delivery, 13*, 1290–1302.

Tiwari, H., Gupta, P., Verma, A., Singh, S., Kumar, R., Gautam, H. K., & Gautam, V. (2024). Advancing era and rising concerns in nanotechnology-based cancer treatment. *ACS Chemical Health & Safety, 31*, 153–161.

Truong-Dinh Tran, T., Ha-Lien Tran, P., Wang, Y., Li, P., & Kong, L. (2016). Nanoparticulate drug delivery to colorectal cancer: Formulation strategies and surface engineering. *Current Pharmaceutical Design, 22*, 2904–2912.

Udompornmongkol, P., & Chiang, B.-H. (2015). Curcumin-loaded polymeric nanoparticles for enhanced anti-colorectal cancer applications. *Journal of Biomaterials Applications, 30*, 537–546.

Van Vlerken, L. E., & Amiji, M. M. (2006). Multi-functional polymeric nanoparticles for tumour-targeted drug delivery. *Expert Opinion on Drug Delivery, 3*, 205–216.

Vashishat, A., Singh, A., Kurmi, B. D., Gupta, G. D., & Singh, D. (2023). A short appraisal of polylactic-co-glycolic acid based polymer nanotechnology for colon cancer: Recent advances and literature evidences. *Therapeutic Delivery, 14*, 459–472.

Venkatesan, P., Puvvada, N., Dash, R., Kumar, B. P., Sarkar, D., Azab, B., Pathak, A., Kundu, S. C., Fisher, P. B., & Mandal, M. (2011). The potential of celecoxib-loaded hydroxyapatite-chitosan nanocomposite for the treatment of colon cancer. *Biomaterials, 32*, 3794–3806.

Wei, Y., Gu, X., Sun, Y., Meng, F., Storm, G., & Zhong, Z. (2020). Transferrin-binding peptide functionalized polymersomes mediate targeted doxorubicin delivery to colorectal cancer in vivo. *Journal of Controlled Release, 319*, 407–415.

Wilkes, G. M. (2005). Therapeutic options in the management of colon cancer: 2005 update. *Clinical Journal of Oncology Nursing, 9*, 31.

Wong, K. E., Ngai, S. C., Chan, K.-G., Lee, L.-H., Goh, B.-H., & Chuah, L.-H. (2019). Curcumin nanoformulations for colorectal cancer: A review. *Frontiers in Pharmacology, 10*, 152.

Xu, G., Shi, H., Ren, L., Gou, H., Gong, D., Gao, X., & Huang, N. (2015). Enhancing the anti-colon cancer activity of quercetin by self-assembled micelles. *International Journal of Nanomedicine,* 2051–2063.

Yakati, V., Vangala, S., Madamsetty, V. S., Banerjee, R., & Moku, G. (2022). Enhancing the anticancer effect of paclitaxel by using polymeric nanoparticles decorated with colorectal cancer targeting CPKSNNGVC-peptide. *Journal of Drug Delivery Science and Technology, 68*, 103125.

Yamamoto, D. S., Viale, P. H., Roesser, K., & Lin, A. (2005). *The clinical use of tumor markers in select cancers: Are you confident enough to discuss them with your patients? Oncology nursing forum.* Oncology Nursing Society, 1013.

Yang, X., Li, Z., Wang, N., Li, L., Song, L., He, T., Sun, L., Wang, Z., Wu, Q., & Luo, N. (2015). Curcumin-encapsulated polymeric micelles suppress the development of colon cancer in vitro and in vivo. *Scientific Reports, 5*, 10322.

Yıldırım, M., Sessevmez, M., Poyraz, S., & Düzgüneş, N. (2023). Recent strategies for cancer therapy: polymer nanoparticles carrying medicinally important phytochemicals and their cellular targets. *Pharmaceutics, 15*, 2566.

Yoon, S. M., Myung, S.-J., Kim, I.-W., Do, E.-J., Ye, B. D., Ryu, J. H., Park, K., Kim, K., Kwon, I. C., & Kim, M. J. (2011). Application of near-infrared fluorescence imaging using a polymeric nanoparticle-based probe for the diagnosis and therapeutic monitoring of colon cancer. *Digestive Diseases and Sciences, 56*, 3005–3013.

You, X., Kang, Y., Hollett, G., Chen, X., Zhao, W., Gu, Z., & Wu, J. (2016). Polymeric nanoparticles for colon cancer therapy: Overview and perspectives. *Journal of Materials Chemistry B, 4*, 7779–7792.

Younis, N. K., Roumieh, R., Bassil, E. P., Ghoubaira, J. A., Kobeissy, F., & Eid, A. H. (2022). Nanoparticles: Attractive tools to treat colorectal cancer. In *Seminars in cancer biology*. Elsevier, 1–13.

Yu, M. K., Park, J., & Jon, S. (2012). Targeting strategies for multifunctional nanoparticles in cancer imaging and therapy. *Theranostics, 2*, 3.

Yu, T., Huang, X., Liu, J., Fu, Q., Wang, B., & Qian, Z. (2019). Polymeric nanoparticles encapsulating α-mangostin inhibit the growth and metastasis in colorectal cancer. *Applied Materials Today, 16*, 351–366.

Zhang, N., Zhu, L., Liu, Y., Chen, X., Zhang, B., Wen, C., Zhang, H., Tang, Q., & Zhang, M. (2024). Case report: Successful treatment of advanced colon cancer in an eighty-year-old man with long-term and multi-stage endoscopic minimally invasive therapy. *Frontiers in Oncology, 14*, 1367173.

Zheng, B., Chen, L., Pan, C.-C., Wang, J.-Z., Lu, G.-R., Yang, S.-X., Xue, Z.-X., Wang, F.-Y., & Xu, C.-L. (2018). Targeted delivery of miRNA-204-5p by PEGylated polymer nanoparticles for colon cancer therapy. *Nanomedicine: Nanotechnology, Biology, and Medicine, 13*, 769–785.

Zhong, Y., Meng, F., Deng, C., & Zhong, Z. (2014). Ligand-directed active tumor-targeting polymeric nanoparticles for cancer chemotherapy. *Biomacromolecules, 15*, 1955–1969.

CHAPTER

# 13 Nanomicelles-based drug delivery systems as for targeted therapy of colon cancer

**Ankit Paul[1], Adil Ali[1], Pranav Kumar Prabhakar[2] and Tarun Kumar Upadhyay[1]**

[1]*Department of Biotechnology, Parul Institute of Applied Sciences and Research and Development Cell, Parul University, Vadodara, Gujarat, India,*

[2]*Department of Biotechnology, School of Engineering and Technology, Nagaland University, Kohima, Nagaland, India*

## 13.1 Overview of colon cancer and its challenges

The colon and rectum, parts of the large intestine, are the sites where colorectal cancer (CRC) initially progresses. The colon represents the large intestine of the human body and has been the prominent site where CRC primarily develops (Singh et al., 2023). CRC is one of the most prevalent malignancies after lung carcinoma, prostate neoplasms, and breast carcinoma. According to the WHO GLOBOCAN 2022, it ranks fourth globally with over 1.1 million incidences and 0.5 million mortalities (Sung et al., 2021; Wang et al., 2022). It is also the third and second most prevalent cancer in men and women, respectively. It is projected that CRC incidences will reach 2.4 million, and CRC will cause 1.3 million deaths worldwide by 2035 (Bose et al., 2021; Center et al., 2009; Dey et al., 2023; Khan et al., 2022). Colon-associated neoplasms are more common than rectal neoplasms in developed countries; the ratio of colon incidences is twice that of rectum incidences (slightly more in females). Around 0.25 million new colon cases in Europe are identified yearly, accounting for around 9% of all malignancies. It is more prevalent in developed countries, yet its occurrence is rising in middle and developing nations (Labianca et al., 2010). In India, the CRC ranked sixth, with over 70 thousand incidences and 50 thousand mortalities. Generally, CRC can be caused by many factors, including low intake of fruits and veggies, changes in diet and lifestyle, obesity, smoking, and physical inactivity (Center et al., 2009).

The expansion of CRC involves multiple steps that include a continuous loss of genes. Rapid progress in molecular and cellular biology methods has enabled the identification of genetic alterations driving this multistep process (Labianca et al., 2010).

CRC is characterized by a multifaceted process involving various somatic and molecular alterations that are influenced by dietary intake, environmental conditions, pathogenic bacteria, and host immune responses. Not all individuals consuming a high-fat diet, low intake of fiber-based foods, or high-meat diet foster CRC; similarly, a sustainable diet abundant in green veggies, coarse grains, and fresh, juicy fruits does not guarantee protection against the disease (Tjalsma et al., 2012). CRC is not hereditary but sporadic, and an individual's diet can significantly influence neoplasm initiation,

Nano Drug Delivery Systems for Colon Cancer. DOI: https://doi.org/10.1016/B978-0-443-40443-6.00018-0

progression, and promotion. The progression of malignancy is affected by the duration of genetic defects and the intensity of environmental variables (Beyer-Sehlmeyer et al., 2003).

Substantial shreds of evidence suggest that CRC arises from step-by-step disruptions in the gut microbial constitution caused by dietary elements or food, as well as genetic alterations in proto-oncogenes and tumor-inhibitor genes/proteins (Verma, 2013). For instance, maximum CRC incidences harbor alterations in adenomatous polyposis coli (APC) coding sequence, a tumor inhibitor that modulates Wnt signaling by regulating β-catenin intracellular concentrations, controlling subsequent signaling of inflammation-mediated molecules, and cellular differentiation and proliferative signal transduction pathways (Beyer-Sehlmeyer et al., 2003).

CRC arises from sequential disruptions in gut microbiota makeup caused by dietary components, along with genetic modifications in oncogenes and tumor suppressor genes. Most CRCs exhibit alterations in the APC gene, a tumor suppressor that modulates Wnt signaling by regulating β-catenin levels, which subsequently influence downstream-mediated inflammation and its related molecules, cellular cycle, and proliferative signal transduction pathways (Tarapore et al., 2012).

However, several investigations have revealed that microorganism-based fermented goods and their phytoconstituent derivatives have an extensive spectrum of antineoplastic impacts that respond to a neoplastic signaling cascade and escalate epigenetic processes such as suppression of histone deacetylase, which helps in suppressing cell growth, halting neoplastic transformation, and apoptosis (O'keefe, 2016).

## 13.2 Colorectal cancer progression

Most CRC neoplasms are adenocarcinomas that initiate from benign polyps, explicitly adenomas and notched polyps. CRC progression is characterized by four distinct stages. At stages 0 and 1, the tumor develops within the inner layers of the colon, referred to as a "polyp."

In stage 2, cancer extends beyond the colon's walls and begins to infiltrate the outer layers, yet it does not progress to the lymph nodes, being classified as "adenomas." In stage 3, the tumor invades the lymph nodes and is classified as "carcinoma"; stage 4 indicates metastasis within the body (Dey et al., 2023). The accumulation of alterations in proto-oncogenes, tumor-inhibitor genes, and genomic instability genes causes most neoplasms. The advancement from healthy colorectal epithelia to adenoma culminated in a metastatic, aggressive neoplasm (Singh et al., 2023), shown in Fig. 13.1.

## 13.3 Targeted therapy for colorectal cancer

The US FDA has sanctioned various targeted therapies for the treatment of CRC, notably including bevacizumab (anti-VEGF Ab), panitumumab and cetuximab (anti-EGFR Ab), and regorafenib (multikinase inhibitor). These therapeutic molecules primarily act by inhibiting the RAS-RAF-MEK-ERK and PI3K-AKT signal transduction pathways, leading to temporary remissions; however, they rarely result in complete cures (Brown et al., 2015). The limited efficacy can be explained by the upregulation of survival pathways and the downregulation of apoptotic pathways. These mechanisms contribute to a drug-tolerant phenotype and, consequently, therapeutic resistance. Notably, first-line

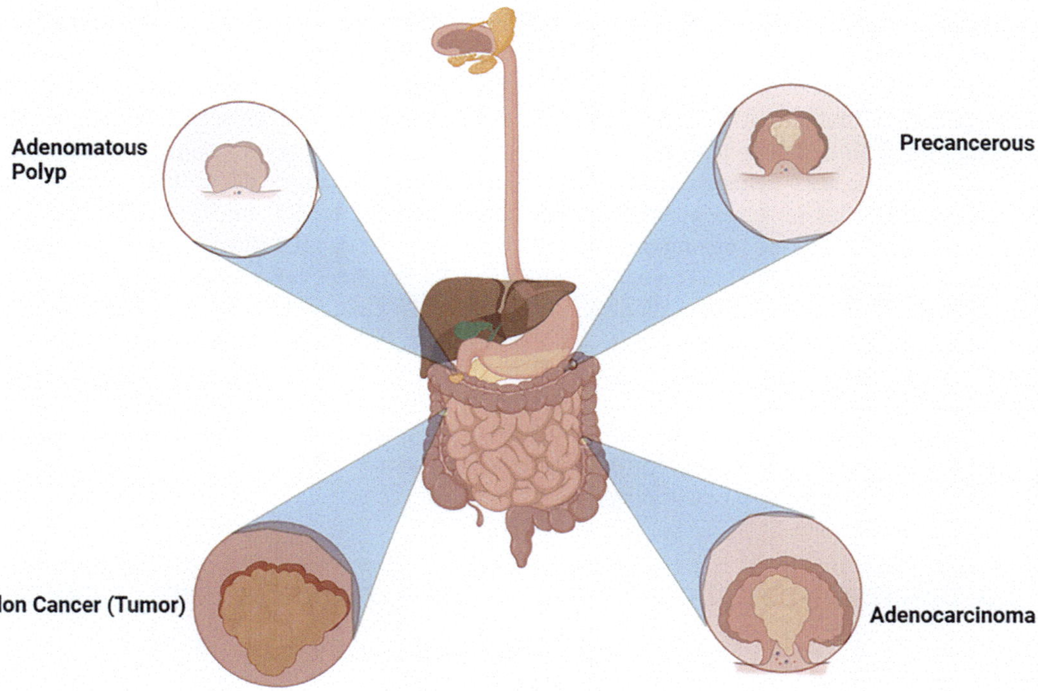


**FIGURE 13.1**

A diagram shows the stages of colon tissue from adenomatous polyp to adenocarcinoma around the large intestine.

treatment with bevacizumab has been exhibited to augment T and B cellular populations in patients with advanced metastatic CRC. Furthermore, the combination of cetuximab with chemotherapy has been associated with increased immunity via $CD8^+$ T-cell activity (Ruan et al., 2020).

## 13.4 Nanomaterials in colorectal cancer diagnosis and treatment

Nanoparticles (NPs) are valuable therapeutic molecules for neoplasms. The most important characteristics of NPs are high biocompatibility and high bioavailability, which make them ideal candidates for drug delivery systems (DDS), diagnosis, and neoplastic therapeutics. Currently, NPs have attracted worldwide interest due to their potent roles in the diagnosis and treatment of CRC. The primary challenge in contemporary drug treatment is the efficient delivery of drugs into neoplastic cells, a task that is unfeasible for larger MW compounds. Active pharmaceutics must be small in size to effectively penetrate neoplastic cells. Nanoparticles possess the advantage of nanoscale dimensions, facilitating their penetration into neoplastic cells (Khan et al., 2022). The morphology and dimensions of nanoparticles are critical factors influencing the effective delivery and permeation of anticancer therapeutics. Effective delivery of pharmaceutics and therapy can be achieved via the synthesis of NPs in various shapes, including nanotubes, nanoflowers, nanoprism, spherical forms, nanostars, and nanorods. The morphology of nanoparticles is a crucial determinant for effective targeting and infiltration within neoplastic cells (Schütz et al., 2014). The spherical morphology of nanoparticles facilitates deep penetration into neoplastic cells. In addition to shape, nanoparticle size is crucial for effective therapeutic response, as it directly influences the cellular internalization process, which relies on endocytosis. NPs vacillating from 10 to 400 nm in size are capable of penetrating the neoplastic membrane of cells (Navarro & de ILarduya, 2009). Furthermore, nanoparticles exceeding 500 nm can enter neoplastic cells via endocytosis. The size of nanoparticles significantly influences the pharmacokinetic and pharmacodynamic profile of drug release and cellular internalization. Iron oxide, quantum dots, Au-NPs, Ag-NPs, Pt-NPs, and $CeO_2$-NPs ranging from 10 to 100 nm can efficiently penetrate neoplastic cells (Ma et al., 2017). Polypeptide nanoparticles and dendrimers with particle sizes vacillating from 150 to 350 nm can readily penetrate neoplastic cells (Pelka et al., 2009). Nanoformulations, which consist of various NPs, are employed for CRC diagnosis and therapy (Khan et al., 2022).

## 13.5 Different types of nanocarriers used in colorectal cancer treatment

Numerous nanocarriers are being employed for the drug distribution in CRC treatment. For instance, nano-DDS utilized for CRC therapeutics include dendrimers, polymeric nanoparticles, metallic-based carriers, lipid-based carriers, micelles, peptides, and hydrogels (Singh et al., 2023).

## 13.6 From micelles to nanomicelles

Micelles (M) are amphiphilic colloidal structures with diameters ranging between 5 and 100 nm. Micelles are composed of molecules with two distinct regions that exhibit varying hydrophilicity. Amphiphilic molecules constitute aggregation of M at critical concentrations and temperatures (Vadlapudi & Mitra, 2013). The critical micellar concentration (CMC) refers to the concentration at which accretion initiates, and micellar molecules are designed, whereas the critical micellar temperature (CMT) is defined as the temperature at which M molecules combine; below this temperature, M do not form and exist solely as monomers. The aggregation of amphiphilic molecules resulting in micelle formation is led by the exclusion of hydrophobic components from the aqueous conditions and the establishment of H-bonds in an aqueous solvent, which collectively reduce the system's free energy (Bose et al., 2021). Nanomicelles serve a large number of advantages over other forms of micelles (Table 13.1).

Micelles serve as drug carriers in pharmaceutical applications, effectively encapsulating lipophilic drugs in their core while their surface interacts with polar molecules. There are various subtypes of micelles, which include regular micelles, reverse micelles, and polymeric micelles (PMM) (Bertrand et al., 2014). Regular micelles are amphiphilic molecules with a hydrophobic tail and hydrophilic head that self-assemble in an aqueous environment (i.e., polyethylene glycol [PEG], poly (lactic-co-glycolic acid) [PLGA]); reverse micelles are just contrary, having an exposed hydrophobic tail and

**Table 13.1 Advantages of nanomicelles.**

| Advantages | Description |
|---|---|
| Enhanced drug solubility | Improve the solubility of antineoplastic drugs, enabling efficient delivery to neoplastic cells. |
| Targeted drug delivery | The drug's ligands enable active targeting, enhancing specificity for colorectal cancer cells. |
| Improved bioavailability | Prevent drug degradation, improve systemic circulation, and enhance bioavailability. |
| Controlled drug release | Sustained drug release minimizes side effects while improving therapeutic efficacy in treatment applications. |
| Reduced toxicity | Minimize side effects and toxicity by delivering drugs targeted to tumor sites. |
| Enhanced cellular uptake | Help efficient drugs penetrate neoplastic cells through endocytosis mechanisms. |
| Overcome multidrug resistance (MDR) | Encapsulation of drugs in nanomicelles can bypass efflux pumps, combating MDR in neoplastic cells. |
| Biodegradability and biocompatibility | Often made from biodegradable materials, reducing potential toxicity for long term. |
| Combination therapy potential | Codeliver multiple drugs or therapeutic agents, allowing synergistic effects for better outcomes. |
| Tumor microenvironment | Targeted drug delivery system responds to tumor microenvironment stimuli, enhancing therapeutic efficacy and precision. |
| Improved penetration | Small size enhances tumor tissue penetration through improved permeability and retention effects. |

hydrophilic head as the interior core self-assembled in a nonaqueous environment (i.e., phosphazene micelles in chloroform); and PMM are colloids of amphiphilic copolymers that self-assemble in an aqueous solvent, having an interior hydrophobic core and a hydrophilic exterior round structure, linearly arranged in a 2-block structure (A-B), 3-block (A-B-A), or 5-block (A-A-B-A-A) with multiple branches, where A and B can be any polymer such as polylactic acid (PLA), Poly(D-lactic acid) (PDLA), poly(L-lysine) (PLL), PEG, PLGA, and Poly(L-amino acid) (PLAA) (Trinh et al., 2017; Bose et al., 2021).

Nanomicelles are prepared from either surfactants or artificial block copolymers. These amphiphilic subunits carry positive, negative, or no net charge, or exist in a zwitterionic state. Negatively or anionically charged surfactants include sodium dodecyl sulphate (SDS), whereas positively or cationically charged surfactants include dodecyl-trimethyl-ammonium-bromide. Nonionic surfactants carry no net charge, including n-dodecyl tetra(ethylene oxide). In contrast, zwitterionic-based surfactants hold both anionic and cationic charges, such as di-octanoyl phosphatidylcholine (Bose et al., 2021).

Nonsoluble pharmaceutics can be encapsulated in regular nanomicelles, while solubilized pharmaceutics can be encapsulated in reverse nanomicelles (Trivedi & Kompella, 2010). Nanomicelles act as a shield by directly diminishing the interaction of pharmaceutics with the in vivo setting;

consequently, they improve drug bioavailability and reduce antagonistic complications (Tawfik et al., 2020; Xu et al., 2013). Nanomicelles facilitate targeted drug delivery, enhancing tissue penetration and increasing drug bioavailability within the organism (Bose et al., 2021).

## 13.7 Targeting nanomicelles

Targeted ligands can be combined to create various targeting nanomicelles, optimizing biodistribution and reducing side effects. Activated nanomicelles can selectively target cell bodies through interactions with specific target receptors and coupling with local active signaling proteins (Sutton et al., 2007; Li et al., 2020), as shown in Fig. 13.2.

Tao Wang et al. developed paclitaxel drug-encapsulated polyethylene glycol-phosphatidyl ethanolamine (PEG-PE) M, tailored with MCF-7-selective phage-fusing proteins aimed at countering neoplastic cells. The study demonstrated that countered phage nanomicelles exhibited enhanced tumor specificity in neoplastic cells compared to healthy cells. This countering technique can augment the antineoplastic efficacy of paclitaxel drugs in mouse models. Additionally, alternative neoplasm-

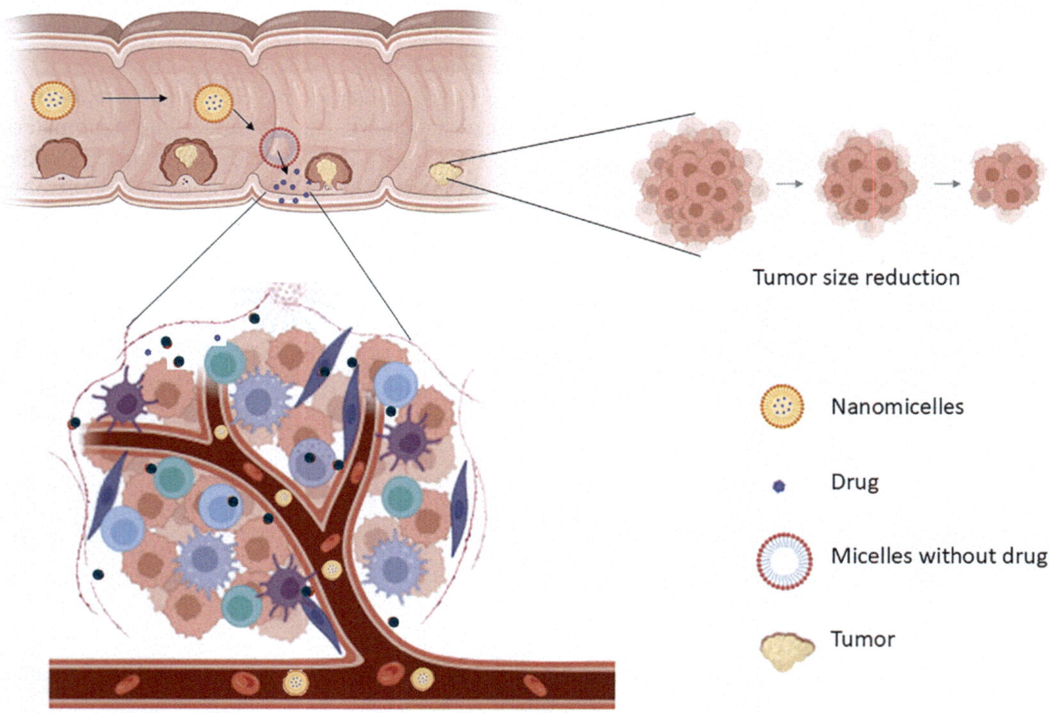


**FIGURE 13.2**

Diagram shows nanomicelles delivering drug actively and passively to tumor leading to tumor size reduction.

countering agents, including antibodies, peptides, and aptamers, were examined (Li et al., 2020; Wang et al., 2014).

Ahn et al. have developed novel conjugated nanomicelles with antitissue factor antibodies laden with platinum drugs to counter pancreatic neoplasm. These nanomicelles exhibit a 15-fold increase in cell intake and considerable anticancer effects lasting over seven weeks in pancreatic neoplasm compared to nonspecified micelles (Ahn et al., 2015; Li et al., 2020).

Currently, stearic-g-polyethyleneimine acid amphiphilic nanomicelles were developed and activated with folic acid-derived carbon dots (CDs) to facilitate targeted delivery of the antineoplastic drug doxorubicin, that is, "DOX," and harmonize bioimaging for triple negative breast cancer (TNBC). The fluorescence characteristics of folic acid-based CDs enable CDSP-25 to serve as an effective bioimaging agent for TNBC (Li et al., 2020).

## 13.8 Polymer selection for nanomicelle synthesis

Nanomicelle consists of hydrophilic polymers and hydrophobic polymers; hence, the selection of suitable polymers is of utmost crucial for the synthesis of sustained, stabilized, and biocompatible nanomicelles.

### 13.8.1 Hydrophilic/polar polymers in block copolymers

Hydrophilic and hydrophobic polymers play crucial roles in the development of block copolymers, particularly from the perspective of DDSs. The physicochemical properties of hydrophilic M, including MW and drug entrapment capacity, are vital for their stabilization, circulation duration, and biodistribution in vivo (Li et al., 2020).

Among the various polar polymers, PEG stands out due to its biocompatibility, aqueous solubility, and reduced toxicity, particularly at MW oscillating from 1 to 20 kilo Daltons. PEG is recognized as a standard in nanomedicine, enhancing nanoparticles with antifouling properties and significantly influencing the circulation duration and removal of PMM (Li et al., 2020).

Another notable hydrophilic polar polymer is Polyvinylpyrrolidone (PVP), which is utilized as a substitute for PEG in liposomes and di-block polymer micelles due to its superior biocompatibility. PVP is synthesized through radical polymerization and is characterized by its antifouling properties, although the underlying mechanisms remain to be fully elucidated. Poly(2-oxazoline) (POx) has also garnered attention as an innovative biomaterial, exhibiting enhanced mucus penetration capabilities that are particularly advantageous for oral and intestinal drug delivery applications (Li et al., 2020).

### 13.8.2 Hydrophobic/nonpolar polymers in block copolymers

On the other hand, nonpolar polymers typically form the core of block copolymers and include polyesters, poly-ethers, and poly-amino acids. Polyesters, such as PLA, Poly(ε-caprolactone) (PCL), and PLGA, are biodegradable polymers used in applications and are particularly favored for biomedical uses because of their biodegradability and biocompatibility. These hydrophobic blocks are often conjugated with hydrophilic blocks like PEG to form M, which can encapsulate polar drugs.

Polyanhydrides, such as poly-sebacic-anhydride (PSA), also exhibit promising bioavailability and low toxicity, enabling them to be suitable for DDS applications (Li et al., 2020).

Advancements in di- and tri-block copolymer synthesis enhance long-circulating DDSs. These auto-assembling copolymers are integrated through anionic and ring-opening polymerization techniques, allowing for variations in molecular weight and hydrophilic-hydrophobic ratios. The stretch of the hydrophobic sequence directly influences the size of the hydrophobic core, enhancing the capacity for drug encapsulation. The efficiency of DDSs is determined by various factors, including surface activity, stability, targeting capabilities, morphology, and size. Smaller micelles, particularly those around 30 nm, have demonstrated improved tumor penetration and increased anticancer activity (Li et al., 2020).

Poly-ethers, synthesized via ring-opening anionic polymerization, serve as core interior-forming components for loading hydrophobic drugs. Poloxamers, which are Poly(ethylene oxide) and Poly (propylene oxide) (PEO-PPO-PEO) copolymers, have been widely utilized for dissolving nonpolar pharmaceutics and integrating PMM formulations. These block copolymers were among the pioneering materials employed for DDS in PMMs (Li et al., 2020).

Hydrophobic/nonpolar poly-amino acids, produced through the active polymerizing activity of α-amino acid-N-carboxy anhydrides, also contribute to the formation of amphiphilic block copolymers. These polymers facilitate the dissolution of sparsely bioavailable drugs. For instance, Poly(β-benzyl-L-aspartate (PBLA) and Poly(γ-benzyl-L-glutamate (PBLG) can form micelles capable of encapsulating drugs like camptothecin and doxorubicin, demonstrating significant drug-loading capacities and favorable size distributions (Li et al., 2020).

In conclusion, the integration of polar and nonpolar polymers in block copolymers presents an encouraging approach to enhancing DDS. The careful selection and modification of these polymers can lead to improved stability, biocompatibility, and therapeutic efficacy, ultimately advancing the field of nanomedicine (Li et al., 2020).

## 13.9 Preparation of nanomicelles

The preparation of nanomicelles is influenced by the characteristics of polymer stretch. Two general methods are primarily employed for the synthesis of nanomicelles:

1. **Direct dissolution:** Moderately hydrophobic block copolymers typically self-assemble into nanomicelles via direct dissolution or by simple equilibrium method. This technique involves the instantaneous suspension of the copolymer as well as the pharmaceutical drug in a hydrous solution at the appropriate ratio, followed by heating to initiate the formation of nanomicelles. High temperature induces dehydration in the internal core structure, resulting in nanomicellar formation (Li et al., 2020).
2. **Solvent casting:** The solvent casting type encompasses three techniques, that is, dialysis, oil-in-water (o/w) dispersion, and solution forming.
3. The dialysis method is effective for isolating nanomicelles from the water-insoluble copolymer. Drugs and copolymers are solubilized in suitable organic solvents that are water-miscible and possess a high b.p. A copolymer and drug solution are dialyzed against an aqueous solvent overnight. During dialysis, the organic-solution is gradually vaporized, leading to the formation of

drug-loaded nanomicelles. This technique addresses the challenge of removing organic-solutions with high boiling points; however, it is limited by issues such as drug loss resulting from low drug entrapment ability (Li et al., 2020).

4. Secondly, the o/w dispersion technique includes the physical encapsulation of the drugs. The entrapment method involves dissolving pharmaceuticals and polymers in an immiscible organic-solution containing a minimal amount of aqueous solvent. The evaporation of the solution leads to the physical drug encapsulation within the interior core of the synthesized nanomicelles (La et al., 1996).
5. Solution or solvent casting involves the use of organic as well as aqueous solutions. Polymers and drugs are solubilized in an organic-solution to yield a clear solution. Drug-encapsulated nanomicelles can be produced as a thin film following the elimination of organic-solutions under high vacuum conditions (Li et al., 2020).

## 13.10 Ideal characteristics of nanomicelles for colorectal cancer therapy

**Structural stability:** Nanomicelles exhibit kinetic as well as thermodynamic stability, attributed to the structural complexity and entanglement of polymer stretches within the micelle's inner core. Nanomicelles attain thermodynamic stability when the concentration of their copolymer exceeds the CMC, which is influenced by the polar-lipophilic balance of the copolymer (Koo et al., 2005). CMC values for nanomicelles typically range from $10^{-6}$ to $10^{-7}$ molar, which is lower than those of micelles designed from low-MW constituents, enabling nanomicelles to preserve their structure during dilution (Wiradharma et al., 2009). Kinetic stability becomes crucial in drug delivery (DD) scenarios where conditions are not in balance, particularly when copolymer concentration falls below its CMC. Nanomicelles exhibit slower disassembly rates below the CMC due to kinetic stability. The gradual dissociation resulting from structural stability enables nanomicelles to preserve their integrity and drug content until they arrive at the target site, thereby enhancing the drug's bioavailability (Adams et al., 2003; Bose et al., 2021).

**Hydrophobic drug solubilization ability:** Nanomicelles serve as effective pharmaceutical carriers for nonpolar drugs, addressing the challenge of formulating clear aqueous solutions with sufficient drug concentrations for therapeutic efficacy. Their structure consists of a nonpolar core that encapsulates the drug, surrounded by a polar shell formed by extended chains. This configuration facilitates the solubilization of polar compounds, thereby enhancing their bioavailability and therapeutic potential in aqueous environments (Lavasanifar et al., 2002; Velagaleti et al., 2010).

**Physical and chemical attributes:** Nanomicelles, formed from amphiphilic block copolymers, are categorized into di-block and tri-block types (Velagaleti et al., 2010). Di-block copolymers consist of polar (A) and nonpolar (B) segments, while tri-block copolymers can be classified as ABA or accelerated blood clearance (ABC) types. The AB and ABA configurations are particularly effective for synthesizing nanomicelles, which enhance drug solubility and loading efficiency (Kabanov et al., 2002; Xu et al., 2013). The nonpolar blocks create an inner core stabilized by nonpolar and electrostatic interactions among oppositely charged macromolecules, resulting in the formation of a Polyion Complex. In addition to these, it is also formed through metal-ligand binding and hydrogen ion bonding. The hydrophilic outer shell exhibits steric stability and is significant in in vivo conditions due to its interactions with cells. Micelles self-assemble in water, enhance drug solubility, and maintain stability in the gastrointestinal tract due to their structure (Bose et al., 2021).

**pH sensitivity:** Drug release is contingent upon intracellular signals, highlighting the significance of pH-responsive systems. Micelles, polybases, and polyacids serve as fundamental components that confer pH sensitivity to drug release (Yoncheva et al., 2012). The amine core remains unchanged, exhibiting hydrophobic properties at high pH and transitioning to hydrophilic upon protonation at low pH. While, the carboxylic acid core is usually uncharged at acidic pH but becomes negatively charged upon protonation at alkaline pH. Thus, protonation can be responsible for destabilizing micelles. pH must be above the pKa of the protonable group for the formation of the nanomicelles. When the pH drops below the pKa value, there is an increase in hydrophilicity and electrostatic repulsion due to the ionization of the polymers. This results in the destabilization of the micelle, facilitating controlled drug release (Lee et al., 2003). Therefore, the absorption of drugs in the intestine can be enhanced by utilizing this pH gradient (Bromberg, 2008).

**Mucoadhesive property:** Nanoparticles used in oral medication typically adhere to mucosal surfaces and penetrate the mucosal layer, where they disintegrate, leaving the drug attached to the mucosa. This attachment is subsequently removed by the mucus clearance mechanism, resulting in reduced bioavailability of the drug (Lee and Na, 2003). Mucosal retention prolongs transit time in the gastrointestinal tract, thereby enhancing the duration of drug release. Mucoadhesive nano micellar polymers swell and occupy the cervices of the mucous membrane, thereby enhancing the effective surface area and resulting in a higher local concentration of drugs (Tang et al., 2009). Finally, bioadhesion enhances the drug concentration gradient, facilitating the interaction of particles with the mucosal surface (Smart, 2005). Research indicates that the intestinal mucosa, characterized by a negative charge from glycocalyx, attracts positively charged poly micelles. Particle mobility is contingent upon surface charges. Crater and Carrier demonstrated that anionic particles diffuse 20–30 times faster than cationic particles (Ponchel & Irache, 1998).

**Precise binding ability:** Targeted drug delivery can be enhanced through various strategies, including the attachment of specific targeting ligands to micelle surfaces, the development of stimuli-responsive amphiphilic block copolymer micelles, and leveraging the enhanced permeability and retention effect. A notable approach involves the synthesis of immuno-micelles, where monoclonal antibodies are conjugated to p-nitrophenyl carbonyl groups (Bose et al., 2021; Crater & Carrier, 2010).

**Light sensitivity:** According to studies, light exposure can change a substance's hydrophilicity and hydrophobicity. Light can therefore cause nanomicelles to break down, releasing medications (Bose et al., 2021; Crater & Carrier, 2010; Torchilin, 2007).

## 13.11 Delivery of anticancer drugs using different nanomicelles in colorectal cancer therapy

The PMM's properties and mechanisms for drug delivery in CRC therapy are discussed in Table 13.2.

**Table 13.2 Current status of different nanomicelles employed in vitro and in vivo for colorectal cancer (*CRC*) treatment.**

| Block copolymer | Therapeutic substance | In vitro/in vivo | Micelles facilitate the targeted delivery of therapeutic substances at a cellular level | References |
|---|---|---|---|---|
| Polyethylene glycol (PEG)-based Dox-functionalized copolymer | Adriamycin | CT-26 mice's CRC cells; BALB/c mice implanted cutaneously with CT-26-derived tumors | (1) The intervention does not lead to a substantial elevation in the levels of blood markers associated with organ toxicity, specifically aspartate aminotransferase (AST); and (2) increases the alanine aminotransferase (ALT) and lactate dehydrogenase levels | Brunato et al. (2021); Fatfat et al. (2022); Men et al. (2018); Sanati et al. (2022) |
| PEG-Poly(L-lysine hydrazide (PLLHA) and FA-PEG-polymer | Taxotere | CT26 cells; female BALB/c mice implanted the cells via subcutaneous injection | (1) Induce neoplastic cell demise compared to nonspecific counterparts; (2) erythrocyte agglutination is not caused; (3) histological damage does not induce levels; and | Li et al. (2020); Sanati et al. (2022); Su et al. (2020) |
| Hyaluronic acid conjugated to D-α-Tocopheryl Succinate (TOS) | Taxol | CT26 murine CRC cells and various human CRC cell lines, utilizing BALB/c mice for subcutaneous injection | (1) Apoptosis induction was observed in HT29 and Lovo neoplastic cells; and (2) cell demise induces proliferation | Lee et al. (2011); Sanati et al. (2022); Zhu et al. (2018) |
| Methoxy poly(ethylene glycol)-poly(ε-caprolactone) (mPEG-PCL) and DOTAP | Plasmid pVAX1-mIL22BP, which encodes murine IL-22 binding protein | C26 CRC cells injected into BALB/c mice; 293 T cells | (1) Induce cellular demise under artificial cell conditions; (2) decrease in microvessel density indicated by CD31+ve staining; and (3) lymphocyte infiltration influences tumor microenvironment and immune response | Chen et al. (2020); Sanati et al. (2022) |

(*Continued*)

**Table 13.2 Current status of different nanomicelles employed in vitro and in vivo for colorectal cancer (*CRC*) treatment.** ***Continued***

| Block copolymer | Therapeutic substance | In vitro/in vivo | Micelles facilitate the targeted delivery of therapeutic substances at a cellular level | References |
|---|---|---|---|---|
| PEI-poly (DL-lactic acid) | Survivin shRNA combined with camptothecin | C-26 and CHO cell lines; female BALB/c mice implanted with C26 cells subcutaneously | (1) Induce cellular demise under artificial cell conditions evaluated with their nonspecific complements; and (2) reduced in vivo accumulation linked to nonspecific counterparts | Kim et al. (2010); Sanati et al. (2022) |
| PEI-based poly (D, L-lactide) and 1,2-Distearoyl-sn-glycero-3-phosphoethanolamine-polyethylene glycol (DSPE-PEG) conjugates | miRNA-34a enhances irinotecan's efficacy in neoplastic therapy | CT-26 cells were injected into female BALB/c mice | (1) miR-34a expression reduces Bcl-2 and mammalian target of rapamycin (mTOR) phosphorylation levels | Lu et al. (2018); Sanati et al. (2022) |
| PNMDS for Poly (N-methyldietheneamine sebacate) | Adriamycin and TNF-related apoptosis-inducing ligand | SW480 human CRC cells; WI38 human lung connective tissue cells | (1) Caspase-dependent apoptosis induces | Sanati et al. (2022); Talelli et al. (2015) |
| FA-dextran-paclitaxel | Adjudin and Taxol | HCT-8 and HCT-8/PTX cultures were used in mouse model studies | (1) Decrease mitochondrial membrane potential and adenosine triphosphate concentrations | Ciardiello et al. (2014); Sanati et al. (2022) |

## 13.12 Advantages of nanomicelles as prodrugs

Nanomicelles provide several advantages as prodrugs in DDSs.

### 13.12.1 Factors influencing nano micelle stability

At low concentrations of micelle solutions ($<10^{-5}$ millimolar), the constituent level is inadequate to facilitate the auto-organization of M. They typically accumulate at the hydrous medium interface, disintegrating the M. A lower dose of surfactant, known as the CMC, is necessary to sustain the conformation of M (Kim et al., 2010; Lu et al., 2018). The intrinsic properties of micelles influence the CMC and are essential for assessing their stability at diluted concentrations.

The conformational constancy of M is affected by dilution and the complex biological conditions, including factors such as salt dose, temperature, pH, and solvents. Subsequently, through intravenous administration, M may experience various environmental changes, such as substantial dilution, alterations in pH, and interactions with a range of peptides, sterols, and cells. The polar and nonpolar areas of specific M are connected through amide, ester, and various functional groups. The M conformation may be interrupted by linker hydrolysis in response to substantial pH variations. Additional disruptors encompass protein absorption via nontargeted interactions or electric charge-based interactions. The early release of encapsulated healing contents may result in drug accretion in healthy organs, potentially leading to significant adjacent effects and a reduction in the therapeutic efficacy of the drug (Lu et al., 2018; Talelli et al., 2015). Enhancing intra-micellar interactions is a fundamental strategy for improving micelle stability, typically indicated by a reduced CMC. Consequently, assessing the CMC value provides a direct evaluation of the effectiveness of these strategies (Lu et al., 2018).

### 13.12.2 Pros and cons of personalized medicine

Personalized medicine (PM) enhances treatment efficacy by tailoring drug prescriptions to an individual's genetic profile, contrasting with conventional therapies that use a standardized approach. This individualized strategy increases the likelihood of therapeutic success, as it considers the unique hereditary characteristics of patients, thereby improving treatment outcomes compared to traditional methods that may not be effective for all individuals.

Neoplastic treatments often exhibit toxicity and adverse effects. Personalized therapy addresses this by prescribing medications based on genetic factors, allowing for the administration of targeted drugs. This approach minimizes unnecessary treatments, thereby reducing potential side effects and overall treatment costs, as only essential medications are used. Personalized therapy represents a significant advancement in cancer care, enhancing patient outcomes (Ciardiello et al., 2014).

PM enhances traditional therapies by improving treatment options and optimizing strategies within healthcare systems, behavior, lifestyle, and surgical interventions, ultimately leading to better individual outcomes.

PMs are contingent upon the distinct circumstances of individuals, which may influence their health. Currently, individuals are accountable for therapeutic decisions, as physicians do not consistently possess knowledge regarding the outcomes of specific treatments or their effects on particular patients. By utilizing the precision of PM, healthcare providers can devise individualized treatment plans, thereby enhancing the probability of patient recovery.

To operate effectively, PM requires a comprehensive genomic data collection. However, regulatory barriers, such as FDA restrictions on hereditary information access and lack of governmental accountability, pose significant challenges (Rehman et al., 2016). Additionally, privacy concerns regarding the management of extensive data collections may lead to substantial opposition against the implementation of such initiatives, complicating progress in the field.

PM has the potential to affect healthcare significantly. However, it will require a significant amount of time for execution and necessitate substantial infrastructure development. Essential advancements are needed in infrastructure, as well as in systems for data exchange, collection, and storage, to enable personalized therapy.

Personalized treatment is revolutionizing medicine by enhancing prevention and prediction tailored to individuals. Integrating evidence-based actions, scientific knowledge, and foundational education maximizes opportunities for health improvement while mitigating potential adverse outcomes. This approach signifies a significant shift toward more effective and individualized healthcare strategies (Dey et al., 2023).

## 13.13 Conclusion and future perspectives

Recent studies on nanomicelles have highlighted their potential applications in CRC diagnostics and therapeutics. Researchers are focusing on synthesizing novel drug conjugate-nanomicelles to enhance CRC treatment efficacy. These nanomicelles serve as effective carriers for antineoplastic drugs, facilitating their encapsulation and targeted delivery to neoplastic cells. The development of optimized synthetic methods is crucial to meet the increasing demand for these DDS. Particularly, nanomicelles exhibit biocompatibility and low toxicity, making them suitable for clinical applications. Their ability to provide a controlled release of antineoplastic agents is essential for maximizing therapeutic outcomes. Additionally, the nanoscale size of these micelles enables efficient penetration into neoplastic cells, forming stable nanocomplexes that protect the drugs from nuclease degradation and ensure targeted delivery. Cohort research is needed for more advanced and precise drug delivery into the tumor cells.

## Acknowledgment

The authors express their gratitude to Geetika Madan Patel, Vice President and Medical Director, and the Research and Development Cell team of Parul University, Vadodara, for providing the necessary facilities and motivation to conduct this multidisciplinary collaborative publication.

## References

Adams, M. L., Lavasanifar, A., & Kwon, G. S. (2003). Amphiphilic block copolymers for drug delivery. *Journal of Pharmaceutical Sciences, 92*, 1343–1355.

Ahn, J., Miura, Y., Yamada, N., Chida, T., Liu, X., Kim, A., & Kataoka, K. (2015). Antibody fragment-conjugated polymeric micelles incorporating platinum drugs for targeted therapy of pancreatic cancer. *Biomaterials, 39*, 23–30.

Bertrand, N., Wu, J., Xu, X., Kamaly, N., & Farokhzad, O. C. (2014). Cancer nanotechnology: the impact of passive and active targeting in the era of modern cancer biology. *Advanced Drug Delivery Reviews, 66*, 2–25.

Beyer-Sehlmeyer, G., Glei, M., Hartmann, E., Hughes, R., Persin, C., Böhm, V., & Pool-Zobel, B. L. (2003). Butyrate is only one of several growth inhibitors produced during gut flora-mediated fermentation of dietary fibre sources. *British Journal of Nutrition, 90*, 1057–1070.

Bose, A., Roy Burman, D., Sikdar, B., & Patra, P. (2021). Nanomicelles: types, properties and applications in drug delivery. *IET Nanobiotechnology, 15*, 19–27.

Bromberg, L. (2008). Polymeric micelles in oral chemotherapy. *Journal of Controlled Release, 128*, 99–112.

Brown, M. F., Leibowitz, B. J., Chen, D., He, K., Zou, F., Sobol, R. W., & Yu, J. (2015). Loss of caspase-3 sensitizes colon cancer cells to genotoxic stress via RIP1-dependent necrosis. *Cell Death & Disease, 6*, e1729.

Brunato, S., Mastrotto, F., Bellato, F., Bastiancich, C., Travanut, A., Garofalo, M., & Caliceti, P. (2021). PEG-polyaminoacid based micelles for controlled release of doxorubicin: rational design, safety and efficacy study. *Journal of Controlled Release, 335*, 21–37.

Center, M. M., Jemal, A., Smith, R. A., & Ward, E. (2009). Worldwide variations in colorectal cancer. *CA: A Cancer Journal for Clinicians, 59*, 366–378.

Chen, D., Ge, S., Zuo, L., Wang, S., Liu, M., & Li, S. (2020). Adjudin-loaded redox-sensitive paclitaxel-prodrug micelles for overcoming multidrug resistance with efficient targeted Colon cancer therapy. *Drug Delivery, 27*, 1094–1105.

Ciardiello, F., Arnold, D., Casali, P. G., Cervantes, A., Douillard, J. Y., Eggermont, A., & Stahel, R. (2014). Delivering precision medicine in oncology today and in future—the promise and challenges of personalised cancer medicine: a position paper by the European Society for Medical Oncology (ESMO). *Annals of Oncology, 25*, 1673–1678.

Crater, J. S., & Carrier, R. L. (2010). Barrier properties of gastrointestinal mucus to nanoparticle transport. *Macromolecular Bioscience, 10*, 1473–1483.

Dey, A., Mitra, A., Pathak, S., Prasad, S., Zhang, A. S., Zhang, H., & Banerjee, A. (2023). Recent advancements, limitations, and future perspectives of the use of personalized medicine in treatment of colon cancer. *Technology in Cancer Research & Treatment, 22*, 15330338231178403.

Fatfat, Z., Fatfat, M., & Gali-Muhtasib, H. (2022). Micelles as potential drug delivery systems for colorectal cancer treatment. *World Journal of Gastroenterology, 28*, 2867.

Kabanov, A. V., Batrakova, E. V., & Alakhov, V. Y. (2002). Pluronic® block copolymers as novel polymer therapeutics for drug and gene delivery. *Journal of Controlled Release, 82*, 189–212.

Khan, F. A., Albalawi, R., & Pottoo, F. H. (2022). Trends in targeted delivery of nanomaterials in colon cancer diagnosis and treatment. *Medicinal Research Reviews, 42*, 227–258.

Kim, S., Shi, Y., Kim, J. Y., Park, K., & Cheng, J. X. (2010). Overcoming the barriers in micellar drug delivery: loading efficiency, in vivo stability, and micelle–cell interaction. *Expert Opinion on Drug Delivery, 7*, 49–62.

Koo, O. M., Rubinstein, I., & Onyuksel, H. (2005). Camptothecin in sterically stabilized phospholipid micelles: a novel nanomedicine. *Nanomedicine: Nanotechnology, Biology and Medicine, 1*, 77–84.

La, S. B., Okano, T., & Kataoka, K. (1996). Preparation and characterization of the micelle-forming polymeric drug indomethacin-incorporated poly (ethylene oxide)–poly (β-benzyl L-aspartate) block copolymer micelles. *Journal of Pharmaceutical Sciences, 85*, 85–90.

Labianca, R., Beretta, G. D., Kildani, B., Milesi, L., Merlin, F., Mosconi, S., & Wils, J. (2010). Colon cancer. *Critical Reviews in Oncology/Hematology, 74*, 106–133.

Lavasanifar, A., Samuel, J., & Kwon, G. S. (2002). Poly (ethylene oxide)-block-poly (L-amino acid) micelles for drug delivery. *Advanced Drug Delivery Reviews, 54*, 169–190.

Lee, E. S., Shin, H. J., Na, K., & Bae, Y. H. (2003). Poly (l-histidine)–PEG block copolymer micelles and pH-induced destabilization. *Journal of Controlled Release, 90*, 363–374.

Lee, E. S., Na, K., & Bae, Y. H. (2003). Polymeric micelle for tumor pH and folate-mediated targeting. *Journal of Controlled Release, 91*, 103–113.

Lee, A. L., Dhillon, S. H., Wang, Y., Pervaiz, S., Fan, W., & Yang, Y. Y. (2011). Synergistic anti-cancer effects via co-delivery of TNF-related apoptosis-inducing ligand (TRAIL/Apo2L) and doxorubicin using micellar nanoparticles. *Molecular Biosystems, 7*, 1512–1522.

Li, Y., Jia, F., Deng, X., Wang, X., Lu, J., Shao, L., & Wu, Y. (2020). Combinatorial miRNA-34a replenishment and irinotecan delivery via auto-fluorescent polymeric hybrid micelles for synchronous colorectal cancer theranostics. *Biomaterials Science, 8*, 7132–7144.

Lu, Y., Zhang, E., Yang, J., & Cao, Z. (2018). Strategies to improve micelle stability for drug delivery. *Nano Research, 11*, 4985–4998.

Ma, L., Chen, Q., Ma, P., Han, M. K., Xu, Z., Kang, Y., & Merlin, D. (2017). iRGD-functionalized PEGylated nanoparticles for enhanced colon tumor accumulation and targeted drug delivery. *Nanomedicine: Nanotechnology, Biology, and Medicine, 12*, 1991–2006.

Men, K., Huang, R., Zhang, X., Zhang, R., Zhang, Y., Peng, Y., & Duan, X. (2018). Delivery of interleukin-22 binding protein (IL-22BP) gene by cationic micelle for colon cancer gene therapy. *RSC Advances, 8*, 16537–16548.

Navarro, G., & de ILarduya, C. T. (2009). Activated and non-activated PAMAM dendrimers for gene delivery *in vitro* and *in vivo*. *Nanomedicine: Nanotechnology, Biology and Medicine, 5*, 287–297.

O'keefe, S. J. (2016). Diet, microorganisms and their metabolites, and colon cancer. *Nature Reviews Gastroenterology & Hepatology, 13*, 691–706.

Pelka, J., Gehrke, H., Esselen, M., Türk, M., Crone, M., Bräse, S.,, & Marko, D. (2009). Cellular uptake of platinum nanoparticles in human colon carcinoma cells and their impact on cellular redox systems and DNA integrity. *Chemical Research in Toxicology, 22*, 649–659.

Ponchel, G., & Irache, J. M. (1998). Specific and non-specific bioadhesive particulate systems for oral delivery to the gastrointestinal tract. *Advanced Drug Delivery Reviews, 34*, 191–219.

Rehman, A., Awais, M., & Baloch, N. U. A. (2016). Precision medicine and low-to middle-income countries. *JAMA Oncology, 2*, 293–294.

Ruan, H., Leibowitz, B. J., Zhang, L., & Yu, J. (2020). Immunogenic cell death in colon cancer prevention and therapy. *Molecular Carcinogenesis, 59*, 783–793.

Sanati, S., Taghavi, S., Abnous, K., Taghdisi, S. M., Babaei, M., Ramezani, M., & Alibolandi, M. (2022). Fabrication of anionic dextran-coated micelles for aptamer targeted delivery of camptothecin and survivin-shRNA to colon adenocarcinoma. *Gene Therapy, 29*, 55–68.

Schütz, C. A., Staedler, D., Crosbie-Staunton, K., Movia, D., Chapuis Bernasconi, C., Kenzaoui, B. H., & Juillerat-Jeanneret, L. (2014). Differential stress reaction of human colon cells to oleic-acid-stabilized and unstabilized ultrasmall iron oxide nanoparticles. *International Journal of Nanomedicine,* 3481–3498.

Singh, S. K., Solanki, R., & Patel, S. (2023). Nano pharmaceutical delivery in combating colorectal cancer. *Medicine in Drug Discovery*100173.

Smart, J. D. (2005). The basics and underlying mechanisms of mucoadhesion. *Advanced Drug Delivery Reviews, 57*, 1556–1568.

Su, M., Xie, J., Zeng, Q., Shu, M., Liu, J., & Jiang, Z. (2020). Enzymatic synthesis of PEGylated lactide-diester-diol copolyesters for highly efficient targeted anticancer drug delivery. *Materials Science and Engineering: C, 115*, 111125.

Sung, H., Ferlay, J., Siegel, R. L., Laversanne, M., Soerjomataram, I., Jemal, A., & Bray, F. (2021). Global cancer statistics 2020: GLOBOCAN estimates of incidence and mortality worldwide for 36 cancers in 185 countries. *CA: A Cancer Journal for Clinicians, 71*, 209–249.

Sutton, D., Nasongkla, N., Blanco, E., & Gao, J. (2007). Functionalized micellar systems for cancer targeted drug delivery. *Pharmaceutical Research, 24*, 1029–1046.

Talelli, M., Barz, M., Rijcken, C. J., Kiessling, F., Hennink, W. E., & Lammers, T. (2015). Core-crosslinked polymeric micelles: principles, preparation, biomedical applications and clinical translation. *Nano today, 10*, 93–117.

Tang, B. C., Dawson, M., Lai, S. K., Wang, Y. Y., Suk, J. S., Yang, M., & Hanes, J. (2009). Biodegradable polymer nanoparticles that rapidly penetrate the human mucus barrier. *Proceedings of the National Academy of Sciences, 106*, 19268–19273.

Tarapore, R. S., Siddiqui, I. A., & Mukhtar, H. (2012). Modulation of Wnt/β-catenin signaling pathway by bioactive food components. *Carcinogenesis, 33*, 483–491.

Tawfik, S. M., Azizov, S., Elmasry, M. R., Sharipov, M., & Lee, Y. I. (2020). Recent advances in nanomicelles delivery systems. *Nanomaterials, 11*, 70.

Tjalsma, H., Boleij, A., Marchesi, J. R., & Dutilh, B. E. (2012). A bacterial driver–passenger model for colorectal cancer: beyond the usual suspects. *Nature Reviews. Microbiology, 10*, 575–582.

Torchilin, V. P. (2007). Micellar nanocarriers: pharmaceutical perspectives. *Pharmaceutical Research, 24*, 1–16.

Trinh, H. M., Joseph, M., Cholkar, K., Mitra, R., & Mitra, A. K. (2017). Nanomicelles in diagnosis and drug delivery. In *Emerging Nanotechnologies for Diagnostics, Drug Delivery and Medical Devices*. Elsevier, 45–58.

Trivedi, R., & Kompella, U. B. (2010). Nanomicellar formulations for sustained drug delivery: strategies and underlying principles. *Nanomedicine: Nanotechnology, Biology, and Medicine, 5*, 485–505.

Vadlapudi, A. D., & Mitra, A. K. (2013). Nanomicelles: an emerging platform for drug delivery to the eye. *Therapeutic Delivery, 4*(1), 1–3.

Velagaleti, P. R., Anglade, E., Khan, I. J., Gilger, B. C., & Mitra, A. K. (2010). Topical delivery of hydrophobic drugs using a novel mixed nanomicellar technology to treat diseases of the anterior and posterior segments of the eye. *Drug Delivery Technology, 10*, 42–47.

Verma, M. (2013). Cancer control and prevention: nutrition and epigenetics. *Current Opinion in Clinical Nutrition & Metabolic Care, 16*, 376–384.

Wang, K., Shen, R., Meng, T., Hu, F., & Yuan, H. (2022). Nano-drug delivery systems based on different targeting mechanisms in the targeted therapy of colorectal cancer. *Molecules (Basel, Switzerland), 27*, 2981.

Wang, T., Yang, S., Mei, L. A., Parmar, C. K., Gillespie, J. W., Praveen, K. P., & Torchilin, V. P. (2014). paclitaxel-loaded PEG-PE–based micellar nanopreparations targeted with tumor-specific landscape phage fusion protein enhance apoptosis and efficiently reduce tumors. *Molecular Cancer Therapeutics, 13*, 2864–2875.

Wiradharma, N., Zhang, Y., Venkataraman, S., Hedrick, J. L., & Yang, Y. Y. (2009). Self-assembled polymer nanostructures for delivery of anticancer therapeutics. *Nano Today, 4*, 302–317.

Xu, W., Ling, P., & Zhang, T. (2013). Polymeric micelles, a promising drug delivery system to enhance bioavailability of poorly water-soluble drugs. *Journal of Drug Delivery, 2013*, 340315.

Yoncheva, K., Calleja, P., Agüeros, M., Petrov, P., Miladinova, I., Tsvetanov, C., & Irache, J. M. (2012). Stabilized micelles as delivery vehicles for paclitaxel. *International Journal of Pharmaceutics, 436*, 258–264.

Zhu, C., Zhang, H., Li, W., Luo, L., Guo, X., Wang, Z., & You, J. (2018). Suppress orthotopic colon cancer and its metastasis through exact targeting and highly selective drug release by a smart nanomicelle. *Biomaterials, 161*, 144–153.

CHAPTER

# Regulatory framework in nanomedicine targeting colon cancer

14

**Hitesh V. Shahare[1], Deepak B. Somavanshi[2], Shweta H. Shahare[3], Dipti G. Phadtare[3], Mrunalini Kulkarni[4] and Rajesh Gavit[5]**

[1]*Deparment of Pharmaceutical Chemistry, SNJB's Shriman Sureshdada Jain College of Pharmacy, Nashik, Maharashtra, India,*
[2]*Dept. of Pharmacognosy, K. B. H. S. S. Trust Institute of Pharmacy, Malegaon, Maharashtra, India,*
[3]*Department of Pharmaceutics & Pharmaceutical Chemistry, KCT, R. G. Sapkal Institute of Pharmacy, Nashik, Maharashtra, India,*
[4]*Deparment of Pharmaceutical Chemistry, Vishwakarma University, Pune, Maharashtra, India,*
[5]*Herbanext Laboratories Inc, Bacolod, Negros Occidental, Philippines*

## 14.1 Background

Nanotechnology has gained major research attention over the past three decades because materials at the nanoscale possess distinctive electrical, optical, chemical, magnetic and biological characteristics (Nel et al., 2009). A relatively recent development in science and technology is nanomedicine. The phrase can be interpreted differently, especially between Europe and the United States, and at times it seems to be poorly defined. The synthesis and characterization of a wide range of nanomaterials have advanced significantly, with many demonstrating encouraging preclinical outcomes. These developments are currently being translated into practical applications of nanomedicine that are useful for patients. The unique characteristics of nanomedicine, the state of the field, and the difficulties in attaining effective clinical translation are all highlighted in this chapter (Ramos & Almeida, 2022).

Designing and applying materials and technologies at the 1–100 nm scale is known as nanotechnology. A subfield of this study, called nanomedicine, focuses on using these nanoscale technologies in healthcare with the goal of improving patient health overall as well as illness detection, treatment, and monitoring (Logothetidis, 2011).

The dimensions that are within the nanoscale (1–100 nm) can be utilized for categorizing nanomaterials. This classification, illustrated in Fig. 14.1, includes the following categories:

1. Three-dimensional nanomaterials (3-ND) nanoscale incorporates all three dimensions (x, y, and z). Nanoparticles, quantum dots, and fullerenes are a few examples.
2. Two-dimensional nanomaterials (2-ND)nanoscale encompasses just two dimensions, whereas the third dimension is bigger. Nanofibers, nanotubes, nanorods, and nanowires are all included in this category.
3. One-dimensional nanomaterials (1-ND) are flat, sheet-like structures like nanosheets, nanowalls, and nanolayers that have only one dimension inside the nanoscale (Kairdolf et al., 2013).

Nano Drug Delivery Systems for Colon Cancer. DOI: https://doi.org/10.1016/B978-0-443-40443-6.00001-5

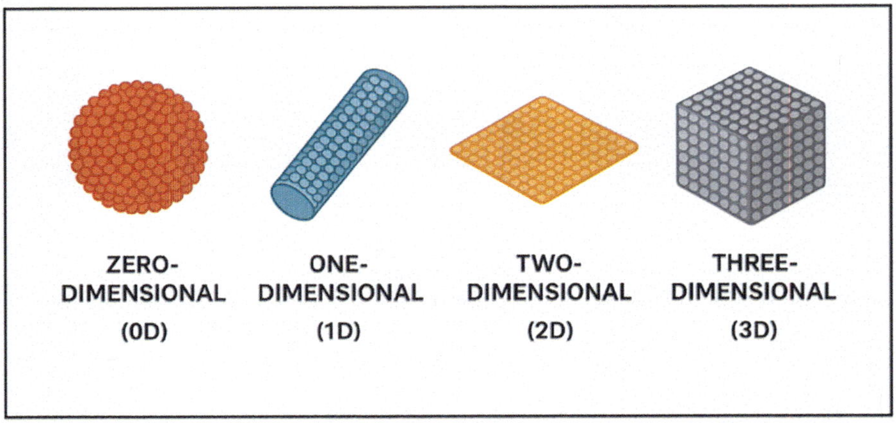


**FIGURE 14.1**

Different kinds of nanomaterials according to dimensions.

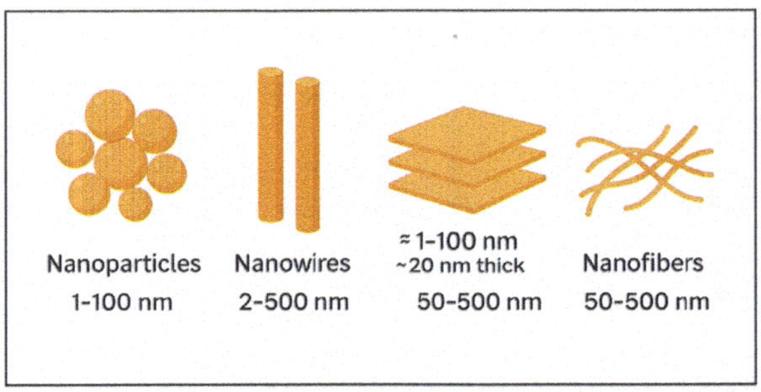


**FIGURE 14.2**

Various nanomaterials with increasing surface area.

### 14.1.1 Need for regulations

Nanomaterials can be constructed to exhibit a variety of physicochemical and biological properties on their surface and within due to their minuscule size (Mekuye & Abera,2023; Navya & Daima, 2016). By altering their size, shape, or chemical composition, quantum dots—tiny semiconductor particles that emit fluorescence—can have their emission characteristics changed (Vallabani & Singh, 2018) (Fig. 14.2).

By encapsulating them with biocompatible polymers and antibodies, these nanomaterials can be used as probes for both in vitro and in vivo microscopy. Additionally, gold-based rod-shaped nanomaterials may generate heat from their surface when exposed to laser light (Akinc et al., 2019).

De Jong et al. (2008) found that intravenously administered gold nanoparticles in rats exhibited organ-specific accumulation depending on particle size. Gold nanoparticles of 10 nm were widely distributed, with higher uptake in the kidneys, testes, thymus, heart, brain, spleen, and liver. In contrast, 50 nm particles localized mainly to the lungs, spleen, and liver, whereas particles larger than

100 nm were found only in the spleen and liver (Cho et al., 2009). Similarly, a separate study on intravenously injected silica nanoparticles in mice revealed that 50 and 100 nm particles accumulated in both the liver and spleen, while 200 nm particles were restricted to the spleen. A similar pattern of enhanced splenic accumulation for nanoparticles exceeding 200 nm has also been observed for other nanocarriers, including liposomal formulations (Vallabani & Singh, 2018). Integrating multiple types of nanoparticles into one platform enables the creation of a nanosystem capable of providing both therapeutic and diagnostic functions at the same time (Sun et al., 2025).

### 14.1.2 History

PEGylated doxorubicin formulation (Doxil) was approved as the first liposomal nanomedicine in 1995. Currently, it is utilized as a second-line treatment for Kaposi's sarcoma and ovarian cancer. Approximately 15 nanomedicines, mostly liposomal formulations intended to encapsulate chemotherapeutic medications, have been authorized for use in cancer therapy worldwide (He et al., 2019).

DaunoXome (laden with daunorubicin), Myocet (a non-PEGylated equivalent of Doxil approved only in Canada), MARQIBO (vincristine), MEPACT (mifamurtide), ONIVYDE (irinotecan), and Vyxeous (combining cytarabine and daunorubicin in a 5:1 molar ratio) are notable examples of approved nanomedicines. Vyxeous stands out as the first nanomedicine in a Phase III research study to considerably increase survival in acute myeloid leukemia (AML), extending it from 5.9 to 9.6 months, despite the fact that many similar formulations have not resulted in significant improvements in patient survival. Additionally, it is the first nanomedicine with a perfectly optimized combination therapy that has been approved. This illustrates one of the main advantages of nanomedicines, which is their capacity to transport and administer several therapeutic chemicals to the same location more successfully than traditional small-molecule medications (Anselmo & Mitragotri, 2019).

### 14.1.3 Nanomedicine in cancer therapy

Oncology is just one aspect of nanomedicine. Numerous treatments based on the nanotechnology have been approved previously or are undergoing clinical trials for a variety of ailments, such as iron deficiency anemia, uncommon liver illnesses, nucleic acid-based vaccine delivery, and different fungal infections (Utkarsh et al., 2025).

Atherosclerosis, Alzheimer's disease, skin problems, toxic dosages, and inflammatory bowel disease are just a few of the ailments for which researchers are currently developing nanomaterials for diagnosis and treatment (Duan et al., 2021).

Additionally, they are investigating how to modify gut microbiome in preclinical models using nanoparticles. Furthermore, the delivery of CRISPR gene-editing components is being investigated with nanoparticles (Ruberg & Berk, 2012).

The development of nanoparticle-based probes for sophisticated imaging methods like magnetic resonance imaging (MRI), computed tomography (CT), and positron emission tomography (PET) is also making significant strides. Several of the most advanced instances of these cutting-edge nanotechnologies will be covered in this chapter (Hawkins et al., 2015).

Alnylam Pharmaceuticals' Onpattro (patisiran), which was approved in 2018, is the first nanomedicine that has been successfully translated to transport nucleic acids. It makes use of a lipid nanoparticle (LNP) system with an aqueous center that contains negatively charged nucleic acids and an exterior shell made of pH-responsive cationic lipids. To improve stability and maintain near-neutral charge, a polyethylene glycol (PEG) is applied to the particle surface. This LNP, which is less than 100 nm, is used to treat hereditary transthyretin amyloidosis (hATTR) by delivering short-interfering RNA (siRNA) (Stern et al., 2010).

### 14.1.4 Nanomedicine in colon cancer

The use of engineered nanoparticles in therapeutic applications for cancer and other disorders is growing. Despite these developments, researchers and regulatory bodies are still investigating the specific characteristics of nanomaterials that affect their safety and therapeutic efficacy (Hare et al., 2017) (Table 14.1).

One of the most researched applications of nanomedicine is the use of nanoparticles to treat cancer; this field is known as cancer nanomedicine (Giri et al., 2023). In order to achieve more effective delivery directly to tumors, nanomaterials were primarily designed to convey either diagnostic or anticancer medications. This strategy aimed to solve the problems associated with conventional small-molecule chemotherapeutics and imaging agents, such as poor accumulation at target areas and unexpected consequences. Research conducted in the 1980s revealed that injected proteins and dyes tended to collect particularly in xenograft tumors rather than in the skin, which gave rise to the idea of employing nanomaterials (Adams et al., 2018).

## 14.2 Global regulatory overview

Governments of India have put in place a number of complex legislative, administrative, and technical requirements known as health product regulations to ensure that medications and medical equipment are safe, effective, and of high quality. Additionally, these rules guarantee the accuracy and dependability of the information supplied regarding such products (Shi et al., 2017). Particularly significant are the legal systems of the United States and the European Union (EU), which frequently act as models for international regulations. The FDA is the main body in charge of overseeing important health product laws in the United States, whereas the European Commission (EC) creates the fundamental legislative framework in the EU. These areas are often used as benchmarks for global regulatory procedures. Nanotechnology-enabled health products (NHPs) are typically categorized as either medical devices or medicinal products, depending on how the product primarily accomplishes

**Table 14.1 Clinical trial and commercial examples of anticancer nanomedicines.**

| Drug | Product name/company | Indication | Phase |
|---|---|---|---|
| Irinotecan | HA-irinotecan HyACT/Alchemia | Colorectal cancer (CRC) | Phase II |
| Oxaliplatin | MBP-426/Mebiopharm | Gastrointestinal adenocarcinoma | Phase II |
| Camptothecin | XMT1001 (Fleximer)/Mersana | CRC | Phase II |
| Doxorubicin | ThermoDox/Celsion | Colorectal metastases | Phase III |

its desired outcome. Compared to other health products, NHPs in contact with the human body present special technical and safety challenges due to their intended purpose (Mangla et al., 2025).

### 14.2.1 European Union

Directive 2001/83/Ec defines The legal structure regarding pharmaceuticals in the EU, with Article 1(2) offering a thorough explanation of what a medicinal product (European Commission, 2001). Before being sold or given to patients in the EU, all pharmaceutical products—including those that use nanotechnologies—must obtain marketing authorization. Because of their complex physicochemical characteristics, several nanomedicines are informally categorized as nonbiological complex pharmaceuticals (NBCDs). Depending on the product's regulatory category, a marketing authorisation application (MAA) needs to be submitted in order to obtain marketing authorization. Three methods are available for evaluating the MAA: a centralized approach at the level of the EU community, a decentralized mechanism involving several member states, or a national procedure within a single member state (Rodríguez-Gómez et al., 2025).

The European Medicines Agency (EMA) is in charge of the evaluation under the centralized approach, whilst the corresponding National Competent Authorities (NCAs) oversee the review of the MAA in national and decentralized procedures (Lezotre, 2013). Because of their physicochemical characteristics, nanomedicines—which are categorized as NBCDs—are extremely complex and lack a precise legal categorization in the EU. As a result, they lack a specific regulatory mechanism and are evaluated on an individual basis. Biotechnology-derived pharmaceuticals, on the other hand, need to go through the centralized process. It is unclear whether to pursue a "generic application" as per Article 10(1) or a "hybrid application" as per Article 10(3) due to the ambiguous legal environment for follow-on NBCDs; research indicates a rising preference for the hybrid approach (Klein et al., 2019).

#### *14.2.1.1 The European Union market for nanomedicines*

Nanomedicines have been commercially available for more than 20 years. In Europe, regulatory authorities assess each medicinal product prior to approval and make detailed information on its use and properties publicly accessible. The first nanomedicine received authorization from the EC in 1996. For every product it reviews, the EMA releases a European Public Assessment Report (EPAR) on its website, outlining the drug's characteristics and approved indications. Several nanomedicines are already available on the market, as illustrated by the examples below (Nanomedicine on the EU Market, 2025) (Table 14.2).

The primary legal frameworks governing the filing and assessment of MAAs in the EU are Directive 2001/83/EC and Regulation 726/2004/EC. These rules guarantee that every pharmaceutical product, including nanomedicines categorized as NBCDs, is thoroughly examined for quality, safety, and efficacy. Candidates must be legally established within the European Community, and applications must be submitted through the eSubmission channel in the electronic Common Technical Document (eCTD) format, in accordance with ICH M4 regulations (Gaspar et al., 2020).

Moreover, national agencies, not the EMA, are in charge of approving clinical trials. Therefore, the appropriate NCAs should be contacted with questions about the appropriateness or completeness of documents for clinical trial authorization. On the other hand, the EMA can provide advice

**Table 14.2 Marketed nanomedicines on the European Union market.**

| Medicine | Active substance and platform/technology | Use | Advantages of nanoformulations |
|---|---|---|---|
| Abraxane | Colloidal suspension of albumin-bound paclitaxel (active component) in spherical nanoparticle form without the use of a solvent | Therapy for nonsmall cell lung cancer, pancreatic adenocarcinoma, and breast cancer | The active principle's solubility issues have been resolved (paclitaxel is insoluble in water). Compared to earlier formulations, there were less frequent and severe toxicity (nausea, vomiting, exhaustion, arthralgia, myalgia, baldness). |
| Caelyx | Pegylated liposomal doxorubicin | Therapy for breast cancer that has spread, ovarian cancer, Kaposi's sarcoma (a cancer of the blood vessels), multiple myeloma | Greater accumulation within tumor cells and increased circulation time–reduction of cardiotoxicity of doxorubicin. |
| CosmoFer/ INFeD/ Ferrisat* | Iron nanoparticles coated with dextran for parenteral delivery (injection) | Management of anemia | Compared to nonnano formulations, this iron-carbohydrate combination is more stable and has less toxicity (nephrotoxicity, intracellular iron buildup). |
| Rapamune | Sirolimus particles stabilized to decrease particle size in a nanocrystal colloidal nanodispersion | Preventing organ rejection in kidney transplant recipients | Enhanced bioavailability and stability. |

*Nanomedicine on the EU Market, European Union Observatory For Nanomaterials https://euon.echa.europa.eu/nanomedicines-on-the-eu-market. Accessed on 02 Dec 2025.*

regarding whether a clinical trial's overall design is sufficient to support a MAA (Bernabe et al., 2019).

### 14.2.2 United States

In the United States, the definition of a medicinal product is set forth in section 201(g) of the Federal Food, Drug, and Cosmetic Act (FD&C Act) ICH M4, 2016):

The term "drug" means:

1. Articles included in the official national formulary, official homeopathic pharmacopeia of the United States, official US pharmacopeia, or any addition to any of these;

2. Articles designed to be used in the diagnosis, treatment, prevention, mitigation, or cure of a disease in humans or other animals;
3. Articles (apart from food) meant to alter a human or other animal's body structure or function; and
4. Articles meant to be used in conjunction with any article mentioned in clauses (1), (2), or (3). A food or dietary supplement for which a claim is made in compliance with the requirements of section 343(r) of this title, subject to sections 343(r)(1)(B) and 343(r)(3) of this title or sections 343(r)(1)(B) and 343(r)(5)(D) of this title, is not a drug just because the label or the labeling contains such a claim.

### 14.2.3 Food and Drug Administration activities

FDA personnel engage in a wide range of initiatives aimed at advancing regulatory science, strengthening collaborations with stakeholders, and promoting a flexible, science-driven, product-specific regulatory approach for technologies involving nanomaterials or nanotechnology applications. These efforts support innovation, improve timely and transparent communication about the FDA's work in nanotechnology, and uphold the agency's commitment to protecting public health.

Further details on specific initiatives can be found in the following resources:

- Nanotechnology Task Force
- Nanotechnology Task Force Report 2020
- Nanotechnology Task Force Report 2007
- Nanotechnology Guidance Documents
- Nanotechnology Partnerships at FDA.

According to USFDA, Content current as of 05/15/2024:

The FDA has issued new guidance titled *"Drug Products, Including Biological Products, that Contain Nanomaterials,"* outlining how to identify and manage risks associated with nanomaterials in drug products. Although these products must meet the same safety, efficacy, and quality standards as all other drugs, the guidance helps address uncertainties unique to nanotechnology. It covers characterization, testing, controls, and qualification of intentionally engineered nanoscale materials, excluding naturally nanoscale biological substances and incidental nanoparticles.

Nanomaterial-containing drug products are diverse, treating conditions ranging from cancer to inflammation and infections. A 2017 analysis showed that most engineered nanomaterials submitted to the FDA were under 300 nm. Due to their extremely small size, nanomaterials require specialized analytical methods, as standard microscopes cannot resolve them. Proper characterization—including particle size distribution—is essential because these features can significantly influence a product's performance (FDA, 2025).

The primary agency responsible for evaluating and approving drugs in the United States is the FDA. This large government organization has several offices, each of which is devoted to a particular regulatory product category. The FDA's Center for Biologics Evaluation and Research (CBER) and the Center for Drug Evaluation and Research (CDER) are principally responsible for drug regulation. The CDER's Office of Drug Safety (ODS) monitors the safety of drugs once they are approved, while its Office of New Drugs (OND) evaluates new medications before they are put on the market. Meanwhile, the Office of Tissues and Advanced Medicines (OTAT) within CBER oversees advanced medicines like gene and cell treatments. These offices work together to guarantee that drugs fulfill the fundamental requirements of safety and effectiveness, which form the basis of pharmaceutical regulation (CFR. Code of Federal Regulations - title 21 - food & drugs, 1996 and Darrow et al., 2020).

### 14.2.4 Innovation Task Force meeting

Experts in science, law, and regulation comprise the multidisciplinary Innovation Task Force (ITF) of the EMA. ITF meetings give applicants and the EMA an opportunity to discuss novel areas of product development early on. These gatherings provide a venue for early interaction to assist in the development of new therapeutic strategies and research. They enable businesses to communicate with EMA specialists early on and to have broad, informal conversations where opinions can be openly shared (Vereen & Ngnitewe, 2025 and Dabrowska & Thaul, 2018).

During these sessions, companies typically present their development plans, and the EMA provides general feedback—such as comments on the suitability of proposed animal models or other development strategies. In recent years, nanotechnology-related topics, including nanomedicines, have been among the top ten subjects discussed in ITF meetings, appearing in about 5% of procedures between 2019 and 2022 (Uster et al., 2024).

ITF meetings, which provide applicants with early feedback from EMA experts, are generally considered a beneficial preparatory tool (Ferrell, 2014).

It is crucial to remember, nonetheless, that the ITF Secretariat has the final say over whether to approve meetings based on requests for a meeting and a succinct draft briefing document (Uster et al., 2024). Furthermore, the Secretariat may only choose particular subjects for debate based on whether they are within the purview of the ITF (Vass et al., 2023).

### 14.2.5 ICH guidelines

While ICH standards are typically applicable to medicines, they often lack specific provisions for nanotechnology-based health products (NHPs) and may not always be acceptable. For example, by yielding false-negative results, conventional genotoxicity testing may underestimate the mutagenic risk of nanomaterials. It has been demonstrated that the commonly used in vitro Ames test is unreliable for nanomaterials since some particles cannot enter bacterial cells, making the assay inappropriate for assessing mutagenicity. Carcinogenicity studies, which are laborious and typically take two years, are only necessary for long-term drug use. The necessity for precise and trustworthy testing methods is highlighted by the fact that the causes of carcinogenicity for some nanomaterials, like those based on nickel, are still unclear (Henry et al., 2024).

In terms of immunotoxicity, it is well recognized that nanomaterials influence the immune system and may cause immunological reactions that might impact a product's effectiveness or safety. Cases of CARPA syndrome, a pseudo-allergy linked to complement activation, have been documented across several nanomaterial platforms. There is currently no particular regulatory structure in place for these specific difficulties. The necessity of creating new testing models that could lead to revised regulatory requirements is evident, highlighting the significance of developing guidelines specific to the unique features of products enabled by nanotechnology (Halamoda-Kenzaoui et al., 2021).

### 14.2.6 Other agencies

Standards-developing bodies like ISO and American Society for Testing and Materials (ASTM International) play a crucial role in creating and refining guidelines relevant to the medical device sector (Oualikene-Gonin et al., 2023). Quality management, risk assessment, preclinical and clinical testing, and many other aspects of medical device development are covered by the standards provided by both organizations. These organizations have specialized committees that concentrate on nanotechnology-enabled health products (NHPs), such as ASTM Committee E56.08 on nano-enabled medical goods and ISO/TC 229 on nanotechnologies (Daryanani et al., 2025).

The FDA in the United States considers and adopts standards from other organizations, including ISO and ASTM, prior to recognizing them as universal guidelines. Declarations from manufacturers that they comply with these standards are recognized as proof that the necessary Fundamental and Specific Controls have been fulfilled. An acute, non-IgE-mediated hypersensitivity reaction known as Complement Activation-Related Pseudoallergy (CARPA) develops when nanomedicines, liposomes, and biological products stimulate the complement system. Flushing, fever, dyspnea, and cardiopulmonary symptoms are some of the symptoms of CARPA, which can happen following the initial administration of a medication. Regulatory organizations like the FDA take complement activation and CARPA risk into account when examining innovative drug delivery systems and nanomedicines because of its significance to the safety evaluation of nanotechnology-based treatments (Bernal-Chávez et al., 2021). Two important documents pertinent to nanotechnology-enabled health products (NHPs) are ISO/TR 13014:2012(E), which offers guidance on physicochemical characterization of engineered nanoscale materials for toxicological assessment, and ISO/TR 10993–22:2017, which offers guidance on using nanoparticles to assess medical devices biologically, despite not being formally harmonized or designated as consensus standards. In order to facilitate toxicological assessments and aid in differentiating between seemingly similar nano-objects, these articles describe techniques for characterizing physicochemical features. ISO/TR 10993–22:2017 also offers instructions for sample preparation and toxicokinetic research concerning distribution, absorption, metabolism, and excretion or elimination (Daryanani et al., 2025).

The Pharmaceuticals and Medical Devices Agency (PMDA) is Japan's regulatory body in charge of reviewing and approving sophisticated therapeutic items, medical devices, pharmaceuticals, and regenerative medicines. PMDA assesses product safety, efficacy, and quality in close collaboration with Japan's Ministry of Health, Labour, and Welfare (MHLW), making sure that novel treatments satisfy legal requirements prior to being administered to patients. Sterility remains a major challenge, especially for nanomedicines intended for sterile routes of administration. The size and structural characteristics of nanoparticles can limit the suitability of certain sterilization techniques. For example, filtration becomes problematic for rigid nanoparticles that resemble the typical pore size of filtering membranes. Environmental safety is another important consideration during nanoparticle production. Extra precautions are required when working with dry nanomaterials to avoid aerosol formation, which could lead to inhalation hazards. Additionally, appropriate measures must be taken to prevent dermal exposure and protect personnel involved in the manufacturing process (Simeonidis et al., 2019).

## 14.3 Regulatory challenges specific to nanomedicines for colon cancer

Because of their intricate physicochemical characteristics, erratic biological interactions, and requirement for solid proof of safety, efficacy, and quality, nanomedicines intended for colon cancer treatment pose particular regulatory obstacles. Before approval, regulatory bodies including the FDA, EMA, and NCAs want thorough characterization and interdisciplinary assessment. The key challenges include:

**(1)** (Complexity in physicochemical characterization
**(2)** Unpredictable biodistribution and organ accumulation
**(3)** Safety assessment and long-term toxicity
**(4)** Manufacturing and quality control (critical micellar concentration [CMC] challenges)
**(5)** Lack of harmonized global regulatory frameworks
**(6)** Challenges in demonstrating bioequivalence and substitutability (Vetten et al., 2014)

## 14.4 Future directions in regulation

Future regulation of nanomedicines for colon cancer will focus on creating clearer, harmonized, and science-based frameworks to address the unique challenges of nanoscale materials. International agencies are expected to develop unified guidelines for nanoparticle characterization, toxicity testing, and clinical evaluation, along with more precise definitions tailored to nano-oncology. Advanced tools such as AI-based models and physiologically based pharmacokinetic (PBPK) simulations will increasingly support predictions of biodistribution, toxicity, and therapeutic outcomes. Regulators will work toward establishing standardized nanotoxicology methods to assess long-term safety, biodegradation, GI tract interactions, and microbiome effects. Adaptive regulatory pathways, including accelerated approvals and real-world evidence, will help bring innovative colon-targeted nanotherapies to patients more efficiently. Strengthened Quality-by-Design approaches and updated frameworks for combination nanomedicines will further ensure consistent quality, safety, and efficacy as the field continues to evolve (Krasteva & Georgieva, 2022).

## 14.5 Discussion

Nanomedicine has grown from fundamental studies on nanoscale materials into a highly specialized field with applications spanning oncology, diagnostics, vaccine delivery, imaging, and gene editing. Materials with one, two, or three dimensions in the nanoscale range demonstrate unique behaviors that differ substantially from their bulk counterparts. For example, quantum dots allow tunable fluorescence through modifications in size and composition, whereas gold nanorods can convert light energy into heat when irradiated. These properties enable nanomaterials to function as imaging probes and drug carriers, thereby addressing limitations of conventional therapeutics such as poor bioavailability, rapid clearance, and off-target toxicity.

In oncology, early discoveries from the 1980s revealed the accumulation of macromolecules in tumor tissues, giving rise to the concept of using nanoparticles to enhance tumor-directed drug delivery. Several landmark products—such as Doxil, Vyxeous, and Abraxane—demonstrate the clinical utility of nanoscale formulations in improving drug solubility, prolonging circulation, and reducing systemic toxicity. Colon cancer, in particular, has seen growing interest in nanoparticle-based therapies, as evidenced by nanocarriers containing irinotecan, oxaliplatin, camptothecin, and doxorubicin that are currently in clinical trials. However, despite promising therapeutic outcomes, nanoparticle biodistribution studies show that size, shape, and surface chemistry significantly influence organ accumulation, clearance, and toxicity. For instance, gold and silica nanoparticles show size-dependent uptake in liver, spleen, lungs, and kidneys, highlighting the need for precise nanoparticle engineering and safety evaluation.

Regulatory agencies worldwide recognize these complexities and have adapted their structures to guarantee the effectiveness and safety of nanomedicine products. In the EU, nanomedicines fall under Directive 2001/83/EC and are often treated as NBCDs due to their structural and functional complexity. Since no specific nanomedicine regulation exists, each product is evaluated on a case-by-case basis under centralized, decentralized, or national procedures. The EMA provides scientific assessment for centralized MAAs, while national authorities approve clinical trials and assess documentation quality. Similar regulatory patterns are seen in other regions, where classification, manufacturing controls, and clinical trial design require close coordination between developers and regulatory bodies. As the field advances, the integration of multifunctional nanoparticles, gene-editing platforms, and imaging agents further complicates regulatory pathways, emphasizing the need for clearer standards and global harmonization.

## 14.6 Conclusion

Nanomedicine continues to redefine modern healthcare by enabling precise, efficient, and innovative approaches to disease diagnosis and therapy. With several nanomedicines already in clinical use and many more in development for colon cancer and other diseases, the field has demonstrated significant therapeutic potential. However, the translation of nanotechnology into clinical practice depends heavily on rigorous scientific characterization, robust toxicity evaluation, and harmonized regulatory frameworks. Current regulatory systems, particularly in the EU, classify nanomedicines within broader medicinal product categories, often treating them as nonbiological complex drugs that require detailed, case-specific assessments. This approach reflects both the potential of nanomedicine and the challenges posed by its complexity. Moving forward, advances in nanoparticle design, improved understanding of biodistribution, and clearer regulatory guidance will be essential to accelerate clinical translation. Ultimately, continued collaboration among researchers, clinicians, regulatory authorities, and industry stakeholders will shape the effective and safe integration of nanomedicine into future healthcare systems.

## References


Adams, D., Gonzalez-Duarte, A., O'Riordan, W. D., Yang, C. C., Ueda, M., Kristen, A. V., Tournev, I., Schmidt, H. H., Coelho, T., Berk, J. L., & Lin, K. P. (2018). Patisiran, an RNAi therapeutic, for hereditary transthyretin amyloidosis. *New England Journal of Medicine, 379*(1), 11–21.

Akinc, A., Maier, M. A., Manoharan, M., Fitzgerald, K., Jayaraman, M., Barros, S., et al. (2019). The Onpattro story and the clinical translation of nanomedicines containing nucleic acid-based drugs. *Nature Nanotechnol, 14*, 1084–1087.

Anselmo, A. C., & Mitragotri, S. (2019). Nanoparticles in the clinic: An update. *Bioengineering & Translational Medicine, 4*(3), e10143.

Bernabe, R. D., van Thiel, G. J., Breekveldt, N. S., Gispen-de Wied, C. C., & Van Delden, J. J. (2019). Ethics in clinical trial regulation: Ethically relevant issues from EMA inspection reports. *Current Medical Research and Opinion, 35*(4), 637–645.

Bernal-Chávez, S. A., Del Prado-Audelo, M. L., Caballero-Florán, I. H., Giraldo-Gomez, D. M., Figueroa-Gonzalez, G., Reyes-Hernandez, O. D., González-Del Carmen, M., González-Torres, M., Cortés, H., & Leyva-Gómez, G. (2021). Insights into terminal sterilization processes of nanoparticles for biomedical applications. *Molecules (Basel, Switzerland), 26*(7), 2068.

CFR. Code of Federal Regulations - title 21 - food and drugs. (1996). Federal Register, Washington, Dc: Federal Register.

Cho, M., Cho, W. S., Choi, M., Kim, S. J., Han, B. S., Kim, S. H., Kim, H. O., Sheen, Y. Y., & Jeong, J. (2009). The impact of size on tissue distribution and elimination by single intravenous injection of silica nanoparticles. *Toxicology Letters, 189*, 177–183. https://doi.org/10.1016/j.toxlet.2009.04.017.

Dabrowska, A., & Thaul, S. (2018). How FDA approves drugs and regulates their safety and effectiveness. *Washington: Congressional Research Service, 8*, 1–25.

Darrow, J. J., Avorn, J., & Kesselheim, A. S. (2020). FDA approval and regulation of pharmaceuticals, 1983-2018. *JAMA the Journal of the American Medical Association, 323*(2), 164–176.

Daryanani, A. E., Maduekwe, U. N., Baird, P., & Ehrenfeld, J. M. (2025). Ensuring medical device safety: The role of standards organizations and regulatory bodies. *Journal of Medical Systems, 49*(1), 16.

De Jong, W. H., Hagens, W. I., Krystek, P., Burger, M. C., Sips, A. I., & Geertsma, R. E. (2008). Particle size-dependent organ distribution of gold nanoparticles after intravenous administration. *Biomaterials, 29*, 1912–1919. https://doi.org/10.1016/j.biomaterials.2007.12.037.

Directive 2001/83/Ec. (2001). Directive 2001/83/Ec of the European Parliament and of the council of 6 November 2001 on the community code relating to medicinal products for human use. Brussels, Belgium: European Commission (Ec).

Duan, L., Ouyang, K., Xu, X., Xu, L., Wen, C., Zhou, X., Qin, Z., Xu, Z., Sun, W., & Liang, Y. (2021). Nanoparticle delivery of CRISPR/Cas9 for genome editing. *Frontiers in Genetics, 12*, 673286.

Ferrell, G. (2014). Electronic management of assessment (EMA); a landscape review. Bristol: JISC http://repository.jisc.ac.uk/5599/1/EMA_REPORT.pdf

Gaspar, R. S., Silva-Lima, B., Magro, F., Alcobia, A., da Costa, F. L., & Feio, J. (2020). Non-biological complex drugs (NBCDs): Complex pharmaceuticals in need of individual robust clinical assessment before any therapeutic equivalence decision. *Frontiers in Medicine, 7*, 590527.

Giri, P. M., Banerjee, A., & Layek, B. (2023). A recent review on cancer nanomedicine. *Cancers, 15*(8), 2256.

Halamoda-Kenzaoui, B., Vandebriel, R. J., Howarth, A., Siccardi, M., David, C. A., Liptrott, N. J., Santin, M., Borgos, S. E., Bremer-Hoffmann, S., & Caputo, F. (2021). Methodological needs in the quality and safety characterisation of nanotechnology-based health products: Priorities for method development and standardisation. *Journal of Controlled Release, 336*, 192–206.

Hare, J. I., Lammers, T., Ashford, M. B., et al. (2017). Challenges and strategies in anti-cancer nanomedicine development: An industry perspective. *Advanced Drug Delivery Reviews, 108*, 25–38.

Hawkins, P. N., Ando, Y., Dispenzeri, A., Gonzalez-Duarte, A., Adams, D., & Suhr, O. B. (2015). Evolving landscape in the management of transthyretin amyloidosis. *Annals of Medicine, 47*, 625–638.

He, H., Liu, L., Morin, E. E., Liu, M., & Schwendeman, A. (2019). Survey of clinical translation of cancer nanomedicines—Lessons learned from successes and failures. *Accounts of Chemical Research, 52*(9), 2445–2461.

Henry, L. M., Hansen, E., Chimoff, J., Pokstis, K., Kiderman, M., Naim, R., Kossowsky, J., Byrne, M. E., Lopez-Guzman, S., Kircanski, K., & Pine, D. S. (2024). Selecting an ecological momentary assessment platform: Tutorial for researchers. *Journal of Medical Internet Research, 26*, e51125.

https://www.fda.gov/science-research/science-and-research-special-topics/nanotechnology-programs-fda Accessed on 05 Dec 2025.

ICH M4. (2016). Organisation of the Common Technical Document for the Registration of Pharmaceuticals for Human Use. Retrieved February 22, 2017 from https://database.ich.org/ites/default/files/M4_R4__Guideline.pdf

Kairdolf, B. A., Smith, A. M., Stokes, T. H., Wang, M. D., Young, A. N., & Nie, S. (2013). Semiconductor quantum dots for bioimaging and biodiagnostic applications. *Annual Review of Analytical Chemistry, 6*, 143–162.

Klein, K., Stolk, P., De Bruin, M. L., Leufkens, H. G., Crommelin, D. J., & De Vlieger, J. S. (2019). The EU regulatory landscape of non-biological complex drugs (NBCDs) follow-on products: Observations and recommendations. *European Journal of Pharmaceutical Sciences, 133*, 228–235. https://euon.echa.europa.eu/nanomedicines-on-the-eu-market accessed on 21 Nov 2025.

Krasteva, N., & Georgieva, M. (2022). Promising therapeutic strategies for colorectal cancer treatment based on nanomaterials. *Pharmaceutics, 14*(6), 1213.

Lezotre, P. L. (2013). *International cooperation, convergence and harmonization of pharmaceutical regulations: A global perspective*. Academic Press.

Logothetidis, S. (2011). Nanomedicine: The medicine of tomorrow. In *Nanomedicine and Nanobiotechnology, 21*, Berlin, Heidelberg: Springer Berlin Heidelberg, 1–26.

Mangla, B., Kumar, P., Javed, S., Pathan, T., Ahsan, W., & Aggarwal, G. (2025). Regulating nanomedicines: Challenges, opportunities, and the path forward. *Nanomedicine Nanotechnology, Biology, and Medicine, 20*(15), 1911–1927. https://doi.org/10.1080/17435889.2025.2533107.

Mekuye, B., & Abera, B. (2023). Nanomaterials: An overview of synthesis, classification, characterization, and applications. *Nano Select, 4*(8), 486–501.

Nanomedicine on the EU Market, European Union Observatory For Nanomaterials. Retrieved December 2, 2025 from https://euon.echa.europa.eu/nanomedicines-on-the-eu-market

Navya, P. N., & Daima, H. K. (2016). Rational engineering of physicochemical properties of nanomaterials for biomedical applications with nanotoxicological perspectives. *Nano Convergence, 3*(1), 1. https://doi.org/10.1186/s40580-016-0064-z.

Nel, A. E., Madler, L., Velegol, D., Xia, T., Hoek, E. M. V., Somasundaran, P., et al. (2009). Understanding biophysicochemical interactions at the nano-bio interface. *Nature Materials, 8*, 543–557.

Oualikene-Gonin, W., Sautou, V., Ezan, E., Bastos, H., Bellissant, E., Belgodère, L., Maison, P., & Ankri, J. (2023). Scientific advisory board of the ANTM. Regulatory assessment of nano-enabled health products in public health interest. Position of the scientific advisory board of the French national agency for the safety of medicines and health products. *Frontiers in Public Health, 11*, 1125577.

Ramos, D., & Almeida, L. (2022). Overview of standards related to the occupational risk and safety of nanotechnologies. *Standards, 2*(1), 83–89.

Rodríguez-Gómez, F. D., Monferrer, D., Penon, O., & Rivera-Gil, P. (2025). Regulatory pathways and guidelines for nanotechnology-enabled health products: A comparative review of EU and US frameworks. *Frontiers in Medicine (Lausanne), 12*. https://doi.org/10.3389/fmed.2025.1544393 PMID: 40109724; PMCID: PMC11919859.

Ruberg, F. L., & Berk, J. L. (2012). Transthyretin (TTR) cardiac amyloidosis. *Circulation, 126*, 1286–1300.

Shi, J., Kantoff, P. W., Wooster, R., & Farokhzad, O. C. (2017). Cancer nanomedicine: Progress, challenges and opportunities. *Nature Reviews Cancer, 17*(1), 20–37.

Simeonidis, K., Martinez-Boubeta, C., Zamora-Pérez, P., Rivera-Gil, P., Kaprara, E., Kokkinos, E., & Mitrakas, M. (2019). Implementing nanoparticles for competitive drinking water purification. *Environmental Chemistry Letters, 17*(2), 705–719.

Stern, S. T., Hall, J. B., Yu, L. L., et al. (2010). Translational considerations for cancer nanomedicine. *Journal of Controlled Release, 146*, 164–174.

Sun, W., Zheng, L., Gao, X., Li, H., Fang, Y., Zhang, M., Nan, D., Wang, Z. L., & Chen, B. (2025). Gasochromic-triboelectric-photodynamic multimodal patch for wound sensing and healing with congenital insensitivity to pain. *Nano Materials Science, 6*, 4.

Uster, D. W., Cordo, V., Cormier, E., & Ehmann, F. (2024). Insights into early interactions on innovative developments with European regulators. *Therapeutic Innovation & Regulatory Science, 58*(6), 1108–1119.

Utkarsh, K., Kumar, S., Khan, A. R., Alam, M. Z., & Khan, A. (2025). Nanomaterials application for diagnosis and treatment of common inflammatory disorders. In *Emergence of Sustainable Biomaterials in Tackling Inflammatory Diseases, 18*, Singapore: Springer Nature Singapore, 251–276.

Vallabani, N. S., & Singh, S. (2018). Recent advances and future prospects of iron oxide nanoparticles in biomedicine and diagnostics. *Biotechnology, 8*(6), 279.

Vass, P., Akdag, D. S., Broholm, G. E., Kjaer, J., Humphreys, A. J., & Ehmann, F. (2023). Enabling technologies driving drug research and development. *Frontiers in Medicine, 10*, 1122405.

Vereen, A., & Ngnitewe, G. (2025). FDA overview. In *InTranslational Cardiology*, *1*, Academic Press, 387–391.

Vetten, M. A., Yah, C. S., Singh, T., & Gulumian, M. (2014). Challenges facing sterilization and depyrogenation of nanoparticles: Effects on structural stability and biomedical applications. *Nanomedicine: Nanotechnology, Biology and Medicine, 10*(7), 1391–1399.

CHAPTER

# 15 Current and future perspectives of nanomedicine in colon cancer treatment

**Pranav Kumar Prabhakar**[1], **Ashok Kumar Sah**[2] **and Satyajit Mohanty**[3]

[1]*Department of Biotechnology, School of Engineering and Technology, Nagaland University, Kohima, Nagaland, India,*

[2]*Department of Medical Laboratory Sciences, College of Applied & Health Sciences, A' Sharqiyah University, Ibra, North Sharqiyah, Oman,*

[3]*Department of Pharmacy, School of Medical and Allied Science, Galgotias University, Greater Noida, Uttar Pradesh, India*

## 15.1 Introduction

Colorectal cancer (CRC), which includes malignancies arising from the colon and rectum, remains one of the most commonly diagnosed cancers globally and a major cause of cancer-related morbidity and mortality. Advances in diagnostic techniques, surgical strategies, radiotherapy, and chemotherapy have improved patient outcomes over the past few decades, yet the burden of colon cancer remains substantial, particularly in low- and middle-income countries where screening programs and early diagnostic infrastructure are limited. Conventional treatment modalities, although effective to an extent, are associated with significant limitations including poor specificity, systemic toxicity, multidrug resistance (MDR), and suboptimal therapeutic indices. These challenges have ignited interest in innovative treatment approaches, among which nanomedicine has emerged as one of the most promising. Nanomedicine—defined as the application of nanoscale materials for diagnosis, imaging, and therapy—offers new avenues for targeted drug delivery, controlled drug release, and improved therapeutic efficacy. Nanotechnology-based interventions aim to address many of the shortcomings of conventional colon cancer treatments by enhancing drug bioavailability, minimizing systemic exposure, enabling site-specific delivery, and exploiting unique features of the tumor microenvironment (TME). In recent years, nanocarriers such as liposomes, polymeric nanoparticles (PNPs), dendrimers, metal-based nanostructures, micelles, and stimuli-responsive platforms have demonstrated considerable potential in preclinical and early clinical studies. This chapter provides an overview of the global burden of colon cancer, the inherent limitations of traditional treatment strategies, and the emergence of nanomedicine as a transformative tool in colon cancer management. These discussions establish the foundation for understanding how nanotechnology can contribute to more effective, personalized, and less toxic therapeutic approaches.

Nano Drug Delivery Systems for Colon Cancer. DOI: https://doi.org/10.1016/B978-0-443-40443-6.00012-X

### 15.1.1 Global burden of colon cancer

Colon cancer is a major public health concern worldwide. According to the Global Cancer Observatory, CRC is the third most commonly diagnosed cancer and the second leading cause of cancer-related deaths globally (Sung et al., 2021). Its incidence continues to rise in many regions due to aging populations, lifestyle changes, dietary patterns high in processed foods, sedentary behaviors, and increasing rates of obesity. Although colon cancer historically affected older adults, recent epidemiological trends show a worrying increase in early-onset cases among individuals under 50 years of age (Siegel et al., 2020). Geographically, patterns of colon cancer incidence and mortality vary significantly across regions. High-income countries such as the United States, Japan, Australia, and several European nations have traditionally reported higher incidence rates, largely attributed to lifestyle factors and better detection practices. However, improvements in screening and early detection have contributed to declining mortality in these regions. Conversely, many developing nations—particularly in Asia, Africa, and Latin America—are experiencing increasing incidence and higher mortality due to limited access to screening programs, diagnostic tools, and advanced treatment modalities (Arnold et al., 2017). The economic burden of colon cancer is equally significant. Costs associated with diagnosis, treatment, hospitalization, and long-term care are substantial, posing major financial pressures on healthcare systems globally. The burden is compounded by indirect costs such as productivity loss, caregiver burden, and reduced quality of life among survivors. As the global population continues to age, and as lifestyle-related risk factors persist, projections suggest that the incidence and mortality of colon cancer may increase further in the coming decades unless effective prevention and therapeutic interventions are implemented (Bray et al., 2018). Despite widespread public health campaigns, colon cancer screening rates remain suboptimal in many regions. Standard screening tools—including colonoscopy, fecal occult blood tests, and sigmoidoscopy—have proven effective in reducing mortality but face barriers such as lack of access, limited awareness, stigma, and cost. As a result, many patients present at advanced stages when the prognosis is poor. This growing burden highlights the need for improved diagnostic and therapeutic strategies, including advanced drug delivery technologies that can enhance treatment outcomes.

### 15.1.2 Limitations of conventional therapies

Current colon cancer treatments typically include surgery, chemotherapy, radiotherapy, targeted therapy, and immunotherapy. While these modalities have improved survival rates, they also present significant limitations.

1. **Surgical limitations:** Surgery remains the gold standard for early-stage colon cancer, yet its success depends heavily on tumor location, stage, and patient health. Advanced cancers that have metastasized to distant organs are often not curable with surgery alone. Moreover, postoperative complications, prolonged recovery times, and recurrence remain major concerns (Benson et al., 2021).
2. **Chemotherapy constraints:** Chemotherapy using agents such as 5-fluorouracil, oxaliplatin, and irinotecan has been the cornerstone of systemic treatment. However, traditional chemotherapeutics lack specificity, allowing cytotoxic agents to damage both cancerous and healthy tissues. This

leads to severe side effects, including gastrointestinal toxicity, neuropathy, immunosuppression, and hematological disorders (Longley et al., 2003). Additionally, the development of MDR significantly reduces the effectiveness of chemotherapy, often resulting in treatment failure.

3. **Radiotherapy challenges:** Radiation therapy is often used in rectal cancer but has limited application in colon cancer due to anatomical constraints. Even when used, it carries risks of collateral tissue damage, inflammation, and long-term gastrointestinal complications.
4. **Targeted therapy and immunotherapy limitations:** Targeted therapies such as epidermal growth factor receptor (EGFR) inhibitors and vascular endothelial growth factor (VEGF) inhibitors have improved outcomes for some patients but are limited by the development of resistance mutations, high cost, and varying responses across patient populations (Vardy et al., 2019). Immunotherapies such as checkpoint inhibitors have shown impressive results in a subset of patients with mismatch repair-deficient tumors, but the majority of colon cancer cases do not respond to these therapies due to immunologically "cold" TMEs (Ganesh et al., 2019).
5. **Pharmacokinetic barriers:** Many anticancer drugs exhibit poor solubility, rapid metabolism, and inadequate tumor accumulation, leading to suboptimal therapeutic concentrations at the tumor site. The TME—characterized by abnormal vasculature, hypoxia, and high interstitial fluid pressure (IFP)—further limits drug penetration and efficacy.

These limitations collectively underscore the need for more effective, targeted, and less toxic treatment strategies. Nanotechnology offers promising solutions to many of these challenges.

### 15.1.3 Emergence of nanomedicine

Nanomedicine has rapidly emerged as a promising field aimed at improving cancer diagnosis, imaging, and therapy by utilizing nanoscale materials and carriers. Its potential stems from unique physicochemical properties such as small particle size, large surface area-to-volume ratio, tunable surface chemistry, and the ability to load multiple bioactive molecules. These characteristics enable nanoparticles to overcome biological barriers that impede conventional drugs. One of the primary advantages of nanomedicine is the ability to deliver drugs specifically to tumor tissues through passive and active targeting. Passive targeting exploits the enhanced permeability and retention (EPR) effect caused by leaky tumor vasculature, allowing nanoparticles to accumulate more readily in tumor tissue than in normal tissues (Maeda et al., 2013). Active targeting involves the functionalization of nanoparticles with ligands such as antibodies, peptides, or small molecules that bind specifically to overexpressed receptors on cancer cells, enhancing the precision of drug delivery. Nanocarriers also offer controlled and sustained release capabilities, enabling consistent therapeutic levels while minimizing systemic toxicity. Platforms such as liposomes, PNPs, micelles, dendrimers, gold nanoparticles (AuNPs), and mesoporous silica nanoparticles (MSNs) have been extensively studied for colon cancer therapy. Many of these systems can encapsulate hydrophobic drugs, protect them from degradation, and enhance biodistribution. Additionally, the integration of imaging agents within therapeutic nanoparticles has led to the development of *theranostic* systems capable of simultaneous diagnosis and therapy. These multifunctional platforms allow real-time monitoring of drug release, biodistribution, and tumor response, supporting personalized treatment strategies.

Recent advances also highlight the use of nanoparticles for delivering genetic materials such as siRNA, miRNA, and CRISPR/Cas components. These technologies allow modulation of oncogenes, tumor suppressor genes, and signaling pathways involved in colon cancer progression (Wang et al., 2021). Overall, nanomedicine offers transformative possibilities for addressing the long-standing limitations of colon cancer therapy. Its capacity for precise delivery, reduced toxicity, and enhanced efficacy positions it as a key innovation in the future of oncological treatment.

## 15.2 Nanotechnology in colon cancer: fundamental concepts

Nanotechnology has emerged as a transformative approach in modern oncology, providing innovative strategies for enhancing the diagnosis and treatment of malignancies such as colon cancer. The fundamental advantage of nanomedicine lies in its ability to manipulate materials at the nanoscale, typically between 1 and 100 nm, allowing for novel physicochemical features not present in bulk materials. These attributes—such as high surface-area-to-volume ratio, tunable charge, controlled drug release, and modifiable surfaces—enable nanoparticles to overcome biological barriers, improve pharmacokinetics, enhance tumor accumulation, and reduce systemic toxicity (Peer et al., 2007). Colon cancer, characterized by its heterogeneous tumor biology, complex metastatic pathways, and variable TME, presents multiple therapeutic challenges that nanotechnology is uniquely poised to address. This section elaborates on the types of nanocarriers used in colon cancer therapy, the principles of passive and active targeting that guide nanoparticle biodistribution, and the interactions between nanoparticles and the TME that influence treatment outcomes.

### 15.2.1 Types of nanocarriers

Nanocarriers have emerged as transformative tools in colon cancer therapy due to their ability to enhance drug solubility, protect therapeutic agents from degradation, improve pharmacokinetics, and achieve site-specific delivery. Their physicochemical diversity allows researchers to tailor drug delivery systems to meet specific therapeutic needs such as sustained release, reduced toxicity, targeted accumulation, or multimodal therapy. The major classes of nanocarriers include lipid-based carriers, PNPs, dendrimers, inorganic nanoparticles, polymeric micelles, nanogels, and carbon-based systems. Each of these carriers offers distinct structural, functional, and pharmacological advantages.

1. **Lipid-based nanocarriers:** Lipid-based drug carriers are among the most mature and clinically successful nanomedicines due to their biocompatibility and structural similarity to biological membranes. Liposomes, the earliest developed lipid nanoparticles (LNPs), consist of one or more phospholipid bilayers surrounding an aqueous core capable of encapsulating both hydrophilic and lipophilic agents. Their structural flexibility allows them to solubilize a wide range of anticancer drugs, protect them from enzymatic degradation, and reduce systemic toxicity by restricting drug release into healthy tissues. These attributes have been demonstrated in colon cancer therapy, where liposomal formulations of drugs such as 5-fluorouracil, irinotecan, and doxorubicin achieve enhanced tumor accumulation and reduced side effects compared to free drugs (Akbarzadeh et al., 2013). Liposome PEGylation further prolongs systemic circulation by preventing opsonization and clearance by the mononuclear phagocyte system. Beyond conventional liposomes, solid lipid nanoparticles (SLNs) and nanostructured lipid carriers (NLCs) represent next-generation lipid

carriers designed to overcome limitations such as low drug-loading capacity and potential drug expulsion during storage. SLNs comprise solid lipids that remain in a crystalline form at body temperature, providing a stable matrix for sustained drug release. NLCs introduce a mixture of solid and liquid lipids, creating imperfections in the lipid matrix that increase drug-entrapment efficiency and minimize expulsion. Their biodegradability, stability in gastrointestinal conditions, and ability to modify surface chemistry for colon targeting make them particularly attractive for oral colon cancer therapy.

2. **PNPs:** PNPs are among the most versatile nanocarriers owing to the diversity of natural and synthetic polymers used in their fabrication. Polymers such as poly lactic-co-glycolic acid (PLGA), polylactic acid (PLA), polyethylene glycol (PEG), and chitosan offer biodegradable and biocompatible platforms that allow fine control over drug release kinetics, degradation rates, and surface functionality. Their tunable physicochemical properties enable the design of nanoparticles that respond to specific physiological triggers, such as pH or enzymatic activity, which are highly relevant in colon-selective drug delivery. The ability of PNPs to encapsulate hydrophobic chemotherapeutics (e.g., paclitaxel, camptothecin) while preserving systemic stability has made them promising candidates for colon cancer therapy (Danhier et al., 2012). Additionally, PNPs are widely explored for nucleic acid delivery—including siRNA, miRNA, and DNA—due to their capacity to protect fragile genetic cargo and facilitate cellular uptake. The incorporation of targeting ligands such as folic acid, hyaluronic acid, or antibodies onto polymeric surfaces further improves specificity toward colon cancer cells overexpressing corresponding receptors. Their controlled release and targeting abilities directly address challenges associated with conventional chemotherapy.
3. **Dendrimers:** Dendrimers are three-dimensional, monodisperse, highly branched nanostructures characterized by internal cavities and numerous peripheral functional groups. Their precise molecular architecture allows exceptional control over size, surface chemistry, and drug-loading mechanisms. Polyamidoamine (PAMAM) dendrimers are the most widely studied dendrimers for colon cancer therapy, especially for gene and nucleic acid delivery (Sherje et al., 2018). The interior voids can encapsulate small drugs through hydrophobic interactions, while surface groups can conjugate larger biomolecules or targeting ligands. The high density of surface groups enables multivalent interactions with cancer cell receptors, improving cellular uptake and therapeutic efficacy. The ability to modify dendrimers with PEG, peptides, or antibodies allows fine-tuning of biodistribution and reduction of cytotoxicity associated with unmodified cationic dendrimers. Their unique "tree-like" structure provides advantages in codelivering drugs and nucleic acids, making dendrimers a versatile platform for combination therapy in colon cancer.
4. **Inorganic nanoparticles:** Inorganic nanocarriers, including AuNPs, MSNs, and iron oxide nanoparticles, offer unique optical, magnetic, and structural properties conducive to both therapy and imaging, making them ideal for theranostics. AuNPs stand out due to their favorable biocompatibility, ease of synthesis, and tunable surface chemistry. Their optical properties, especially surface plasmon resonance, allow applications in photothermal therapy, whereby nanoparticles convert absorbed light into heat to induce cancer cell apoptosis. These nanoparticles can also be conjugated with chemotherapeutics, peptides, or nucleic acids for combined treatment approaches (Dreaden, Alkilany, et al., 2012; Dreaden, Austin, et al., 2012). MSNs possess a high surface area, uniform pore structure, and excellent drug-loading capacity. Their porous framework allows controlled release profiles, and the silanol groups on the surface facilitate easy

functionalization with targeting ligands or stimuli-responsive gates. These attributes have positioned MSNs as promising carriers for colon cancer drugs such as camptothecin, curcumin, and 5-FU. Magnetic nanoparticles, primarily iron oxide-based, provide capabilities for magnetic resonance imaging (MRI) and magnetically guided drug delivery. Their responsiveness to external magnetic fields offers an additional mechanism to direct nanoparticles to tumor sites, enhancing specificity and reducing off-target effects.

5. **Polymeric micelles:** Polymeric micelles are nanosized, core–shell structures formed by the self-assembly of amphiphilic block copolymers in aqueous environments. Their hydrophobic cores solubilize poorly water-soluble drugs, a common challenge in delivering chemotherapeutics such as paclitaxel, docetaxel, and curcumin. Meanwhile, the hydrophilic shell—often made of PEG—provides steric stabilization and prolongs circulation by reducing protein adsorption and clearance (Kwon, 2003). These micelles typically fall within an optimal size range (10–100 nm) that favors tumor accumulation via the EPR effect. In colon cancer therapy, polymeric micelles have shown promise in overcoming MDR, a significant barrier to effective chemotherapy. Micelles can achieve high intracellular drug concentrations by bypassing efflux pumps and enhancing endocytic uptake. Their physicochemical properties can also be engineered for pH-responsive drug release, exploiting the mildly acidic TME to trigger drug deposition specifically within cancerous tissues.
6. **Nanogels:** Nanogels are nanoscale hydrogel particles composed of crosslinked hydrophilic polymers that can swell in water and encapsulate a variety of therapeutic agents. Their high water content, biocompatibility, and mechanical softness mimic biological tissues, allowing improved interaction with the colon's mucosal environment. Nanogels can encapsulate both small-molecule chemotherapeutics and macromolecules such as proteins, peptides, or nucleic acids. Their responsiveness to environmental stimuli—such as temperature, pH, redox gradients, or enzymatic activity—enables controlled and site-specific drug release (Vinogradov, 2010). Nanogels have shown particular promise in colon-targeted delivery because their structural properties can be tailored to remain stable in the acidic stomach environment and release payloads preferentially in the neutral-to-alkaline conditions of the colon. Additionally, surface modification with targeting moieties enhances tumor-cell uptake. Nanogels are increasingly explored for combination therapies and for delivering immunomodulators that can synergize with chemotherapy.
7. **Carbon-based nanocarriers:** Carbon-based nanomaterials, including carbon nanotubes (CNTs), fullerenes, and graphene oxide (GO), are valued for their high surface area, mechanical strength, and ability to adsorb aromatic or hydrophobic molecules. CNTs, with their tubular morphology, can penetrate biological membranes more easily than many other nanoparticles, enhancing intracellular drug delivery. GO offers abundant oxygen-containing functional groups that allow for efficient drug conjugation or loading. Despite their advantages, the major challenge restricting their clinical translation is concern over biopersistence, cytotoxicity, and potential induction of oxidative stress or inflammation. Studies continue to investigate strategies to improve their biocompatibility through surface functionalization, PEGylation, or coating with biopolymers. While promising for imaging and drug delivery, carbon-based nanocarriers require extensive safety evaluation before widespread adoption in colon cancer therapy.

Together, these nanocarrier platforms provide a diverse technological foundation for improving colon cancer treatment through enhanced bioavailability, targeted accumulation, controlled release, and multifunctional capabilities. Their versatility allows for the incorporation of chemotherapeutics, genetic materials, diagnostic agents, and immunotherapeutics, paving the way for more precise and effective nanomedicine-based interventions.

### 15.2.2 Nanoparticle–tumor microenvironment interactions

Understanding the interactions between nanoparticles and the TME is critical for optimizing nanomedicine design. The TME of colon cancer is a dynamic and complex ecosystem composed of cancer cells, cancer-associated fibroblasts, immune cells, endothelial cells, extracellular matrix (ECM) components, and microbiota. Each component influences drug transport, nanoparticle accumulation, and therapeutic response. One of the most significant features of the TME is the abnormal tumor vasculature. Colon tumors exhibit chaotic, tortuous, and hyperpermeable blood vessels, which, on the one hand, facilitate passive nanoparticle accumulation but, on the other hand, limit deep penetration due to uneven perfusion. High IFP within tumors further restricts nanoparticle movement from blood vessels into deeper tumor regions (Jain, 2014). To overcome these barriers, researchers are developing nanoparticles with optimized sizes, shapes, and surface characteristics that allow improved intratumoral transport. Smaller nanoparticles, typically within the range of 20–80 nm, have demonstrated enhanced penetration into densely packed tumor tissues. Hypoxia is another characteristic feature of the colon cancer TME. Low oxygen levels induce angiogenesis, promote metastasis, and contribute to therapeutic resistance. Nanoparticles can exploit hypoxia by delivering hypoxia-activated prodrugs or agents that modify the tumor vasculature. Hypoxia-responsive nanocarriers have gained attention for their ability to release drugs selectively in hypoxic zones, thereby improving therapeutic index (Brown & Wilson, 2004). Variations in pH within the colon cancer microenvironment also influence nanoparticle performance. Tumors typically exhibit a mildly acidic extracellular pH due to the Warburg effect, where cancer cells preferentially utilize glycolysis. pH-responsive nanoparticles can exploit this gradient, remaining stable under normal physiological conditions but releasing their cargo in the acidic tumor milieu (Bae et al., 2013). This approach minimizes off-target effects while maximizing drug efficacy at the tumor site. The immune microenvironment in colon cancer significantly affects nanoparticle behavior. Tumor-associated macrophages (TAMs), regulatory T cells, and myeloid-derived suppressor cells contribute to an immunosuppressive milieu that supports tumor progression. Nanoparticles can modulate this immune landscape by delivering immunostimulatory molecules or reprogramming immune cell phenotypes to enhance antitumor immunity (Fan et al., 2020). Additionally, nanoparticles may be opsonized by serum proteins and cleared by macrophages, influencing biodistribution and efficacy. PEGylation and other surface modifications help reduce immune clearance and prolong circulation time.

The ECM, composed of collagen, proteoglycans, and glycoproteins, forms a physical barrier that limits nanoparticle diffusion. Dense ECM structures are common in colon cancer and often correlate with poor penetration of both nanoparticles and conventional drugs. Strategies such as incorporating ECM-degrading enzymes or using nanoparticles that alter ECM architecture are being explored to enhance drug distribution. Finally, the gut microbiota plays a unique role in colon cancer progression and nanoparticle behavior. Microbial enzymes such as azoreductases, nitroreductases, and glycosidases can interact with nanoparticle coatings or drug conjugates, leading to colon-specific activation or degradation (Prasad & Tyagi, 2015). Harnessing these microbial interactions allows the design of microbiota-responsive nanoparticles that release drugs specifically in the colon, minimizing systemic exposure. Altogether, the nanoparticle–TME interplay is central to the success of nanomedicine in colon cancer, guiding the development of carriers that navigate biological barriers, exploit tumor-specific features, and achieve high therapeutic efficacy with minimal toxicity.

## 15.3 Colon-specific nanodrug delivery strategies

The development of colon-specific nanodrug delivery systems has emerged as a transformative approach for enhancing the precision, safety, and therapeutic outcomes of colon cancer treatment. Unlike conventional oral chemotherapy, where drugs are subjected to extensive degradation in the stomach and small intestine, colon-targeted nanocarriers are engineered to navigate the gastrointestinal (GI) tract and release their therapeutic payloads in the large intestine with spatial and temporal precision. Such targeted delivery minimizes systemic toxicity, enhances localized drug concentration at tumor sites, and improves patient compliance compared to parenteral or systemic routes. Colon-targeting nanoplatforms exploit distinctive physiological features of the colon, including variations in pH, enzyme distribution, microbial composition, and mucosal architecture. This section provides an in-depth discussion of four principal strategies: pH-responsive systems, enzyme-triggered delivery, microbiota-activated nanocarriers, and mucoadhesive or colon-retentive formulations. Together, these technologies exemplify how nanotechnology can be harnessed to improve drug delivery under complex GI conditions.

### 15.3.1 pH-responsive systems

pH-responsive nanocarriers are among the most widely explored systems for colon-specific drug delivery, primarily because the gastrointestinal tract exhibits a predictable gradient of pH values. The stomach maintains a highly acidic environment (pH 1.5–3.5), which gradually increases in the small intestine (pH 5.5–7.0) and reaches near-neutral to slightly alkaline values in the colon (pH 7.0–7.5). This pH progression serves as a physiological trigger that can be strategically utilized to achieve site-specific drug release.

**Polymeric coatings and pH-sensitive materials:** A common strategy is to coat nanocarriers with pH-sensitive polymers that remain intact under acidic and mildly acidic conditions but dissolve when exposed to the neutral-to-alkaline pH of the colon. Polymers such as Eudragit S100, Eudragit L100, hydroxypropyl methylcellulose phthalate (HPMCP), cellulose acetate phthalate (CAP), and poly (methacrylic acid) derivatives are frequently used to fabricate enteric coatings or shell layers around nanoparticles. These polymers prevent premature degradation or drug leakage in the stomach and small intestine. For example, Eudragit S100 dissolves at pH values above 7, making it well suited for colon-targeted delivery of chemotherapeutics like 5-fluorouracil (5-FU), curcumin, or irinotecan. Researchers have developed Eudragit-coated PNPs and liposomes that remain stable throughout the upper GI tract and release drugs upon encountering colonic pH conditions. Such systems significantly improve therapeutic indices by enhancing colonic deposition and reducing gastric irritation.

**pH-sensitive hydrogels and nanogels:** Another approach employs pH-sensitive hydrogels or nanogels, which swell or degrade selectively in response to pH changes. These hydrogels are typically composed of poly(acrylic acid), chitosan derivatives, or poly(β-amino esters). In the acidic stomach, the hydrogels remain compact, restricting drug release. Upon reaching the colon, their network swells due to ionization of carboxyl or amino groups, allowing drug diffusion. Such swelling-responsive behavior enables prolonged retention and controlled release profiles. Nanogels encapsulating doxorubicin, capecitabine, or natural compounds like quercetin have shown improved antitumor efficacy in preclinical colon cancer models.

#### 15.3.1.1 Advantages and limitations

pH-responsive systems are relatively simple to fabricate, cost-effective, and compatible with a variety of nanocarriers. However, variability in colonic pH among patients—especially in pathological states like inflammation or diarrhea—may affect release profiles. Supplementing pH-responsive systems with additional targeting triggers (e.g., enzyme- or microbiota-responsive elements) is a growing trend to improve reliability.

### 15.3.2 Enzyme-triggered delivery

Enzyme-responsive nanocarriers take advantage of the differential enzymatic activity along the GI tract, especially the distinct abundance of pancreatic, brush border, and colonic bacterial enzymes. Because certain enzymes are primarily present in the colon, they can act as highly specific biological triggers for drug release.

**Polysaccharide-based enzyme-sensitive systems:** The most prominent enzyme-responsive nanocarriers are those built from natural polysaccharides such as pectin, chitosan, dextran, guar gum, inulin, and amylose. These polymers are resistant to degradation in the stomach and small intestine but can be selectively broken down by colonic enzymes such as pectinase, dextranase, β-glucuronidase, and azoreductase. For instance, dextran-based nanoparticles remain intact throughout the upper GI tract but degrade rapidly when exposed to colonic dextranase, leading to controlled drug release. Pectin-coated liposomes and nanoparticles exhibit similar colon-specific degradation due to the action of pectinolytic enzymes.

**Azoreductase-triggered systems:** Azoreductase, produced predominantly by colonic bacteria, cleaves azo bonds (–N=N–). This property has been extensively exploited in designing prodrugs and azo-linked nanoparticle systems. Mesalamine azo-prodrugs for inflammatory bowel disease (IBD) serve as classical examples, and researchers have extended this principle to create azo-linked polymeric nanogels, dendrimers, and micelles for colon cancer therapy.

**Protein- and peptide-based enzyme-responsive nanocarriers:** Protease-responsive platforms incorporate peptide linkers that can be cleaved by colonic or tumor-associated enzymes such as matrix metalloproteinases (MMPs), cathepsins, or elastases. These systems are particularly beneficial in colon cancer, where tumor tissues overexpress MMPs. For example, MMP-responsive PNPs that release doxorubicin or siRNA only within TMEs minimize systemic toxicity and enhance localized efficacy.

#### 15.3.2.1 Advantages and challenges

Enzyme-responsive systems demonstrate high specificity and reduced variability compared to pH-responsive systems. However, their performance depends on the presence and activity of colonic enzymes, which may vary with diet, disease state, or microbiota composition. Combining enzyme sensitivity with other targeting modalities enhances delivery precision.

### 15.3.3 Microbiota-activated nanocarriers

The colon harbors a dense and metabolically active microbiota comprising trillions of bacteria, primarily from the genera *Bacteroides*, *Firmicutes*, *Bifidobacterium*, and *Clostridium*. This microbial ecosystem provides unique enzymatic and metabolic activities that can be exploited for drug delivery. Unlike the upper GI tract, the colon contains high concentrations of reductive, hydrolytic, and fermentative enzymes, making microbiota-triggered nanocarriers highly specific and efficient.

**Mechanisms of microbial activation:** Microbial enzymes such as azoreductase, nitroreductase, β-glucosidase, β-galactosidase, and β-xylosidase can degrade specific polymeric substrates or chemical linkers incorporated into nanocarrier systems. When nanoparticles reach the colon, these enzymes break down the outer protective layers or linker molecules, releasing the encapsulated drug. For example, nanoparticles coated with polysaccharides like inulin, guar gum, or resistant starch are selectively digested by colonic bacteria, facilitating site-specific drug release. Similarly, nanoparticles functionalized with azo bonds or nitroaromatic linkers are activated by microbial reductases.

**Microbiota-modulating nanocarriers:** An emerging strategy involves designing nanocarriers that interact therapeutically with the microbiome. Certain nanoparticles restore microbial balance disrupted during cancer progression or chemotherapy. For instance, curcumin- or resveratrol-loaded nanoparticles not only exert anticancer effects but also promote beneficial microbial populations, improving overall colon health.

#### *15.3.3.1 Advantages over other strategies*

Microbiota-activated nanocarriers offer enhanced site specificity due to the high density and unique enzymatic repertoire of colonic bacteria. Moreover, microbial activity remains relatively robust across patient populations, making this approach more reliable than strategies dependent solely on pH or GI transit time.

#### *15.3.3.2 Limitations*

Variations in microbiota composition due to antibiotics, diet, or diseases such as IBD may affect the degradation rate of microbiota-sensitive nanocarriers. Nevertheless, the growing understanding of microbiome–drug interactions continues to guide the design of resilient, adaptive delivery systems.

### 15.3.4 Mucoadhesive and colon-retentive systems

The colon possesses a relatively thick mucus layer composed of mucin glycoproteins, which can act either as a barrier or an anchor depending on nanocarrier design. Mucoadhesive and colon-retentive nanocarriers are engineered to prolong residence time in the colon, enhance contact with tumor tissues, and improve the absorption or local deposition of therapeutic agents.

**Mucoadhesive polymers:** Common mucoadhesive polymers include chitosan, pectin, carbopol, hyaluronic acid, thiolated polymers, and lectins. These materials form hydrogen bonds, electrostatic interactions, or covalent disulfide linkages with mucin proteins, enabling the nanocarriers to adhere to the mucosal surface (Peppas & Huang, 2004). Chitosan-based nanoparticles exhibit strong mucoadhesion due to their cationic properties, which facilitate electrostatic attraction with negatively charged mucin. Thiolated polymers, known as thiomers, form stable covalent interactions with mucin cysteine residues, significantly enhancing retention time.

**Colon-retentive strategies:** Beyond mucoadhesion, colon-retentive nanocarriers are designed to resist peristaltic clearance. Approaches include:

- **Bioadhesive microspheres and nanoparticles** that anchor to mucosal folds.
- **Floating or gas-generating nanoparticles** that maintain buoyancy within the colon.
- **Swelling hydrogels** that expand in size once they reach the colon, reducing elimination.
- **Magnetically retentive nanoparticles**, where external magnetic fields hold magnetic nanocarriers at tumor sites.

These strategies ensure prolonged exposure of tumor tissues to therapeutic payloads, increasing uptake and cytotoxicity.

#### *15.3.4.1 Tumor-penetrating mucoadhesive nanocarriers*

Recent innovations combine mucoadhesion with tumor-penetrating functionalities. For instance, nanoparticles decorated with RGD peptides (Arginine-Glycine-Aspartic acid (Arg-Gly-Asp) tripeptide) or *Internalizing RGD (*iRGD) sequences adhere to mucus and subsequently penetrate tumor stroma, improving intratumoral drug distribution. Such multifunctional platforms enhance the therapeutic outcomes of chemotherapeutics, siRNA, or immunomodulators.

#### *15.3.4.2 Advantages and challenges*

Mucoadhesive systems significantly enhance drug residence time, reduce dosing frequency, and improve bioavailability. However, the protective role of mucus may hinder deep tumor penetration. To address this, researchers are increasingly using mucus-penetrating coatings, such as PEG or zwitterionic polymers, to complement mucoadhesive strategies, creating dual-functional systems.

Colon-specific nanodrug delivery strategies represent a cornerstone of modern nanomedicine for colon cancer therapy. By exploiting physiological cues such as pH gradients, enzymatic signatures, microbial enzymatic activity, and mucosal architecture, these nanocarriers enable highly localized, efficient, and safe drug delivery. As research advances, hybrid platforms incorporating multiple targeting triggers—such as pH- and enzyme-responsive, or mucoadhesive and microbiota-activated systems—are emerging as next-generation solutions. These sophisticated systems hold great promise for improving therapeutic outcomes, minimizing adverse effects, and paving the way toward precision nanomedicine in colon cancer treatment.

## 15.4 Applications of nanomedicine in colon cancer treatment

Nanomedicine has emerged as a transformative platform in the treatment of colon cancer due to its capacity to overcome major clinical challenges such as drug resistance, systemic toxicity, limited tumor specificity, and heterogeneous tumor biology. By engineering nanoscale systems that can encapsulate therapeutics, respond to physiological signals, or engage immune and genetic mechanisms, nanotechnology offers unique advantages over conventional colon cancer therapies. Nanocarriers can improve drug solubility, extend circulation time, facilitate targeted delivery to tumor tissues, and allow combination or multimodal therapy within a single platform (Wilhelm et al., 2016). The following sections provide a detailed overview of the major therapeutic applications of nanomedicine in colon cancer management.

### 15.4.1 Chemotherapy-loaded nanoparticles

Chemotherapy remains one of the frontline treatments for colon cancer; however, the clinical use of chemotherapeutics such as 5-fluorouracil (5-FU), oxaliplatin, and irinotecan is often limited by rapid clearance, off-target toxicity, and MDR. Nanoparticles have been widely explored as vehicles for delivering cytotoxic agents more effectively to tumor tissues while reducing systemic side effects. PNPs, liposomes, dendrimers, and inorganic platforms have been engineered to encapsulate single or multiple chemotherapeutic molecules and release them in a controlled or stimuli-responsive manner (Danhier et al., 2010). Liposome-based formulations, such as PEGylated liposomal irinotecan, demonstrate improved pharmacokinetics and tumor accumulation due to EPR effects (Torchilin, 2014). PNPs fabricated from PLGA or chitosan offer sustained drug release and improved bioavailability. 5-FU-loaded nanoparticles, for example, have shown significantly higher cytotoxicity in colon cancer xenograft models than free drug, attributable to better intracellular uptake and prolonged retention in tumor tissues (Li et al., 2020). Nanocarriers also help bypass resistance by codelivering efflux pump inhibitors or by promoting endosomal escape. Another advantage of chemotherapy-loaded nanoparticles is their potential for colon-specific release. pH-sensitive polymeric systems can protect drugs during gastrointestinal transit and release them selectively in the higher pH of the colon, thereby increasing local drug concentration while reducing systemic exposure (Laroui et al., 2012). These advancements collectively highlight how nanotechnology-enabled chemotherapy platforms are reshaping the therapeutic landscape for colon cancer.

### 15.4.2 Nanocarrier-based gene therapy (siRNA, miRNA, CRISPR)

Gene therapy offers highly targeted treatment avenues by altering gene expression patterns responsible for tumor progression, invasion, metastasis, and chemoresistance. However, the major challenge with nucleic acids such as siRNA, miRNA, and CRISPR complexes is their instability, susceptibility to enzymatic degradation, and poor cellular uptake. Nanocarriers provide essential protection and facilitate safe and efficient gene delivery to colon cancer cells (Chen et al., 2021).

**siRNA delivery:** Small interfering RNAs can silence oncogenes such as KRAS, BRAF, and VEGF, which are commonly dysregulated in colon cancer. LNPs and polymeric carriers encapsulating siRNA have demonstrated robust knockdown efficiency and tumor growth inhibition in preclinical models. For example, KRAS-targeted siRNA nanoparticles significantly reduced tumor viability and enhanced sensitivity to chemotherapeutics.

**miRNA replacement or inhibition:** Many miRNAs function as tumor suppressors or oncogenes. Nanocarriers enable replacement of downregulated miRNAs (e.g., miR-34a) or inhibition of oncogenic miRNAs (e.g., miR-21). Delivery of miR-34a via lipid-based nanoparticles has restored apoptotic signaling and suppressed colon cancer metastasis (Bader et al., 2010).

**CRISPR-based gene editing:** The CRISPR/Cas9 system holds promise for permanent genetic modifications. Nanoparticle systems are being engineered to deliver Cas9 ribonucleoproteins or plasmids directly to colon tumor cells. AuNPs and polymer-based vectors have demonstrated high editing efficiency for genes involved in DNA repair and tumor proliferation (Liu et al., 2020). These platforms avoid the immunogenicity associated with viral vectors and permit controlled and transient expression. Overall, nanocarrier-mediated gene therapy provides precision modulation of oncogenic pathways and is a rapidly growing direction in advanced colon cancer therapeutics.

### 15.4.3 Immunomodulatory nanoplatforms

Immunotherapy has transformed cancer treatment, but its success in colon cancer—especially in microsatellite-stable tumors—remains limited due to immunosuppressive TMEs and inadequate immune activation. Immunomodulatory nanoplatforms are designed to enhance antitumor immunity, improve the presentation of tumor antigens, and reverse immune evasion mechanisms.

**Nanoparticle-based delivery of immune adjuvants:** Nanocarriers have been employed to deliver Toll-like receptor agonists, cytokines, or (STimulator of INterferon Genes) (STING) activators directly to antigen-presenting cells. This strategy increases dendritic cell maturation and enhances T-cell priming within the TME (Korshoej et al., 2021). For example, PNPs carrying CpG (cytosine-guanine) oligonucleotides have demonstrated strong antitumor responses in colon cancer models.

**Checkpoint inhibitor nanoparticles:** While PD-1/PD-L1 inhibitors show limited response in most CRCs, nanoparticle-enabled delivery improves local concentration and reduces systemic toxicity. PLGA nanoparticles conjugated with anti-PD-L1 antibodies or siRNA against PD-L1 have enhanced T-cell infiltration and tumor rejection in murine models (Havel et al., 2019).

**Immune cell–targeted nanocarriers:** Macrophages, dendritic cells, and cytotoxic T cells can be selectively targeted using ligand-modified nanoparticles. Reprogramming TAMs from an M2 to an M1 phenotype using iron oxide nanoparticles, for instance, has enhanced antitumor immunity and reduced tumor burden (Zanganeh et al., 2016). Immunomodulatory nanoplatforms thus offer a potent means to enhance the immune recognition and destruction of colon cancer cells.

### 15.4.4 Combination and synergistic nanotherapeutics

Combination therapy is essential for overcoming drug resistance and targeting multiple cancer hallmarks. Nanocarriers allow coloading or codelivery of drugs, genes, immune ligands, or imaging agents, ensuring synchronized delivery to tumor cells.

**Chemo-gene combinations:** Codelivery of 5-FU with siRNA targeting antiapoptotic genes such as BCL-2 enhances apoptosis and reduces drug resistance (Zhang et al., 2019). Nanocarriers enable controlled ratio maintenance and enhance synergistic interactions.

**Chemo-immunotherapy platforms:** Nanoparticles can simultaneously deliver chemotherapeutics and immune adjuvants. For example, oxaliplatin-loaded nanoparticles combined with CpG oligonucleotides have triggered strong immunogenic cell death, increasing T-cell activation and tumor clearance.

**Photothermal and photodynamic therapy combinations:** AuNPs or porphyrin-based nanocarriers can produce localized heat or reactive oxygen species upon irradiation. When combined with chemotherapy, these platforms enhance tumor permeability and drug uptake, providing synergistic tumor regression.

Combination nanotherapeutics unify multiple treatment modalities into a single formulation, enhancing therapeutic intensity while minimizing systemic toxicity.

### 15.4.5 Theranostic nanoparticles

Theranostic nanomedicine integrates diagnostic and therapeutic capabilities into one platform, enabling real-time monitoring of drug delivery, tumor response, and treatment efficacy. This is particularly valuable for colon cancer because of its heterogeneous progression and unpredictable metastatic behavior.

**Imaging-guided nanotherapy:** Iron oxide nanoparticles provide MRI contrast, while simultaneously carrying chemotherapeutics. Quantum dots, AuNPs, and upconversion nanoparticles allow fluorescence, computed tomography (CT), or photoacoustic imaging, guiding clinicians to the tumor site and enabling precise treatment planning (Xie et al., 2020).

**Stimuli-responsive theranostics:** Some nanocarriers release drugs in response to endogenous cues (pH, enzymes, glutathione) while simultaneously signaling the release event through imaging changes. For example, pH-sensitive polymeric micelles tagged with fluorescent dyes allow real-time visualization of drug release within acidic TMEs.

**Radio-theranostics:** Nanoparticles labeled with radionuclides such as $^{64}Cu$ or $^{99m}Tc$ offer PET/SPECT imaging and radiotherapeutic potential. This dual functionality improves early detection of metastasis and delivers localized radiation to tumor tissues.

Theranostic nanoparticles bring personalized medicine closer to clinical application by integrating diagnosis, targeted therapy, and treatment monitoring.

## 15.5 Clinical advances and translational status

The clinical translation of nanomedicine for colon cancer has progressed from promising preclinical results toward several early-phase and some later-phase human studies, but widespread clinical adoption remains limited by biological complexity, manufacturing hurdles, and regulatory scrutiny. This section summarizes the key preclinical findings that justified clinical testing, reviews the most relevant trials underway or completed to date, and discusses the regulatory and practical considerations that shape the path from bench to bedside.

### 15.5.1 Preclinical findings

Preclinical studies—spanning in vitro assays, orthotopic models, and patient-derived xenografts—have repeatedly shown that nanocarriers can improve drug delivery to colorectal tumors, enhance antitumor efficacy, and mitigate systemic toxicity compared with free drugs. Liposomal formulations of camptothecin derivatives (SN-38, irinotecan) and cyclodextrin-based nanoparticle–drug conjugates (e.g., CRLX101/NLG207) demonstrate superior tumor accumulation, prolonged circulation, and improved tumor control in multiple animal models. For example, CRLX101 showed enhanced antitumor activity and radiosensitizing effects in colorectal xenografts, providing the rationale for early-phase trials combining CRLX101 with chemoradiation in rectal cancer. Liposomal irinotecan formulations and other SN-38 delivery systems (LE-SN38, NK012, Irinophore C/CPX-1) improved the pharmacokinetics and therapeutic index of camptothecin family drugs in preclinical colorectal models. Such studies often reported increased tumor drug exposure, reduced gastrointestinal and hematologic toxicity at equivalent or greater efficacy, and the potential to overcome certain mechanisms of drug resistance (Batist et al., 2009). Beyond small-molecule chemotherapeutics, nanoparticle vectors have demonstrated promise as carriers for nucleic acids (siRNA, miRNA) and gene-editing complexes in preclinical colon cancer studies—protecting cargo from nuclease degradation, enhancing uptake, and achieving meaningful gene knockdown or editing in tumor models. Preclinical immuno-nanomedicines (e.g., nanoparticle delivery of adjuvants, checkpoint inhibitors, or payloads that reprogram TAMs) have shown improved tumor immunogenicity in murine CRC models, supporting translation into combined chemo-immunotherapy approaches. Reviews synthesizing these findings emphasize reproducible advantages—improved biodistribution, controlled release, and multifunctionality—while noting variability across models and tumor types that complicates prediction of clinical performance.

### 15.5.2 Current clinical trials

Clinical testing of nanomedicines for CRC has concentrated on liposomal and polymer-based carriers for camptothecin analogs, nanoparticle–drug conjugates, mucoadhesive/mucosal delivery strategies, lymph node tracers, and a small but growing number of nucleic acid delivery trials. Several representative programs and trial outcomes illustrate the field's status. CRLX101 (NLG207)—a cyclodextrin-polymer nanoparticle conjugated to camptothecin—progressed from robust preclinical efficacy into phase I/II testing. Early clinical studies focused on safety, pharmacokinetics, and tumor uptake; subsequent trials evaluated CRLX101 in combination with capecitabine and radiotherapy as neoadjuvant therapy for locally advanced rectal cancer National Clinical Trial (NCT02010567), with pilot data indicating tumor preferential uptake and tolerability that supported further study. Liposomal irinotecan and SN-38 platforms have been intensively evaluated. CPX-1 (a fixed-ratio liposomal formulation of irinotecan and floxuridine) completed phase I–II studies in advanced CRC and showed disease control in a subset of heavily pretreated patients, justifying continued development (NCT00361842). Separate liposome-encapsulated SN-38 (LE-SN38) and polymeric micelle SN-38 formulations such as NK012 have also entered clinical testing, including phase II studies in unresectable or metastatic CRC, demonstrating feasibility and some activity in selected cohorts. More recent trials continue to explore liposomal irinotecan schedules and combinations (including with Poly (ADP-ribose) polymerase (PARP) inhibitors or bevacizumab) in metastatic settings (multiple NCT entries). Other clinical applications include the use of carbon nanoparticle tracers for sentinel lymph node mapping in rectal cancer and theranostic agents to combine imaging with therapy. Multiple ClinicalTrials.gov records document active or completed studies evaluating nanoparticle tracers and image-guided nanotherapies for CRC staging and intraoperative guidance. Landscape and systematic reviews indicate that while a number of nanomedicines have reached clinical trials in CRC, most remain in phase I–II testing, with heterogeneous endpoints and small cohorts. Positive signals—improved pharmacokinetics, manageable safety profiles, and occasional objective responses—exist, but definitive phase III evidence establishing clear survival benefits over standard regimens remains sparse. Comprehensive systematic reviews summarize a growing but still nascent clinical evidence base and call for better-designed, biomarker-stratified trials.

### 15.5.3 Regulatory considerations

Regulatory agencies have acknowledged the potential of nanomedicines but emphasize that products containing nanomaterials must meet the same standards of safety, efficacy, and quality as other drug products, with additional attention to unique physicochemical attributes and manufacturing controls. The US Food and Drug Administration (FDA) has issued guidance documents to help sponsors determine whether a product involves nanotechnology and to outline expectations for characterization, control strategies, pharmacokinetic evaluation, and product labeling (FDA, 2024). These guidances stress thorough characterization of size, morphology, surface chemistry, drug release, and lot-to-lot consistency, and encourage early engagement with regulators. Translational bottlenecks extend beyond regulatory paperwork. Key practical challenges include: (1) scale-up and manufacturing reproducibility, since many promising nanoparticle syntheses are optimized at milligram laboratory scale but must be adapted to kilogram Good Manufacturing Practice (GMP) production while preserving critical quality attributes (CQAs); (2) complex biodistribution and safety assessment, because nanomaterials can have unexpected organ accumulation, complement activation, or long-term persistence requiring expanded toxicology packages; (3) heterogeneity of preclinical models, which complicates clinical prediction—animal xenografts often fail to recapitulate human TME and immune interactions; and (4) clinical trial design—the need for robust biomarker strategies to select patient subsets likely to benefit (e.g., lesions with enhanced permeability or specific receptor expression) and standardized endpoints that capture nanomedicine advantages (pharmacokinetics, intratumoral drug levels, immune modulation). These translational issues are widely discussed in contemporary reviews and regulatory analyses. Regulators also highlight manufacturing changes as potentially "significant" events; alterations in particle size distribution, surface coating, or solvent residuals can change a product's safety or efficacy profile and may trigger additional comparability studies. Sponsors are therefore advised to define CQAs early, develop robust analytical methods, and prepare comprehensive control strategies that encompass raw materials, in-process testing, and final product release criteria. International regulatory harmonization—between FDA, European Medicines Agency (EMA), Pharmaceuticals and Medical Devices Agency (PMDA), and others—remains incomplete, adding complexity for global development programs. Finally, ethical and economic considerations affect translation. Nanomedicines frequently require complex manufacturing and specialized cold-chain logistics, increasing cost; payers will demand clear evidence of clinical benefit to justify higher prices. Long-term safety monitoring (postmarketing surveillance) is particularly important for novel nanomaterials due to potential delayed adverse effects or environmental concerns. Together, scientific, regulatory, and economic realities mean that translating promising nanomedicines for colon cancer into standard clinical practice requires interdisciplinary collaboration, early regulatory engagement, well-powered clinical studies with meaningful endpoints, and scalable manufacturing strategies.

Preclinical work has convincingly demonstrated that nanocarriers can change the pharmacology of established drugs, improve tumor targeting, and enable new therapeutic modalities (gene editing, theranostics, immune modulation). Several nanomedicines—particularly liposomal and polymeric camptothecin formulations and nanoparticle–drug conjugates—have advanced into clinical trials for CRC, yielding encouraging pharmacokinetic and safety results. However, large randomized phase III data proving unequivocal survival benefits are limited, and major translational challenges remain: scalable GMP manufacturing, comprehensive safety evaluation, optimal patient selection, and regulatory alignment. Overcoming these hurdles will require rigorous translational pipelines, adaptive clinical trial designs, and sustained dialog between scientists, clinicians, industry, and regulators to realize the promise of nanomedicine for colon cancer patients.

## 15.6 Challenges and safety considerations

Nanomedicine has shown tremendous potential in improving the delivery, specificity, and efficacy of therapeutics for colon cancer. However, its translation into routine clinical practice remains limited by multiple scientific, technical, and regulatory challenges. Understanding the safety concerns associated with nanomaterials, manufacturing constraints, and interpatient variability is essential for designing clinically viable nanoformulations. This section discusses three major issues shaping the risk–benefit profile of nanomedicine: toxicity and biodistribution, scale-up and manufacturing challenges, and variability in patient response.

### 15.6.1 Toxicity and biodistribution

Toxicity and biodistribution represent the most critical considerations in the development of nanomedicines for colon cancer. Unlike small-molecule drugs that follow relatively predictable pharmacokinetic patterns, nanoparticles exhibit complex interactions with biological systems. Their distribution is influenced by size, shape, surface charge, hydrophobicity, protein corona formation, and interactions with immune cells (Blanco et al., 2015). These physicochemical characteristics determine whether nanoparticles accumulate within tumors, remain in circulation, or become sequestered in the liver, spleen, or kidneys. One major concern is off-target accumulation, particularly in organs of the reticuloendothelial system (RES). Studies show that nanoparticles larger than 50 nm are rapidly captured by macrophages in the liver and spleen, reducing their therapeutic availability and increasing the risk of organ-specific toxicities (Wilhelm et al., 2016). Surface modifications such as PEGylation reduce immune recognition, but prolonged circulation also raises concerns about long-term persistence and unknown chronic toxicity. Inflammatory and immunogenic responses are another safety consideration. Nanoparticles may activate complement pathways, induce cytokine release, or trigger hypersensitivity reactions. Cationic liposomes and dendrimers are particularly prone to causing membrane disruption and pro-inflammatory signaling (Mukherjee et al., 2019). For colon cancer patients undergoing combination chemotherapy, immune dysregulation caused by nanoparticles could exacerbate treatment-related side effects. A related challenge is the formation of a protein corona, a dynamic layer of plasma proteins adsorbed onto the nanoparticle surface. This corona alters nanoparticle identity, affecting biodistribution, cellular uptake, and toxicity (Monopoli et al., 2012). For example, opsonins within the corona accelerate clearance, while dysopsonins may prolong circulation. The protein corona composition varies widely among individuals, introducing unpredictability in nanoparticle behavior. Furthermore, genotoxicity and oxidative stress have been reported with metal-based nanoparticles such as gold, silver, and iron oxide. Although generally considered biocompatible, these particles may catalyze reactive oxygen species formation, leading to DNA damage, mitochondrial dysfunction, or apoptosis (Ahamed et al., 2017). While conventional chemotherapeutics intentionally induce cytotoxicity, unintended effects on healthy tissues remain a critical safety barrier. Long-term toxicity studies are particularly important for nanoparticles with slower degradation rates. Some inorganic nanoparticles may persist for months, and the consequences of chronic exposure are not fully understood. Regulatory agencies increasingly require extended monitoring of nanoparticle metabolism and clearance mechanisms, including renal or fecal elimination pathways. In colon cancer therapy, the TME adds complexity. Hypoxia, acidity, and impaired vasculature may alter nanoparticle degradation or trap particles within necrotic tissue. While this can enhance therapeutic retention, it may also impede clearance and complicate toxicity predictions (Shi et al., 2017).

Therefore, comprehensive toxicological profiling—covering acute, subchronic, chronic, immunotoxicity, and genotoxicity—is vital for safe clinical translation. Advanced in vitro systems (e.g., organ-on-a-chip), physiologically based pharmacokinetic (PBPK) modeling, and longitudinal in vivo studies can improve predictability and reduce clinical risk.

### 15.6.2 Scale-up and manufacturing challenges

Even when nanomedicines show significant therapeutic promise, their translation is frequently hindered by difficulties in scaling up production while maintaining quality, reproducibility, and regulatory compliance. Laboratory-scale synthesis often involves precise control over reaction conditions, mixing rates, solvents, and purification steps that are not easily replicated at industrial scale (He et al., 2020). One fundamental issue is achieving batch-to-batch consistency. Nanoparticles are highly sensitive to minor variations in temperature, pH, mixing times, or surfactant concentrations. Small deviations can lead to changes in particle size distribution, drug loading, surface charge, or morphology—each of which significantly affects safety and efficacy. Regulatory agencies classify many of these parameters as CQAs, requiring stringent analytical testing and validation (FDA, 2024). Manufacturing technologies such as microfluidic mixing, high-pressure homogenization, and scalable polymerization systems aim to address these issues; however, each method has limitations. Microfluidics ensures precise particle formation but is difficult to implement at large volumes. High-pressure homogenization is widely used but may degrade sensitive biological payloads such as siRNA or CRISPR complexes. Polymer-based nanoparticles require solvent removal processes that must be carefully controlled to avoid residual toxicity. Purification remains another challenge. Techniques such as ultracentrifugation, chromatography, or tangential flow filtration must be adapted to GMP conditions, often increasing production cost and complexity. Additionally, sterilization of nanoparticles is nontrivial; methods like filtration may disrupt particle integrity, while gamma irradiation can alter chemical structure.

Stability during storage and transport is a major concern. Nanoparticles may aggregate, fuse, precipitate, or degrade over time, especially under temperature fluctuations. Stabilizers, cryoprotectants, or lyophilization techniques can mitigate these issues, but they require careful optimization to avoid altering pharmacokinetics or therapeutic activity (Ventola, 2017). Another critical barrier is cost.

Production of complex nanomedicines—particularly those incorporating targeting ligands, biomaterials, or gene-editing components—remains expensive. High production costs limit commercial interest unless clear clinical superiority over conventional treatments is demonstrated. Supply chain considerations also impact scalability. Many nanomedicines rely on specialized lipids, polymers, or biological targeting moieties that require high-purity synthesis and may have limited suppliers. Ensuring a consistent supply of materials that meet GMP standards is essential for commercial viability. Ultimately, successful translation requires integrated collaboration among material scientists, engineers, pharmacologists, clinicians, and regulatory experts. Development of standardized manufacturing platforms, automation, and real-time monitoring technologies (e.g., Process Analytical Technology tools) will be crucial for scaling nanomedicine production to clinically meaningful levels.

### 15.6.3 Variability in patient response

Even when nanomedicines are optimally designed and manufactured, their clinical performance varies widely among patients due to biological, genetic, and microenvironmental heterogeneity. This variability complicates dosing, efficacy prediction, and safety assessment. A major source of variability stems from differences in the TME. The EPR effect, historically considered the cornerstone of passive nanoparticle targeting, is now known to vary drastically among patients and tumor types. Some colon tumors exhibit leaky vasculature that favors nanoparticle accumulation, while others have dense stroma or poor perfusion that limits penetration (Danhier, 2016). This inconsistency affects therapeutic index and contributes to unpredictable outcomes. Genetic and molecular heterogeneity further influence response to nanomedicine. Mutations in KRAS, BRAF, or PI3K pathways may alter nanoparticle uptake or sensitivity to delivered therapeutics. For gene-based nanotherapy (e.g., siRNA, miRNA, CRISPR), variability in RNA expression, nuclease activity, and cellular endocytosis pathways can critically affect therapeutic efficacy. Patient physiology—including liver and kidney function, immune status, metabolic rate, and gut microbiota composition—also shapes nanoparticle behavior. For instance, immune-mediated clearance through macrophages or complement proteins differs significantly between individuals, affecting nanoparticle half-life and biodistribution (Zhang et al., 2018). Differences in protein corona formation represent another important cause of variability. Because the corona composition depends on the individual's plasma proteome, the same nanoparticle formulation may behave differently across patients, altering targeting ability, opsonization, and toxicity (Monopoli et al., 2012). The gut microbiome plays a particularly important role in colon cancer. Microbiota may degrade, activate, or chemically modify nanoparticles intended for colon-specific delivery, thereby influencing therapeutic availability and efficacy (Gao et al., 2020). Sociodemographic factors—including age, diet, comorbidities, and ethnicity—can indirectly affect response by shaping metabolic and immune processes. Elderly patients, for example, may exhibit reduced clearance or enhanced susceptibility to toxicity compared to younger individuals.

## 15.7 Future perspectives

Nanomedicine continues to evolve as a transformative platform in colon cancer therapy, driven by advances in biomaterials science, computational modeling, artificial intelligence (AI), personalized medicine, and translational strategies. While current nanocarrier systems have demonstrated significant improvements in drug delivery efficiency, tumor penetration, and therapeutic outcomes, next-generation technologies are poised to push these boundaries further. The future of nanomedicine in colon cancer is likely to be shaped by innovations that integrate multifunctionality, biological intelligence, adaptive behavior, and precision engineering. These advancements will not only expand the therapeutic arsenal but also help overcome persistent limitations such as off-target toxicity, poor clinical predictability, and manufacturing bottlenecks.

## 15.7.1 Biomaterial innovations

Advances in biomaterials are central to the next phase of nanomedicine. First-generation nanocarriers—such as liposomes, PNPs, and dendrimers—laid the foundation for improved pharmacokinetics and targeted delivery. However, their performance can be limited by issues such as premature drug leakage, immune clearance, and restricted tumor penetration. Emerging biomaterials aim to address these shortcomings through enhanced biocompatibility, programmability, and responsiveness to tumor-specific cues.

### *15.7.1.1 Smart and stimuli-responsive biomaterials*

Future nanocarriers are increasingly being designed to respond to multiple stimuli, including pH, redox gradients, enzymes, hypoxia, and temperature. Multistimuli-responsive nanocarriers offer stricter spatial and temporal control over drug release, improving targeting specificity. For example, dual pH- and enzyme-responsive polymers degrade selectively in the colon TME, enhancing intratumoral drug deposition (Mura et al., 2013). Redox-sensitive nanocarriers incorporating disulfide linkages enable rapid payload release in the high-glutathione environment of cancer cells (Meng et al., 2021).

### *15.7.1.2 Biomimetic and cell-membrane-coated nanoparticles*

Biomimetic nanocarriers represent a major leap forward by leveraging natural cell membranes—such as those derived from erythrocytes, platelets, macrophages, or cancer cells themselves—to camouflage nanoparticles. These cell membrane coatings extend circulation time, facilitate immune evasion, and allow homotypic tumor targeting (Fang et al., 2018). In colon cancer, cancer cell–coated nanoparticles demonstrate enhanced tumor accumulation through self-recognition, improving therapeutic efficacy.

### *15.7.1.3 Hybrid and multifunctional nanoplatforms*

Hybrid nanocarriers combining lipid–polymer, inorganic–organic, or hydrogel–nanoparticle systems allow researchers to exploit the strengths of each material. For example, combining mesoporous silica with a lipid shell improves drug-loading flexibility while minimizing cytotoxicity (Zhao et al., 2020). Multifunctional nanoplatforms may integrate therapeutic payloads, imaging agents, and targeting ligands, enabling simultaneous diagnosis, therapy, and real-time monitoring.

#### 15.7.1.4 Injectable and in situ-forming biomaterial systems

Another promising direction is the development of injectable nanogels or in situ self-assembling scaffolds that release nanoparticles over time. These systems can be applied postsurgery to prevent recurrence by maintaining localized therapeutic concentrations (Liu et al., 2022). Their biodegradability and tunable mechanical properties make them ideal adjuncts to systemic treatment.

#### 15.7.1.5 Gene-activated biomaterials

Emerging biomaterials are being engineered to carry genetic elements—such as plasmids, CRISPR components, or microRNA mimics—in tissue-responsive formats. These systems enable selective activation of tumor-suppressive pathways or gene editing within colon cancer cells (Wang et al., 2022). Integrating gene delivery within biodegradable scaffolds or hydrogels provides controlled, long-term therapeutic effects. Overall, biomaterial innovations will continue to improve targeting precision, reduce toxicity, enhance intratumoral diffusion, and support multifunctional therapy in colon cancer.

### 15.7.2 AI-enabled nanoparticle design

AI and machine learning (ML) are revolutionizing nanoparticle design by enabling predictive modeling, automation, and optimization. Traditional experimental methods for nanoparticle formulation rely heavily on trial-and-error approaches, which are time-consuming, expensive, and poorly scalable. AI-driven methods accelerate this process by identifying optimal compositions, predicting biological behavior, and adjusting design parameters with unprecedented accuracy.

#### 15.7.2.1 Predictive modeling of nanoparticle behavior

AI algorithms can predict critical nanoparticle parameters such as size, zeta potential, drug-loading efficiency, and release kinetics. Deep learning models trained on experimental datasets enable researchers to simulate how nanoparticles interact with biological environments, including protein corona formation, immune clearance, and tumor penetration (Zhu et al., 2021). This predictive capacity significantly reduces the number of required laboratory experiments.

#### 15.7.2.2 Optimization of targeting ligands and surface chemistry

ML tools help identify optimal targeting ligands for colon cancer biomarkers such as EGFR, epithelial cell adhesion molecule (EpCAM), Vascular Endothelial Growth Factor Receptor (VEGFR), and integrins. By analyzing thousands of ligand–receptor interactions, AI can recommend combinations with the strongest binding affinity and lowest immunogenicity. This is especially useful in developing next-generation active-targeting nanoparticles.

#### 15.7.2.3 AI in nanotoxicology assessment

Nanotoxicology is one of the primary barriers to clinical translation. AI-enabled toxicity prediction models can analyze patterns in chemical composition, surface charge, and morphology to forecast adverse effects and off-target organ accumulation (Sun et al., 2022). Predictive toxicity tools enable safer initial designs and reduce preclinical failure rates.

#### *15.7.2.4 Automated nanoparticle synthesis platforms*

Robotic automation combined with AI algorithms enables high-throughput nanoparticle synthesis. These automated platforms adjust formulation conditions in real-time based on predictive outputs and experimental feedback. The iterative loop drastically accelerates screening of materials, drug combinations, and environmental triggers.

#### *15.7.2.5 AI for personalized therapy planning*

AI algorithms can analyze patient-specific genomic, proteomic, and imaging data to recommend individualized nanoparticle formulations. Such approaches enable precision nanotherapy by tailoring drug ratios, release profiles, and targeting ligands to the unique tumor biology of each colon cancer patient. Overall, AI is expected to fundamentally transform the landscape of nanomedicine research, enabling rational design, reduced development timelines, and personalized treatment paradigms.

### 15.7.3 Personalized nanotherapeutics

Personalized nanomedicine represents the future of colon cancer treatment, focusing on tailoring therapeutic interventions to individual patients' tumor biology, immune profile, and pharmacogenomics. This shift from "one-size-fits-all" approaches to precision-guided nanotherapy has profound implications for improving efficacy and reducing toxicity.

#### *15.7.3.1 Tumor genomics and targeted nanotherapy*

Advances in genomics and transcriptomics have revealed significant heterogeneity among colon tumors, including KRAS mutations, microsatellite instability (MSI) status, and adenomatous polyposis coli (APC) or TP53 alterations. Nanocarriers can be engineered to deliver drugs or gene-editing tools targeting specific genetic defects. For instance, KRAS-mutant colon cancer—which resists standard therapy—may benefit from siRNA-loaded nanoparticles designed to silence oncogenic KRAS (Ely et al., 2021).

#### *15.7.3.2 Pharmacogenomic-driven nanocarrier customization*

Patient-specific variations in drug metabolism influence therapeutic outcomes. Nanomedicine can circumvent pharmacogenomic limitations by bypassing metabolic pathways or altering drug distribution patterns. Personalized nanocarriers may adjust drug-loading amounts or release kinetics based on pharmacogenomic profiles.

#### *15.7.3.3 Personalized biomarker-responsive nanoparticles*

Next-generation nanocarriers may incorporate biomarker-sensitive triggers such as overexpressed enzymes (e.g., MMP-2, COX-2), specific metabolites, or tumor-specific lipids. Such personalized, stimuli-responsive systems ensure selective drug activation in individual tumor environments.

#### 15.7.3.4 Immunoprofiling for personalized nanovaccines

Advances in immunoprofiling allow nanocarriers to be designed to stimulate immune responses tailored to individual patients. Personalized nanovaccines incorporating tumor neoantigens can induce potent cytotoxic responses against colon cancer cells (Ott et al., 2020).

#### 15.7.3.5 Patient-specific organoids and ex vivo models

Ex vivo colon cancer organoids—derived from patient tissue—are increasingly used to screen nanoparticle formulations. These organoids retain the phenotypic and genetic characteristics of the original tumor and serve as platforms to identify the most effective nanotherapeutics for each patient. Personalized nanotherapy is expected to bridge the gap between precision oncology and advanced drug delivery, marking a pivotal shift toward individualized treatment regimens.

### 15.7.4 Roadmap for clinical translation

Despite significant advancements in preclinical research, the translation of nanomedicine to clinical practice remains limited. Establishing a clear roadmap is critical to ensure that new nanotechnologies progress successfully from laboratory to clinic.

#### 15.7.4.1 Standardization of preclinical evaluation

A major obstacle is the lack of standardized protocols for evaluating nanoparticle biodistribution, toxicity, and pharmacodynamics. Harmonizing these protocols—along with advanced in vitro and in vivo colon cancer models—will enhance reproducibility and regulatory acceptance (Wilhelm et al., 2016).

#### 15.7.4.2 Scalable and Good Manufacturing Practice–compliant manufacturing

Clinical translation requires nanocarriers to be produced at large scale under GMP. Novel microfluidic reactors and automated synthesis systems are emerging to facilitate reproducible, scalable, and sterile nanoparticle fabrication (Khan et al., 2019).

#### 15.7.4.3 Regulatory alignment and framework development

Regulatory agencies are still developing guidelines for nanomedicines. Enhanced communication between researchers and regulators, along with adaptive regulatory frameworks, will support smoother translation. Agencies such as the FDA and EMA have begun issuing guidance documents for nanotechnology-based therapeutics, but further clarity is needed (Syed et al., 2023).

#### 15.7.4.4 Long-term safety and postmarket surveillance

Nanoparticles can exhibit long-term accumulation or unexpected interactions within biological systems. Integrating long-term toxicity studies and establishing postmarket monitoring programs are essential to ensure safety and build patient trust.

#### 15.7.4.5 Clinical trial design adapted for nanomedicine

Nanomedicines often exhibit nonlinear pharmacokinetics and unique biodistribution patterns. Future clinical trials must adopt adaptive designs that consider nanoparticle-specific behavior, patient stratification, and biomarker-based outcome measures.

#### 15.7.4.6 Multistakeholder collaboration

Effective translation requires collaboration among scientists, clinicians, regulatory agencies, manufacturing experts, and industry. Public–private partnerships and consortia can accelerate development and commercialization pathways.

A coordinated roadmap will significantly enhance the likelihood of successful nanomedicine integration into colon cancer treatment.

## 15.8 Conclusion

Nanomedicine holds extraordinary promise for revolutionizing colon cancer treatment by enabling targeted delivery, reduced toxicity, enhanced tumor penetration, and multifunctional therapy. Over the past two decades, substantial progress has been made in developing advanced nanocarriers, including liposomes, PNPs, dendrimers, micelles, and hybrid systems. Their integration with gene therapy, immunotherapy, combination therapy, and theranostics has further expanded treatment possibilities. However, despite impressive preclinical data, only a limited number of nanomedicines have reached clinical use, largely due to challenges related to toxicity, scalability, regulatory acceptance, and patient heterogeneity.

Future progress will depend heavily on next-generation biomaterials, AI-guided nanoparticle engineering, personalized nanotherapeutics, and a clear translational roadmap that addresses regulatory, manufacturing, and safety concerns. Innovations in biomimetic systems, multistimuli responsiveness, gene-activated biomaterials, and adaptive nanoplatforms will lead to more precise and effective therapies. Personalized nanomedicine—powered by genomics, immunoprofiling, and patient-derived organoids—will ensure that treatment regimens are tailored to individual tumor biology. AI will streamline nanoparticle design, predict toxicity, and support clinical decision-making.

Ultimately, the successful clinical integration of nanomedicine in colon cancer will require rigorous standardization, multistakeholder collaboration, and transparent regulatory pathways. With these advancements, nanomedicine has the potential to fundamentally reshape colon cancer management, ushering in a new era of precise, effective, and patient-centered therapy.

## References

Ahamed, M., Akhtar, M. J., Alhadlaq, H. A., & Alshamsan, A. (2017). Nanotoxicity and mechanisms of oxidative stress. *Journal of Clinical Toxicology, 7*(3), 1–6.

Akbarzadeh, A., et al. (2013). Liposome: Classification, preparation, and applications. *Nanoscale Research Letters, 8*(1), 102.

Arnold, M., Sierra, M. S., Laversanne, M., Soerjomataram, I., Jemal, A., & Bray, F. (2017). Global patterns and trends in colorectal cancer incidence and mortality. *Gut, 66*(4), 683–691.

Bader, A. G., Brown, D., & Winkler, M. (2010). The promise of microRNA replacement therapy. *Cancer Research, 70*(18), 7027–7030.

Bae, Y. H., Fukushima, S., Harada, A., & Kataoka, K. (2013). Design of environment-sensitive supramolecular assemblies for drug delivery. *Advanced Drug Delivery Reviews, 65*(1), 3–18.

Batist, G., Sawyer, M. B., Yacoub, N., Qu, X., Xie, H., & Chapman, T. (2009). Safety, pharmacokinetics, and efficacy of CPX-1 liposome injection in patients with advanced colorectal cancer. *Clinical Cancer Research, 15*(2), 692–702 PubMed.

Benson, A. B., et al. (2021). Colon cancer, version 2.2021, NCCN clinical practice guidelines in oncology. *Journal of the National Comprehensive Cancer Network, 19*(3), 329–359.

Blanco, E., Shen, H., & Ferrari, M. (2015). Principles of nanoparticle design for overcoming biological barriers to drug delivery. *Nature Biotechnology, 33*(9), 941–951.

Bray, F., Ferlay, J., Soerjomataram, I., Siegel, R. L., Torre, L. A., & Jemal, A. (2018). Global cancer statistics 2018: GLOBOCAN estimates. *CA: A Cancer Journal for Clinicians, 68*(6), 394–424.

Brown, J. M., & Wilson, W. R. (2004). Exploiting tumour hypoxia in cancer treatment. *Nature Reviews. Cancer, 4*(6), 437–447.

Chen, Y., Gao, D., & Li, Y. (2021). Nanocarrier-mediated nucleic acid delivery for colon cancer therapy. *Advanced Drug Delivery Reviews, 169*, 1–21.

Danhier, F. (2016). To exploit the tumor microenvironment: Since the EPR effect fails in the clinic, what is the future of nanomedicine? *Journal of Controlled Release, 244*, 108–121.

Danhier, F., Ansorena, E., Silva, J. M., Coco, R., Le Breton, A., & Préat, V. (2012). PLGA-based nanoparticles: An overview of biomedical applications. *Journal of Controlled Release, 161*(2), 505–522.

Danhier, F., Feron, O., & Préat, V. (2010). To exploit the tumor microenvironment: Passive and active tumor targeting of nanocarriers. *Journal of Controlled Release, 148*(2), 135–146.

Dreaden, E. C., Alkilany, A. M., Huang, X., Murphy, C. J., & El-Sayed, M. A. (2012). The golden age: Gold nanoparticles for biomedicine. *Chemical Society Reviews, 41*(7), 2740–2779.

Dreaden, E. C., Austin, L. A., Mackey, M. A., & El-Sayed, M. A. (2012). Size matters: Gold nanoparticles in targeted cancer drug delivery. *Therapeutic Delivery, 3*(4), 457–478.

Ely, Z., et al. (2021). KRAS-targeted nanoparticle gene silencing in colorectal cancer. *Molecular Therapy, 29*(3), 1123–1134.

Fan, Y., Moon, J. J., & Nanomedicine (2020). Nanoparticle drug delivery systems for cancer immunotherapy. *Advanced Drug Delivery Reviews, 156*, 104–121.

Fang, R. H., et al. (2018). Cell membrane-coated nanotechnology. *Advanced Materials, 30*(23), 1706759.

FDA. (2024). *Drug products that contain nanomaterials: Guidance for industry*. U.S. Food and Drug Administration.

Ganesh, K., Stadler, Z. K., Cercek, A., Mendelsohn, R. B., Shia, J., Segal, N. H., & Diaz, L. A. (2019). Immunotherapy in colorectal cancer: Rationale, challenges, and potential. *Cancer Journal, 25*(3), 225–236.

Gao, R., Gao, Z., Huang, L., & Qin, H. (2020). Gut microbiota and colorectal cancer: Opportunities and strategies for clinical interventions. *Biochimica et Biophysica Acta (BBA) - Reviews on Cancer, 1873*(1), 188314.

Havel, J. J., Chowell, D., & Chan, T. A. (2019). The evolving landscape of biomarkers for checkpoint inhibitor immunotherapy. *Nature Reviews. Cancer, 19*(3), 133–150.

He, H., Liu, L., Morin, E. E., Liu, M., & Schwendeman, A. (2020). Survey of clinical translation of cancer nanomedicines—Lessons learned from successes and failures. *Accounts of Chemical Research, 52*(9), 2445–2461.

Jain, R. K. (2014). Antiangiogenesis strategies revisited: From starving tumors to alleviating hypoxia. *Cancer Cell, 26*(5), 605–622.

Khan, S., et al. (2019). Microfluidic synthesis of nanomedicine. *Advanced Drug Delivery Reviews, 144*, 38–58.

Korshoej, M., Hansen, M., & Madsen, M. (2021). Nanoparticle immune modulation in cancer therapy. *Nature Nanotechnology, 16*(2), 195–207.

Kwon, G. S. (2003). Polymeric micelles for delivery of poorly water-soluble compounds. *Critical Reviews in Therapeutic Drug Carrier Systems, 20*(5), 357–403.

Laroui, H., Wilson, D. W., & Greene, M. W. (2012). Targeted nanoparticles for colon cancer therapy. *Biomaterials, 33*(28), 7221–7230.

Li, J., Wang, Y., & Liang, R. (2020). 5-FU-loaded polymeric nanoparticles for colon cancer treatment. *International Journal of Pharmaceutics, 583*, 119470.

Liu, C., Zhang, L., & Sun, Y. (2020). Nanocarrier-based delivery of CRISPR/Cas9 for cancer therapy. *Advanced Functional Materials, 30*(29), 2002086.

Liu, Y., et al. (2022). Injectable hydrogels for localized cancer therapy. *Biomaterials, 285*, 121522.

Longley, D. B., Harkin, D. P., & Johnston, P. G. (2003). 5-fluorouracil: Mechanisms of action and clinical strategies. *Nature Reviews. Cancer, 3*(5), 330–338.

Maeda, H., Nakamura, H., & Fang, J. (2013). The EPR effect for macromolecular drug delivery to solid tumors. *Advanced Drug Delivery Reviews, 65*(1), 71–79.

Meng, X., et al. (2021). Redox-responsive nanocarriers. *Advanced Functional Materials, 31*(3), 2004970.

Monopoli, M. P., Åberg, C., Salvati, A., & Dawson, K. A. (2012). Biomolecular coronas provide the biological identity of nanosized materials. *Nature Nanotechnology, 7*(12), 779–786.

Mukherjee, S., Ray, S., & Thakur, R. S. (2019). Solid lipid nanoparticles: A modern formulation approach in drug delivery system. *Indian Journal of Pharmaceutical Sciences, 71*(4), 349–358.

Mura, S., Nicolas, J., & Couvreur, P. (2013). Stimuli-responsive nanocarriers for drug delivery. *Nature Materials, 12*(11), 991–1003.

Ott, P. A., et al. (2020). Personalized neoantigen cancer vaccines. *Nature Reviews Clinical Oncology, 17*(10), 591–602.

Peer, D., Karp, J. M., Hong, S., Farokhzad, O. C., Margalit, R., & Langer, R. (2007). Nanocarriers as an emerging platform for cancer therapy. *Nature Nanotechnology, 2*(12), 751–760.

Peppas, N. A., & Huang, Y. (2004). Nanoscale technology of mucoadhesive interactions. *Advanced Drug Delivery Reviews, 56*(11), 1675–1687.

Prasad, S., & Tyagi, A. K. (2015). Colonic delivery of bioactives through polysaccharide-based nanoparticles: A perspective. *Carbohydrate Polymers, 132*, 79–89.

Sherje, A. P., Jadhav, M., Dravyakar, B. R., & Kadam, D. (2018). Dendrimers: A versatile nanocarrier for drug delivery and targeting. *International Journal of Pharmaceutics, 548*(1), 707–720.

Shi, J., Kantoff, P. W., Wooster, R., & Farokhzad, O. C. (2017). Cancer nanomedicine: Progress, challenges and opportunities. *Nature Reviews. Cancer, 17*(1), 20–37.

Siegel, R. L., Miller, K. D., & Jemal, A. (2020). Colorectal cancer statistics. *CA: A Cancer Journal for Clinicians, 70*(3), 145–164.

Sun, D., et al. (2022). AI in nanotoxicology prediction. *Nano Today, 42*, 101367.

Sung, H., Ferlay, J., Siegel, R. L., Laversanne, M., Soerjomataram, I., Jemal, A., & Bray, F. (2021). Global cancer statistics 2020. *CA: A Cancer Journal for Clinicians, 71*(3), 209–249.

Syed, A. M., et al. (2023). Regulatory challenges of nanomedicine. *Nature Nanotechnology, 18*(2), 85–98.

Torchilin, V. (2014). Multifunctional, stimuli-sensitive nanoparticulate systems for drug delivery. *Nature Reviews. Drug Discovery, 13*(11), 813–827.

Vardy, J. L., Dhillon, H. M., & Fardell, J. E. (2019). Pharmacological and non-pharmacological interventions for cognitive impairment after chemotherapy. *The Lancet Oncology, 20*(2), e47–e57.

Ventola, C. L. (2017). Progress in nanomedicine: Approved and investigational nanodrugs. *P&T, 42*(12), 742–755.

Vinogradov, S. V. (2010). Nanogels in the race for drug delivery. *Nanomedicine: Nanotechnology, Biology, and Medicine, 5*(2), 165–168.

Wang, H., Agarwal, P., & Zhao, S. (2021). Advances in nanomedicine for colorectal cancer therapy. *Cancer Letters, 504*, 97–109.

Wang, M., et al. (2022). Gene-activated biomaterials in cancer therapy. *Biomaterials Science, 10*(6), 1432–1447.

Wilhelm, S., et al. (2016). Nanoparticle delivery efficiency. *Nature Reviews Materials, 1*(5), 16014.

Xie, J., Lee, S., & Chen, X. (2020). Nanoparticle-based theranostic agents. *Advanced Drug Delivery Reviews, 134*, 3–19.

Zanganeh, S., et al. (2016). Iron oxide nanoparticles inhibit tumor growth by inducing pro-inflammatory macrophage polarization. *Nature Nanotechnology, 11*(11), 986–994.

Zhang, Y., Hong, H., & Cai, W. (2019). Dual-functional nanocarriers for chemo-gene therapy. *Biomaterials Science, 7*, 1710–1723.

Zhang, Y. N., et al. (2018). Nanoparticle–liver interactions: Cellular uptake and hepatobiliary elimination. *Journal of Controlled Release, 240*, 332–348.

Zhao, Y., et al. (2020). Hybrid lipid–silica nanocarriers. *ACS Nano, 14*(8), 9171–9182.

Zhu, F., et al. (2021). Machine learning in nanoparticle design. *Small (Weinheim an der Bergstrasse, Germany), 17*(15), 2008131.

# Index

*Note*: Page numbers followed by "*f*" and "*t*" refer to figures and tables, respectively.

# D



# E

## S



## T



## U



## V



## W



## Z

Printed and bound by CPI Group (UK) Ltd, Croydon, CR0 4YY
28/07/2026
02179060-0001